AF556956

Earth Science Mapping

for Planning, Development and Conservation

Earth Science Mapping

for Planning, Development and Conservation

Edited by

Joseph McCall and Brian Marker

Graham & Trotman
A member of the Kluwer Academic Publishers Group
LONDON/DORDRECHT/BOSTON

Graham and Trotman Limited
Sterling House
66 Wilton Road
London
SW1V 1DE
UK

Graham and Trotman Inc.
Kluwer Academic Publishers Group
101 Philip Drive
Assinippi Park
Norwell, MA 02061
USA

ISBN 0 86010 989 5

First published 1989

British Library Cataloguing in Publication Data

Earth science mapping for planning, development and conservation.
1. Geological features. Mapping
I. McCall, Joseph II. Marker, Brian
551.8

Library of Congress Cataloging-in-Publication Data

Earth science mapping for planning, development, and conservation/ edited by Joseph McCall and Brian Marker.
p. cm.
Includes bibliographical references.
1. Geological mapping. I. McCall, G.J.H. (Gerald Joseph Home).
II. Marker, Brian.
QE36.E25 1989
550'.22'3—dc20 89-37622
CIP

Typeset in Garamond and printed and bound in Malta by Interprint Ltd.

Contents

Contributors viii
Acknowledgements ix

PART I – INTRODUCTION

Chapter 1 **The Scope of Earth Science Mapping** 3
Brian Marker and *Joseph McCall*

Introduction 3
The development of mapping 4
Land use planning and the earth sciences 5
Mapping techniques 7
Data handling and data systems 8
Water 8
Minerals 8
Soils 9
Hazards 9
Construction 9
Environmental geology maps 9
Liabilities and responsibilities 9
The various types of earth science maps 9
References 10

PART II – TECHNIQUES

Chapter 2 **Techniques of Map Presentation** 15
John C. Doornkamp

Earth science maps and plans 15
Aims of mapping 15
The nature of earth science data 16
Point data 16
Line data 20
Area data 25
The third dimension 26
The role of computers 26
Map production 27
Cartographic considerations 27
Using earth science maps 28
References 28

Chapter 3 **Data Handling and Examples of Data Systems** 29
Joseph McCall

Introduction 29
The nature of information 29
Locating data 30
The essential steps in making and using a data system 30
Spatial characteristics of observations and measurements 34
Examples of data systems 34
Conclusions 48
References 49

PART III – EARTH SCIENCE TOPICS FOR PLANNING, DEVELOPMENT AND CONSERVATION

Chapter 4 **Water** 53
Melinda Anne Lewis

Introduction 53
The hydrological cycle 53
Factors affecting regional water resources 55
The power, uses and problems of water 57
Water and land uses 60
Water and health 62
Water-related maps 63
References 76

Chapter 5 **Mineral Resources** 79
Joseph McCall

General discussion 79
Case studies 90
Conclusion 113
References 117

Chapter 6 **Soils** 119
Malcolm J. Reeve

Introduction 119
Soil formation 119
Soil classification 120
Mapping soil distribution 122
Interpretive and specialist soil maps and their use 132
Soil mapping in integrated studies 151
Conclusions 154
References 154

Chapter 7 **Hazards** 157
John C. Doornkamp

Introduction 157
Map scales 158
Aids to hazard mapping 158
Mapping and classification systems 159
Hazard mapping and perception 160
Cartographical approaches to hazard mapping 160
Hazard mapping and plotting accuracy 160
Landslips and avalanches 160
Natural ground subsidence 163
Quarrying and mining 163
Erosion and deposition 164
Flooding 165
Saline soils 167
Permafrost 168
Seismological hazards 170
Volcanic hazards 170
Application of hazard maps 171
Wider responses 171
Conclusion 172
References 172

Chapter 8 **Construction** 175
E.F.J. de Mulder and *J.G. Bakker*

Introduction 175
Ground conditions—problems and responses 175
Site investigation 180
Site preparation 183
Design of structures 183
Special considerations relating to excavations and tunnels 183
Presentation of information 184
Case histories 186
Conclusions 197
References 197

PART IV – CURRENT STATUS AND FUTURE PROSPECTS

Chapter 9 **Environmental Geology Mapping** 201
Brian Marker and *Joseph McCall*

Introduction 201
Definitions 202
Reasons for undertaking environmental geology mapping 202
Planning and development objectives 202
Levels of detail in presentation of results 204
Commissioning agencies and contractors 204
Data and updating 205
Limitations and liabilities 207
Implementation 208
Examples of environmental geology exercises 209
Conclusions 233
Acknowledgement 234
References 234

Chapter 10 **Potential and Future Prospects – Earth Science Mapping for Rational Management of Natural Resources and the Environment** 237
G. Lüttig

Situation and problems 237
Expected requirements of geosphere management 238
The potential of the earth sciences information 239
The technical steps necessary 239
Contributions of geoscience to solving the problems of regional and environmental planning and management 241
Education 244
The global perspective 244
Strategies and problems 245
Conclusion 247
References 248

Epilogue 249
Glossary 251
Additional Bibliography 261
Index 265

Contributors

J.G. Bakker, Rijks Geologische Dienst, Spaarne 17, Haarlem, Postbus 157, 2000 AD Haarlem, The Netherlands

J.C. Doornkamp, Department of Geography, The University of Nottingham, University Park, Nottingham NG7 2RD, UK

M.A. Lewis, British Geological Survey (Hydrogeological Group), Crowmarsh Gifford, Wallingford, Oxon OX10 8BB, UK

G. Lüttig, Institut für Geologie und Mineralogie, Lehrstuhl für Angewandte Geologie, Friedrich-Alexander-Universität Erlangen-Nürnberg, 8520 Erlangen, Schlossgarten 5, West Germany

B.R. Marker, Minerals and Land Reclamation Division, Department of the Environment, 2 Marsham Street, London SW1P 3EB, UK

G.J.H. McCall, GAPS Geological Consultants, 13 Deodar Road, Putney, London SW15 2NP, UK; Fellow, Liverpool University

E.F.J. de Mulder, Rijks Geologische Dienst, Spaarne 17, Haarlem, Postbus 157, 2000 AD Haarlem, The Netherlands

M.J. Reeve, Soil Survey and Land Research Centre, Shardlow Hall, Shardlow, Derby, DE7 2GN, UK

Acknowledgements

The authors and editors of this volume would like to express their thanks to all who made this publication possible through granting their permission to reproduce figures and tables. The following is a list of acknowledgements. (The numbers in brackets indicate the chapter number.)

Figure 6(6) reproduced courtesy of Agriculture Canada and by permission of the Ministry of Supply and Services, Canada, from *Soils of the Lac La Hache-Clinton Area, British Columbia, map, B.C. Soil Survey Report No. 25,* 1980.
Figures 3(6), 11(6) and 26(6) reproduced by permission of Blackwell Scientific Publications Ltd.
Figures 6(2), 7(3), 8(3), 6(5), 9(5) and 17(5) reproduced by permission of the Director, British Geological Survey: NERC Copyright reserved.
Figure 13(4) reproduced from Geological Survey geothermal map of the UK by permission of the Director, British Geological Survey: NERC Copyright reserved.
Figure 6(4) reproduced from Geological Survey hydrogeological map of the area between Cambridge and Maidenhead by permission of the Director, British Geological Survey: NERC Copyright.
Figure 7(4) reproduced from Geological Survey hydrogeological map of the Permo-Trias and other minor aquifers of South West England by permission of the Director, British Geological Survey: NERC Copyright.
Figure 23(6) reproduced by permission of the Central Electricity Generating Board.
Figures 9(4), 10(4) and 11(4) reproduced by permission of the Commission of the European Communities.
Figure 9(7) reproduced by permission of M. Cross.
Figure 5(6) reproduced by permission of CSIRO, Melbourne.
Figure 14(6) reproduced by permission of the Division of Land and Soil Sciences, New Zealand, Department of Scientific and Industrial Research.
Figures 4(6), 13(6) and 20(6) reproduced by permission of Elsevier Applied Science Publishers.
Figure 4(7) reproduced by permission of Professor C. Embleton.
Figure 10(5) reproduced by permission of Engineering Geology Ltd/Geomorphological Services Ltd.
Figure 12(9) reproduced by permission of the *Geographical Journal*.
Figures 6(2) and 7(5) reproduced by permission of the Geological Society, London.
Figures 1, 2, 3, 4, 5, 6, 7, 8, 9, (all 8) reproduced by permission of the Geological Survey of the Netherlands (Rijks Geologische Dienst) and Delft Geotechnics.
Figures 3(7) and 1(9) reproduced by permission of Geomorphological Services Ltd.
Figures 20(5), 21(5), 22(5), 23(5) reproduced by permission of R. McQuillin.
Figure 4(2) simplified from mapping by J.C. Griffiths.
Figures 6(2), 2(3), 4(3), 5(3), 6(3), 4(4), 8(4), 10(5), 11(5), 12(5), 13(5), 14(5), 1(9), 4(9) and 7(9) reproduced by kind permission of Her Britannic Majesty's Stationery Office, © Crown Copyright, 1989.
Figure 5b(4) reproduced by permission of the Institute of Hydrology (© NERC copyright).
Figure 1(3) reproduced by permission of the Institute of Mining and Metallurgy.
Figures 1(2), 15(5), 16(5) and 6(9) reproduced by permission of the International Association of Engineering Geologists (AIGI/IAEG), Paris.
Figure 4(5) reproduced by permission of David Jarvis Associates, Swindon.
Figure 8(2) reproduced after JEBCO, London.
Figure 17(6) reproduced by permission of MAFTech, Palmerston North and Division of Land and Soil Sciences, DSIR.
Figure 1a & b (5) reproduced by permission of Norges Geologische Undersøkelse.
Figure 21(6) reproduced by permission of Ohio Department of Transportation, Columbus, Ohio.
Figure 5a(4) reproduced from the Ordnance Survey map of Northern Ireland with the sanction of the controller of H.M.Stationery Office, Crown Copyright reserved. Permit number 224.
Figure 2(3) reproduced by permission of Ove Arup and Partners.
Figures 7(7) and 8(7) reproduced by permission of Oxford University Press, from *Geomorphology in Environmental Management* (2nd ed.), (Cooke and Doornkamp, eds.).
Figures 5(9) and 11(9) reproduced by permission of E. Schweitzerbart'sche Verlagsbuchhandlung.
Figure 24(5) reproduced courtesy of Shell Petroleum, London.
Figure 19(6) reproduced by permission of the Soil Society of America Inc. and the American Society of Agronomy Inc.
Figures 12(4), 1(6), 7(6), 12(6), 22(6) and 25b(6) reproduced by permission of the Soil Survey and Land Research Centre.
Figure 5(7) reproduced by permission of M. Stocking.
Figure 2(2) redrawn from a field sketch by P.H. Temple.
Figure 2(9) from *Engineering Geology Maps*, © UNESCO 1985. Reproduced by permission of UNESCO.
Figure 1c(5) reproduced after the United States Department of the Interior, Bureau of Mines.
Figures 2(5), 3(5), 5(5), 8(9), 9(9) and 10(9) reproduced after the United States Geological Survey.

Figures 17(5) and 18(5) reproduced by permission of the Warrington and Runcorn Development Corporation.
Figure 5(2) reproduced by permission of John Wiley and Sons Ltd, from *International Geomorphology 1986* (Gardiner, V., ed.).
Tables VIII, IX and X(5) are reproduced after the Geological Society, London, and Rendel Geotechnics, London.

Chapter 4 by Melinda Lewis is published with the approval of the Director, British Geological Survey (NERC).

Part I

Introduction

Chapter 1

The Scope of Earth Science Mapping

Brian Marker and *Joseph McCall*

INTRODUCTION

In an increasingly technological world with a rapidly increasing population, the demand for earth resources is rising. These resources include water for agriculture; minerals for energy production, industry and construction; soils for agriculture; and stocks of good land for building on. It is important both that sufficient supplies of resources should be maintained and that resources which are not yet needed should not be unnecessarily sterilised, for example, by building on top of scarce mineral deposits. Ground stability may also influence the safe and cost-effective placing of constructions and a number of other types of geological hazards, such as earthquakes and volcanic eruptions, may affect the safety even of large areas. Environmental geology is as much about surface processes (geomorphology) as it is about geological materials.

A wide range of social, economical and political pressures affect the ways in which development of land takes place. In many parts of the world, development is now constrained in some locations by attempts to conserve the natural environment. Planning of land use and control of the ways in which development takes place attempts to find the best possible balance between the social, economic and environmental concerns. The manner in which this is done varies greatly from country to country, depending on the national legal and advisory provisions and the economic pressures faced. There may be a strong political element in some systems, whereas in others more directive regulations exist.

Earth Science Mapping for Planning, Development and Conservation

Earth science information may be relevant to good decision-making, for instance:

- building development might sterilise minerals;
- waste disposal might affect water resources within the ground;
- construction might initiate landsliding;
- buildings might be susceptible to subsidence risks.

It is important to realize that most problems concerned with earth materials or geological processes can be overcome provided their existence is appreciated. Planners may benefit therefore from earth science advice which is material to decision making, although not necessarily of overriding importance.

Provision of the correct information in forms which can be properly applied to decision making requires planners and developers to appreciate when and where earth science advice is needed, what information is required and how it should be presented. One effective way of communicating earth science information for these purposes is by means of carefully designed maps and diagrams. Indeed, maps have long been designed for many other matters on which planning and development decisions are based, such as population distribution and transport patterns.

Over the past two decades, great advances have been made in presenting earth science information for these purposes. Equally, much greater note is being taken of these factors. Even so, the intensifying competition for use and reuse of land still frequently leads to losses of, and damage to, earth resources; and damage to development, because it is poorly located, or inadequate precautionary or remedial works have been undertaken. The costs of losses and remedial measures must be considerable, but have not generally been systematically studied or assessed. Equally, very few studies of cost and benefits of applying earth science maps have been undertaken. Recent work by de Mulder (1988) has demonstrated very well, however, the savings resulting from Engineering Geology Mapping of selected areas in the Netherlands, and it is likely that similar relationships exist in many other areas.

Other earth science issues have been little addressed until recently; for example, the relationship between geochemistry and health (Bowie and Webb, 1980), although concern is growing about both natural and man-made geochemical damage to health—e.g. the growing concern about radon emissions.

The purpose of this book is to review the types of earth science mapping that are now available, and that are relevant to planning of land use and to development of land. This review is aimed at earth scientists who wish to broaden their perspective of the subject, and at planners and developers who might need to consult earth science information. It is a defect of many reviews that they are too general for specialists and too complicated for non-specialists. The contributors to this volume have deliberately tried to provide sufficient introductory material for non-specialists, but hope that specialists will still find here much that is of interest to them.

The book is organized into two main sections: the first covers techniques, and the second various relevant earth science topics. Within these sections are chapters which present background information on each topic, outlines of case studies and examples, various conclusions and a list of references. A short concluding section discusses the likely future development of earth science mapping related to planning and development. Appendices at the end of the book contain a supplementary bibliography, pointing to some further reading, and a glossary of terms, other than those defined in the text.

Because various branches of the earth sciences partly consider the same material from different angles, there is inevitably a degree of overlap in the Chapters. This has been intentionally left by the editors so that readers can consult self-contained chapters as well as read the book as a whole.

Because planning and development is at present mainly aimed at land areas, the sea is by and large omitted from this volume, though the writers can visualize a time, not far off, when the seas and oceans will also need to be taken into account. Although earth science mapping also includes the atmosphere, this book is aimed mainly at the land and water areas, though it makes some passing reference to atmospheric contamination, another growing concern.

'Land use planning' is a term which refers to the attempt to produce a framework for decisions on development and conservation which resolves competing demands for the use of land at the best balance of social, economic and environmental costs.

The nature of that balance will vary between differently endowed territories. Small countries with high population densities and considerable wealth may face, for example, severe conflict between the needs of society and the aspirations of those same people for an improved environment. Conversely, a large but relatively poor country with growing population density may find it difficult to justify anything beyond maximizing the potential for economic survival (even if, in the long term, the environment may suffer greatly and quickly gained benefits may be repaid with long-term problems).

In both cases, the starting point of planning will be to consider population trends, housing, industry, transport and infrastructure, waste disposal, agriculture, forestry, fisheries, minerals and water supply—the natural resources and activities on which wealth creation is based, and the meeting of the material needs of society. Where circumstances permit, activities such as recreation will be taken into account; and, even in countries with limited resources, tourism may be an important source of revenue. A corollary to this may be the need to protect the landscape and the coastline, where economic circumstances permit.

The legal frameworks for, and levels of, decision making vary greatly but commonly include:

(1) general or strategic planning policies;
(2) local or more specific planning policies;
(3) regulations to control construction, mineral extraction, waste disposal and other activities.

Systems may be directive or, at the other extreme, they may be advisory. Although earth science factors have long been taken into account in minerals and waste disposal planning, the emphasis is often on specific sites rather than on the broader range of options that might be available. In the case of potential natural and man-made hazards, the level of consideration has been much more variable. Indeed, the position on these has often been one of reaction in the form of emergency planning before an event or withdrawal of investment and insurance after an event, rather than land use planning and economic planning beforehand. In many parts of the world efforts are now being made to ensure that earth science information is taken into account, when appropriate, at the planning stage.

Since planners, developers and earth scientists have a common currency in maps, these are a useful means of conveying the necessary information. It is important to note that this does not necessarily mean using earth science information as an overriding influence in decision-making; rather it is a matter of properly considering and giving weight to relevant earth science information alongside planning factors and development proposals. Equally, to obtain results in the most useful form is a task for co-operation between the various disciplines involved. There is a need to generalize information and to strip it down to the essential facts and interpretations, and to present these in plain, jargon-free terms, whilst still preserving the scientific and technological credibility and standing of the study. Equally, it is important that the objectives of studies should be drawn up in planning terms rather than in earth science terms.

THE DEVELOPMENT OF MAPPING

The earliest known maps concerned land holdings in Babylon and Egypt some 5000 years ago, although maps for navigational purposes may, perhaps, have a greater antiquity. Certainly, examples survived amongst tribal societies in Polynesia until recent times. Mapping undoubtedly became diverse during the ancient Greek civilization, but examples have not survived. However, it is likely that in addition to features such as lakes, rivers and wells, early property maps may have made reference to some mines and quarries. The great expansion of mapping in Western Europe came in the late 16th century and was linked to the dynamic world exploration going on at that time. Many such maps are remarkably accurate (for example, those of John Speede, *c.* 1690). Some of these maps carried general notes on the natural produce, including broad mineral bearing localities. However, the principal purposes were still financial, legal, political, military and topographical. It was the military requirements for accurate maps that led to the first triangulation survey of France in 1740—a technique which greatly improved accuracy and was quickly adopted by many other countries. The National Topographic Survey of Great Britain (later the Ordnance Survey), for example, was set up in 1791. From the mid-eighteenth century onwards, specialist thematic maps began to be prepared to show economic and scientific aspects, as well as topography, routes and settlements. These thematic maps included the first earth science maps (Slade and Ferguson, 1911). The need for such maps was expressed by John Aubrey at some time between 1656 and 1691:

> I have often times wished for a mappe of England coloured according to the colours of the earth, with markes for the fossils and minerals.

It was not until 1751, however, that J.E. Guettard produced his map and memoir of the minerals and rocks of France. This was later extended to a mineralogical map of Northern Europe. Subsequently, in 1780, a more detailed map of France, commenced by Guettard but finished by Monnet, included rocks, minerals, quarries, fossil localities, cold and hot springs and minerals: an early example of applied geological mapping. In 1799, William Smith, prepared a geological map of the country around Bath, England. This was the first of a number of sets of coloured geological maps of parts of England, issued between 1801 and 1824, based on Smith's realization that various layers of sedimentary rocks contained fossils characteristic of specific periods of time (Eyles, 1969). These were not merely academic exercises, they were of practical application, for William Smith was an engineer of canals, coal mines, coastal defences, drainage and navigation schemes. He well appreciated the applied value of his stratigraphical maps, but this appreciation still eludes many non-geologists. The early nineteenth century saw a rapid expansion of stratigraphical mapping in many countries; for example, William Maclure's work on the eastern United States (1809), Elie de Beaumont's work in France (1840–1848) and A.H. Dumont's work in Belgium (1836–1854).

The burgeoning of properly triangulated topographical maps in the late eighteenth century provided accurate base maps for recording geological observations. It is perhaps not surprising that the two activities became linked. Thus, Colonel T.F. Colby, superintendent of the Trigonometrical Survey of Great Britain, became a member of the Geological Society of London in 1814, and appointed J. Maccullough as a geologist. In 1832, the work of Henri de la Beche on Devon and Cornwall came to Colby's notice and, by 1835, de la Beche had become the first Director of the Geological Survey of Great Britain (McCartney, 1977). However, the World's first public geological survey had been commenced in 1830 by Edward Hitchcock in Massachusetts, USA (Woodward, 1911).

Such systematic surveys formed the basis of many advances of geological knowledge over the next 150 years and involved repeated development of more accurate surveys of the more populated parts of Western Europe, North America and some other regions at scales generally ranging from 6 inches to the mile of 1:10 000 to one inch to one mile or 1:50 000. Coverage of other areas at smaller scales (for example 1:125 000 in Kenya and 1:250 000 in Australia), usually a preliminary to mineral resource exploration, has gone on until the present day in successively remoter parts of the world.

Soils have been commented upon in connection with agriculture in a wide variety of topographical and travellers' documents. Even the mapping of soils has a long history. In the period from 1790–1810, for example, the English Board of Agriculture produced a series of small-scale county maps of soils: these divided the soils in terms of topsoil structure (e.g., rich loam, wet clay, poor sand) and made some reference to agricultural land use to give a general guide to agricultural potential of land.

However, the science of pedology progressed slowly in the first half of the nineteenth century and it was only in the later years that the understanding of soil structure and formation was carried forward, principally in Germany, Russia and the United States (Cruickshank, 1972). The new forms of understanding and classification formed the basis for mapping. Thus a Soil Survey was set up in 1898 in the United States, and, for example, a State Survey was set up in Missouri soon after, leading to the publication of a soil map of the State in 1904 (Soil Science Society of America, 1942).

In subsequent years, soil maps were used as the basis for a variety of agricultural potential and land use maps; thus, by the 1930s, soil maps formed parts of publications such as the Atlas of American Agriculture (1935), and soil mapping of agricultural land grade was systematically carried out in Britain.

Recognition of the significance of soils mapping for land use planning, land valuation, engineering, forestry and agriculture led to a diversification of thematic and derived maps for specific purposes which have been addressed in symposia (e.g., Baird *et al.*, 1966) and illustrative publications such as *Subsoil Uncovered* (Rijks Geologische Dienst, undated).

LAND USE PLANNING AND THE EARTH SCIENCES

While the relationship between geological surveying and minerals was established early on, it was very much a matter for the scientist and mineral operator. It was not until the late nineteenth century that stirrings of the need for planning of land use began to appear, with, for example, the design of 'model villages' in England and the formation of conservation bodies such as the National Trust in England and the Sierra Club in the United States. The significance of the earth sciences for planning and development were indicated as early as the 1890s when the Reverend Charles Kingsley, the author of the *Water Babies*, lectured regularly on 'Town Geology' (Kingsley, 1887; Legget, 1987). A decade later, an 'eccentric palaeontologist' named Ami reported on the *Geology of the Principal Cities of Canada* (Ami, 1900; Legget, 1987). He suggested that 'what the drill has to penetrate in any of the larger city centres of Canada...is a matter not only of interest, but of economic value'. This prophetic statement predicts the importance of studying any record of subsurface penetration, made for whatever purpose, in environmental geology.

The long delay in incorporating consideration of geological and other scientific information into land use planning is puzzling to the present-day environmental geologist. For nothing can more simply state this essential need than the axiom of Legget (1987):

> Land is after all but the surface expression of the underlying geology.

Worth (1987) in a thoughtful essay considering the reason for this delay, attributed it possibly to the fact that the early planners considered only physical planning and ignored socio-economic issues. This may have led to a swing towards social factors at the expense of physical considerations. He also suspected a deficiency in the training of the planners at that time, and a sort of technocratic arrogance that 'the technocrat knows best'. There is also an interesting discussion in Dubrovolny and Schmoll (1968) on the reasons for planners ignoring the stated dangers (Miller and Dubrovolny, 1959) of the Bootlegger Cove Clay Formation under earthquake shock at Anchorage, Alaska, prior to the 1964 catastrophe; there were no geologists capable of interpreting the geological map on the planning staff. Planners and geologists 'went on their separate ways essentialy in ignorance of one another, largely because there was no common meeting ground'. This discussion also stressed the need for earth science information to be prepared in a manner that can be readily appreciated by planners and those not trained in the earth sciences.

The change brought about in the mid-1960s was triggered by socio-economic pressures related to the need to promote economic development and improve living conditions in a world increasingly troubled by higher population densities, land use conflicts and environmental constraints. Another important factor was the involvement of the general public in the planning process, once the undisputed domain of planning authorities. Muskie (1976) believed that the new approach to land use planning was triggered in the United States by the Santa Barbara oil blow-out, a catastrophe that threatened the coast of California with between one and three million barrels of 'black tide': this single event aroused public concern which made possible tougher environmental laws and changed the attitudes to land use planning. The Alaska earthquake of 1964 (Dubrovolny and Schmoll, 1968) also had a similar triggering effect.

Environmental geology as we know it became an important subdiscipline of geology in the early 1960s and thus virtually all the literature relevant to our topic dates from 25 years ago or less. Despite this short history of publication, the literature is immense. In the following pages, the key literature covering the development of environmental geology related to planning and development as a subdiscipline is traced, citing the more important contributions.

The practical applications of geoscience mapping were given a wide airing at about the same time in the United States and Europe. In the United States, Professor McHarg, Professor of Regional Planning at the University of Pennsylvania, led the way with a book entitled *Design with Nature*, wherein he emphasized the important role of geology in planning, and developed plastic overlays covering 'Physiographic Constraints' (slopes, drainage, bedrock geology and soils) in order to study their effects on planning (McHarg 1969a, b; Legget, 1987).

Legget (1973) published a comprehensive study of *Cities and Geology*, a pioneer treatment of urban geology well worth reading today, and later produced an equally important book on *Geology in Civil Engineering* (Legget, 1982).

On the other hand, it is surprising that even as late as 1965, Chapin, in an account of 'land use planning' did not significantly discuss the importance of geological maps at all (Chapin, 1965; Frye, 1976). However, the message of the importance of geology in land use planning and environmental protection was apparent to several other contemporary authors. Among the earliest and most far-sighted writers on this subject must be Cloud (e.g., Cloud, 1969), who has returned again and again to the subject of the fragility of our planetary environment. Flawn (1970) published a book entitled *Environmental Geology—Conservation, Land Use Planning and Resource Management*; Strahler and Strahler (1973) published an account of *Environmental Geoscience* (the interaction between natural systems and Man); Menard (1974) published *Geology Resources and Society*; Young (1975) published *Geology, the Paradox of Earth and Man*; Marsh (1977) published a manual for planners and developers; and Utgard *et al.* (1978) published a book entitled *Geology and the Urban Environment.*

These publications educated a wide audience and engendered a wider appreciation of the role of the earth sciences in planning and development. Perhaps the most persuasive of all was the outsize publication *Nature to be Commanded* (Robinson and Spieker, 1978), a glossy USGS Professional Paper with liberal use of colour diagrams (the bane of all librarians because of its user-unfriendly shelf characteristics), it is, however, a deceptive volume, and provides a number of excellent case studies of geological investigations and mapping for planning and development purposes.

A more recent publication by Coates (1985) entitled *Geology and Society* is an excellent statement of the importance of the earth sciences in the entire human range of activities.

In Europe, the importance of extending geological mapping to planning and development appears to have been first appreciated in Czechoslovakia and Spain. In Czechoslovakia, a very early study comprising a 1 : 25 000 engineering geology map was produced by Pasek and Rybar (1961), and other regional engineering geology studies were initiated by Matula (1965, 1969); Matula, however, did not publish general review treatments on the subject of engineering geology mapping for planning purposes until later (Matula, 1976; 1979). Echevarria (1967) published a general treatment of geotechnical mapping in Spain. Lüttig, in West Germany, at a slightly later date produced a series of publications relating geoscience to planning and development, and pioneered the practice in West Germany of geotechnical mapping (Lüttig, 1971; 1972; 1975; 1977; 1978; 1979a and b; 1984; 1987a and b; Becker-Platen and Lüttig, 1980; Daniels and Lüttig, 1982; Lüttig and Pfeiffer, 1974).

Among the more important general treatments of the subject published elsewhere in Europe during the last two decades are those of: Abad *et al.* (1974); Arnould *et al.* (1980); Champetier de Ribes (1979); Dearman and Fookes (1974); Dearman and Matula (1976); De Beer *et al.* (1980); De Mulder (1988); De Mulder and Hillen, (in press); Lee and Wood (1978); Letko (1977); Mazeas and du Mouza (1979); Melidoro (1970); Olivier and Camboly (1979); Rivoire (1979); Ronai (1979); Sanejouand (1972); UNESCO (1979); Viberg and Adestam (1980); and Radbruch-Hall et al. (1987).

In a newly published symposium volume (Wolff, 1988) there are further papers by Follestaad and Neeb, Forster and Culshaw, and Suette, covering the present approach to this form of mapping in Norway, Britain and Austria.

General treatments by Cendrero *et al.* (1976), Cendrero and Teran (1988), and Deveughele and Usseglio Polaterra (1979) covered the application of numerical quantification to comparison of environmental and geotechnical variables for planning purposes.

A publication of the Geological Society of London (1972) covered in detail the preparation of maps and plans in the Engineering Geology context. A later conference led to the symposium volume *Planning and Engineering Geology* (Culshaw *et al.*, 1987), which contains a number of review papers that bring up to date this application of geological mapping and investigation very concisely. Bell *et al.* (1987) stated that 'the ultimate objective of planning is to determine a particular course of action and it thus involves attempting to resolve perceived problems... land use planning is an attempt to reduce the number of conflicts and adverse environmental impacts both in relation to society and nature. In the first instance land use planning involves the collection and evaluation of relevant data from which plans can be formulated'.

This emphasized the need for planning before development, and for full geological information to be available before planning. It also reflected the change in attitudes to land use planning noted by Worth (1987) in the same volume in a review of governmental and legal systems covering land use planning in Britain. Worth noted that the new pressures brought about the need for economic growth and the land use conflicts, together with the modern trend towards rehabilitation of old sites and derelict land.

In another paper, a planner (McDonic, 1987) noted that the need for geological information for planners was never so great as at present, and likewise the need to establish a working relationship between geologists and planners. Land use planning was a political process with its basis the positive role of development plans and the negative constraints of development control. Technical advice was essential to this process, though just one of the many factors influencing decision-making. He regarded minerals planning as one area where earth science had the greatest value and related the growing interest in earth science to new legislation regarding working, after-care and restoration, and also to the careful watch on minerals planning by the public and the demand for higher standards. He listed the information that the planner requires from the geologist:

(1) Basic identification of mineral resources, either for their immediate extraction or their safeguarding in the future.
(2) Identification of geological hazards which may restrict land from being developed.
(3) Identification of key geological features that may merit conservation.
(4) Information which establishes criteria for the classification of the value of land used or to be used for agriculture and forestry, the definition of land liable to flooding, the possibility of serving an area with water and sewerage, and the possibility of building stable roads, railways and pipeline routes.

Brook and Marker (1987) took the view that thematic geological mapping is an essential tool for land use planning. They believed that the density of population in Britain has introduced more and more competing requirements for land use, and that options are constrained by existing development, economics, the need to safeguard the environment, the need to reclaim damaged land, and pressures for urban renewal. Geology, geotechnics, geomorphology and hydrology can, with the related sciences, provide additional information, both on the constraints on development and the availability of resources for safeguarding and development. Recognition of these factors permits safer and more cost-effective planning, and also allows rational decisions to be taken on the exploitation or sterilization of the Earth's resources. These authors, like McDonic, stressed the need to improve communication, especially by clear and easily understandable presentation for non-specialists such as new and improved methods of map presentation.

Other papers in the same volume include a world-wide review of examples of geological mapping for land use planning (Legget, 1987), and an international review of environmental geology mapping (Doornkamp *et al.*, 1987).

McCall (1988, in press), in a paper presented at an International Conference on Geosciences in Development, noted the emergence of a subdiscipline of geoscience related to land use planning in the last 25 years: the previous lack of communication between earth scientists and planners and the role of natural and man-made disasters such as the Santa Barbara Oil Spill, the Anchorage earthquake and the Aberfan Coal Tip Disaster in arousing public concern, and so necessitating better practices. He emphasized the inevitability of growth of this subdiscipline because of the increasing pressures on land use and resultant conflicts, the fragility of the planetary biosphere, and its imminent risk of irreversible damage due to unwise human activities. He noted that the problem of increasing population was the biggest challenge facing mankind.

This book draws on these and many other sources in order to outline the present scope and state of applied earth science mapping in the hope that workers in various disciplines and subdisciplines may gain a broader perspective of the potential and significance of these activities. Because of the widely varying audience that this volume is aimed at, and an awareness that that review volumes often give insufficient background information, each chapter has been deliberately prefaced by an outline of the subject which it deals with. As noted above, the topics are related by common factors and thus some overlap of the chapters has been incorporated to allow them to be free-standing as well as contributory to the whole review. These topics are:

- mapping and data handling techniques;
- mineral and water resources;
- soils;
- construction;
- hazards.

These topics are followed by a chapter on environmental geology maps in general which acts as a summary and conclusion.

The examples taken are, it must be emphasized, only a sample of the available material, and many other equally meritorious studies could be cited, but for the constraints of space.

MAPPING TECHNIQUES

Mapping is concerned with the display and communication of information which shows lateral variations. The approach taken is conditioned by:

(a) the nature, quantity and accuracy of the basic information;

(b) the purpose to which the map is to be put.

Most mapping is necessarily selective—a set of generalizations from large quantities of information—and, to that extent, also interpretative. In absolute terms, however, the amount of 'factual data' and inference will vary greatly. Important issues which are discussed in the chapter on mapping techniques include:

- *Significance of scale*: large-scale maps based on relatively inaccurate data, for example, may give a spurious sense of accuracy. Conversely, maps for general planning purposes are often best shown at a moderately small-scale, even if the base data is abundant and accurate, because the broad patterns rather than local details are what are important.
- *Target audience*: the technical complexity and language of the maps must be aimed at the audience, though it is important not to oversimplify maps aimed at lay audiences to the point where the technical integrity of the exercise is destroyed.
- *Types of data*: data may represent points on the mapped surface, lines or areas, though a point on one scale may become an area at a large scale.
- *Selective representation*: the more information that is put onto a map, the more crowded and difficult to read it becomes; therefore a degree of selectivity is essential. Care is needed, since selectivity may reveal more about the bias of the map maker than about the mapped distribution.
- *Clarity of presentation*: the avoidance of clutter is only part of a wider issue of clarity which relates to choice of symbols, use of colours and the nature of the linework and ornamentation.
- *Methods of representation*: information may be represented as points on the map, either annotated or coded by means of symbols, or may be generalized, for example, by contouring or shading/ornamenting areas with similar characteristics of measured values.

While all these factors are important, a further one which needs most careful consideration is the 'legend' or 'key' to any map. If this is clear and lucid, and the classification is not overcomplicated, the map is likely to communicate successfully. For this reason, the authors of sections of this volume have found that it is often better to illustrate the legends of example maps rather than the faces of the maps themselves.

DATA HANDLING AND DATA SYSTEMS

Earth science data may be collected in primary surveys such as direct field mapping or indirect ground evaluation by interpreting satellite images or air photographs (photogrammetry), or geophysical surveys. Samples may be collected and subjected to laboratory examination and tests to determine compositions or physical properties. Geochemical analyses and engineering testing may be carried out in both the field and the laboratory. In addition to this, there may be, in countries with a long legacy of resource evaluation and development, considerable amounts of secondary data—both published and unpublished. The tracing, collection, collation and interpretation of such information may supplement primary surveys and provide information which would be prohibitively expensive to obtain by new ground investigations. Making use of such data often depends on the goodwill and assistance of those holding or owing the original records.

The assembly of data relevant to specific maps depends on careful, systematic indexing and, often, tabulation of data. This may be handled manually by card indexes and paper copies, but in the case of larger exercises computerized data bases are more cost effective. Some of these give the possibility of direct digital map production from the stored data, and, with it, the chance to continuously update maps as new data become available and are fed into the system, or to produce additional themes when the need arises.

In addition, there is the potential for presenting earth science information on graphical displays of digital terrain models to give a three-dimensional picture of relationships in the ground; a topic which has been catered for in the past by less flexible cross-sections, block diagrams and solid models. Such three-dimensional displays may be critical in presenting information to non-specialists in a way which emphasizes the true extent of features or factors of interest.

WATER

Water is essential to living organisms and a precious resource for human society. The impact of drought on society can lead to disasters such as the famine in recent years affecting the Sahel Region of Africa. In many areas groundwater provides an important resource for agriculture and industry, and, more obviously, fish farming. Over the last twenty years there has been increasing concern about the pollution of ground and surface waters by industrial and other wastes, fertilizers and pesticides, which may become damaging to crops, people and wildlife. Protection of water resources may constrain types of development and agricultural practices.

Control of ground and surface water is equally important in relation to oversupply. Catastrophic flooding such as that recently experienced in Bangladesh and the rise of water tables threatening cities, as is being experienced in Bangkok, are examples of the close relationship between water and some hazards.

Water may also be harnessed for energy production. Examples of this are hydroelectricity and the use of geothermal energy to produce electricity, either from hot water or steam deep within the ground.

MINERALS

Energy production, many industrial processes and much construction depend on supplies of minerals from the Earth. Increasing populations and advancing technology have led to larger demands for minerals from surface pits, and quarries, underground mines and wells, and the seafloor by dredging. In some cases supplies may be of international and strategic significance—for example, uranium and chromium ores, or petroleum, the production of which strongly influences the buoyancy of the world economy.

It is important to ensure supplies and to avoid unnecessary sterilization of mineral resources. There are penalties to be paid for mineral extraction such as residual voids in the ground, tips of mineral waste, and effects of extraction and processing on the local environment. In areas where land is at a premium, careful restoration and after-care of worked ground is becoming increasingly important, and, where groundwater conditions permit, quarries are commonly used for disposal of domestic, industrial and constructional wastes.

SOILS

Soils are an essential medium for growth of most terrestrial plants and are therefore of great significance for agriculture, forestry, and conservation. They form from weathering of rocks and their characteristics are influenced by the parent rock materials and the climate. The characteristics of soils are pointers not only to the suitability of land for cropping, grazing and other uses, but indicate the risks of soil erosion if the existing vegetation is removed—either temporarily, as in seasonal ploughing, or more permanently, as when deforestation occurs. Not surprisingly, soil maps link to climatic information, water resource data and topographical maps. The results of chemical analyses may be relevant, as for example, when acidification of rain-water is taking place, a matter of great current concern in parts of North America and Europe.

HAZARDS

Earth resources are only one side of the coin. A wide range of problems may arise from geological or related causes, and these range from large-scale disasters due to earthquakes, volcanic eruptions and floods, to more localized catastrophes resulting from major mudflows or landslides. Such events may be terrible in their magnitude and costs (the recent major earthquake in Armenia cost tens of thousands of lives and billions of pounds worth of damage, and this only shortly after the loss of thousands of lives due to catastrophic flooding of the Ganges delta in Bangladesh) but they are only part of a wide range of physical and chemical constraints on planning and development. Many are unspectacular but lead to financial losses and sometimes risks to life and health. These include subsidence of ground into natural underground cavities and old mine cavities, rock falls and avalanches, compaction of ground under load causing damage to foundations, chemical reactivity of groundwater with construction materials, and emissions from the ground of explosive and radioactive gases. Problems may also be caused by erosion of land by wind, water or ice, and by the deposition of the resulting sediment of clog harbours, channels and underground pipes.

Though natural hazards predominate, man-made hazards, such as the catastrophic failure of dams, are becoming increasingly threatening. The total costs of all these problems are poorly known both on a world and a national scale, but are undoubtedly extremely large. A better knowledge of these costs would help to demonstrate the value of earth science investigations into planning and development, and, more particularly, to define the extent, frequency, magnitude and risk of the various problems.

CONSTRUCTION

Many geological problems have a direct bearing on safe, cost-effective construction. A knowledge of the ground conditions is required not only for adequate design of foundations and precautionary work against hazards, but to help assess the costs of excavation and surface works. All development requires proper investigations of the ground in the site and general earth science maps showing factors such as ground susceptible to compaction or mass movement, mined ground, depth to beds suitable for foundations and ease of excavation. These can aid in:

(a) preparing preliminary desk studies;
(b) design of site investigations to take account of the likely full range of problems which may be encountered;
(c) tracing original documents and reports which need to be consulted.

The information can aid, but not replace, site investigations; if the problems are appreciated, an engineering solution can generally be found for most of them, although at a price.

ENVIRONMENTAL GEOLOGY MAPS

Although the term 'environmental geology maps' has been used in a great variety of senses, it is employed in this volume to denote sets of maps covering a wide range of earth science issues in a single geographical area, and particularly those which contain summary maps depicting: (a) resources for development; and (b) constraints to development.

Such sets may, therefore, consist of amalgams of types of maps discussed in the earlier chapters, or in some cases may include an even greater variety of themes such as biological, ecological and cultural factors. They tend, therefore, to go beyond studies intended to meet restricted objectives into the preparation of a more comprehensive information base for planning and development. Because of the character of such map sets it is important that appropriate multi-disciplinary teams should be assembled to produce them. Many diverse methods have been adopted to produce them and a working group of the Directors of the Western European Geological Surveys is currently considering possible standardization of methods of presentation (Lüttig, G., personal communication).

LIABILITIES AND RESPONSIBILITIES

Many types of maps may affect the status of land. Discovery of valuable minerals enhances land values. Definition of a high risk of land instability may conversely cause loss of value, withdrawal of insurance cover or refusal of financial loans—an effect sometimes referred to as 'blight'. The preparation of such maps therefore carries very real responsibilities to be accurate, thorough and to carry out all interpretations of data in terms of the best current state of knowledge. Reciprocally, there are potential liabilities if work is carried out negligently. It is in the best interests of both the map maker and the map user to indicate, preferably on the map, the limitations of the study and the way in which the results should be used. In this way misinterpretations and misuse are less likely, and map makers are less vulnerable to possible legal actions.

THE VARIOUS TYPES OF EARTH SCIENCE MAPS

Traditional geological maps may be 'Solid Geology Maps', which show only the boundaries of the rock formations

Table I Some names used in conjunction with applied earth science mapping

Name	*Description*
Applied earth science maps, geoscience maps, environmental geology maps	Geological, geomorphological, hydrogeological, geochemical, geophysical and other maps, aimed at assisting in the solution of practical problems; usually presented as sets, sometimes with summary maps for specific purposes such as land use planning. The sets commonly include maps showing non-geological aspects and constraints
Geopotential maps	Maps demonstrating the resource and development potential of land
Engineering geology maps	A subset of applied earth science maps aimed at dealing with engineering problems
Thematic maps	Maps, earth science or otherwise, devoted to specific specialist topics
Element map	Thematic map showing observational or factual data on a single theme
Derived map	Interpretative map based on one or more element maps, synthesizing several types of information relevant to a single issue
Potential map	A category of derived map which demonstrates potential uses of land or the potential for processes to occur
Constraint map	A thematic map which indicates limitations on the use of land
Hazard map	A thematic map which shows the known extent and types of hazards in an area
Risk map	A map which attempts to quantify the likelihood of a damaging occurrence of given type and size
Resource map	A thematic map indicating the nature, extent and quality of resources, on and under the ground

beneath the veneer of geological recent drift or superficial cover such as sands, gravels, clays and peat, 'Drift Maps', which show only the drift (superficial cover), or combined 'Solid Geology and Drift Maps', which portray both together on a single map sheet. The drift or superficial cover deposits are very important in applied mapping related to planning and development. Because of the prevalence of locally quite thick glacial drift cover in parts of Britain affected by the Quaternary ice sheets, there is a strong tradition of drift mapping in Britain. In some other parts of the world, such as Ontario, Canada, drift deposits are if anything more important, geologically, than the bedrock—they may be more than 50 m thick. Giant eskers (sand and gravel ridges deposited by glacial melt water) traverse the countryside for tens of kilometres, and much of the countryside is mantled by the deposits of a vast, ice-barrier-bounded ancient lake. Such deposits require very special mapping techniques. Equally important are river deposits and coastal beach deposits, and the extensive sand deposits of the desert terrains. Loess, a wind blown sand deposit related to glaciation, provides a special problem in China and other parts of Eastern Asia.

Besides these traditional types of basic geological maps there are a number of commonly used applied earth science maps. However, these are referred to by a wide, and sometimes confusing, variety of terms. The main types are listed in Table I.

REFERENCES

Abad, J., Ayala, F. and Pena, J.L. (1974). Mapa geotécnico de ordinación territorial. *Jornados Minero, Metalúrgica*, Bilbao, Sections 1–4.

Ami, H.M. (1900). On the geology of the principle cities of Canada. *Trans. Roy. Soc. Canada* (second series), **6**, Section 55, 125–173.

Arnould, M., Deveughele, M. and Usseglio Polaterra, J.M. (1980). Utilisation complémentaire de fichiers et des cartes géotechniques en milieu urbain. *Bull. Int. Assoc. Eng. Geol.*, **22**, 117–123.

Baird, J.V., Bartelli, L.J., Heddleson, M.R. and Klingebeil, A.A. (1966). *Soil surveys and land use planning*. Soil Science Society of America, Madison, Wisconsin.

Becker-Platen, J.D. and Lüttig, G. (1980). Naturraumpotentialkarten als Unterlagen für Raumordnung und Landesplanung. *Arbeitsmat. Akad. Raumforsch. Landesplang.*, **27**, 1–60, Hannover.

Bell, F.G., Cripps, J.C., Culshaw, M.G. and O'Hara, M. (1987). Aspects of geology in planning. In *Planning and Engineering Geology* (Culshaw, M.G. *et al.*, eds.), Geol. Soc. London, Eng. Geol. Spec. Publ. No. 4, 1–38.

Bowie, S.H.U. and Webb, J.S. (1980). *Environmental geochemistry and health—a Royal Society discussion*. The Royal Society, London, 216 pp.

Brook, D. and Marker, B.R. (1987). Thematic geological mapping as an essential tool in land use planning. In *Planning and Engineering Geology* (Culshaw, M.G. *et al.*, eds.), Geol. Soc. London, Eng. Geol. Spec. Publ. No. 4, 211–214.

Champetier de Ribes, G. (1979). Données géologiques et plans d'occupation des sols. *Bull. Liaison Laboratoire des Ponts et Chaussées*, Ref. 2277–2278, 99.

Cendrero, A., Díaz De Teran, J.R. and Saiz de Omenaca, J. (1976). A technique for the definition of environmental geologic units and for evaluating their environmental value. *Landscape Planning*, **3**, 35–66.

Cendrero, A. and Díaz de Teran, J.R. (1988). The environmental map system of the University of Cantabria, Spain. In 'Mineral Resource Extraction, Environmental Protection, and Land-Use Planning' (Arndt, P. and Lüttig, G., eds.), Schweitsserbart Verlag, Stuttgart, *in press*.

Chapin, F.S. (1965). *Urban Land-Use Planning*. Urbana Univ., Illinois Press, 498 pp.

Cloud, P.E. (1969). *Resources and Man: a study and recommendations*. W.H. Freeman and Co., San Francisco, California, 259 pp.

Coates, D.R. (1985). *Geology and Society*. Chapman and Hall, New York and London, 406 pp.

Cruickshank, J.G. (1972). *Soil Geography*. David and Charles, Newton Abbot, Devon, 256 pp.

Culshaw, M.G., Bell, F.G., Cripps, J.C. and O'Hara, M. (1987).

Planning and Engineering Geology, Geol. Soc. London. Spec. Publ. No. 4, 641 pp.

Daniels, C.H. von and Lüttig, G. (1982). Geowissenschaftliche Karten des Naturraumpotentials als Unterlagen fur Raumordnung und Landesplanung. In *Eneergierohstoffe im Alpen-Adria-Raum*, Graz, 151–168.

Dearman, W.R. and Fookes, P.G. (1974). Engineering geological mapping for civil engineering practice in the United Kingdom. *Quart. J. Eng. Geol.*, 7, 223–256.

Dearman, W.R. and Matula, M. (1976). Environmental aspects of engineering geology mapping. *Bull. Int. Assoc. Eng. Geol.*, **14**, 141–146.

De Beer, E., Fagnoul, A., Lousberg, E., Nuyens, J. and Maertens, J. (1980). A review of engineering geology mapping in Belgium. *Bull. Int. Assoc. Eng. Geol.*, **21**, 91–98.

De Mulder, E.F.J. (1988). Thematic applied Quaternary mapping—a profitable investment or expensive wallpaper? In *Applied Quaternary* (De Mulder, E.F.J. and Hageman, B.P., eds.), Balkema, Rotterdam.

De Mulder, E.F.J. and Hillen, R. (1988). Preparation and application of thematic engineering and environmental geology maps in the Netherlands. *Geol. en Mijnbouw, in press.*

Department of the Environment (1976). *Research Report II, Environmental Impact Analysis*, London.

Deveughele, M. and Usseglio Polaterra, J.M. (1979). Utilisation de la théorie des variables régionalisees pour l'interpretation geotechnique d'espersées. *Symp. de AIGI sur la Cartographie Géotechnique*, Session 4, Newcastle upon Tyne, 1979.

Doornkamp, J.C., Brunsden, D., Cooke, R.U., Jones, D.K.C. and Griffiths, J.S., 1987. Environmental geology mapping: an international review. In *Planning and Engineering Geology* (Culshaw, M.G. *et al.*, eds.), Geol. Soc. London, Eng. Geol. Spec. Publ. No. 4, 215–219.

Dubrovolny, E. and Schmoll, H.R. (1968). Geology as applied to urban planning: an example from the Greater Anchorage Borough, Alaska, *Proc. 23rd Int. Geol. Congress*, **12**, 39–56.

Echevarria, C.M.R. (1967). Geotecnica y mapas geotecnicos. *Not. Y Com. del Instituto Geologico y Minero Espana*, Madrid, 97–98.

Eyles, J.M. (1969). William Smith (1769–1839): a chronology of signficant dates in his life. In *International Field Symposium on the British Jurassic* (Torrens, H.S., ed.), Keele University, Staffordshire, UK, 2–7.

Flawn, P.T. (1970). *Environmental Geology—Conservation, Landuse Planning and Resource Management.* Harper and Row, New York, 313 pp.

Forster, A. and Culshaw, M. G. (1987). Engineering geological maps as an aid to planning. In *Geology for Environmental Planning* (Wolff, F.C., ed.), Norges Geologiske Undersøkelse, Trondheim.

Frye, J.C. (1976). A geologist views the environment. In *Focus on Environmental Geology* (Tank, R., ed.), Oxford University Press, New York, London, Toronto, 3–10.

Geological Society of London (1972). The preparation of maps and plans in terms of engineering geology. *Quart. J. Eng. Geol.* **5**(4), 294–302.

Hugdal, H. and Wolff, F.C. (1987). Development of geotype maps in Norway. In *Geology for Environmental Planning* (Wolff, F.C., ed.), Norges Geologiske Undersøkelse, Trondheim.

Kingsley, C. (1887). *Town Geology*. Daldy, Isbister and Co., London.

Lee, N. and Wood, C. (1978). Environmental impact assessment of projects in EEC countries. *J. Engineering Management*, **6**, 57–71.

Legget, R.F. (1973). *Cities and Geology*, McGraw-Hill, New York.

Legget, R.F. (1982). *Geology in Civil Engineering*, McGraw-Hill, New York.

Legget, R.F. (1987). The value of geology in planning. In *Planning and Engineering Geology* (Culshaw, M.G. *et al.*, eds.), Geol. Soc. London., Eng. Geol. Spec. Publ. No. 4, 53–58.

Letko, V. (1977). Engineering geological evaluation of the environment for planning. Thesis (in Slovak), Cornelius University, 153 pp.

Look, E.G. (1987). Land use mapping, an attempt to demonstrate conflicts in different forms of areal utilisation. In *Geology for Environmental Planning* (Wolff, F.C., ed.), Norges Geologiske Undersøkelse, Trondheim.

Lüttig, G. (1971). Die Bodenschatze des Nordsee-Kustenraumes und ihre Bedeuting fur Landesplanung und Raumordnung. *N. Arch. f. Nieders.*, **21**, 1:13–25, Göttingen.

Lüttig, G. (1972). Naturraumliches Potential I, II und III. In *Niedersachsen, Industrieland mit Zukunft*, 9–10 Hannover.

Lüttig, G. (1975). Geoscience and the potential of the natural environment. Geoscientific Studies and the Potential of the Natural Environment, 29–40, Köln (Deutsche UNESCO—Komm).

Lüttig, G. (1977). Die Rolle der geowissenschaftlichen Kartographie in der verausschauenden Umweltforschung. *Kartogr. Nachr.*, **27**, 3, 81–89, Bonn-Bad Godesberg.

Lüttig, G. (1978). Von den Moor- und Bodenkarten zur Kartierung des Naturraum-Potentials. *Z.f. Kulturtechn. u Flurbereing*, **19**, 65–76, Berlin-Hamburg.

Lüttig, G. (1979a). Geoscientific maps of the environment as an essential tool in planning. *Geol. en Mijnbouw*, **57**, 527–532.

Lüttig, G. (1979b). Geoscientific maps as a basis for land use planning. *Geol. Foren. Stockh. Forh.*, **101**, 65–69, Stockholm.

Lüttig, G. (1982). Contributions of the geoscientific cartography to the solution of conflicts arising from utilisation of natural resources and the protection of the environment. In *Resources, Environment and the Future* (Fisher, W.B. and Kent, P.W., eds.), German Acad. Exch. Serv., London, 203–227.

Lüttig, G. (1984). Study of aggregates requirements and resources. *Intern. Symp. Granulats. Gen. Rep. Theme 1*, 1–19, Nice.

Lüttig, G. (1987a). Keynote address. In *Geology for Environmental Planning* (Wolff, F.C., ed.), Norges Geologiske Undersøkelse, Trondheim, 5–10.

Lüttig, G. (1987b). Large scale maps for detailed environmental planning. In *Geology for Environmental Planning* (Wolff, F.C., ed.), Norges Geologiske Undersøkelse, Trondheim, 71–76.

Lüttig, G. and Pfeiffer, D. (1974). Die Karte des Naturraum-Potentials. Ein neues Ausdrucksmittel geowissenchaftlicher Forschung for Landesplanung aund Raumordnung. *N. Arch. F. Nieders.*, **23**, 1:3–13, Gottingen.

McCall, G.J.H. (1988). The 'alternative' Earth Science mapping: mapping for planning and development. 'Geosciences in Development', Proc. of international conference on the application of geology in developing countries, Nottingham University (AGID, Geological Society of London, Institution of Geologists, Institution of Mining and Metallurgy), 26–30 Sept. 1988, *in press.*

McCartney, P.J. (1977). *Henry de la Beche: Observations on an Observer* (Bassett, D.A., ed.), National Museum of Wales, Cardiff, 77 pp.

McDonic, G.F. (1987). Planning and geology. In *Planning and Engineering Geology* (Culshaw, M.G. *et al.*, eds.), Geol. Soc. London, Eng. Geol. Spec. Publ. No. 4, 49–52.

McHarg, J. (1969a). *Design with Nature*. Nat. Hist. Press (American Museum of Natural History).

McHarg, J. (1969b). Ecological values and regional planning. *Civil Engineering*, New York, **40**, 40.

Matula, M. (1965). Regional engineering geological characteristics of the Zvolen Basin. *Acta Geol. et Geograph. UC Geologica*, 10, Bratislava.

Matula, M. (1969). Regional engineering geology of Czechoslovak Carpathians. *Slovenskej Akademievied*, Bratislava, 225.

Matula, M. (1976). Principles and types of engineering geological zoning. *Mem. Soc. Geol. Italiana*, **14**, 327–336, Roma.

Matula, M. (1979). Regional engineering geological evaluation for planning purposes. *Bull. Int. Assoc. Eng. Geol.*, **19**, 18–24.

Mazeas, H. and de Mouza, J. (1979). Cartographie géotechnique de formations superficielles en zones non urbanisées. *Bull. Int. Assoc. Eng. Geol.*, **19**, 47–52.

Melidoro, E.S. (1970). La cartografia geotecnica, principi, experience e reflession. *Geol. Appl. Idrog.*, **5**, 27–47.

Menard, H. W. (1974). *Geology, Resources and Society.* W.H. Freeman and Co., San Francisco, California, 621 pp.

Miller, R.D. and Dubrovolny, E. (1959). Surface geology of Anchorage and vicinity. *USGS Bull. 1093*, 128 pp.

Moldan, B. (1988). Environmental geology mapping in Czechoslovakia. In Geology for Environmental Planning (Wolff, F.C., ed), Norske Geologiske Undersøkelse, Trondheim, *in press.*

Munro, S.K. and Hull, J.H. (1986). Environmental geology in Great Britain. In *Geo-Resources and Environment* (Bender, E.F., ed.), E. Scheitzerbart'sche Verlagsbuchhandling (Nagele e Obermiller), Stuttgart, 107–124.

Muskie, E.S. (1976). Foreword. In *Focus on Environmental Geology* (Tank, R., ed.), Oxford University Press, N.Y., 538 pp.

National Academy of Sciences—Committee on Geological Sciences (1972). *The Earth and Human Affairs.* Canfield Press, San Francisco, California, 142 pp.

Olivier, G. and Camboly, D. (1979). Cartographie géotechnique pour l'amènagement. *Bull de Liaison de Laboratoire des Pontes et Chaussées*, 2278, 55–64.

Pasek, J. and Rybar, J. (1961). An illustration of engineering-geological conditions in the map 1 : 25,000. *Geol. Exploration*, **3**, 45–47, Prague.

Pryde, P.R. (1972). The quest for environmental quality in the USSR. *Am. Scientist*, **60**, 739–745.

Radbruch-Hall, D.H. *et al.* (1987). Experimental engineering geologic and environmental geologic maps of the conterminous United States. *Bull. U.S. Geol. Surv. No. 1610.*

Rivoire, M. (1979). Cartographie géotechnique pour l'amènagement. *Bull. de Liaison Laboratoire des Ponts et Chaussées, 99 ref. 2278*, 55–56.

Rijks Geologische Dienst (undated). *Subsoil Uncovered.*

Robinson, G.D. and Spieker, A.M. (1978). Nature to be Commanded: Earth Science Maps Applied to Land and Water Management. *USGS Prof. Paper 950*, 95 pp.

Robinson, A.H. and Sale, R.D. (1969). *Elements of cartography.* Wiley International Edition, John Wiley and Sons, New York, 415 pp.

Ronai, A. (1979). Fundamentals of engineering geology maps. *Bull. Int. Assoc. Eng. Geol.*, **19**, 62–68.

Sanejouand, R. (1972). La cartographie géotechnique en France. *Min. de l'Equipement et des Lodgement*, Ivry-sur-Seine, 96 pp.

Slade, H.E. and Ferguson, W.E. (1911). *The science history of the universe. Vol. II, Geology.* Current Literature Publishing Company and Waverley Book Company, London, 296 pp.

Soil Science Society of America (1942). *Life and work of C.F. Marbut*, Columbia, Missouri.

Strahler, A.N. and Strahler, A.H. (1973). *Environmental Geoscience: Interaction between Natural Systems and Man.* Hamilton Publishing Co., California, 511 pp.

Suette, G. (1987). Environmental geology in Styria, Austria. In *Geology for Environmental Planning* (Wolff, F.C., ed.), Norges Geologiske Undersøkelse, Trondheim. p. 113.

Turk, L.J. (1975). *Environmental Geology.* Springer-Verlag, NY. Inc., New York.

UNESCO (1976). *Engineering Geology Maps—A Guide to Their Preparation.* Publication No. 15, Earth Sciences, 79 pp.

Utgard, R.O., McKenzie, G.D. and Foley, D. (1978). *Geology and the Urban Environment*, Burgess Publishing Company, Minneapolis, Minnesota.

Viberg, L. and Adestam, L. (1980). Geotechnical terrain classification for physical planning—a Swedish research project. *Bull. Int. Assoc. Eng. Geol.*, **21**, 174–178.

Wermund, E.G. (1974). Approaches to environmental geology. *Bureau of Economic Geology Report of Investigations, No. 81*, Univ. of Texas, Austin, Texas, 286 pp.

Wiltshire, S.E., Cowley, J. and Doornkamp, J.C. (1987). Environmental impact assessment: the background and the potential for the geologist. In *Planning and Engineering Geology* (Culshaw, M.G. *et al.*, eds.), Geol. Soc. London, Eng. Geol. Spec. Publ. No. 4, 607–610.

Wolff, F.C. (1987). *Geology for Environmental Planning*, Norges Geologiske Undersøkelse, Trondheim, 121 pp.

Woodward, H.B. (1911). *History of Geology.* Watts and Co., London, 154 pp.

Worth, D.H. (1987). Planning for Engineering Geologists. In *Planning and Engineering Geology*, (Culshaw, M.G. *et al.*, eds.), Geol. Soc. London, Eng. Geol. Spec. Publ. No. 4, 39–46.

Young, K. (1975). *Geology, the Paradox of Earth and Man.* Houghton, Mifflin and Co., Boston, 526 pp.

Part II

Techniques

Chapter 2

Techniques of Map Presentation

John C. Doornkamp

EARTH SCIENCE MAPS AND PLANS

The most common earth science maps are those of geology and soils. In some instances systematic maps may also be found showing geomorphology, hydrology, or geotechnical conditions. In addition, there are many examples of maps that show a restricted set of information within these general subject areas (e.g. maps of geological structure, landslides, clay soils, flood lands or depth to bedrock, to name but a few examples).

There is a distinction that needs to be drawn between maps and plans. This is a function of scale. In general, if the document scale is larger than 1 : 2000 then it is a plan, at smaller scales it is a map (UNESCO, 1976). A commonly adopted classification of maps is that shown in Table I.

Table I Maps and plans classified by scale

Synoptic maps	1 : 100 000 and smaller
Medium-scale maps	1 : 25 000 – 1 : 50 000
Large-scale maps	1 : 10 000 – 1 : 25 000
Detailed maps	1 : 2 000 – 1 : 5 000
Site plans	larger than 1 : 2 000

Since at every scale there is a balance between the size of area that can reasonably be represented and the amount of detail that can be shown, it is necessary to select map scale very carefully to suit the purpose for which the map is to be used.

It is usual, therefore, for site investigations (e.g. for an engineering project) to use scales larger than 1 : 5000. This allows the necessary levels of detail to be portrayed.

Earth Science Mapping for Planning, Development and Conservation

Local planning maps do not require this same level of detail and generally need to show larger areas; for this reason they tend to be in the range 1 : 25 000 – 1 : 50 000. Regional planning maps tend to be at smaller scales still, and may be accompanied by a simplification of the original data set in order to display its dominant regional characteristics.

Local and regional planning maps may carry similar information to that shown at larger scales, but this is accompanied by various amounts of generalization. Indeed, this generalization may be necessary in order to suit a more general planning purpose, and is therefore entirely appropriate for that purpose.

AIMS OF MAPPING

The ultimate aim of any map is to provide information. If it does not inform then it serves no useful purpose. This process of transmitting information is dependent both on map content and on the manner of presentation. In the case of earth science data a map or plan is used to communicate information about one or more of the following aspects of ground conditions:

- their nature,
- their distribution,
- their extent,
- the spatial relationships between them.

In all earth science mapping there is a perennial problem in that data and interpretation are sometimes confused, or at least shown on the same map. This makes it difficult, at times, for the map user to distinguish between information and interpretation, especially if the user is not an earth scientist. In recent years this difficulty has led to the production of different styles of maps: one style is intended for the specialist in the subject (e.g. geotechnical maps of soil and rock properties), and another style for those who come from other disciplines (e.g. map showing areas of 'poor' ground conditions for development purposes). Other professions which need to use and refer to earth science maps include planners, engineers, developers, farmers, foresters and others in the general field of environmental management.

While general maps may be available to a large 'public' (e.g. the geological maps produced by a national geological survey organization), they may actually be of limited value to many potential users because they may be too technical and require interpretation by a specialist. For many users it would be better if the map provided data in a form, and using terminology, which addresses a particular problem or problem site. When this is done the map tends to be of value to a more restricted 'public' and its unit production costs thereby may become prohibitive. This is a genuine dilemma.

THE NATURE OF EARTH SCIENCE DATA

The types of earth science data most frequently displayed on maps include information on geology, geomorphology, soils, hydrology and vegetation. For present purposes the last of these is not included because of its rather different nature to the rest. Under the general heading of geology there are included various specialist topics such as geotechnical conditions, geophysics and geochemistry. Standard geological maps sometimes appear in two versions, one showing 'solid geology' (including bedrock structures) and the other showing 'drift geology' and thereby including superficial deposits such as those of glacial and periglacial origin. The latter provides a natural overlap with geomorphological information. Hydrogeological maps portray those geological conditions which have a bearing on ground water behaviour. Consequently, they provide an overlap with maps showing hydrology. Thus while it is convenient to refer to various systematic disciplines there is in reality a considerably overlap in the information which they include.

There have been a number of publications which describe mapping systems and appropriate mapping legends. These include:

- *Manual of Detailed Geomorphological Mapping* (Demek, 1972)
- *Guide to Medium-scale Geomorphological Mapping* (Demek and Embleton, 1978)
- *The Preparation of Maps and Plans in Terms of Engineering Geology* (Engineering Group Working Party, 1972)
- *Land Surface Evaluation for Engineering Practice* (Engineering Group Working Party, 1982)
- *International Legend for Hydrological Maps* (UNESCO, 1970)
- *Engineering Geology Maps: A Guide to their Preparation* (UNESCO, 1976)

While this chapter concentrates on the techniques of map presentation, it also makes frequent reference to the type of information which is to be portrayed. Reference will therefore be made to geological, geomorphological, hydrological, pedological, geophysical and geotechnical data, but not so much in terms of singular themes as to the manner and style of their cartographic representation. For example, a basic cartographic distinction has to be made between point, line and area data. But in doing so a number of important consequences arise that have a strong bearing on the value of the map to the end-user.

POINT DATA

Data acquisition

Point data in the earth sciences may relate to the location of specific objects, or true point data (e.g. the distribution of volcanoes; the location of mine shafts), or they may comprise small samples of a continuous feature (e.g. thickness of soils obtained at borehole sites). Statistically, these latter provide point samples of a continuous feature (population), and they are subject, therefore, to the principles of statistics and are amenable to statistical analyses. The fundamental importance of this distinction will be elaborated below. Within geology, sampling involves the collection of point data either to fill a data gap, such as with a borehole (e.g. to find the position of stratigraphic horizons, or to test their thickness) or to assess the properties of a 'representative' sample (e.g. a rock fragment).

Information about the location of specific objects which are themselves discrete and of limited extent (such as volcanoes) only pose the problems of identification and the mapping of location. In the case of volcanoes there may be little difficulty in identification, but this is not so with all earth science data. Landslides, protalus ramparts and fossil pingos all cause identification problems. There must be an expectation, therefore, that point data for singular objects will only be as complete as the skill of the investigator allows. Point samples, by definition, are not 'complete'; they are extracts from reality (a sample of the whole population) which may represent the character of that reality with varying degrees of success. Quite how well they do so may never be known. The nearest approach to such knowledge comes through statistics, whereby levels of confidence about a sample's representativeness are built up as the number of samples increases. In particular, a greater understanding of variability in the whole population comes from an increase in the number of samples.

In the case of spatial data, confidence in the representativeness of the sample increases as the spatial distribution improves. In other words small clusters of data do little to increase confidence, or indeed information about the whole population, whereas the same number of points distributed widely, and more or less evenly across the surface, will give a better impression of the nature and variability of the parameter being sampled.

In the earth sciences, point data which have a spatial component are often subjected to various forms of analyses such as contouring or trend surface analysis. Of importance in these cases are both the amount and the spatial distribution of the data. In particular it is necessary to know if data are collected on a regular grid or at random, as this affects subsequent analyses and the degree of reliance that can be placed upon them.

Data representation

Point data may either be plotted true to scale or be indicated by a symbol. The choice is essentially related to the message which it is intended to convey. Figure 1

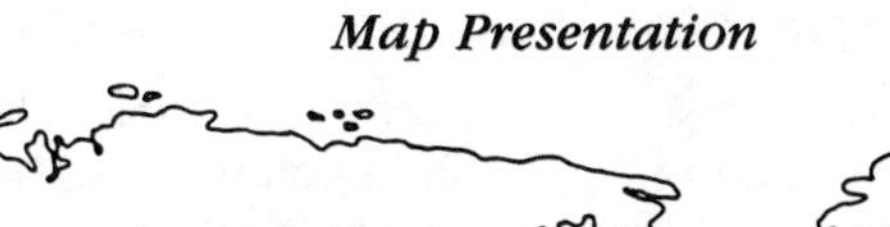

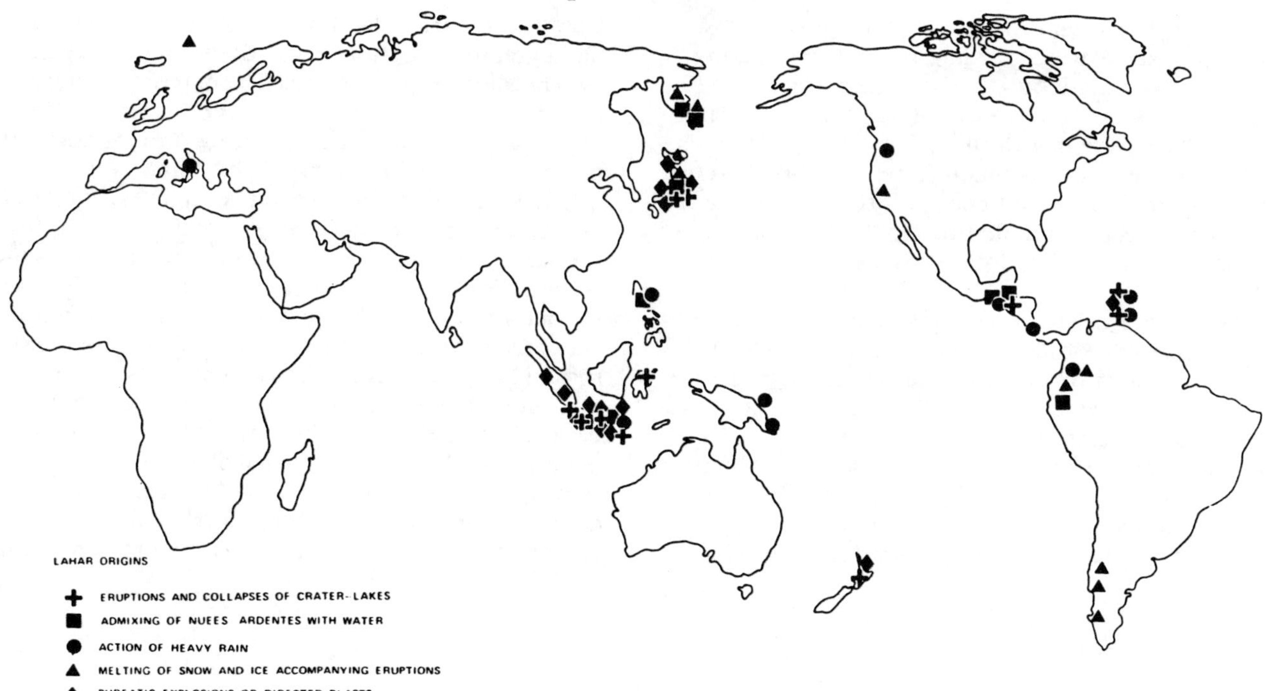

Fig. 1. Global distribution of historical lahars (from Neall, 1976).

shows the global distribution of historical lahars for which the initiating mechanisms are recorded. Only symbolic representation is possible at this scale, but the symbols allow a classification to be shown, concentrations to be identified, and the general world pattern to be appreciated. Figure 2, on the other hand, shows the distribution of some lahar characteristics for a lahar in Tanzania. Point data have become area data because of the change of scale. Internal details of the lahar can now be included, and boundaries shown in identifiable loca-

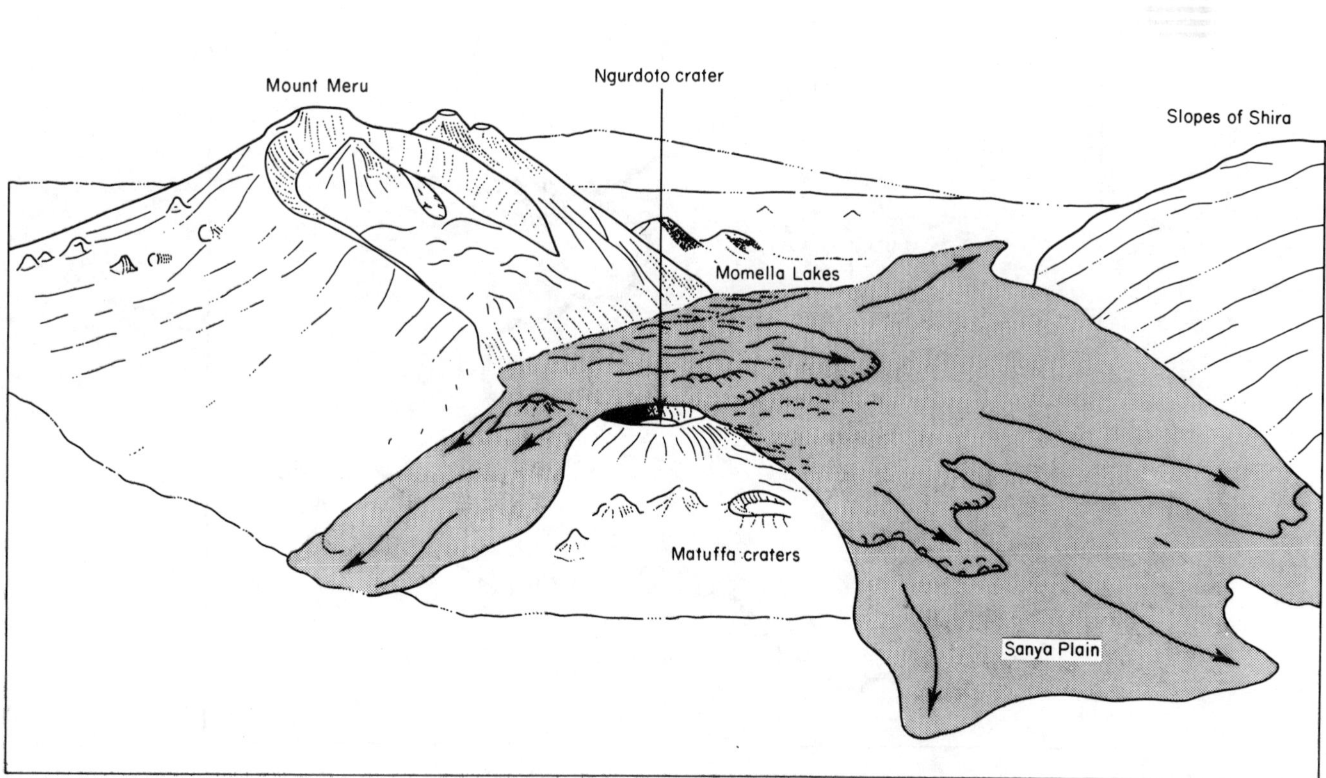

Fig. 2. The Main Meru Lahar (from a field sketch by P.H. Temple).

tions. Since the origin of this lahar is known (it was caused by excess water turning the volcanic deposits into a form of mudflow) it should feature in Fig. 1, yet it does not. Moreover, this is not an isolated example. There are many other lahars in the same area. In other words Fig. 1 is incomplete, and the sense of global distributions would change if these East African examples were to be added to Fig. 1. This illustrates the danger of accepting dot distribution maps as comprehensive.

Point data representation can take a number of forms. At its simplest, a dot distribution map which is neither true to scale nor provides a classification of the feature mapped only concentrates the mind on general locations and regional variations in density. For example, Fig. 3 shows an extract from a national landslide distribution map of Britain. It appears to suggest that the Torbay area of Devon contains a restricted number of landslides. However, the dots represent *recorded* landslides, or at least the recorded landslides *actually traced* during a bibliographic search. The map does not claim to portray *all* landslides within the area. The extent to which it fails to represent reality is shown by comparing Fig. 3 with Fig. 4 which was compiled during an air-photo interpretation and field survey of the same area after the original study (Fig. 3) had been completed. The original clearly under-represents reality, and yet prior to the preparation of Fig. 4, it was the best available statement on landslide distribution within the area.

Since it was based on published records which were accepted unverified, Fig. 3 also includes the assumption that the original record was correct in its identification of a landslide (and this is known to be in doubt in some cases), and that the position of the feature observed has been correctly plotted, about which there may also be some doubt. Thus both veracity of the data and accuracy of the record become critical to the production of a meaningful map. If we add to this the problems of data

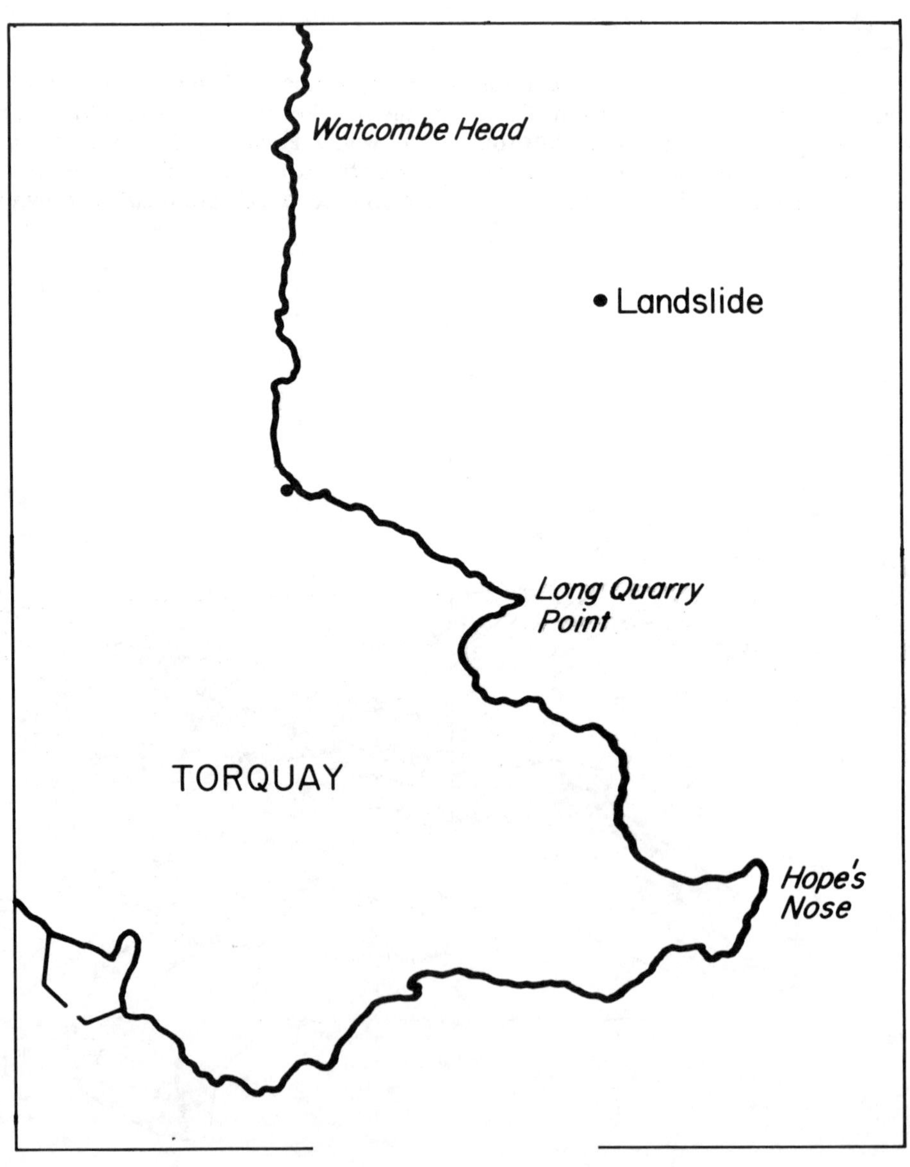

Fig. 3. Extract from a map showing recorded landslides in Britain (area shown is that of Torbay).

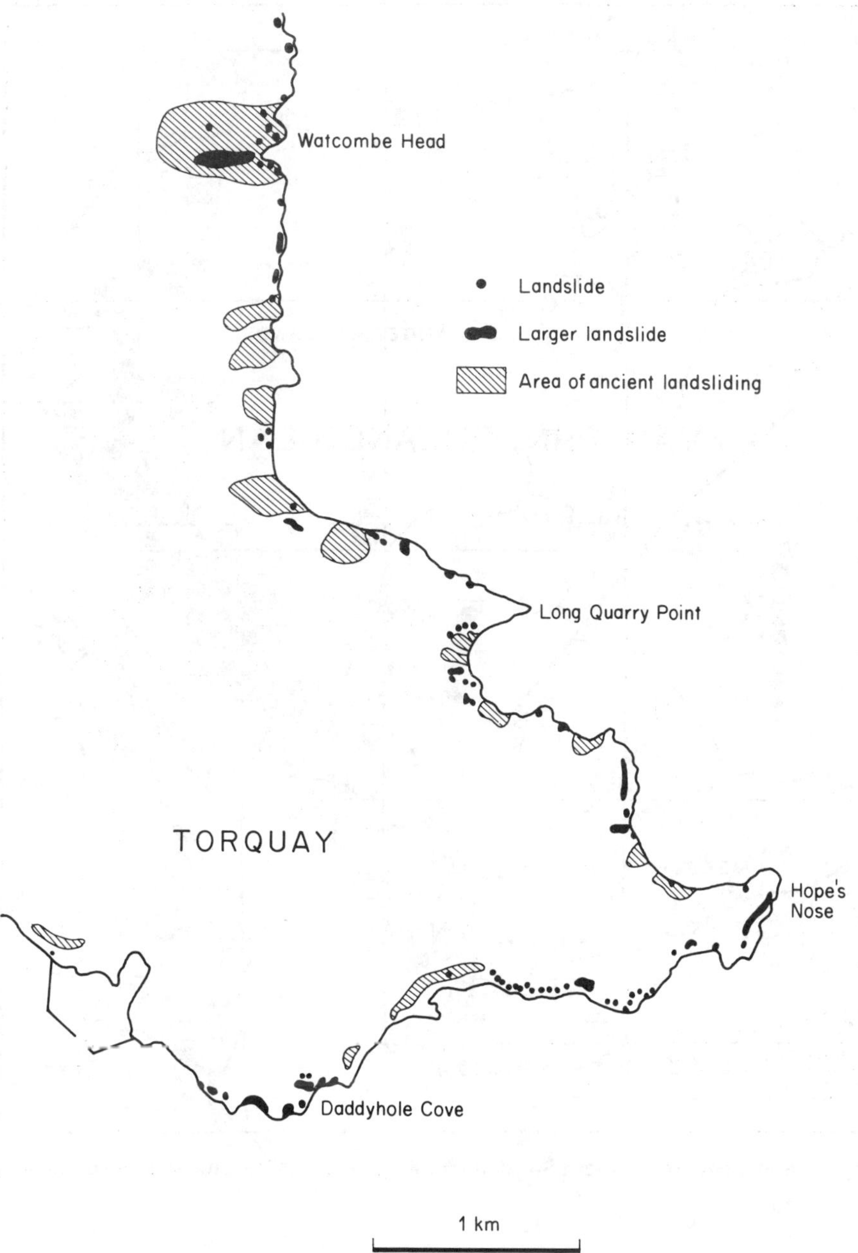

Fig. 4. Landslides of a portion of the Torbay area (simplified from original field and air-photo mapping by J.S. Griffiths).

completeness and map scale, the map user then comes closer to an understanding of some of the limitations attached to the use of point data maps.

An alternative form of showing point data is illustrated by Fig. 5, in which the distribution of caves in a part of northern Sweden is shown symbolically, with symbols chosen to represent cave density. The problem is that the symbols chosen carry no sense of increasing cave density with style or weight of symbol. If each of the symbols in the legend were to be cut out and shaken out of order, who would put them back into the original order knowing that they were supposed to represent different densities of cave concentration? The answer is probably no one. In fact, the case is made worse because the particular choice of symbols does not allow a logical sequence to be established. In the representation of point data cartographic design clearly has an important role. The style of representation should display a sympathy with the nature of the information being displayed.

Data analysis

The analysis of point data, where the points represent single observations (samples) of a continuous distribution, usually involves some form of contouring, and this is itself dependent on some form of interpolation between the data points.

Rockhead contours form a typical case, and many maps have the characteristics of the example in Fig. 6. Control points are unevenly distributed, and over quite large areas (e.g. northwest of Aldridge Shopping Centre in Fig. 6) none are shown. Closer examination reveals that crenulations in the contour pattern (i.e. an uneven rockhead surface) only occur in the vicinity of boreholes. The smoothness of the 'rockhead surface'

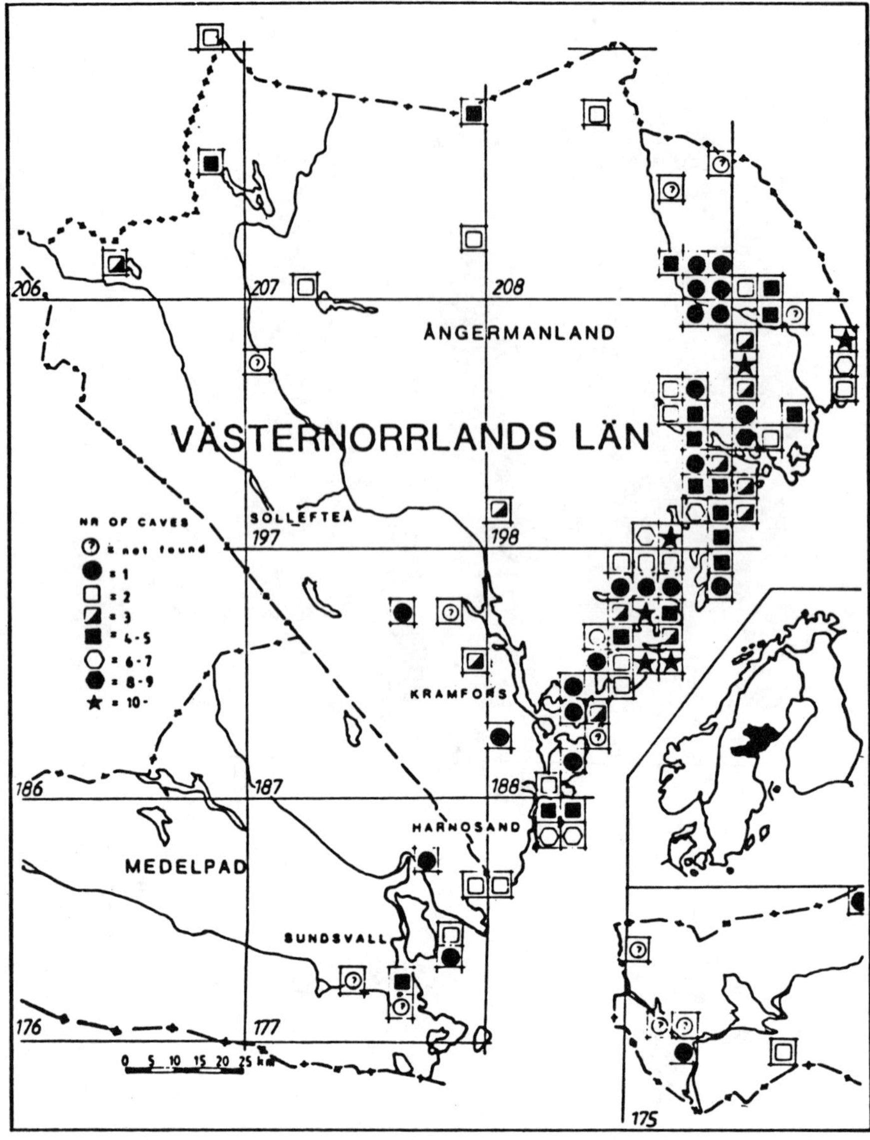

Fig. 5. Cave distribution in a part of northern Sweden (from Sjoberg, 1987). The area of the small squares is 25 km^2.

northwest of the shopping centre, where there are no boreholes, seems particularly 'unreal', and clearly reflects the absence of data points. In such cases the contours drawn between the data points (control points) tend to say more about the distribution of the sample points than about the surface which they are supposed to represent. Yet, as a map user's mind is drawn towards the contour pattern, the significance of the data point limitations may become lost. In some published examples the contours are shown and the original data points have been 'edited out', leaving the user without any objective basis for judging the control available in interpolation between data points.

This problem can be overcome when data are contoured automatically using a prescribed algorithm through a computer. The algorithm used may also allow a statistical confidence surface to be drawn whereby the user can then judge the statistical probability that the actual value of the parameter represented by the map lies within a prescribed range of the contour value shown.

LINE DATA

Data acquisition

While many earth science data are collected along lines few such data are actually linear in character. The prime exceptions include:

- linear geological features:
  faults, joints, dykes;
- boundaries:
  between strata, soils, land and water,
  around specific landforms;
- linear topographic features:
  rivers, ridge crests.

These all form real linear data sets in contrast to contours (and other iso-lines) which are convenient for representing the nature of a continuous surface (the behaviour of which may only be known at the location of point samples through which the contours have been drawn).

Lines of a topographic nature, such as rivers, may be surveyed to any level of detail and accuracy. At its

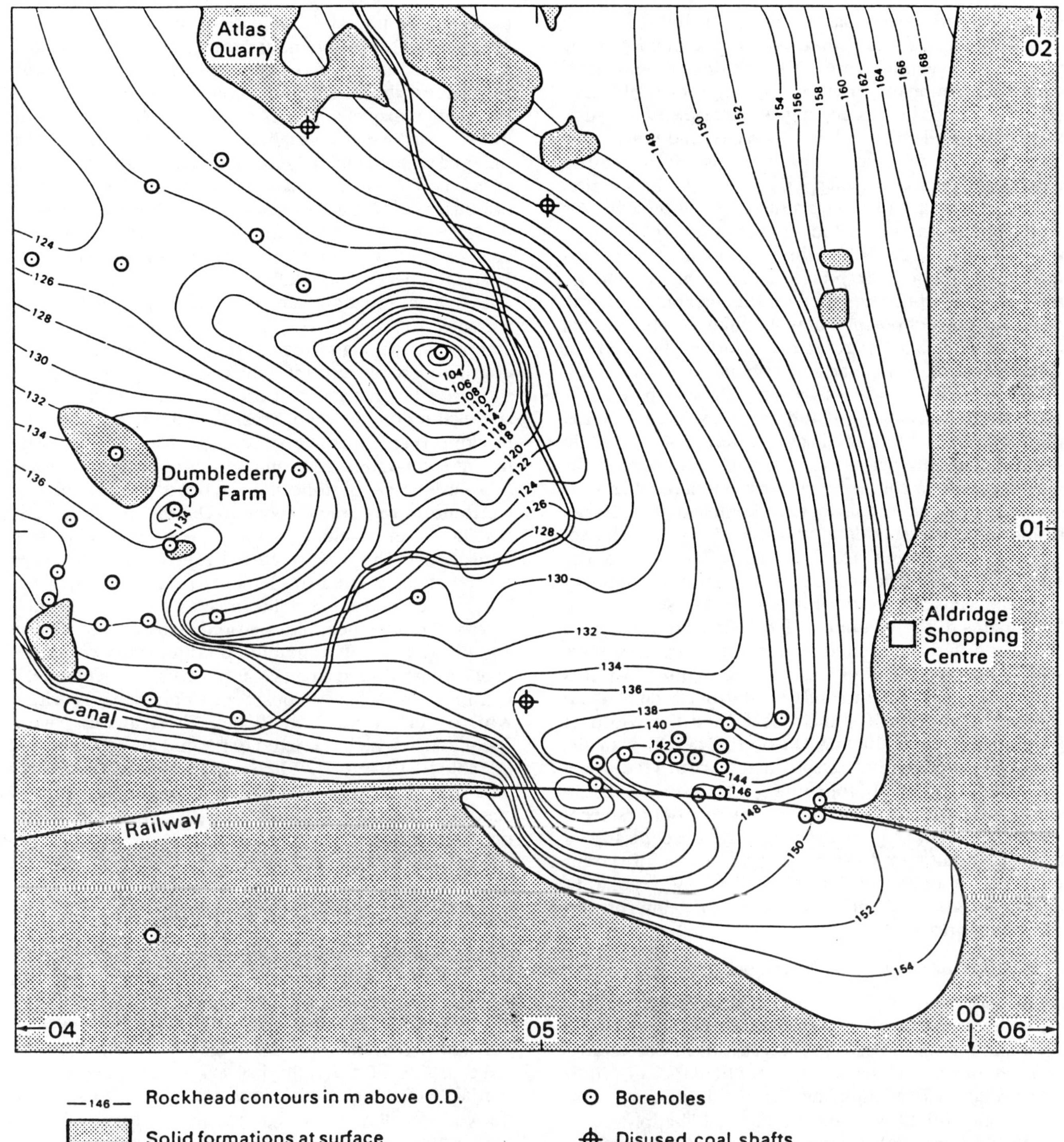

Fig. 6. Rockhead contours—Aldridge Shopping Centre, Birmingham. Reproduced by permission of the Geological Society from 'Geological factors in land use planning at Aldridge–Brownhills, West Midlands' by A.A. Wilson in *Planning and Engineering Geology*, 1986.

ultimate, the level of accuracy in surveying may well exceed the accuracy with which the feature can be plotted on a map. This is why so much survey data may be in a computer data base, from which calculations are made directly, rather than being transferred to a map. Decisions over level of accuracy in a survey may be a function of costs rather than of attainable precision. The tools exist, but the need to use them has to be matched to the financial resources available to employ them. Those interested in more information about surveying techniques should consult any modern textbook that includes a discussion of both 'distomat' instruments and satellite-dependent locational equipment such as satellite navigation systems.

The difficulties that face the earth scientist in mapping linear features is not one of surveying but one of identification and location. For example, the boundary between two strata may either be hard to identify when exposed (which is a geological problem), or obscured (which is an access problem). In the latter case (e.g. when the boundary is buried by soils and covered with vegetation) it is common practice for surrogate information to be used. The best example is that of using relief features and details of slope form to indicate where a geological boundary passes across a hillside. Such 'feature mapping' has been standard practice for many years in well vegetated areas. Clearly, since other causes exist to account for variations in slope form, absolute reliance cannot be placed on 'feature mapping'. In such situations the lines drawn lack veracity, and hence may be plotted out of their true position even if the feature used by the geologist is mapped precisely in the location where he saw it. In such circumstances, another factor which comes into play is the acceptable level of error for the map scale used. While the field scientist may have incorrectly judged the precise location of a boundary, the thickness of the line used on the map may well also cover the correct position of that boundary. The original misjudgement is no longer relevant. Indeed, he need not spend very long being particular about it.

Not all such boundaries are derived (in whole or in part) by field mapping. Other sources include aerial photography and satellite imagery. Here, a second scale which comes into play is the scale of the photography or imagery used. In this sense, as far as satellite imagery is concerned, scale should not be confused with image resolution (the size of detail which the image can show).

When considering aerial photography (or satellite imagery) as a data source for geological boundary mapping the scale factor provides two constraints. The first relates to the amount of detail that can be seen on the image, and the second relates to the accuracy with which observed boundaries can be plotted. In this context Table II shows the relationship between pen line thickness, air-photo (or map) scale, and the width of ground covered by the line. Observable detail on an air-photo is related to photograph characteristics (i.e. scale, lens used, type of photographic paper, altitude of the aeroplane, haze, and cloud cover), whether or not stereoscopes are used, and if so the type of stereoscope and its facilities. Additional factors include the type of terrain involved (especially its ruggedness, and the vegetation cover) and the skill and experience of the interpreter. Thus the phrase:

'map based on air-photo interpretation'

should stimulate questions about all of these factors. They determine how much reliance can be placed on the resulting map. This varies a great deal. Confidence in such maps may be increased when the air-photo interpretation is accompanied by field checking. The limitations of field mapping, however, still apply.

The problems of boundary mapping are further increased when the boundary is itself diffuse. Most of the problems are those of deciding where the boundary really lies, or indeed if a definable boundary really exists. This is a familiar problem for the soil surveyor, especially in cultivated areas where changes in soils may be difficult to relate to abrupt (distinct) boundaries. In many cases the change may take place over a zone. Again map scale plays a part, for at a detailed level the zone should be mapped as such and its outer limits defined; at a smaller scale a broken line may be used to signal 'diffuse boundary'; at a regional scale a solid line can be used to define the location of boundaries regardless of their sharpness in the field if only location, and not boundary character, needs to be defined.

Table II. Pen line thickness and width of ground covered

Line thickness on map (mm)	*Map scale*	*Ground width covered by line (m)*
0.1	1: 5 000	0.5
0.5		2.5
1.0		5.0
0.1	1:10 000	1.0
0.5		5.0
1.0		10.0
0.1	1:25 000	2.5
0.5		12.5
1.0		25.0

A similar situation can arise in the case of maps showing the boundaries between slopes of differing steepness. Although the features being mapped may be continuously present (and observable) in some places, they may be sharp and in others diffuse. This leads to the distinction between 'breaks' of slope and 'changes' of slope (Fig. 7). However, changes of slope at a sufficient level of detail become either convex or concave (upwards) areas of slope and can be mapped as such.

Attention was drawn above to the different scale and resolution in satellite imagery. The former is self-explanatory. The latter refers to the ground area covered by the smallest picture element (pixel) of which the image is composed. Satellite images are made up of a matrix of reflectance values, stored as digital information and printed as a photographic product (after computer processing). If such an image is progressively enlarged it can be seen to be composed of single squares within which there is no more detail. These pixels have a specific size which is related to ground area as indicated in Table III, Chapter 7. When using satellite imagery as a data source, such as in boundary mapping, the level of accuracy (normally) cannot be better than that of the pixel resolution.

Data representation

Linear data are normally represented by lines which almost invariably are not true to scale. Most of the earth science data shown by lines are of limited aerial extent in themselves. For example, the boundary between geological horizons, or the edge of a gorge, or the traverse line of a geophysical survey, all have limited spatial extent and are seldom drawn true to scale. In these circumstances no significance can be attached to the absolute width of a line as drawn on a map. What is more important is that variations in line thickness, far from indicating that a linear feature covers a variable width, are used instead to indicate its status in terms of importance or clarity. Figure 8 is an extract from a map legend showing that both line thickness and line style have been used to distinguished the regional importance of geological lineaments identified from satellite imagery.

Another important method of representing line data is to use continuous, broken, and dotted lines to indicate the degree of confidence held by the mapper in the line information provided (see Fig. 8). Such a system commonly leads to the annotations 'certain', 'inferred', and 'uncertain', or their equivalents. The last two of these in particular are subjective and are features of the map which are direct functions of the mapper's skill, which is itself a function of both ability and experience. The level of mapping skill is not defined on a map.

In some circumstances line data may be embroidered

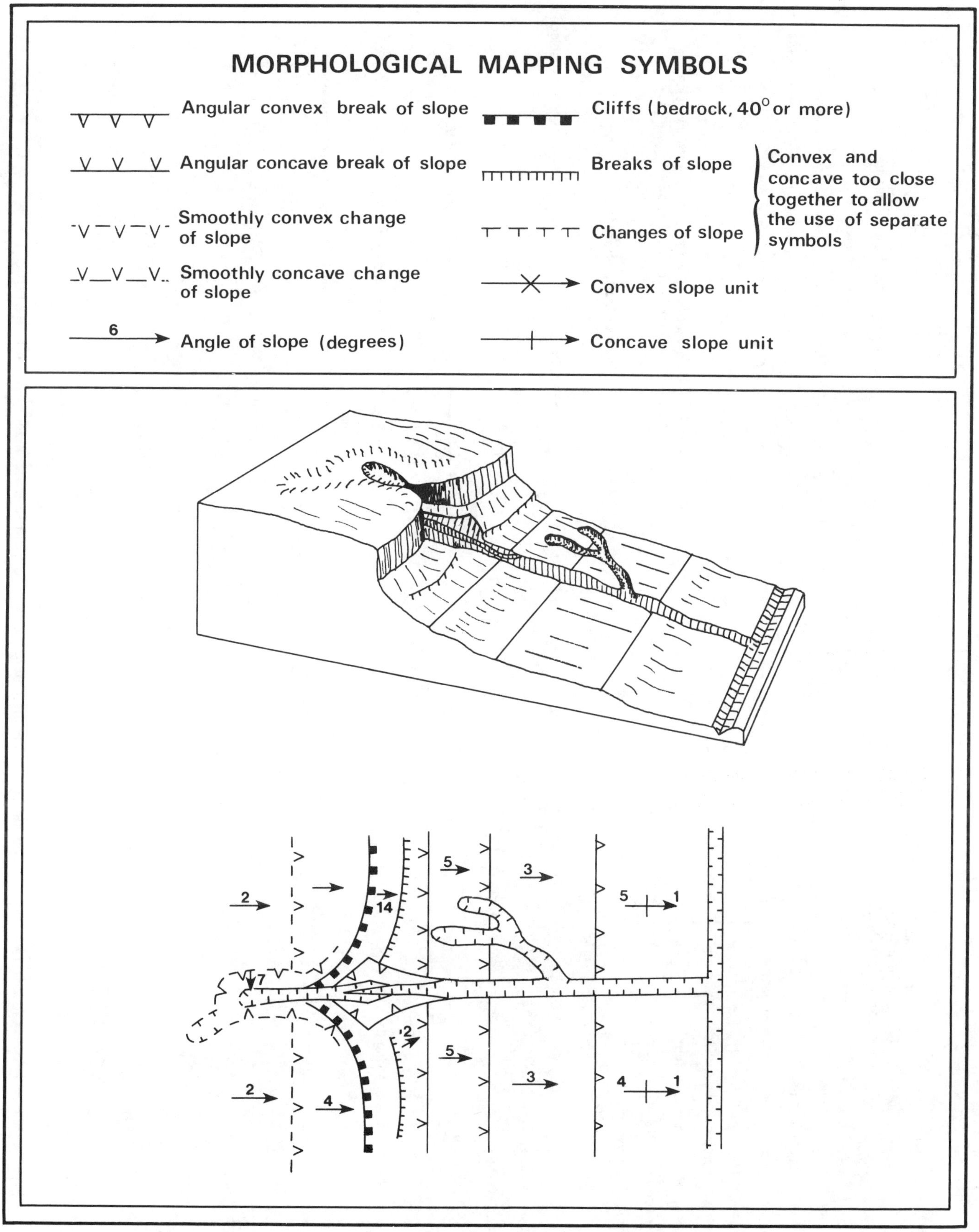

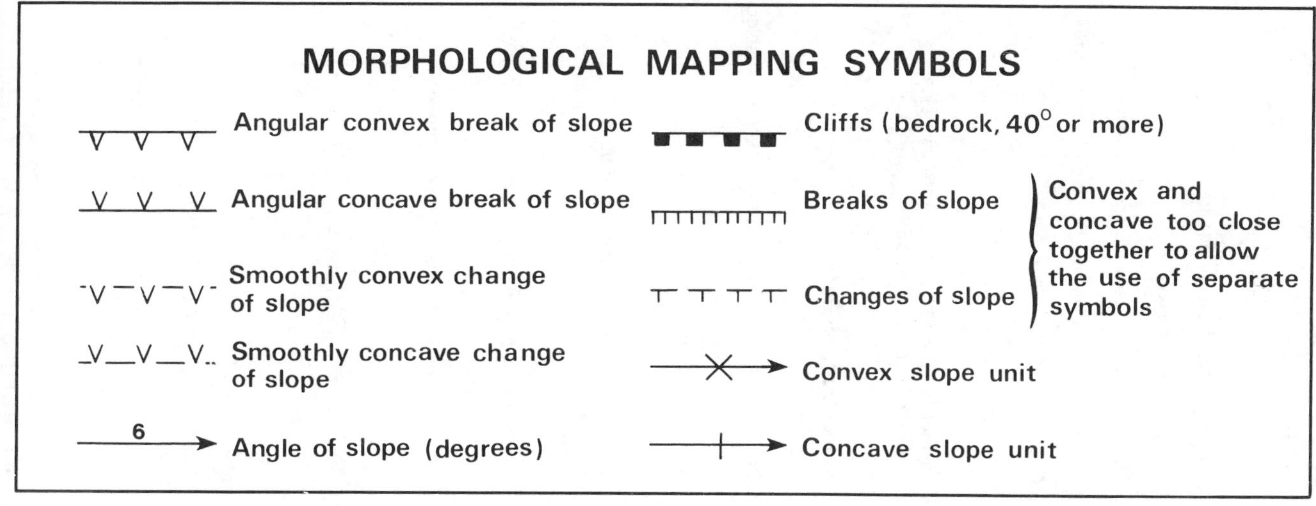

Fig. 7. Morphological mapping symbols.

with symbols. For example, the morphological mapping symbols frequently used in geomorphological mapping (shown in Fig. 7) use a line to show *where* the boundary lies, but a symbol to show what *kind* of boundary it is. Such symbols are seldom, if ever, drawn true to scale, indeed they may not have a dimension at all for they indicate a state rather than form a physical entity. They are a pictorial representation of reality as perceived by the mapper. It is at this point that artistry and map design meet science. There is a considerable skill in developing

LINEAMENTS

Butterknowle L.

Major / Intermediate / Minor — Topographic lineaments broken line indicates degree of uncertainty

Diffuse major tonal lineaments

Diffuse minor tonal lineaments

Landsat lineament coinciding with known fault

Landsat lineament extension of known fault

Known fault with no coinciding Landsat lineament

Landsat lineament subparallel to fault

(With box on downthrow side)

Landsat lineament

Linear glacial features

GEOLOGICAL FOLDS

Monocline

Monocline, inferred

Antiform, inferred

Synform, inferred

Direction of plunge of antiform

Antiform, with vertical axial plane

Antiform, with filled triangle showing dip of axial plane

Antiform, with unknown dip of axial plane

Synform, with vertical axial plane

Synform, with filled triangle showing dip of axial plane

Synform, with unknown dip of axial plane

Regional high axial trace

Regional low axial trace

(With plunge direction if known)

DIPS OF GEOLOGICAL SURFACES

Horizontal bedding

30 Strike and dip of bedding

40 Strike and dip of overturned bedding

Strike of vertical bedding

80 Strike and dip of cleavage or schistosity

90 Strike and dip of vertical cleavage or schistosity

3 Apparent regional dip, offshore

GEOLOGICAL FAULTS

Fault

Normal fault, with box on downthrown side and amount of dip if known

Normal fault, inferred

Strike-slip fault, certain, showing sense of displacement

Strike-slip fault inferred

Thrust or reverse fault, certain, with triangle on upthrown side

Thrust, inferred

Fault, with unknown sense of displacement

AREAS OF HIGH AND LOW MEASURED VALUES

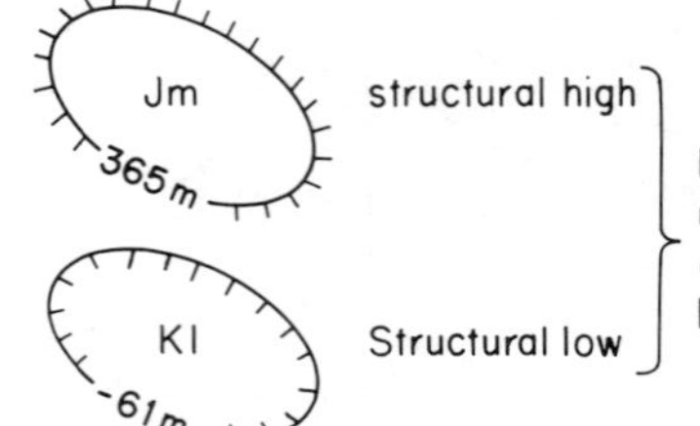

structural high

Structural low

Boundary defined by base of statigraphic unit (or contour in metres above / below O.D. of statigraphic unit) denoted in centre of high / low

Fig. 8. Extract from the legend of a map showing geological lineaments identified from satellite imagery (from JEBCO, 1987).

a set of symbols which are both sensible in terms of what they represent, and clear in terms of their cartographic design. 'Sensible' symbols are those which the map user can directly associate with their intended meaning. Thus the standard symbols for anticlines and synclines, strike and dip, sense of faulting, and the morphological symbols shown in Fig. 7, all fall into this category. The map user 'knows' what is meant by each of these symbols almost without the need to look at the map legend.

Clarity of design involves several distinct aspects of cartography, these are:

> size of symbol, style of symbol, and the distinctiveness of the symbol when compared with others on the map.

Size does not mean 'big' in this context. Big symbols can both obscure other data and over-emphasize the feature being represented. 'Size' refers to the drawing of a symbol such that it can be clearly seen and distinguished from other, perhaps similar, symbols.

Style of symbol is very important. Not only should it be easily associated in the map user's mind with the feature being represented, but its design should allow for ease of recognition.

It is crucial to the map user that one symbol can be easily distinguished from another. Obvious as this may seem, it is a rule that is frequently broken.

Data analysis

The analysis of linear data within the earth sciences tends to fall into the following categories:

- pattern analysis;
- network analysis;
- surface model generation.

Pattern analysis in its visual form usually breaks down into a search for trends within the linear data. In particular it may seek to identify the presence of a characteristic trend, or set of trends, within regional data. For such an analysis to be meaningful, however, it is necessary for the data to be either complete or to be an adequate sample. It is doubtful if these conditions hold in many cases. The biggest danger of all arises when the data set reflects what the interpreter is seeking rather than what is there to be recorded. This is a particular danger when the interpreter is working within the framework of a conceptual model. Such might well be the case, for example, when linears are being interpreted from satellite imagery. The element of subjectivity involved might well lead an individual, if unchecked, to single out a particular trend because of his preference for a particular style of structural evolution for that area.

Pattern analysis is the earth scientist's automatic reaction to contour (iso-line) data. In the simple example of ground surface contours, pattern is used to distinguish:

- highs and lows;
- rate of change from high to low (i.e. slope);
- direction of change;
- surface roughness;
- surface discontinuities.

Contour data, though defining along any one contour those points which have the same value as each other (such as height above sea-level), are often used to represent a continuous surface (i.e. the third dimension is often inferred by the map user). Contours may or may not do this task well. In general, the larger the map scale and the smaller the contour interval, the more precisely the whole surface will be represented. In geological work such maps are often made of surfaces below the ground surface, which are of particular interest. For many practical purposes a second set of lines may be required in order to refine the expression of form. In geomorphology these may be the morphological mapping symbols shown in Fig. 7. They allow discontinuities between slopes of different steepness to be defined regardless of their altitude. In geology, the contours on a stratigraphic horizon may be supplemented with the position of faults that bring about a rapid change in the elevation of that horizon.

Contour data are frequently used to derive other forms of surface representation (e.g. the construction of cross-sections). In so doing, the results can be no more accurate than the original contour data. In fact an element of subjectivity is introduced. While such a cross-section is 'controlled' at its intersection with surveyed contour lines, the behaviour of the surface between these 'control' points is drawn in subjectively. In many practical situations contour maps are used as a basis for deriving slope maps. Slope is determined from the contour spacing and the vertical difference in elevation between adjacent contours. The values derived can only be an average for the area between two adjacent contours; it can make no allowances for surface flexures which do not carry the surface across one of the contour elevations. Such slope maps may, therefore, bear only a modest resemblance to reality when the surface is irregular.

Contouring of the ground surface is a special case in which the surface is usually continuous and (usually) exposed. This may not be true for other 'surfaces'. For example, a lithological contact surface, a water-table, a landslide failure plane. The derivation of 'contours' for such surfaces relies on the analysis of point data as described above.

Networks, which by definition are composed of linear features, may or may not be exposed. Rivers are exposed; piping within soils is not. Yet both are amenable to network analysis. Indeed, physically speaking, the soil pipe system may be an extension of the network system for flows over the surface.

AREA DATA

Data acquisition

Within the earth sciences area data are often closely related to linear data. For example, just as the area of an outcrop is determined by the recognition and mapping of its outer boundaries, similarly with the mapping of soils and many geomorphological features. There is a tacit assumption in such area mapping that the properties (of the rock, soil, or landform) are reasonably uniform within the area identified, and differ in some significant manner from those in adjacent locations, hence the need to identify a boundary. Such an assumption of internal homogeneity may be difficult to justify where exposure is not continuous and sampling (or testing) is based on a limited number of observations (or sites).

This raises the question:

> 'How many samples (observations) are needed in order to define the internal variability of this mapped unit?'

The answer is seldom straightforward. Statistical tests

have been devised which define the extent to which a change in the number of observations is likely to influence the estimate of internal variability within the area defined.

Data density can also be used as a basis for deciding on the appropriate scale of mapping or the degree of generalization to be employed. In some cases the data may be so sparse (or so badly distributed) that the decision should be made not to produce a map at all for it may not carry sufficient veracity to justify its publication.

Within the earth sciences, however, an awareness of the properties of an area may be controlled by a conceptual model. For example, only a few exposures of a reef limestone may soon indicate to the specialist how the nature of the limestone is likely to vary across the whole of its extent. Such an assessment is based on the experience of reef limestones which becomes built into a mental model against which specific observations may be compared. To take another example, the morphology of a landslide may provide indications not only of its type, but also of its extent, and, to the specialist, of the likely depth of its slip surface. In this, and in similar cases where a conceptual model can be established, the process of 'sampling' becomes a process of 'testing' whether or not the model can be justified. This testing process may take relatively few observations to lead to an acceptance, modification, or rejection of the model. Very few geological maps actually indicate the extent to which homogeneity has been tested. They show boundaries based on an adopted classification scheme, but not internal variability.

The basis for most spatial data gathering has to be a classification system. In most mapping situations where the only data readily available are surface data, mapping has to be based on those properties that can be seen. Sometimes this is not the item of interest itself but some surrogate which has to be used in its place. For example, as discussed above, much geological mapping is based on the mapping of relief which is presumed to be associated with a particular geological horizon. In the mapping of soils the same approach is frequently adopted. The assumption of a direct relationship between land form and the underlying rocks or soils is therefore an important one. In East Africa it led to the development of the *catena* concept (Milne, 1935); in Australia it formed the basis of the *land systems* mapping approach (Christian, 1957) used by CSIRO in more than 30 reports, and in South Africa it became the basis for a national geotechnical data bank (Brink *et al.*, 1968). The nine-unit model of landform–soils–process relationships is also based on similar principles (Dalrymple *et al.*, 1968). While these notions of landform–materials relationships are seductive, the particular applications of these general ideas should always be well tested. Bridges and Doornkamp (1962) have shown, for example, that soil properties may vary independently of surface morphology even in an area where in general a close relationship between the two exists.

A further complication arises when not all properties of rock or soil change together. For example, the red soils commonly found over limestones may be mapped as such, and their surface properties may show little variation. However, this uniformity often masks a great variation in soil thickness. For practical purposes, such as engineering work, this variation is even more important than the physical character of the soil. The way that area mapping has accommodated this is to subdivide the material concerned into its significant component properties and to map each of these in turn. The scale of such a task, both in terms of time and cost, is so large that it has been done for very few areas, and published examples are rare. A practical way of coming to terms with these problems is for the map *user* to accept that the information provided by area mapping is essentially a function of map scale and the classification scheme adopted in the map legend. At one extreme the user should not expect a world map of soils, published at a scale of 1 : 200 mln to show detailed patterns, nor should he expect the classification used to be any other than general. At the other extreme he can expect a plan at 1 : 500 to be locationally precise and to carry a detailed classification. The map producer should meet these expectations.

THE THIRD DIMENSION

Since many geological features extend down into the ground, as well as across the ground surface, there is a need to represent this third (depth) dimension graphically. This is usually achieved through the drawing of vertical sections (sometimes in the form or ribbon diagrams), block diagrams, or through the portrayal on a map of data from depth as though it was a surface. Indeed, it is often very difficult for those not trained in Earth sciences to appreciate such vertical variations from maps and careful illustration is often especially beneficial.

Sections: Vertical sections display an interpretation of the behaviour of an earth science property (e.g. rock strata) under the ground. They are an interpretation based on surface observations, possibly backed by borehole or other subsurface data. If geophysics is employed the interpretation will normaly be derived from the reflectance characteristics of sound or shock waves transmitted from the surface. Ribbon diagrams illustrate conditions in a vertical section which follows a sinuous course.

Block diagrams: Block diagrams consist of a combination of three-dimensional views of the ground surface (or some other surface) along whose margins geological sections have also been drawn. This means of visual display is designed to combine in one diagram both the spatial characteristics of the surface and its depth dimensions. This type of representation is especially useful if the intention is to demonstrate the relationship between surface behaviour and subsurface properties.

Map of underground surfaces: In some studies, such as in the structural interpretation of oil-bearing strata, it is useful to portray the nature of the top-surface of a critical stratum. Data control may be provided by boreholes, by seismic surveys, or both. The normal method of portraying such a surface is through contours or three-dimensional (block-diagram) views.

THE ROLE OF COMPUTERS

Introduction

The dominant role of computers within earth science mapping is for data analysis and/or for map printing. Data analysis includes such techniques as the automated contouring of point data, the derivation of trend surfaces, digital terrain modelling from contour data or point data, and the calculation of statistical confidence in a derived (i.e. calculated) surface. Recent ancillary uses of computers include the automated production of borehole logs, the compilation of data acquired by means of electronic monitoring devices (e.g. river gauges, soil tensiometers, strain gauges, inclinometers, piezometers, and so on), but these are not considered here.

AUTOMATIC CONTOURING: The drawing of contours, whether by hand or by computer, involves an interpolation process between data points which establish a value (for that surface) at a fixed point. The greater the point density and the more widespread the data points, the greater the control on the contouring accuracy. The process of contouring by hand involves subjective judgements about the behaviour of the surface between the control points. Automatic contouring, by contrast, takes place in response to mathematically-based algorithms applied without subjectivity by computer. Interestingly, this does not necessarily mean that computer-drawn contours are more 'accurate' than hand-drawn ones. They may be more tightly controlled by the mathematical principles defined in the algorithm, but they frequently show no intelligent knowledge of the way that surface could reasonably be expected to behave in a particular earth science context. Hand-drawn contours do reflect that intelligence; however, this subjectivity may also introduce bias.

TREND SURFACE ANALYSIS: An alternative form of mapping through a set of data points that define a surface (or indeed the behaviour across space of any earth science parameter), is to identify its general trend. This involves computing a two dimensional regression surface through the data. If there are sufficient data points the regression surface may be linear, quadratic, cubic or some higher order surface. The surface can be plotted as a map to show the general trends in the data, and its regression coefficient can be calculated in order to measure the degree of 'fit' of the surface to the data points. In practice, trend surfaces are calculated as much to see which data points depart most from the general trend as they are to establish the trend itself.

STATISTICAL CONFIDENCE SURFACES: Just as a regression line provides an estimate of data values at particular points along a graph with a statistically definable degree of confidence, so trend surfaces provide estimates of parameter (surface) values across an area. The statistical level of confidence which can be associated with those estimates can be shown by the compilation of statistical confidence surfaces. These can also be contoured and displayed overlying the trend surface map.

DIGITAL TERRAIN (SURFACE) MODELS: The data points that represent a surface (or the behaviour of a particular parameter across an area) can be used to produce a three-dimensional view of that surface. This is normally done using a computer.

MAP PRODUCTION

Introduction

Whatever the nature of the earth science data to be recorded in map form, that map always has to pass through a printing process. There are a considerable number of alternative reproduction methods available. These range from the simple (e.g. dye-line printing or photocopying) to the very sophisticated (e.g. computer controlled direct output onto a laser plotter). This account will assume that the former are not used for multiple (or serious) publishing, and that the latter would normally be used to derive material *for* printing rather than being a means *of* printing. This may, however, change in the future. Between these extremes lie the standard means of colour printing—offset litho, which still remains a popular way of printing earth science maps. Approaches to map printing are principally determined by three factors:

- quantity required;
- size of the map;
- the use of colour.

Each of these help to determine the final costs of production, though in real life it may well be the other way round!

CARTOGRAPHIC CONSIDERATIONS

In cartographic terms, it is the *data type* rather than the specific information that affects the techniques of mapping employed. This chapter has concentrated in the main on data arising out of field mapping rather than on the display of statistical data which happen to have a spatial component and involve the earth sciences.

The second important control on cartographic methods is the *scale* of the maps to be produced. Scale greatly influences both the clarity and the content of the final maps, and both of these factors affect the information passed on to the map user. For the purposes of this account the scale distinctions identified in Table I will be used. The choice of map scale is probably the single most important decision to be made when considering map presentation. Not only does scale relate to the size of the area to be portrayed, it also affects the precision with which the location of information can be shown. It also determines how much can reasonably be shown before clutter replaces clarity.

An increase in clarity, and hence effectiveness, is often obtained by the use of *colour*. However, representation in more than one colour increases costs, especially when multiple copies are required. The precise way in which colour can be used, and the costs involved, are determined by the number of copies required, the methods of production used and the method of printing. A distinction also needs to be made between maps produced by traditional methods, using hand-drawing techniques, and those maps resulting from the use of computer-based automated cartography.

All map users need to recognise that there is a fundamental difference between the reliability (veracity) of the data or information shown and the accuracy with which that information is plotted on a map. For example, an incorrect observation may be very precisely located; equally a valid observation may be incorrectly plotted. The hope, of course, is that both the data and the cartography are precise, and that the end-user of the map is thereby properly informed. Naturally the cartographer is also restricted by factors such as the width of line or size of dot which can be used. When compared with map scale, these will in turn determine how precisely the position of a feature can be shown (see Table II).

USING EARTH SCIENCE MAPS

From this account it is clear that both the veracity and the accuracy of earth science maps are dependent upon many factors. These range from the ease with which the data can be obtained to the cartographic methods employed in presenting them. It is also clear that the more aware the user is of the constraints under which some maps have to be produced the more he will know how much reliance he can place upon them.

A series of applied earth science maps were produced in 1988 (commissioned by the Department of the Environment) for planners, developers and engineers in the Torbay area of England (Doornkamp, 1988). These maps, which included each of the topics shown in Table III, and cover about 70 km^2, were compiled from existing data sources, supplemented by the minimum amount of air-photo interpretation and field inspection. The data for these maps were collected over a period of about 12 months. It was never intended that this set of maps should take away the need for site-specific investigations. Indeed, one of the functions of these maps is to warn of ground difficulties that should be investigated within identified areas.

Table III. The map topics included in a suite of applied earth science maps for planning, development and engineering in the Torbay Region, Devon, UK

Bedrock geology
Superficial geology
Geomorphology
Slope steepness
Soils
Geotechnical conditions
Sites of geotechnical investigations
Sites of mineral workings
Land use planning provisions
Ground characteristics for planning and development

These maps are presented at a scale of 1 : 25 000, and use the Ordnance Survey map of the area as a base. This allows individual buildings to be identified. Clearly there is a dilemma. While on the one hand, the stated intention is not to provide site specific data, on the other hand, specific sites can be identified on the map. The intelligent and informed user would understand the context and limitations on the maps. Others are in danger of accepting the maps at face value and without regard to the methods whereby they were compiled. For this reason each of the maps carries an 'earth science health warning' that:

> This map provides only general indications of ground conditions and must not be relied upon as a source of detailed information about specific areas, or as a substitute for site investigations or ground surveys. Users must satisfy themselves, by seeking appropriate professional advice and carrying out ground surveys and site investigations, if necessary, that ground conditions are suitable for any particular land use or development.

Every user of earth science maps might do well to adopt this approach unless there are firm reasons for doing otherwise.

REFERENCES

Bridges, E.M. and Doornkamp, J.C. (1962). Morphological mapping and the study of soil patterns. *Geography*, **48**, 175–181.

Brink, A.B.A., Partridge, T.C., Webster, R. and Williams, A.A.B. (1968). Land classification and data storage for the engineering usage of natural materials. *Proceedings 4th Conf. Australian Road Res. Board*, 1624–47.

Christian, C.S. (1957). The concept of land units and land systems. *Proc. 9th Pacific Sci. Congr.*, **20**, 74–81.

Dalrymple, J.B., Blong, R.J. and Conacher, A.J. (1968). A hypothetical nine-unit land surface model. *Zeitschrijft fur Geomorphologie*, **12**, 60–76.

Doornkamp, J.C. (ed.) (1988). *Planning and Development: Applied Earth Science Background, Torbay*. Geomorphological Services Ltd., Newport Pagnell, UK.

JEBCO (1987). *A structural Atlas of Britain based on Satellite Imagery*. JEBCO, London.

Milne, G. (1935). Composite units for the mapping of complex soil associations. *Trans. 3rd. International Congr. Soil Science*, **1**, 345–7.

Neall, V.E. (1976). Lahars as major geological hazards. *Bull. Int. Assoc. Engng. Geol.* **14**, 233–240.

Sjoberg, R. (1987). An inventory of caves in the County of Vasternorrland, N. Sweden. In *International Geomorphology, 1986*. (Gardiner, V., ed.). Wiley, Chichester, UK, 1191–1198.

UNESCO (1976). *Engineering Geological Maps: A Guide to their Preparation*. UNESCO, Paris.

Wilson A.A. (1986). Geological factors in land use planning at Aldridge–Brownhills, West Midlands. In *Planning and Engineering Geology* (Culshaw, M.G., ed.), Eng. Grp. Geol. Soc., London, 51–64.

Chapter 3

Data Handling and Examples of Data Systems

Joseph McCall

INTRODUCTION

Map preparation depends on the organization, plotting and interpretation of data. The purpose of this chapter is to review the collection of data, various methods of handling the data once collected, and its storage, updating and retrieval. This chapter is related to the previous chapter and includes discussion of data handling of maps, plans, sections, three-dimensional representations (such as block diagrams and ribbon or fence diagrams) and actual solid block models.

Both traditional 'on paper' (manual) methods and also modern computerized methods of data storage, handling and retrieval (data bases) are considered, together with conventional drafting of maps, sections and three-dimensional diagrams, and computerized cartographic systems.

THE NATURE OF INFORMATION

What sort of information may be included in a data bank prepared for earth science studies related to planning and development? The short answer is anything at all that may be relevant to the project, and, besides geological and geotechnical information, information related to planning, such as constraints on land use, should be included.

The information may be:

(1) *Descriptive*: for example, a geologist's description of an outcrop or quarry; a petrographer's description of lithology and microscopic character of a rock sample; or a structural geologist's description of field-observed structures. Such descriptions may involve ***narrative***, ***numerical values*** (for example the angle of dip of the strata), ***physical properties***, and ***mineralogical/chemical compositions***.

(2) *Classificatory*: making use of various widely accepted earth science classification systems (for example, ***rock***, ***mineral***, ***palaeontological*** or ***soil***).

(3) *Interpretative*: such information may consist of the geologist's interpretations of the recorded facts and is a subjective and less stable component of the data base. Geology is a discipline which leans heavily on interpretation (for example, the author of this chapter was one of four successive geologists engaged by Broken Hill Proprietary to map the Carnilya Hill nickel deposit in Western Australia. All four interpretations differed: none was 'correct' or 'incorrect', but each interpretation told the mining company something new and of value. The diversity of interpretation relates to different training, different experience, and simply different thought processes. Interpretative material within data bases should be clearly identified as subjective. It is particularly important that interpretative maps and sections should be clearly labelled as such, as opposed to factual maps and sections.

Primary and secondary information

In many cases it is necessary to rely, partly or wholly, on a newly commissioned 'Primary Survey'. Such surveys may consist of:

(1) *Field surveys*: on the spot site investigations includ-

Earth Science Mapping for Planning, Development and Conservation

ing field testing (for example, measurement of structural surfaces, penetration testing, geophysical tests).

(2) *Laboratory testing*: almost invariably ancilliary to field investigations—field activity of some sort is required just to obtain samples for testing. Examples include lithological/petrographic description, physical tests on samples, chemical analyses and laboratory based engineering geology tests listed in Chapter 8.

It is less costly to rely, partly or wholly, on 'Secondary Sources', which include:

(1) *Published work*.
(2) *Unpublished work*: archives and files, borehole records, site reports, mine records, lists and descriptions of designated areas such as archaeological sites and sites of special scientific interest.

LOCATING DATA

The location of all the potential data sources is a preliminary step in any project. If the project concerns a specific locality, a reconnaissance will be required in the area to derive a list. The local authority, for example, invariably employs staff who have long experience of the area and will know of any engineering and development work, past, present and projected, in the area. They will possess invaluable data on planning and on designated conservation areas such as sites of special scientific interest, nature reserves and archaeological sites. Water, electricity and gas authorities and the local museums are all fruitful areas for research. Some large central governmental bodies (e.g., in England, the Department of Transport and Property Services Agency) are likely to hold subsurface records; the British Geological Survey holds records of all deep boreholes, all hydrocarbon wells (onshore and offshore) and all water boreholes and wells; the Ministry of Agriculture, Fisheries and Food, and the Forestry Commission and the Soil Survey are all valuable information sources. Materials testing laboratories maintained by local authorities to support site investigations are also an invaluable source. In the private sector, engineering consultants, site investigation contractors, developers and borehole/well sinking contractors are all worth approaching for site investigation data.

A data collecting exercise may relate to a small-scale study, such as a site investigation for which such preliminary desk studies are considered essential (see *PSA/CIRIA Site Investigation Manual, Spec. Publ. No. 25,* and *Code of Practice for Site Investigation, BS 5930,* 1981). It may relate to a medium-sized study (for instance, a conurbation or region) or to a very large-scale coverage (for example, the British National Geoscience Datacentre, the USGS Earth Science Directory (Clayton *et al.*, 1987), and the Australian Earth Science Information System (Anon, 1988)). There are even extremely valuable global earth science data banks.

THE ESSENTIAL STEPS IN MAKING AND USING A DATA SYSTEM

The essential steps in making and using a data system are:

(1) collection of data;
(2) storage and organization of data;
(3) retrieval of data in a form suitable for use.

Collection

The collection of all available relevant background information is an essential first step in any earth science study related to planning and development.

The collection of existing data cannot, obviously, be done without financial expenditure. This is true of the smallest-scale coverages and, at the other extreme, of national data banks and global data banks, which may be immensely expensive. However, the cost in real terms of carrying out the investigations which provide the original data was far greater than the cost of collecting and collating the information. Clayton *et al.* (1987) have made this point, and also remarked on the cost in financial terms of the failure to obtain the appropriate information. Where there are no adequate data available, new investigations will have to be initiated, to cover the data requirement of the project in hand, in whole or in part, and this will be far more expensive than utilizing existing high-quality data.

Once a list of potential data holders is obtained, a standard letter can be sent out, followed by a visit if the reply is encouraging. Since one is asking a favour and taking up a data holder's time, the manner and wording of an approach is all important. A diplomatic approach is needed and it is best if the data collector is a reasonably senior person. The donkey work in collecting the data will usually be done by the data collector rather than the holding organization, which is usually not prepared to provide their own staff, though it may prefer to do so. After the initial approach, a junior assistant can collect the data at the source office, but there is a great advantage if an experienced geologist handles this too, in order to evaluate the significance of data fully; otherwise important data may be rejected, or valueless data collected by an inexperienced operator.

Accessibility of data may present a problem. Although a few organizations simply do not want to be bothered, in general most are very co-operative. The main problems lie in confidentiality, possible commercial value of the data, and possible legal risks in supplying data. Most studies of the sort we are concerned with here are carried out either by governmental or local authority agencies (or their contractors) and confidentiality constraints can be surmounted by release on a confidential basis (Clayton *et al.*, 1987). Such data is used in a general way in the course of the study, but request for access by the public must be referred back to the source. This system is particularly applicable to data supplied by commercial companies and from military sources. Possible commercial value may lead to refusal to supply data or the request for an impossibly high price; such data must simply be marked as unobtainable. Some data holding organizations insist that the data collector be a member of a recognized professional body and thus bound by an ethical code.

The legal risk aspect is far too complex to enlarge on here at length, but it does not very frequently arise. One such bar on the supply of data is possible sterilization of property or land (by virtue of releasing some adverse geotechnical information). Also, certain information is unlikely to be released by local authorities to prevent speculation in land (for example, as noted in the case of the Plymouth Project described in Chapter 9, the County Council would not release information on potential land fill sites for waste disposal for this reason). The whole area of mining and quarrying is a very difficult one because all technical information is held by the operators, who may co-operate if they see a benefit for themselves in the study, but otherwise may be uncooperative.

The collection of data is a specialized task and ideally the geologist involved requires both experience of data collection and knowledge of the requirements of the particular project. Data collection should not be regarded as a purely mechanical exercise; it requires discrimination and appreciation of the content of the data *while collecting*. For example, a motorway data bank will have up to a score of volumes and dozens of maps and plans applicable to a single sector a few miles long. From this, the collector must cull, with discrimination, a selection of relevant descriptive sections and quantitative subsurface information (logs of boreholes, trial pits, etc.), and any significant interpretations, comments and recommendations.

Three forms of data collection at source are possible:

(1) taking written or audio dictation notes;
(2) photocopying of selected material;
(3) using a portable microcomputer.

To ensure that data collection is systematic, complete and rapid, it is generaly an advantage to use check sheets. However, these are no substitute for the full source document and may lead to a neglect of relevant aspects not formally listed, which may be uncovered unexpectedly. There is great advantage in using a portable microcomputer for direct input, since the need for reprocessing of data is eliminated.

If geological base mapping or remapping is a requirement of the project specification, a simple single page format with room for additional data and sketches on the blank back page can be adopted (e.g., the system illustrated by McCall and Simonian, 1985, used for a mineral resource related mapping programme in Iran). This system was designed with the data sheets printed in ring bound booklets of 50 pages and had a serial number for each page which was the data point number and specimen serial number. The system was easy to handle in the field, avoided numerical duplications and loss of loose pages, and the books are very easily archived. This system is illustrated in Fig. 1. It was designed for microcomputer storage as well as paper archiving, although computerization was never achieved. A simple system like this could easily be adapted to include required planning and geotechnical information, and one could ring bind more than one leaf per data point, if necessary.

A more specialized and full example, comprising four data sheets is the Check Sheet System adopted by Ove Arup and Partners in a current nationwide study of mining related subsidence, commissioned by the Department of the Environment, London. One of these sheets is illustrated in Fig. 2.

Validation of data

There are two important aspects of validation of data:

(1) discrimination on reliability, quality, etc.;
(2) distinguishing between factual data and interpretation.

Discrimination on reliability, quality, etc.

It is helpful to have some sort of grading system built in to the archive. This is particularly relevant to computer archiving. For example, one can have a numerical system (*1, *2, *3) indicating the collector's grading of reliability quality, and significance to the project. This is important in the construction context, since site reports more than about 10 years old are commonly based on out-of-date testing methods and may be of little value numerically, but the subsurface log of the geological formations encountered is still of real value. It is a peculiar fact that one can encounter geological records of observations in the field that are a century and more old and are as accurate and useful as any that could be produced by a present-day geologist. Geological field observation does not necessarily decay with time, but geotechnical records have become of less value with the passage of time because of the rapid development of new and improved site investigation procedures and testing methods. The validation in this respect will be improved if a space is provided for the collector to write in a concise assessment of the quality and reliability.

It is very important to know the purpose for which the data was collected (which may be quite different to that for which it is now required and may affect its validity): when, how and by whom it was collected and interpreted/calculated.

The numerical system of grading is useful but cannot replace a collector's written assessment, and there is no reason why the latter, though subjective to a degree, cannot at least in summary form, be built into a computer programme.

Distinguishing between factual data and interpretation

From the point of view of manual (paper) systems, such discrimination can only really be made by the user, when retrieving the data. Computerized systems can include not only the facts, but also the subjective interpretations of one or more observers. For example, the formalized system illustrated in Fig. 2 has a section for 'Mineral Valuer's Report'—this is subjective; two mineral valuers might well disagree.

Geologists' interpretations of outcrops or borehole sections/core might show even more marked disagreement. Again, in the system illustrated in Fig. 2, there is a section for 'Brief Summary of Recommendations', another subjective component. It is an obvious advantage if the computer programme discriminates clearly between subjective and factual content, but it is surely not essential; this is something that the trained user should be able to do, and even a planner or developer using engineering geology site reports, for example, should be able to recognize subjective elements and treat them with caution.

Storage

Most data banks collected for purposes relating to planning and development will be designed for permanent storage and updating, but some, related to individual projects, will be compiled, used and ultimately disposed of. This is to be deplored because there is a strong case for comprehensive archival systems, covering all ground and subsurface site investigations. Deep well and borehole information is subject to statutory archiving by the British Geological Survey (water wells and boreholes over 15 m deep, mineral bores and wells over 30 m deep), but so far in Britain only pilot schemes have been carried out covering constructional site

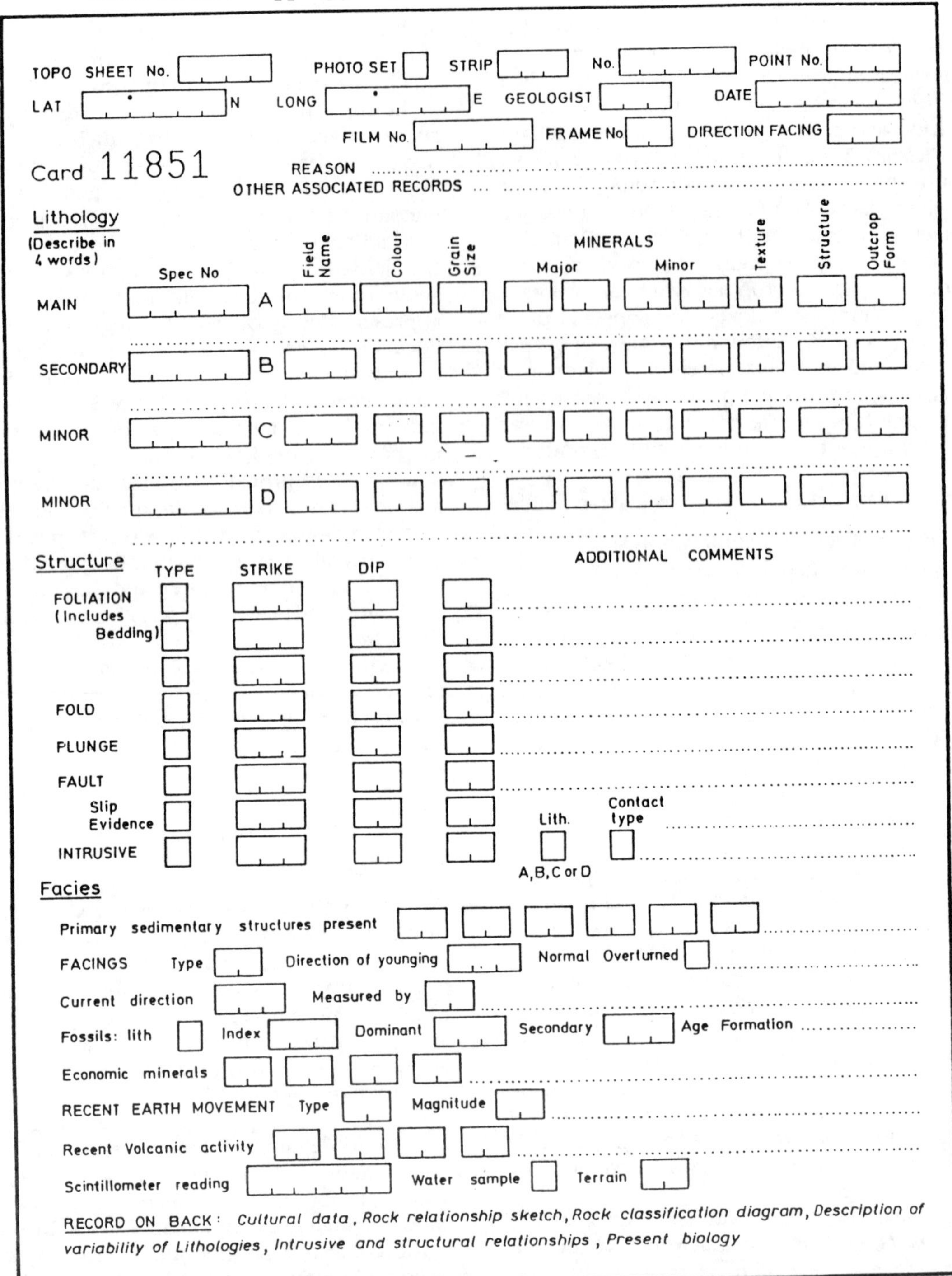

TOPO SHEET No. PHOTO SET STRIP No. POINT No.
LAT N LONG E GEOLOGIST DATE
FILM No FRAME No DIRECTION FACING
Card 11851 REASON ...
OTHER ASSOCIATED RECORDS ...

Lithology
(Describe in 4 words)
Spec No | Field Name | Colour | Grain Size | MINERALS Major | Minor | Texture | Structure | Outcrop Form
MAIN A
SECONDARY B
MINOR C
MINOR D

Structure
TYPE | STRIKE | DIP | ADDITIONAL COMMENTS
FOLIATION (Includes Bedding)
FOLD
PLUNGE
FAULT
Slip Evidence
INTRUSIVE
Lith. Contact type
A,B,C or D

Facies
Primary sedimentary structures present
FACINGS Type Direction of younging Normal Overturned
Current direction Measured by
Fossils: lith Index Dominant Secondary Age Formation
Economic minerals
RECENT EARTH MOVEMENT Type Magnitude
Recent Volcanic activity
Scintillometer reading Water sample Terrain
RECORD ON BACK: *Cultural data, Rock relationship sketch, Rock classification diagram, Description of variability of Lithologies, Intrusive and structural relationships, Present biology*

Fig. 1. Example of a field data sheet as used in Iran (from McCall and Simonian, 1985).

investigation reports (Anon, 1985; Tuckwell and Sadgrove, 1977; British Geological Survey, 1985; Clayton *et al.*, 1987). The permanent archiving of site investigation reports is an obvious ideal to be aimed at for the reasons stated above (under 'Collection'); high-quality geotechnical data can be used again in new site investigations or more general geotechnical investigations. In addition, any subsurface information may contribute to a closer definition of the surface and subsurface geology. In the case of the Plymouth–Plymstock environmental geology mapping study mentioned in Chapter 9, for example, numerous very minor and shallow site investigations allowed a much closer definition of geological boundaries than was possible for the geologist, Ussher, in the 1890s, working almost entirely from surface mapping (McCall *et al.*, 1985). At present deep penetrations are well covered by data banks, but this is not true of shallow penetrations which are of critical importance in the environmental geology context.

For the purpose of this chapter we shall consider that permanent storage and continual updating are in all cases to be aimed at, if at all practicable.

The storage of data presents a major problem of space availability and cost. The first problem may be lessened by the use of computerized systems, though most systems require archival of the original source data in full for future reference. The main requirement is that there is adequate staffing, funding and space to maintain the data bank and update it, and that it is readily accessible to users with easy and not too expensive retrieval systems.

REVIEW OF MINING INSTABILITY IN GREAT BRITAIN for the Department of the Environment	Region 3
	County 4
TECHNICAL REVIEW 10km GRID SQUARE DATA	District 5

SITE INVESTIGATION DATA (Complete a separate sheet for each investigation)

PROJECT DETAILS	GRID REFERENCE 11	LOCATION 12		
	INVESTIGATION FOR	NEW DEVELOPMENT 13	REMEDIAL WORKS 14	
	BRIEF DESCRIPTION 15			
TYPE OF REPORT	DESK STUDY 21	FACTUAL REPORT 22	INTERPRETATIVE REPORT 23	
	MINERAL VALUERS REPORT 24	BRITISH COAL REPORT 25		
LOCAL AUTHORITY	HAS ACTIVE MINING SECTION 311	WAS ACTIVELY INVOLVED 312		
	GAVE GUIDANCE/DIRECTION 313	AT PLANNING APPROVAL STAGE 314	AT BUILDING REGULATION STAGE 315	
	SUPERVISED/INSPECTED THE WORKS 316			
INVESTIGATION TECHNIQUES	CORED BOREHOLES 41	No 411	DEPTH	FROM 412 TO 413
	PROBEHOLES 42	No. 421	DEPTH	FROM 422 TO 423
	TRIAL PITS 43	No 431	DEPTH	FROM 432 TO 433
	GEOPHYSICS 44	TECHNIQUE USED 441	CCTV 45	ULTRASONICS 46
	SURVEY OF MINE BY ENTRY 47	SAFETY MEASURES DURING INVESTIGATION 49		
	ROCK STRENGTH TESTS 48			
	ALTERNATIVES TO INVESTIGATION 50			
FINDINGS 61	MINERAL DEPOSIT FOUND 611	FROM 612	TO	(m)DEPTH
	WORKINGS FOUND 62	FROM 621 TO 622	(m)DEPTH	OPEN (VOIDS) COLLAPSED BACKSTOWED
	MINE ENTRIES FOUND 63	SHAFTS 631	ADITS 632	
BRIEF SUMMARY OF CONCLUSIONS 71				

Fig. 2. Example of one of three data sheets in a set used by Ove Arup and Partners in a nationwide study of mining-related subsidence.

The traditional paper systems were based on files. An example can be found at the Central Engineering Laboratory at Cardington, Bedfordshire, where there is a large collection of government site reports, operated by the Department of the Transport. Such file systems are still necessary even when computerized systems are used, unless the entire source document is put into the computer data bank, something that only very large and expensive systems can handle.

Collation of data

The compilation of data, other than very broad compilations such as a National Archive, is carried out for a purpose of interpretation for the benefit of the particular study in hand. Thus the data base, other than the exceptions noted above, must be tailored to the project. It is pointless and costly to collect, collate and store irrelevant data. There is no standard format for a data base; each one is to a degree *sui generis*. Time spent in careful design of the data, collation, storage and retrieval system is seldom wasted and careful design of the programme will cut the costs substantially.

If the data base is for permanent archival and reference, it will require adequate storage capacity, both on computer, in back-up computer systems (to guard against accidental wiping off) and for paper files, maps, plans, etc., which represent the original source data not accommodated fully by the system. Space must be allowed for the increments of routine or continuous updating.

The purpose of the data collection must be borne in mind in designing storage and retrieval facilities. Access by the potential user and ready retrieval are *all important*, and these considerations may vary depending on the purpose of the collection.

There is a problem of *resolution* facing collation of data: the problem of data compatibility when derived from documents of widely varying scales, and this problem must be surmounted in the design.

Retrieval

A simple retrieval system may consist of a traditional, manual, card index related to the paper file storage of the original documents, or, more commonly nowadays, it may take the form of a computerized retrieval system. Retrieval of actual source documents still commonly involves retrieval of paper files in original or copy form, though in some very complex computer systems even this is eliminated by access through a display screen or print out to actual facsimiles of such original documents. Site reports are commonly lengthy, detailed and of varying format, and contain many diagrams and even handwritten marginal notes. The problem of storage in entirety is an immense one and the funds available will usually allow no more than a computer storage of a summary of the salient aspects, leaving the full report on a paper file, for retrieval if required.

SPATIAL CHARACTERISTICS OF OBSERVATIONS AND MEASUREMENTS

Data can be classified geometrically into:

(1) point data;
(2) linear data;
(3) area data.

These distinctions are very important in relation to map presentation and have been enlarged on in the previous chapter in the context of maps. However, the spatial characteristics of observations and measurements are also of critical significance in constraining the computerization of data, and especially in the design of direct linkages between computerized data storage systems and computerized cartographic systems.

Figure 3 shows the essential distinctions:

(1) Point at surface;
(2) Point below surface;
(3) Line at surface;
(4) Line below surface;
(5) Area (plane) at surface;
(6) Area (plane) below surface.

A point may represent the end of a line where it meets a surface, a line may be the trace of a plane on a surface, and lineaments (lines) may be formed where two surfaces intersect (very important in structural and ore mineral geology). It is critical to realize that a map is only a planar view of an arbitrarily chosen plane (that of the Earth's surface in most cases). Two-dimensional representations such as maps and cross-sections are inadequate in themselves to provide three-dimensional representation, but the two in conjunction do show the three-dimensional picture, though not as well as block diagrams or models. Manual preparation of block diagrams requires time-consuming drafting, and models are even more time-consuming and are also difficult to store.

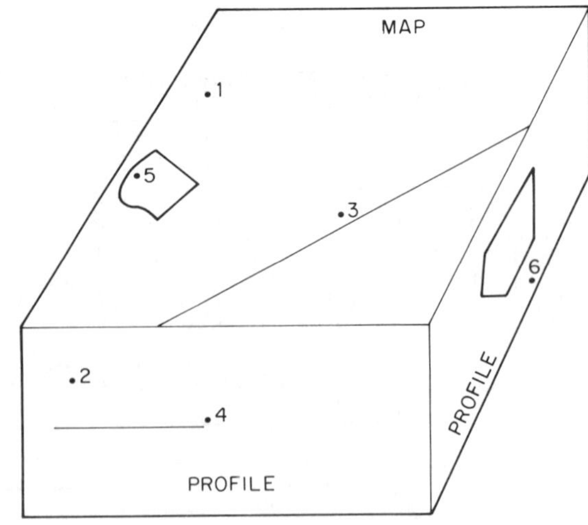

Fig. 3. Spatial types of observation and measurement. (1) Point at surface; (2) Point below surface; (3) Line at surface; (4) Line below surface; (5) Area (plane) at surface; (6) Area (plane) below surface. A point may be the end of a line. A line may be the edge of an area (plane).

Even so, using three-dimensional solid models can be very useful for demonstration purposes. The Shrewsbury sand and gravel study (Crimes *et al.*, 1985) cited in Chapter 5, was supported by a coloured three-dimensional model which opened up into segments both longitudinally and transversely, the profiles being drawn on all sides of individual subsections.

EXAMPLES OF DATA SYSTEMS

The examples given below illustrate the various types of data base system now in common use. These include:

(1) traditional paper systems consisting of card and file archives;
(2) microcomputer data systems;
(3) mainframe computer data systems.

In comparing these three approaches, the following points must be borne in mind:

(a) consideration of the optimum scale and cost effectiveness for the particular project;
(b) design of index and catalogue systems that allow easy retrieval of the information in the form required by users of the data;
(c) potential for graphic linkages to allow the direct production of automatic mapping and three-dimensional modelling systems from computerized data;
(d) capital and equipment required for each system, and subsequent maintenance and running costs;
(e) level of expertise required by the data gatherers, collectors and users.

Traditional paper systems

Environmental geology study of the Plymouth–Plymstock area (GAPS Geological Consultants for the Department of the Environment, London; McCall *et al.*, 1985)

This study used traditional paper systems in a data base

compiled for a planning and development related urban environmental geology mapping study (described in Chapter 9). The objectives of the study were collection, collation and interpretation of all available information on the geology, mineral resources, hydrogeology and geotechnical properties/ground conditions of the area. There was a strong emphasis on engineering geology, slope instability, cliff and quarry face instability and waste disposal.

The data base collection involved a literature search for all published information (reports, papers, maps), and investigation of all likely sources of archival information, particular emphasis being placed on site testing reports and borehole logs. Besides data gathered in a remapping programme (part of the study), all stored cores and samples in museums and other collections were included. No new engineering geology testing was carried out. Thus the entire engineering geology appraisal was based on existing data. As is commonly the case (for example, Edwards *et al.* 1987), the data distribution was geographically uneven, and much reliance had to be placed on modern test results on two major road schemes (Marsh Mills to Tamar Bridge and St Budeaux By-Pass) and the Property Services Agency reports. This data, though unevenly distributed, was adequate to assess the geotechnical properties of the slates which covered most of the area, and their superficial weathered zones, but site reports on the Plymouth Limestone, on the seaward side of the city, were virtually non-existent. Numerous other site reports were of little use for assessing engineering properties due to the outmoded methods of testing used, but were invaluable for assessing the superficial geology subsurface, weathering profiles and bedrock geology.

The major sources are listed in Table I.

Table I List of major sources: Plymouth–Plymstock Study

British Geological Survey (Exeter, Wallingford, Edinburgh)
Central Electricity Generating Board
Central Engineering Laboratory (Department of Transport), Cardington
Department of Transport (London)
Devon County Council (Engineer, Materials Testing Laboratory)
Plymouth City Council (Planning, Health)
Plymouth City Museum (Archives, Environmental Reports)
Property Service Agency
Soil Survey of England and Wales
South West Water Authority
Contractor for the Manadon–Tamar Bridge sector of the A 38 trunk road
British Museum and Library
Public Record Office (Kew, London)
Ordnance Survey (Southampton)

A very useful collection for engineering purposes was a set of old air photographs of various dates showing large areas of the city. The city museum was found to have a remarkable set of maps showing the distribution of bombs from 1939–45, assessing the reprofiling of ground during clearance and making good of bomb damage. In all, 273 data items were collected from these sources.

Collection was made by the project co-ordinator and a technical assistant. Most items were collected at the source. In some cases only summary notes were allowed, but in most cases photocopies were collected and collated in files. Some very valuable rock core material was collected from the two main road investigations, and these were described and sampled for petrographic examination in the laboratory.

Card indexing was carried out under source, locality and subject. A system of coloured spots on cards showed the relevance of the item to general geology, engineering geology, hydrogeology or mining/quarrying (the four main themes of the investigation). These codes were repeated on the paper files in which the full photocopied texts or notes were archived. Some rare older material, mainly from the last century, was not archived in the paper copy, but was indexed with special yellow cards, for reference purposes.

The archive also included maps, published material, and some general technical reports relevant to the interpretation of the data. The archive needs to be regularly updated to retain its value to planners and developers. Some material is classified 'Confidential' for commercial and security reasons and this is clearly marked, and labelled in the computer retrieval system as well as on the cards, so that none is accidentally released. The final printing of the cards was done directly from the subsequent computer programme. A specimen card is shown in Fig. 4.

The advantages of the paper system are its cheapness in terms of capital outlay and its ability to archive and retrieve large files and reports. One road report consisted of 20 large volumes. There are obvious difficulties in fully computerized archival for such material, any part of which could prove useful in a geotechnical data bank covering a city and its fringes. The disadvantages are the requirement of a large number of staff-days to input data, and the possibility of loss of data and mixing-up of the paper records, and also their possible deterioration through wear. Such systems are difficult to duplicate at a large number of data collection and entering sites, with integration of all these in a central data base. They are best reserved for small, compact data bases which are not consulted intensively.

Microcomputer data bases

Retrieval system for Plymouth–Plymstock study described above (GAPS Geological Consultants; McCall *et al.*, 1985)

Although both cost and physical constraints (the bulky nature of the documents) precluded fuller computerization, a retrieval system was subsequently supplied. A separately bound instruction booklet, specially designed for use by planners, developers, and the general public was prepared.

A program for rapid retrieval of archival data using a Sirius 1 microcomputer was held on floppy disks. Keyword data from site reports, technical reports, maps and publications were fed in by the project co-ordinator and technical assistant. This program could be quickly adapted for similar studies. The details of the system are given below.

Hardware

The data base was designed for use on a Victor 9000, 1.2 microcomputer connected to a dot matrix printer with a parallel interface (for example, the C.ITOH 1550). The microcomputer required 128K of memory (128K is standard). A costing for equivalent hardware that would be used in 1989, and which illustrates the rapid techno-

```
Commissioned by DEPT OF TRANSPORT    Consultant FOUNDATION ENG

                                     Date 1979



Title: LONDON TO PENZANCE TRUNK ROAD A38
       MANADON TO TAMAR BRIDGE
       REPORT ON SUPPLEMENTARY SITE INVESTIGATION
       VOLUME 1A BOREHOLE LOGS AND PLANS



Grid refs.
         SX48105770 to SX44005870 and SX          to SX

GAPS ref. DTP 1              Library ref.
GAPS ref.                    Library ref.
```

Fig. 4. Specimen paper data index card used in the Plymouth–Plymstock data base (from McCall *et al.*, 1985).

logical improvement in microcomputer hardware is as follows:

(1) IBM PC XT or compatible with minimum 640K RAM and 20 Mb Hard Disk ... Approx Cost ... £1000 (US$1800)
(2) Epson FX80 dot matrix printer ... Approx Cost ... £400 (US$720)

Software

All the software necessary to use the database is contained on two diskettes: (1) GAPS-DOE PROGRAMME DISC (Curator Version); (2) GAPS-DOE DATA DISC.

Database structure

The data base is arranged at two levels: 'headers' and 'keywords'. The headers contain the document identification information: author, year of publication, commissioning body, consultant, and grid references. Each header has a corresponding list of keywords which refer to the subject matter contained in the document. These are organized at three levels (primary, secondary, tertiary) and are listed in Table II.

Table II Organization of keywords at three levels in the microcomputer retrieval system designed for the Plymouth–Plymstock Environmental Geology Mapping study (from McCall *et al.*, 1985)

GEOLOGY	MINING and QUARRYING	MAPS/PLANS and PHOTOS
Igneous	***Mines***	***Maps***
Petrology	Active	Land use
Dolerites	Disused	Geological
Lavas	Statistics	Topographic
Tuffs	Risks from subsidence	Bomb craters
Gabbros	Shafts	Planning proposals
Felsites		
Granites	***Quarries***	***Plans***
Geochemistry	Active	Planning proposals
	Disused	Topographic
Stratigraphy	Stability	
Quaternary		***Photographs***
Tertiary		Geological
Cretaceous	*ENGINEERING GEOLOGY*	Aerial
Jurassic		Infra red
Triassic	***Stability***	Colour
Permian		
Carboniferous	Cuttings and embankments	
Devonian	Subsidence and settlement	
Tamerton Chert	Slopes	*AREAS and SOURCES*
Cann Wood Slate	Discontinuities	
Plympton Slate	Caverns	***Sources A–D***
Purple and Green Slate		Agricultural Development and Advisory Service
Plymouth Limestone	***Sites***	British Geological Survey
Jennycliff Slate		British Geological Survey (Edinburgh)
Staddon Grit	Soil and rock sequences	British Geological Survey (Exeter)
	Boreholes	British Geological Survey (Wallingford)
Structure	Trial pits	British Museum Library
	Road reports	British Transport Docks Board
Folding	Building and construction site reports	Central Engineering Laboratory (PSA)
Tectonics	Geophysics	CEGB
Cleavage	Dredging reports	Department of Transport (Taunton)
Subsidence	Bridge reports	Devon County Council (Exeter)
Faults		Devon County Council (Materials Lab., Topsham)
Thrusts		

GEOLOGY (Cont.)

Normal faults
Wrench faults
Mylonites
Joints
Seismology

Economic Geology

Arsenic
Tin
Copper
Lead
Zinc
Silver
Wolfram
Iron
China Clay
Aggregates

Sedimentary

Petrology
Limestones
Conglomerates
Slates
Sandstones
Chert
Breccia
Geochemistry
Environment
Stylolites
Solution cavities

Palaeontology

Algae
Ammonoids
Brachiopods
Bryozoa
Corals
Crinoids
Conodonts
Dacryoconarids
Gastropods
Goniatites
Ostracods
Radiolaria
Stromatoporoids
Tentaculitids
Trace fossils
Trilobites
Vertebrates

Geomorphology

Raised beaches
Drowned river valleys
Caves
Fossil erosion surfaces
Glaciation

Superficial

Soil
Head
Colluvium
Alluvium
Estuarine
Tidal flats
Landslip
Beaches

Geophysics

Palaeomagnetism
Seismic
Radiometric dating

ENGINEERING GEOLOGY (Cont.)

Rock Properties

Triaxial strength
Point load strength
Other tests
Degree of fracturing (RQD)
Degree of weathering
Density
Chemistry

Superficial

Made ground
Reclaimed ground
Liquid and plastic limits
Standard penetration tests
Other tests
Density
Chemistry

HYDROLOGY

Hydrogeology

Wells
Boreholes
Thermal linescans
Hydrographs
Interpretations
Groundwater

Water Supply

Surface water
Rainfall and climate
Reservoirs
Licensed extractions
Dams
Flow measurements

Water Quality

Chemical analyses
Pollution
Groundwater chemistry
Marine saline intrusion

Waste Disposal

Rubbish tips
Reclamation
Sewerage
Toxic materials

AREAS and SOURCES (Cont.)

Sources E–Z

Health and Safety Executive, Dept. of Mines and Quarries
Miscellaneous consultancies
Ordnance Survey
Plymouth City Council
Plymouth Polytechnic
Plymouth City Museum
Property Services Agency (Croydon)
Property Services Agency (Devonport)
Public Records Office
Roxburgh, I.S. (Plymouth Polytechnic)
Soil Survey of England and Wales
South West Gas Board
South West Water Authority

Areas A–H

Boringdon Park/Colebrook and Woodford
Brixton
Chaddlewood and Newnham
Chelson Meadow and Saltram Park
City Centre/Stonehouse/The Hoe/Barbican and Drake's Island
Compton/Efford/Laira/Lipson Vale and Longbridge
Crownhill and Manadon
Devonport
Eggbuckland and Hartley Vale
Ernesettle
Estover
Ham/Pennycross/North Prospect and Weston Mill
Hartley and Mannamead
Honicknowle

Areas I–Z

Lipson/St Jude's/Prince Rock and Cattedown
Pennycomequick/Peverell and Mutley
Plym Forest and Elfordleigh
Plympton/Plympton St Maurice and Underwood
Plymstock and Elburton
Pomphlett/Billacombe and Oreston
Roborough and Bickleigh Down
St Budeaux and Kings Tamerton
Southway and Widewell
Stoke/Ford and Keyham
Tamerton Foliot
Turnchapel/Hooe/Mount Batten and Staddiscombe
Whitleigh

Journal titles have standard codes: grid references are eight figures. The area was divided into 27 subareas (Fig. 5). Documents were registered under the subarea which could then be searched for all data items within it by the user of the retrieval package.

Use of the database

The initial (main) menu allows the operator to INPUT DATA or to SEARCH & EDIT DATA. Selection of input gives a screen list ('menu'—academic paper, site report, technical report, map/plan/photo). About 1000 documents can be put on a single diskette. For a larger data base a fixed disk system can be used, or the data base can be divided into smaller units, using one diskette per unit, thus allowing a large volume of data to be handled by a machine of relatively small capacity.

The selection of a primary keyword displays a secondary keyword menu, and likewise selection of a secondary keyword brings up a tertiary menu. The secondary and tertiary menus differ depending on the previous keyword selected.

Once a document has been fully recorded in the data base it can be put into permanent storage by pressing UPDATE DATABASE on the primary keyword menu.

If EDIT or SEARCH are selected from the main menu, the user is able either to:

(1) SEARCH FOR HEADERS & KEYWORDS or

(2) SEARCH ON KEYWORDS

The first option brings up four types of search menu corresponding to the four types of document (Table III).

A search can be based on 'Author', 'Journal', 'Commissioned By', 'Consultant', 'GAPS Ref.' or 'Library Ref.'; by 'Date' or by 'Grid Reference' (SW and NE corners of rectangular areas). A combined headers and keyword search is possible, as is a 'Keyword Search'. In addition an area search is possible using the keyword for the convenient topographic divisions defined in Fig. 5. This degree of flexibility is straightforward on the computer but is cumbersome and very time-consuming to prepare in a card index system.

Systems for compilation, storage and retrieval of data on disused mine openings and workings (Freeman, Fox Ltd., 1988)

The study was commissioned by the Department of the Environment, London, to compare and review methods, scope, contents and costs of existing data banks of information on disused mine openings and workings, and to recommend the best methods for use in areas of varying complexity. This was necessary to aid assessment of the general problems of land development, to help plan specific site investigations and remedial works, and to aid in the work of the emergency services.

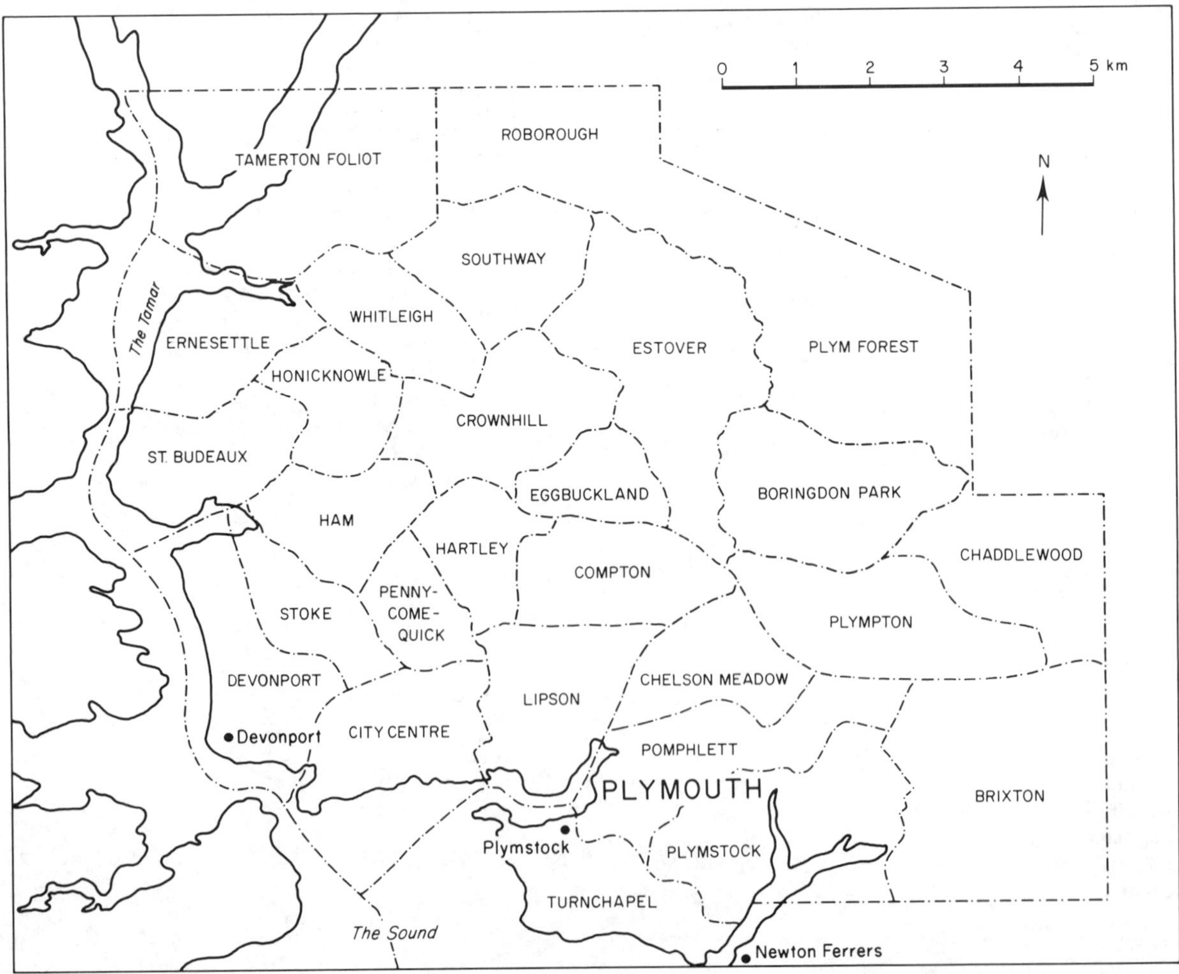

Fig. 5. Data base search areas—Plymouth–Plymstock EGM Study (after McCall *et al.*, 1985).

Table III Menus for the four types of document: Plymouth–Plymstock Programme (after McCall *et al.*, 1985)

Academic paper	*Site report*	*Technical report*	*Map/plan/photo*
Author	Commissioned by	Commissioned by	Commissioned by
Journal	Consultant	Consultant	Consultant
Date	Date	Date	Date
GAPS Ref. No.*	Grid Ref. No.	GAPS Ref. No.	GAPS Ref. No.
Library Ref.	GAPS Ref. No.	Library Ref.	Library Ref.
	Library Ref.		

*The GAPS References Numbers were coded with a letter prefix to indicate source and a serial number within that code set.

This contract arose because of widespread problems in areas of old mining in Great Britain, which has an extensive legacy of underground mining for coal, oil shale, metalliferous and other ore minerals, building stone and others, mainly since the industrial revolution, but sometimes dating from Roman times or earlier. More recent mines are often documented in plans and other records while unrecorded examples are commonly found during site investigations. The data base was developed from one set up for an applied mapping study of the former Chacewater–St Day copper mining area of Cornwall (Freeman, Fox Ltd., 1987). The sources of information are listed in Table IV.

Table IV Sources of information on disused mine workings (after Freeman, Fox Ltd., 1988)

Original Mine Plans and Sections
Abandonment Plans (prepared as a legal requirement on abandonment of a mine since 1872)
Ordnance Survey and other surface maps
Geological Maps
Field Surveys designed specifically to locate evidence of prior mining activity
Field Notes, Memoirs and Reports
Incidental references in a variety of historical records
Aerial Photographs
Site investigation reports recording accidental discoveries in the course of excavations and drilling

There was a considerable resolution problem. The maps and plans have a bewildering variety of scales and very variable accuracy. Some are rare and fragile historic documents which must be kept from frequent handling. In addition to such graphical data there are considerable volumes of descriptive and interpretative data.

The three essential features that had to be included were:

(1) *Shafts*: any feature that can be represented by a symbol at a single point.
(2) *Roadways*: any feature that can be represented by a series straight line segments of specific width. Inclined shafts are special cases of roadways.
(3) *Workings*: any feature that can be represented by an enclosed area. Vertical veins and lodes are a special case of workings that may appear as a series of straight line segments in plan.

It is important to plot the geographic position of all openings (shafts, adits, portals). Roadways need to be defined by their trace plus their depth at each end. Workings need to be shown in plan projection (irregular area shapes or polygons) plus selected point depths or depth contours. In addition, there needs to be a catalogue of source references (with geographical extent definition of workings referred to in the document), and a master plan and register of shafts and openings and of roadways and workings.

Reasons for electing to use microcomputer systems

The main arguments for the use of table-top microcomputers were cheapness, lack of maintenance requirements, lack of need of environment control for operating, and easy operation by non-specialist staff.

Simple microcomputer systems resemble card index system but allow rapid searching through key fields, and simplify the inspection and maintenance of data by portraying a 'card' of data on the screen, which then can be entered, edited or erased with ease. Indexes can easily be copied and used at several locations, and there is no wear or tear, or risk of mixing or losing cards. Electronic 'muddle' can be countered by use of back-up copies.

More complex systems allow users to define the content of data tables, allowing complex simultaneous search systems to produce detailed and structured reports. Such complex processes can be easily programmed so that searches can be carried out by relatively inexperienced computer users. The requirement for data on disused mine openings and workings can easily be accommodated in such systems, but there is an additional requirement for graphic facilities to produce detailed maps, mine working plans and overlays. These can be linked to the microcomputer programme by using CAD (Computer Aided Design) systems, allowing direct input of graphics from digitizers, manipulation of graphics on the screen, and the plotting of drawings, either as draft copies on a computer printer, or as high-quality plots, at chosen scales, on a computer plotter. Such systems can be linked with a variety of microcomputer data base systems.

Recommended storage and retrieval options

Three alternative systems, based on increasing detail and costs, were recommended so that users could consider which was the most suitable.

(1) Catalogue of source documents

The simplest of the three, and a basic element of the other two, was a catalogue arranged in defined areas (e.g., county, mining zone). Search would be by locating a National Grid outline on a displayed tabulation. Such a system is acceptable provided the areas do not contain a large number of documents. Problems arise where documents cover more than one area. Maintenance and searching would be simplified by the use of an elementary data base programme, a card index type of system as described above, with automatic search on a specified search zone which would produce a screen or printer listing. There is no need to group the entries by area. Even the smallest microcomputer could handle this.

(2) Documents catalogue and shafts register

To system 1 this system added detailed information on shafts and openings, and master plans showing the position of the shafts. The register contained each and every source reference to shafts (allowing users to refer back to original source). The register took the form of two tables, A and B:

(A) Listed each shaft shown or referred to in each source document, and a brief set of descriptive data headings.
(B) Master table contained a larger number of data headings (details of ownership, ground conditions, shaft treatments etc.).

This system required standard scaling of shaft/opening positions, conversion to National Grid co-ordinates, and entry onto table A. Each new entry was compared with the current master plan and the master table B.

Interrogation of the master plan for a specific search area was straightforward, though preparation of a summary plan and report (list of shafts and attributes) would not be a minor task except for a very small area. Time-consuming abstraction of shaft positions, conversion to National Grid co-ordinates, and updating of master plan and table were avoided.

There was no difficulty in managing the source document register and shafts register on a data base programme. Graphics were handled by a linked CAD graphics system. There was automatic entry to the register tables with data collection using a map digitizer to achieve automatic transformation of shaft positions from any scale to National Grid co-ordinates.

(3) Source document catalogue with shafts, roadways and workings registers

This system was similar to System 2, but allowed all three types of mine feature to be registered.

Costs

The capital equipment and comparative costs of the first two systems are given below in Table V (adapted from Freeman, Fox Ltd., 1988)

Applicability of the different systems

The simplest system (1) was suitable where only a catalogue was required. The next system (2) added a shaft register, and was applicable to areas where there were relatively few documented references to shafts and openings. The final system (3) was recommended for use wherever extensive past, present or projected mining shows subsidence or other hazards. Table VI (adapted from Freeman, Fox Ltd., 1988) illustrates the data headings for shafts and openings, roadways and workings that would be incorporated in the full data bank system. Figure 6 illustrates a graphic production of a catalogue document.

This third system has the advantage that it is possible to exchange data with other computer data base systems. For manipulation of graphics (digitizing, plan updating and plotting), the AutoCAD system is recommended. Digital OS maps are available in AutoCAD drawing format so that spatial data can be viewed and plotted together with selected OS data if required.

Although the recommended off-the-shelf software systems have the capability to carry out many tasks, in day-to-day operation, many of these tasks would need to be 'streamlined' by pre-programming the commonly used procedures.

Manual versus computer operation

It was concluded that manual operation was practicable for very simple data sets, but otherwise computerization was preferable on account of lower cost (Table VII, adapted from Freeman, Fox Ltd., 1988) and ease of duplication. Such data banks could readily be reproduced for each user location, something virtually impossible in the case of large manual systems. Simple manual systems can however be converted to computer operation at a later date.

Compatibility

Obviously, if systems are to be set up in a number of mining areas, similar hardware and software should be utilized to ensure compatibility. If properly devised, these systems could serve multiple input and retrieval locations or contribute to a unified national data base at a central location.

This system has since been developed by MZ Associates (MZ Associates, 1988) under contract to the Department of the Environment, London, and includes the entire Mining Record Office collection of abandonment plans for 'miscellaneous mines' (mines other than those for coal) in England and Wales.

Summary of microcomputer systems

The two examples above have been used to illustrate the sort of systems that are available to provide data bases for planning and development related earth science programmes. The examples have been drawn from an Environment Geology Mapping Study (an urban coverage with a number of themes), and from an investigation into the most suitable systems for national and regional (integrated) coverage of disused, operative and projected mine openings and workings. By citing these case histories, a picture has been given of the capability of microcomputer data base systems and the way they can form the basis for planning and development related earth science programmes.

Such systems can be readily adapted to the special requirements of other types of environmental geology, hydrogeology, soil survey, mineral resource, construction and hazard-related programmes, which are covered in Chapters 4–8.

Table V Comparison of costs of the three systems (adapted from Freeman, Fox Ltd, 1988). All figures in brackets are US$

SYSTEM 1

	A		B		C	
(i) *Establishment*						
Manual	£ 100	(180)	£ 1 000	(1 800)	£10 000	(18 000)
Computer	£ 1 500	(2 700)	£ 2 400	(4 320)	£11 400	(20 520)
(ii) *Retrieval (per search)*						
Manual	£ 4.02	(7.24)	£ 10.27	(18.49)	£ 72.78	(131.00)
Computer	£ 0.03	(0.05)	£ 0.30	(0.54)	£ 3.00	(5.40)
(iii) *Break Even*						
Number of searches	348		136		16	

A = 100 records, B = 1 000 records, C = 16 records

SYSTEM 2

	A		B		C	
(i) *Establishment of Manual Systems*						
Capital costs, drawing equipment, plan cabinets, card indexes, etc.	£ 1 000	(1 800)	£ 1 500	(2 700)	£ 2 500	(4 500)
Shaft data collection	£ 292	(526)	£ 2 167	(3 900)	£ 2 917	(5 251)
Updating of master plan register	£ 833	(1 500)	£ 8 001	(14 400)	£ 8 333	(15 000)
Total	£ 2 125	(3 826)	£11 668	(21 000)	£15 000	(27 000)
(ii) *Establishment of Computer System*						
Capital costs of computer etc.	£13 000	(23 400)	£13 000	(23 400)	£13 000	(23 400)
Shaft data collection	£ 83	(150)	£ 458	(824)	£ 833	(1 500)
Updating of master plan / register	£ 117	(211)	£ 1 183	(2 130)	£ 1 167	(2 100)
Total	£13 200	(23 761)	£14 641	(26 354)	£15 000	(27 000)
(iii) *Retrieval Costs* (one search of 1 km² area producing 10 shaft references)						
Manual	£ 9.2	(16.60)	£ 13.3	(23.90)	£ 13.3	(23.90)
Computer	£ 0.2	(0.36)	£ 0.3	(0.54)	£ 0.3	(0.54)
(iv) *Break Even*						
Number of retrievals	1 230		229		96	

Data bank A comprises 100 plans and 1 000 shafts
Data bank B comprises 100 plans and 10 000 shafts
Data bank C comprises 1 000 plans and 10 000 shafts

SYSTEM 3

(i) *Establishment*	*Manual*		*Computer*	
Capital costs	£ 2 000	(3 600)	£ 13 000	(23 400)
Shaft data collection	£ 2 167	(3 900)	£ 458	(824)
Roadways data collection	£ 3 125	(5 625)	£ 625	(1 125)
Workings data collection	£ 2 083	(3 750)	£ 417	(751)
Updating				
Shaft master register	£ 8 001	(14 400)	£ 1 183	(2 129)
Roadways master register*	£12 001	(21 600)	£ 1 774	(3 193)
Workings master register*	£ 8 001	(14 400)	£ 1 183	(2 129)
Total	£37 378	(67 280)	£ 18 640	(33 552)

* = additional

Retrieval costs		
One search of 1 km² producing 10 shafts, 3 roadways and 1 working	£20 (36)	£1 (1.80)

Table VI Data headings for openings (e.g. shafts), roadways and workings (adapted from Freeman, Fox Ltd 1988)

Openings	*Roadways*	*Workings*
Identification code	Identification code	Identification code
Name	Name	Name
OS Easting	Line segment positions	Boundary positions
OS Northing	Line segment widths	Boundary and contour depths
Directly surveyed	Line segment endpoint depths	Seam thickness/vein width
Depth (total)	Information sources	Minerals extracted
Date of construction	Function	Method of working
Information sources	Comments	Information sources
Comments		Comments
Additional data headings for directly surveyed openings		
Identification code	Overburden type	Gas problem
Size	Overburden thickness	Adjacent shallow workings
Shape	Bedrock type	Known instability
Inclination	Groundwater level	Existing condition
Bearing	Additional subsurface information	Access
Ground surface profile	Construction information	Owner/tenant
	Comments	Interest groups

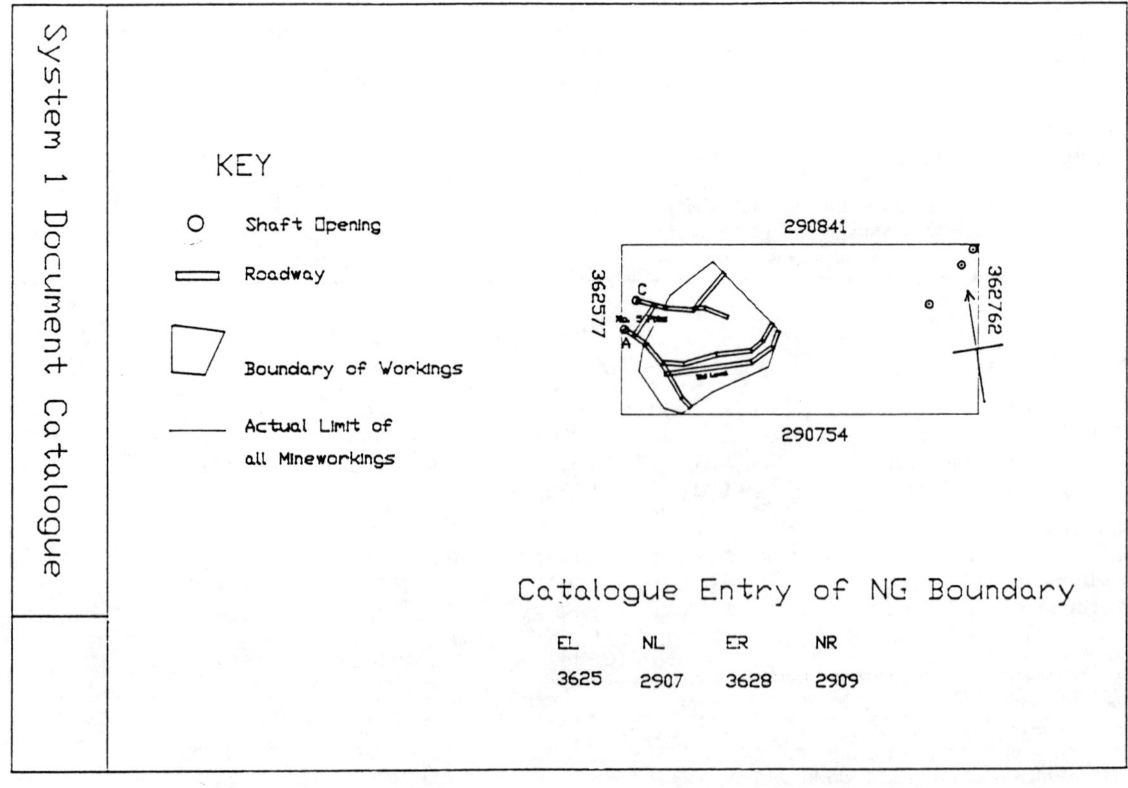

Fig. 6. Plan of Bull Piece Colliery (from Freeman, Fox Ltd., 1988).

Table VII Minimum computer requirements and costs (adapted from Freeman, Fox Ltd 1988). All figures in brackets are US$

Item	*Cost*	
SYSTEM 1		
IBM PC compatible microcomputer with 512K, RAM, twin floppy disk drives, mono screen	£ 600	(1 080)
Printer	£ 300	(540)
Software	£ 500	(900)
Total	£1 400	(2 420)
One document record—101 characters		
1000 document records require 99 K bytes disk storage		
SYSTEM 2		
IBM PC-AT compatible microcomputer with 640 K RAM, 30 Mb hard disk, EGA colour monitor and Mouse interface	£ 3 000	(5 400)
Printer	£ 500	(900)
A1 Digitizer	£ 2 500	(4 500)
A1 Multi-pen plotter	£ 4 000	(7 200)
Software	£ 3 000	(5 400)
Total	£13 000	(23 400)
Data Storage		
One Table A record—71 characters		
One Table B record—70 characters		
Assume 3 'A' records for each 'B' record		
One Master Shaft requires 283 characters		
1000 shafts require 276 K bytes disk storage		

Mainframe data bases

Applied Geological Mapping, Southampton Area (Edwards *et al.*, 1987; Laxton, 1987; Loudon and Mennim, 1987)

The British Geological Survey was commissioned by the Department of the Environment, London, to prepare a set of applied geology maps and descriptive reports for the purpose of land use planning and development in the Southampton area. The study was taken as an opportunity develop computer manipulation of existing geological, geotechnical, hydrogeological and mineral resource data, and to provide an archive of information and computer-plotted applied geology maps (Table VIII).

The results of a field survey by the British Geological Survey from 1973–80 were supplemented by data held in the BGS, including results from an earlier study of sand and gravel resources of the Lymington–Beaulieu area. Commercial seismic data and data from geothermal and hydrocarbons exploration was also available to the BGS. Additional data were collected from various sources including regional, local and water authorities, site investigations, contractors, and consulting engineers. In all, data were recorded from 4549 borehole logs (4501 less than 10 m deep), geotechnical analysis results of 11 893 samples, and particle size analysis of 541 samples. Ordnance Survey data was used in the topographic and cartographic aspects of the data base.

A number of different types of information were assembled in the data base system. Line data, digitized from the geologists' field maps, record the position on the ground of formation boundaries, faults and other geologial features. Borehole descriptions record the positions of the geological boundaries below the surface, together with lithological descriptions of rock units. Geotechnical tests and particle size analysis on samples,

Table VIII List of applied geological maps, Southampton study (after Edwards *et al.*, 1987)

Theme	*Title*	*Scale (and number of sheets)*
Geology	Solid geology	1:25 000 (6)
	Drift geology	1:25 000 (6)
	Drift thickness	1:25 000 (6)
	Rockhead contours	1:25 000 (6)
Mineral resources	Sand and Gravel	1:25 000 (7)
	Clay	1:50 000 (1)
	Distribution, worked ground	1:25 000 (6)
Water resources	Distribution, aquifers	1:50 000 (1)
Engineering geology	Availability of geotechnical test data	1:50 000 (7)
	Engineering geology of solid deposits	1:50 000 (1)
	Engineering geology of superficial deposits	1:50 000 (1)
Slope stability	Distribution of slope angle and landslip	1:50 000 (1)
Landfill and waste disposal	Distribution of made ground	1:25 000 (6)
Data distribution	Distribution of boreholes	1:25 000 (6)
Other	Sites of special scientific interest	1:50 000 (1)

mostly from boreholes, provided another source of relevant data. The contoured topography was also digitized.

For efficient retrieval and processing, organization of data into data sets was necessary. Within each data set, each item had to refer to comparable data, and be held in a uniform format, so that the computer could compare like with like. The organization of the system into data sets is illustrated in Fig. 7. A number of tables hold the borehole data and the results of tests and analyses. This archive is managed by a 'relational data base management system'. The line data from the maps were structured differently, being held in a cartographic work station, which enables the user to call up the data as a screen display and to directly modify the interpretations shown in order to develop geologically credible maps. Contour data from topographic maps were digitized to provide a digital terrain model, a digital record of estimated elevations at points on a 50 m grid across the area. Linking of borehole data with cartographic data was possible. Computer processing and presentation of the data was carried out by a geologist familiar with the area, which was of great value in preparation of the applied geology maps.

Two appendices provide details of the ORACLE Data Base Management System used and possible future improvements. The latter appendix lists the borehole data base as containing information on 5643 boreholes and comprising 11 098 records of reference information and 23 032 records of log data. The number of data items (that is, fields in records containing data) is over 400 000. This second appendix covers the ancilliary use of an IBM XT microcomputer, installed at BGS Exeter Office to input and check data. Two Assistant Scientific Officers were employed on this, and it is noted that 20% of their time was spent on entering reference data, 40% on coding log data on forms and 25% on entering and editing log data. The editing was required to rectify miscodings due to use of inexperienced staff for initial entry and to insert stratigraphical classifications not available at the time of initial ent[illegible].

The use of mainframe computer systems is expensive and requires a big capital cost outlay. This approach was possible because the British Geological Survey is well equipped with such a system to handle National Geoscience Data Base Management. This type of system is feasible for certain types of projects, but in other cases budget restrictions might obviate it as against microcomputer systems.

Laxton (1987) drew the following conclusions:

(1) The volume of data gathered could not have been coped with in a manual system (he did not consider the possibility of using a microcomputerized system).
(2) ORACLE was successful, easy to learn and use, and flexible enough to cope with unforeseen questions and the addition of new data sets.
(3) Entry by a professional data entry operator was more cost-effective than entry by unskilled staff because of the extensive checking and correction needed if unskilled staff were used.
(4) Objectives should be carefully considered before implementing such a data base to avoid waste due to collating unnecessary information.
(5) Considerable expense was involved, but future projects could be less costly because of the lessons learnt. Cost reductions could be made, in particular, by improvement in the coding of geological lines at the time of digitization (this would simplify retrieval) and by closer interaction between the ORACLE and INTERGRAPH systems.
(6) The production of a digital terrain model from the Ordnance Survey Contour Map was successful.
(7) The data base should be kept up to date.
(8) The structure and content of the borehole data base (following conventions developed by BGS in Scotland) were effective and flexible, and were appropriate for future projects of this nature, and could assist in the development of a National Borehole Data Base.

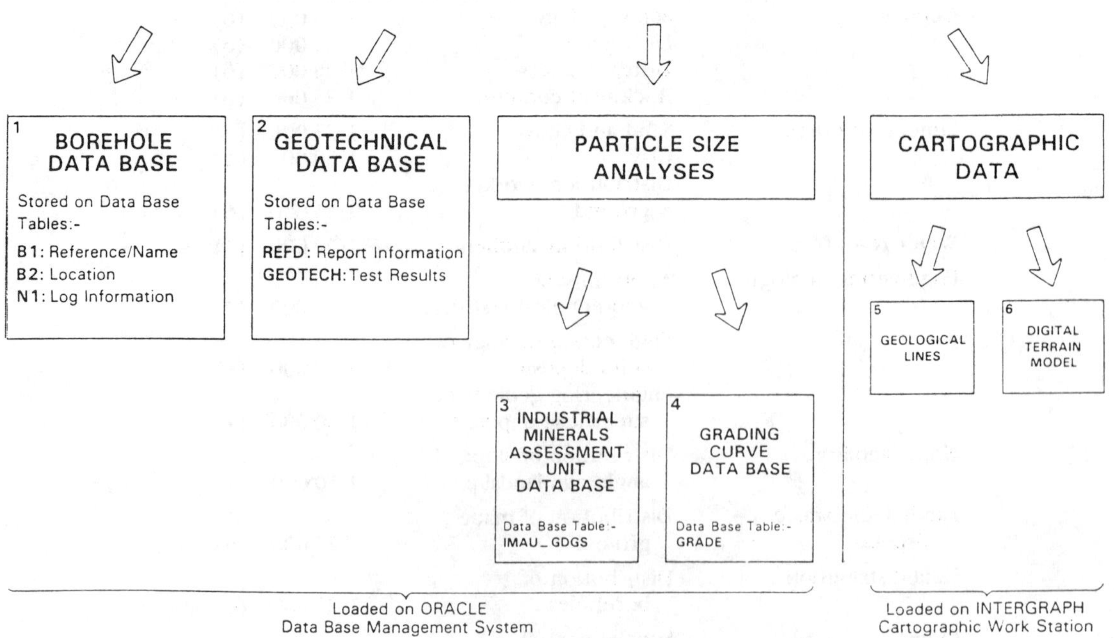

Fig. 7. Southampton project data base system (from Edwards *et al.*, 1987).

Volume 3 (Loudon and Mennim, 1987) covers the mapping techniques. It has recently become practicable, the authors noted, to hold lines and areas from maps in digital form. This can lead to computer systems for direct preparation of geological maps.

Lines are stored as co-ordinates of a string of points spaced along a line ('vector representation'), and more complex patterns, such as the topographic base map can be represented by a 'raster', a set of digits in the computer memory indicating the presence or absence of shading or of colour within each area of a grid of small, square cells covering the map area. The form of the land surface can be shown as a grid of average elevations for each cell or 'Digital Terrain Model' (Fig. 8). Borehole data and the results of geotechnical tests are stored as tables which can be cross-referenced with one another and displayed on maps alongside the cartographic information. Vector data is entered with a cursor on a digitizing table and the raster data (contour data and digital terrain data) is entered from an automatic raster scanning device. Tabular information is entered from a keyboard.

Storage and retrieval present problems for maps; tables were in this case handled by the ORACLE system (see above). The vector data was handled separately on a graphics work station.

It was noted that systems such as that used here can produce coloured maps for small production runs (geotechnical thematic maps have a small print run demand), and could be used to print maps singly on demand by a particular user.

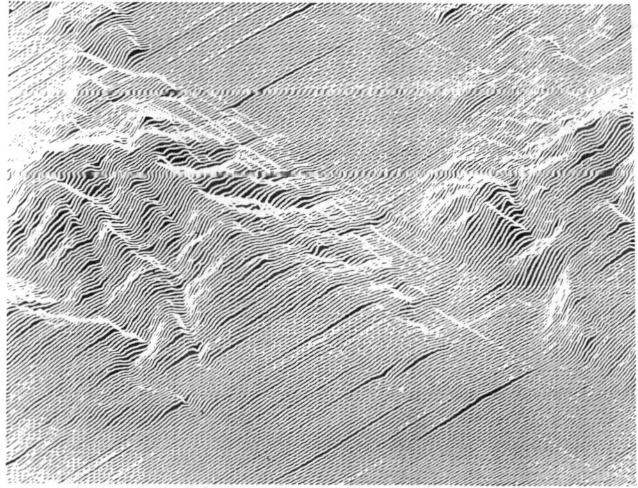

Fig. 8. View from the south–southwest of a digital terrain model, showing the area around the City of Southampton (from Edwards *et al.*, 1987).

The benefits derived from these direct systems of map production using computers are efficiency, flexibility, clarity of presentation, ease and speed of use, ease of updating and revision, cost effectiveness, improvements in the analysis, integration and communication of data (leading to better understanding of its significance). The ideal would be a comprehensive computer system, handling the complete process from capture of data to production of map.

Loudon and Mennim concluded that the available technology has limitations. At present, the combination of traditional and digital methods of map preparation is seen as an interim step. Developing closer links between parallel work on geological information systems and spatial modelling in geology was recommended, and similar projects should be mounted to improve techniques. Follow up work is now under way in the Wrexham and Birmingham areas.

Geological databank pilot study (Clayton *et al.*, 1987)

National Geoscience Data Banks are on the way in several countries (e.g., Netherlands National Geotechnical Data Bank). There are several routes to such data banks. Some depend on legislation for lodging data. Others request donation of data. Whatever the route adopted, it is generally recognized that permanent archival and public access to geoscience data is a real need for planning and development purposes.

While there is in Great Britain a National Geoscience Data Centre operated by the British Geological Survey, there is at present no systematic national earth science data bank. The Department of the Environment and the British Geological Survey jointly commissioned this pilot study which explored the requirements for a computer data base of geological information. Potential users of such information were consulted on their data holdings, and present and future requirements. A demonstration computer programme was developed to illustrate its operation, and a costed implemented programme was then drawn up, although it was qualified by the need to proceed in a phased manner from initial indexing to acquisition of fully digital data.

It was estimated that some 1500 separate groups and organizations in Britain hold site investigation data collections which refer to more than one million boreholes. However, duplication and overlap of information are to be expected. Indeed, actual investigations of individual sites may be repeated, because either original investigation data cannot be easily located or there is a distrust of previous work. Procedures to handle such an enormous data bank are likely to be cumbersome and expensive, but improvements in technology have recently made a stable, cost-effective solution feasible.

One hundred and thirteen organizations participated in the survey of geological data requirements and holdings given in Table IX(a).

The types of data most commonly required and held were:

(1) fundamental geological data (geological maps and boreholes);
(2) land stability data (geotechnical data, site investigation reports, mine plans);
(3) hydrogeological data;
(4) mineral resource, mining and quarrying data.

Tables IX(b) and IX(c) give a perspective on the volume of geological data, held by the organizations consulted in the user study and the holdings of the British Geological Survey alone.

The figures given in Table IX(b) were regarded as conservative and the figures for the 60 organizations covered by this table may in reality be nearer 5.5 million discrete data items.

Three factors prevented estimation of a grand total for Britain:

(1) there is no standard unit of measurement for geological data units, which vary in size and content;
(2) the total number of potential data holders is not known;
(3) the degree of duplication is not known.

However, the total must be immense.

Data sources are spread throughout the country and only national organizations (British Geological Survey, Departments of the Environment, Transport and Energy, etc.) hold data banks covering the entire country. Much of the task of compiling a National Data Bank would involve data collation from scattered local sources (local authorities and water authorities, engineering consultants, civil engineering companies, mining and quarrying companies, etc.). In addition, data may be further dispersed in separate subdivisions of the organization.

Reliability of data is an important consideration, and some form of reliability grading needs to be built in to the system.

Although easy access by the user to relevant information is all important, allowance has to be made for a confidentiality, restricting access to some documents. Whereas most organizations consulted favoured such a data bank, they pointed out the need for such a restriction to safeguard the interests of the owners of data.

Though the BGS holdings (Table IX(c)) are substantial, and most of these are held on computer, these collections

Table IX(a) Geological data requirements and holdings (after Clayton *et al.*, 1987)

Type of geological data	*Number of organizations using data*	*Number of organizations holding data*
Solid and drift geology maps and boreholes etc.	55–60	65–70
Geotechnical engineering geology site investigation	75	60
Hydrogeology	5	45
Mining and quarrying	5	45
Mine Plans	15	n.s.
Mineral resources	30	40
Geophysical data	13	24
Bibliographic information	n.s.	24
Geochemistry	10	19
Photogeology	n.s.	16
Physiography	n.s.	17
Petrology and palaeontology	3	7–12

Total number of organisations participating = 113
n.s. = not specified in user study

Table IX(b) Volume of geological data revealed in used study (after Clayton *et al.*, 1987)

Principal types of data	*Total number of items held*
Individual borehole sections	184 815
Site investigation reports	19 965
Site investigation reports and borehole logs (undifferentiated)	138 570
Seismic sections	300 000
Miscellaneous reports/files	8 000
Chemical analyses	104 000
Core samples	2 000
Combined total (60 organizations)	757 350

Table IX(c) Estimates of British Geological Survey Data Holdings (after Clayton *et al.*, 1987)

Activity or data collection	*Present holding*	*Size of digital holding*
(a) High-level indexes and dictionaries	n/a	30 Mbytes
(b) Library, map library, photographic	6 million (+) documents	300 Mbytes
(c) Publications, maps, memoirs, reports		600 Mbytes
(d) Samples and specimens	16 million + 100 km of core	
(e) Land survey	1 million b/h logs	110 Mbytes
Hydrogeology	1.2 million	500 Mbytes*
Bulk minerals	records and data	
Engineering geology	sheets	
Hydrocarbons		
(f) Marine earth sciences	Included in (d) and (g)	
(g) Regional geophysics	1 250 line-km	20 Mbytes
(h) Global seismology	20 Gbytes	20 Gbytes*
(i) Geomagnetism	4 Gbytes	2 Gbytes 2 Gbytes*
(j) Geochemistry	400 000 sample/results	40 Mbytes

* material archived on magnetic tape
1 Mbyte = 1 million characters
1 Gbyte = 1 000 million characters

were made over a long period from many sources for many different purposes.

The proposal for implementation

A phased approach would be the only practical way to implement this proposal, but collections of data that would be incorporated are in various stages of organization already (in the BGS for example), and it is obviously necessary to take advantage of existing registers. This indicates varying rates of progress for different topics.

A proposal was set out under the following phased objectives:

(1) compilation of a catalogue of sources and collections;
(2) compilation of digital indexes to holdings;
(3) preparation of multi-level indexes;
(4) full digitization of records.

Compilation of a catalogue of sources and collections

This would provide a reference catalogue of data holdings and an outline of their contents. It would also provide a link with other data holdings and encourage co-operation and local use of the computer systems, evaluate quality of the holdings and the extent of duplication, and establish general working principles for data exchange. To be more than an initial browsing facility, spatial reference in a manner that allows specific retrieval of information on holdings for defined geographical areas of interest would be necessary.

The main functions of the directory of holdings would be:

(1) establishment of a computer-based digital directory of geological data, outlining the extent and type of records held in each;
(2) provision of retrieval facilities to allow enquirers to locate geological information pertinent to geographical areas and/or topics;
(3) the initial laying of foundations for extending the introduction of digitally-compiled indexes of the record holdings.

The information would be held in a relational data base management system on a mainframe computer, allowing easy manipulation, checking, updating and retrieval. Initial gathering utilized the ORACLE DBMS on a VAX 8600 computer.

Dissemination of information would be by means of a printed catalogue of listings and summaries made available to interested parties. Selected extracts could be obtained for additional charges, and dissemination of structured data on floppy disks was also envisaged. Such disks would be devised to allow access by topics, geographical areas and organization.

Compilation of digital indexes to data holdings

The aim is to compile indexes to records to enable interactive searching to be conducted on fields of interest to the enquirer. The principal advantage of computerized systems designed to hold information about geological data is to establish by search the existence or absence of relevant records. Ideally, this is followed by access to the records themselves, either by direct screen retrieval or even in traditional paper file storage. The main keys to search for records are geographical parameters, so spatial referencing is critical to this phase.

Individual data holding departments and organizations would need to compile their own digital indexes using a standardized scheme of cataloguing and a common index structure. A common format for data input and update would be essential.

Data could be fed into microcomputers at these departments and organizations. The central co-ordinating body would supply the necessary software to drive the system. Data base design for this function is estimated to require one man-year of effort.

Transfer of the local data indexes to a central system would follow. It is estimated that about £150 000 (US$270 000) would be needed to provide a mini-computer for this and the necessary facilities for a functional central data base management system. Several thousand pounds per annum running costs would be required for operation and maintenance, and a data base manager. The principle attributes at this stage would be:

(1) a common digital format for indexing data collection, the indexing being largely by the holders themselves;
(2) the provision of software to enable local operation on microcomputers, and to search departmental holdings with the aid of user-friendly screen prompts;
(3) transferral of the local digital data indexes to a central storage facility operated by the co-ordinating body responsible for compatibility and integration.

Various means of communicating data between the central and local organizations are discussed by Clayton *et al.*, 1987, including exchange of floppy disks and use of magnetic tape between mainframe operators. However, the aim was to provide a suitable facility for local users demanding little more than a microcomputer and printer. Addition of a *modem* would allow such users to have direct access to the mainframe by telephone. Suitable modems would cost about £300 (US$540), and a charge for appropriate linking software must be added to this. Writing and testing the necessary programme for this are estimated to require about 0.5 man-year effort. Dial-up charges would augment subscribers telephone charges.

The advantages of local stand-alone systems using departmental microcomputers or work-stations is emphasized by the enormous storage potential provided by CD-ROM (Compact Disc Read Only Memory), an optical disk system which can be read by a light beam from a laser diode and can supply directly the on-off pulses required for the computer's processing unit. Disks are moulded from a master and it is claimed that a single 4.7 inch diameter disk can store 600 Mbytes, equivalent to 200 000 pages of text.

Preparation of multi-level indexes

The indexing described above should ideally be followed by fuller indexing, including summary details of data held in the records in a more compact and accessible form than the original documents. For many purposes this would be adequate, possibly even preferable to the documents themselves.

Such things as formalized descriptions of lithological units, cartographic data, analytical and test results are examples of subjects that are very suitable for summary treatment. It was estimated that 2.0 man-years of effort, plus supporting finance for materials and travel, would be needed to investigate the practical implementation of such a facility, the establishment of appropriate standards for data input and exchange, and the retrieval and presentation of data in the manner required by the users.

Full digitization of records

The final objective for a wholly digital data base is conversion of analogue data to machine readable format, to be complemented by direct entry of digital data in raw form as it is acquired. A vast expenditure of effort would be required to keyboard the information held in analogue records to a computer storage medium and this may await the advent of effective optical character readers for automatic conversion of data, if it is ever accomplished or considered warranted. This leaves project specific data bases with a strong advantage for the foreseeable future for mapping exercises, but the indexing stages would greatly ease the task of tracking down some of the data for such studies.

In Table X, (adapted from Clayton *et al.*, 1987) the implementation costs for a five-year rolling programme covering only objectives 1–3 are estimated. Objective 4 would add enormously to the costs.

Table X Implementation costs for five-year rolling programme (objectives 1–3 only) (adapted from Clayton *et al.*, 1987)

Data collections catalogue		
Subsumed within work required for establishing digital indexes to the holdings, as part of the preparatory measures for initiating the scheme		
*Digital indexes to data holdings**		
Staff effort	—3, 3, 2.5, 2 and 1.5 years = 12.0 years	
Travel & subsistence	—£9 000[†]	
Operational costs	—£45 000	
Capital costs	—£167 000	TOTAL £221 000
Central respository of records		
Staff effort	—2, 1, 1, 1, 1 years = 6 years	
Operational costs	—£17 000[‡]	
Capital costs	—£100 000	TOTAL £100 000
Electronic storage of 3 million records		
Scanning (3 per minute)	—1 000 000 minutes	
No. of hours	— 16 600 hours	
No. of 4 hour days	— 4 150 days	
No. of 200 day years	— 20.8 years	
		TOTAL £392 000

OVERALL COSTS
(including staff at appropriate grades)

Total £730 000 over 5 years

(Conversion of the British Geological Survey's own data registers and holdings to digital format was estimated separately and would take the overall cost up to £935 000)

* This assumes that individual collection holders are prepared to compile their own digital indexes to holdings for central integration by the co-ordinating body
† to convert to US$ multiply by 1.8
‡ operational costs to be increasingly covered by charges for access, copies, FAX, transmission, etc. from year 5 onwards

Access to geological records

With dispersed data bases, the local holders must accommodate requests for access to, or copies of, data on request. This may work well if there is capacity to handle all of the requests, and most of these arise locally. If they come from further afield, delays may be incurred due to posting of information. While FAX systems can rapidly transmit information on paper, they cannot cope with high resolution scanning and printing, and the quality of graphic images or multitone illustrations will be poor.

An alternative option is to store copies of the records at a central holding facility, and have all requests for access and copies directed there (removing any obligation on the local holders). This may have the advantage of a back-up system, insuring against loss of data, although back-up local disks may be easily prepared but have the disadvantage that all new data must be sent to the central location for input into the system. Another possible advantage is that indexing can be centrally carried out, rather than locally.

The National Geoscience Data Centre could provide foundation for a British system of this type. There are already immense paper and microfilm archives which, however, are very demanding on staff time and accommodation. Such resources would have to be at least doubled if extension in line with the present proposal was implemented, because there would be a large influx of documents from organizations not at present supplying archival data to the Centre.

The possibility of using the extremely compact Laser-Optical filing system now becoming available was suggested as an alternative to paper and microfilm. One such system now on offer consists of a work-station comprising a document scanner, monitor, IBM PC/AT with 70 MB hard disk drive, an optical disk drive and a laser printer. Documents up to 10 inches wide can be handled (of unlimited length) and can include text, tables, graphs, photographs and handwritten notes. Continuous feeding of up to 30 pages at once will process this number of documents in six minutes. The storage on optical disk provides a permanent record and each disk can store up to 4000 pages of information. Documents can be retrieved at will for viewing on screen or as hard copy by the laser printer. Total cost of such a system is estimated at £50 000. Work-stations can be networked for multiple user access and input.

For the foreseeable future hard copy would be despatched by post to the enquirer, but such systems provide the likely future answer to the critical problem of handling large documents and their direct call-up in full copy of original source document on screen or as hard copy from the archives, with the use of limited space and immediate recall.

CONCLUSIONS

(1) Applied earth science studies require carefully assembled and organized data from: (a) primary surveys, (b) laboratory testing, and (c) secondary data collections.
(2) Secondary data is very dispersed among public and private holders and needs assiduous tracing. A diplomatic approach in securing data and the practice of discretion regarding confidentiality are essential.
(3) There is a need to organize it in a manner best suited for its use for interpretation and mapping.
(4) Three alternative systems of organization exist: (i) card index and paper files, (ii) microcomputer, and (iii) mainframe computer.
(5) Choice of system depends on project complexity, cost effectiveness and the actual amount of funds available

for a particular project. Computer operation should be chosen for any but the smallest data base for two reasons: first, on direct cost grounds, i.e. it will be *more* expensive to operate a manual data base in the long term; secondly, for ease of duplication, where more than one user site is required.

(6) Use of secondary data must be subject to careful validation of the data.

(7) It is sensible to retain and maintain a data base after use in a specific project for possible future use (e.g., site investigation desk studies need reference back to the original data).

(8) Many data sources are located in the areas to which they refer. Data collection should therefore be carried out in or close to these areas, particularly as local knowledge can play an important part in the data collection process.

(9) Small local data bases are best held in the locality and restricted (specialist) topic data bases are best held where the expertise rests. Many users of data banks are interested only in data relating to their local area. Therefore it is useful for data to be available for interrogation locally. Technically there is no problem in providing many interrogation outlets for a single data base.

(10) There is a strong case for local authorities in England, and elsewhere, to add earth science data to their Geographical Information Systems for Land Resource Assessment, if, as is commonly the case, these data are not now included (see Burrough, 1987, for details of these systems).

(11) Although it is recommended that data collection and maintenance be *carried out* 'locally', it is recognized that this is a task requiring specialized expertise. For this reason, and to ensure a constant standard between local data banks, it is recommended that the establishment and maintenance of large data banks covering many regions be *co-ordinated* on a national basis. In particular, a central body could be responsible for developing the system and software, and could operate one or more data collection 'teams' at local sites. This could be on a direct contract basis or the cost could be recouped by licensing the use of the data bank (and suitable hardware/software configurations) to interested users.

(12) There is a strong case for a national geosciences data index (Adlam *et al.* 1988) to provide easy access to local and topic data bases to aid and allow rapidity in future exercises.

(13) There is a case for a national data store, but this requires expensive initial investment, high running costs and perhaps a substantial phasing-in period. Clusters of microcomputer systems may be more cost-effective for local and national data bases than mainframe systems.

(14) An important step for local and national data handling and for information interchange would be to set common standards for data collection, data storage, data retrieval, and compatible equipment.

(15) Whilst all these conclusions are based very much on recent British work, the points made are likely to apply to most other countries, a few of which are already more advanced than Britain in computerization of earth science data.

Acknowledgement

The author of this chapter acknowledges the advice and assistance of Brian Marker and Mark Zytynski.

REFERENCES

Adlam, K.A.M., Clayton, A.R. and Kelk, B. (1988). A demonstrator for the National Geosciences Data Index, *Int. J. Geographical Information Systems*, **2**(2), pp. 161–170.

Anon. (1985). Scottish pilot scheme for ground investigation reports. *CIRIA News*, June 1985, p. 3.

Anon. (1988). Earth science information, *Aus. I.M.M. Bull. and Proc.*, **293**(7), p. 6.

British Geological Survey (1985). *National registry for ground investigation reports: guidance on usage of pilot scheme.* BGS leaflet.

Burrough, P.A. (1987). *Principles of Geographical Information Systems.* Oxford Science Publications, Oxford, 208 pp.

Clayton, A.R., Cannell, B., Adlam, K.A.M., Loudon, T.V. and Robson, P.G. (1987). Geological databank pilot study. Report prepared for the Department of the Environment, BGS, *Research Report ICSO/87/1*, 23 pp.

Crimes, T.P., Marshall, P., Harris, A., McCall, G.J.H., Chester, D.K., Mussett, A.E., Wilson, C.D.V., Brenchley, P.J., Flinn, D. and Hunt, C. (1985). *Assessment of sand and gravel resources to the south of Shrewsbury.* Department of Geological Sciences, University of Liverpool for the Department of the Environment (open file report).

Edwards, R.A. and Freshney, E.C. (1987). The geology of the country around Southampton. *Mem. British Geol. Surv.*, Sheet 315, London, H.M.S.O.

Edwards, R.A., Scrivener, R.C. and Forster, A. (1987). Applied geological mapping, Southampton area. Main Report and Appendix, *BGS Research Report ICSO/87/2*, 69 pp.

Freeman, Fox Ltd. (1987). Investigations of ground characteristics in the area around Chacewater and St. Day, in the County of Cornwall. *Main report (Volume 2) and Computer Database User's Manual, Department of the Environment, London, contract 7/1/224.*

Freeman, Fox Ltd (1988). Methods of compilation, storage and retrieval of data on disused mine openings and workings. *Final Report to Department of the Environment, London, (open file report), contract 7/1/224.*

Laxton, J.L. (1987). Applied geological mapping, Southampton area. Computer database. *BGS Research Report ICSO/87/4*, 22 pp.

Loudon, T.V. and Mennim, K.C. (1987). Applied geology mapping, Southampton area. Mapping techniques. *BGS Research Report ICSO/87/3*, 21 pp.

McCall, G.J.H., Chandler, P., Coote, V.J., Crimes, T.P. and Roxburgh, I.S. (1985). Environmental Geology Study in the Plymouth/Plymstock area. Technical Report and User's Manual for the Computerised Search Programme—Archival Database (Curator Version). GAPS Geological Consultants, for the Department of the Environment, London (open file report).

McCall, G.J.H. and Simonian, G.O. (1985). The Makran project—a case history. In *Prospecting in Areas of Desert Terrain*, Inst. Min. & Metallurgy, London, pp. 31–42.

MZ Associates (Bath, England) (1988). Computer catalogue of miscellaneous mines in England and Wales. *Report to Department of the Environment, London, contract 7/1/293.*

Tuckwell, D.J. and Sadgrove, B.H. (1977). A case for a National Registry of Ground Investigation Reports. *CIRIA Report 70.*

Part III

Earth Science Topics for Planning, Development and Conservation

Chapter 4

Water

Melinda Anne Lewis

INTRODUCTION

Water is essential to all life on earth. It is the one commodity that man is totally unable to survive without. Our bodies comprise about 66% water and we must consume about two and a half litres each day to remain alive. The lack of sufficient water to meet the demands of humans, livestock and crops constitutes a drought. At the other extreme, too much water can lead to extensive flooding. Both drought and floods cause hardship and famine throughout the world, forcing large numbers of people to leave their homes and become refugees. However, around 50 000 people die daily from water-related diseases because they do not have access to a safe and reliable source of water. The quality of drinking water is therefore as important as quantity. Water is also used for many other purposes. Our ancestors used it for defence and transport; even today there are few major cities in the world not located by the sea or on the banks of a perennial river. Water is an essential part of many aspects of our twentieth-century lives and is now used additionally for purposes as varied as power supply and recreation.

Water, like air, is an atypical earth resource—they are both renewable. Although individual water molecules are continually moving, the total volume of water on the earth remains constant and it is only its quality that changes. Water has some unusual properties, the most important being its ability to dissolve many other substances; it is virtually incompressible and it reaches its maximum density at a temperature 4°C higher than it solidifies or freezes.

This chapter will look at the movement of water around the earth (*the hydrological cycle*) and the various factors that affect its distribution and quality. The power of water, its quality and its uses, as well as some of the problems it causes, will then be considered. This is followed by sections dealing with the impact that water has on land use and health. Finally, the various types of maps available, and their applications in relation to water and land use are reviewed. Technical terms mentioned in the text are defined in the glossary.

THE HYDROLOGICAL CYCLE

It has been estimated that the total volume of water on the earth is 1460 million cubic kilometres; about 94% of this occurs in the oceans, which cover nearly 71% of the earth's surface and have an average depth of nearly 3.8 kilometres. The remaining 90.2 million cubic kilometres occur in the atmosphere or are distributed over and within the land's surface. Table I indicates the volumes of water stored at any one time. Apart from the oceans, the two principal locations where water is stored, are as groundwater and in ice-caps and glaciers. The quantity of water in rivers and lakes at any one time is negligible on a global scale, although they discharge 30 000 cubic kilometres of water into the oceans each year. Table I also shows the length of time a molecule of water will remain in any one form. The mechanism by which it moves from one location to another is called the hydrological cycle (Fig. 1). This is driven by two factors: the sun's heat and gravity.

Earth Science Mapping for Planning, Development and Conservation

Table I Estimate of the World's water balance (after Nace, 1971)

	Volume ($km^3 \times 10^6$)	*% of total water*	*Residence time (years)*
Oceans and seas	1 370	94	△ 4 000
Groundwater	60	4	0.04–10 000
Icecaps and glaciers	30	2	10–1 000
Lakes and reservoirs	0.13	⩽0.01	△ 10
Soil moisture	0.07	⩽0.01	0.04–1
Atmospheric water	0.01	⩽0.01	△ 0.03
Swamps	⩽0.01	⩽0.01	1–10
River channels	⩽0.01	⩽0.01	△ 0.04
Biospheric water	⩽0.01	⩽0.01	△ 0.02
Totals	1 460.2	100	

The sun's heat over the oceans causes evaporation to occur. The resulting water vapour rises on air currents and winds. As the air rises it expands due to a decrease in pressure, and as it expands cools, becoming saturated with water vapour which then condenses around any particles suspended in the air. These small water droplets (0.001–0.1 mm in diameter) form clouds, and once they coalesce to produce larger droplets (1 mm in diameter) or ice crystals, precipitation occurs. Due to atmospheric circulation only 80% of this precipitation occurs over the oceans, the remainder falling above land. The uneven distribution of precipitation both spatially and temporally is controlled by the strength and direction of the prevailing winds and by topography. The amount can also vary considerably from year to year; in Britain annual variations are normally between 60% and 150% of the mean, but elsewhere it may range from zero to several hundred per cent of the average.

Some precipitation falling over land will be intercepted by buildings and vegetation and thence evaporate directly back to the atmosphere. The remainder reaches the earth's surface and then follows one of several possible routes. It may remain on the surface in the form of pools or puddles which are eventually evaporated. The rate at which this will occur depends on the climate, texture of the soil and the type and density of any vegetation cover. It may flow over the surface, as overland flow along depressions and channels to become surface runoff in the form of streams, rivers or lakes. At high altitudes or latitudes it may fall as snow, be compacted into ice and form ice-sheets or glaciers. These will move slowly down-gradient under the force of gravity until at their base they either enter a river or the sea as meltwater or chunks break off to form icebergs. Precipitation may infiltrate through the ground surface to join existing soil moisture storage, which may return to the surface as interflow or will percolate downwards to become groundwater.

The rate at which infiltration occurs depends on many factors: the depth, porosity (including the size, number and degree of interconnection of the voids) and permeability (related to texture, structure and clay content) of the soil and near-surface rocks, the antecedent soil moisture conditions, vegetation cover, slope of the ground surface and land use. Once water has penetrated the air/soil or rock interface it enters the vadose zone or zone of aeration. This is where the pore spaces are filled with a mixture of air and water and water movement is mainly downwards; it is shallow in humid regions but in arid areas may reach 100 m in thickness. The upper part forms the soil water zone from which water is readily removed by evaporation or transpiration. The base of the vadose zone is marked by the water-table which forms the boundary between the unsaturated and saturated zones: below this level all the pores are filled with water and the predominant direction of water movement becomes horizontal. The water-table is not a flat surface,

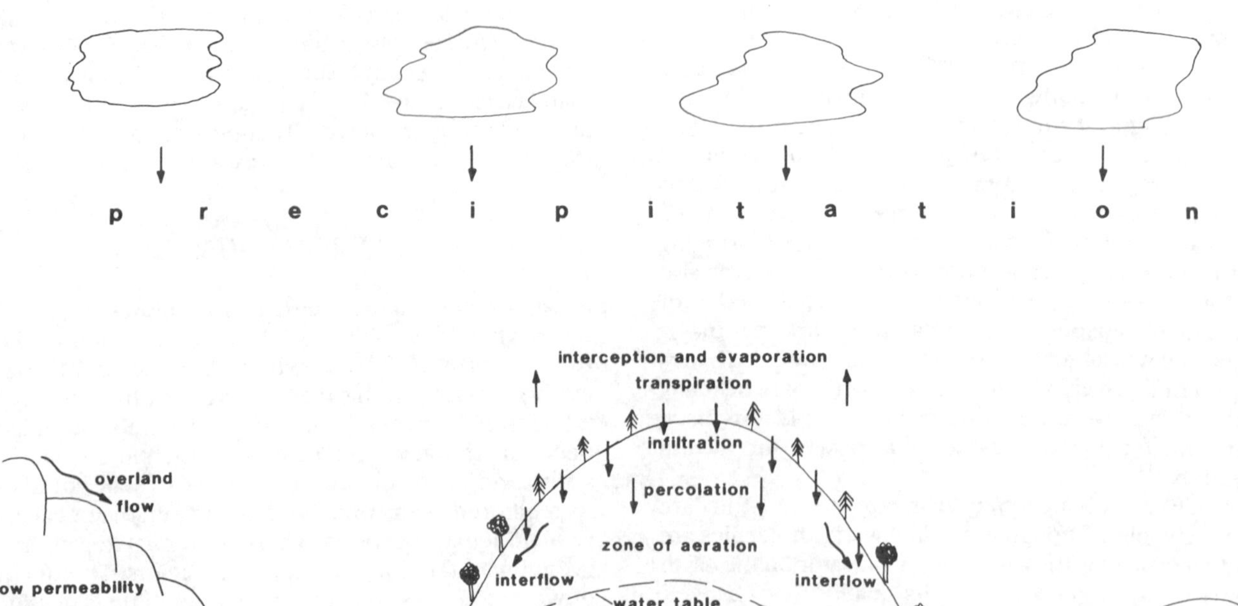

Fig. 1. The hydrological cycle.

but reflects the shape of the topography in a subdued form. Where it intersects the surface of the land, springs and swamps occur. Under constant climatic conditions the water-level would stabilize; however, it rises and falls in response to both seasonal variations and man's interference. The movement of water through a rock is determined by its permeability and interconnected (or linked) porosity. If these are high enough to allow water to flow though the rock at rates at which it is economic to abstract it using wells and boreholes, the rock is called an aquifer. Rocks that do not yield significant quantities of water are referred to as aquicludes. Below depths of between 600 and 1000 m, porosities generally become so low that few rocks remain aquifers.

Once part of the groundwater body, water generally moves slowly down-gradient to discharge as baseflow into rivers or the sea. Hence, most water returns eventually to the oceans to complete the hydrological cycle. Even formation water held within the pores of sedimentary rocks at depths of several hundreds of metres and the considerable quantity of water bound up chemically within the lattices of minerals are part of this cycle. However, this water may be removed from atmospheric circulation for millions of years and it can therefore be ignored.

The hydrological cycle (Fig. 1) can be summarized as follows:

total inflow = total outflow + change in storage
where
inflow = precipitation
outflow = evaporation + transpiration + icebergs + surface runoff + groundwater outflow direct to sea
change in storage = change in amount of water stored as ice, surface water, soil moisture and groundwater

This equation is relevant both on the global scale and also within a given catchment area (if water imports and abstractions are also included) for any given period of time.

FACTORS AFFECTING REGIONAL WATER RESOURCES

The three main factors affecting the water resources of a region are climate, geology and geomorphology.

Climate

Climate affects the water resources of a region in several different ways. Rainfall is obviously the most important factor. Where there is a very low rainfall, surface water will be limited and any groundwater available will either be fossil water that infiltrated when the climate was wetter or recharge from another area. For example the abstraction of groundwater for irrigation from the Nubian Sandstone in Libya is causing a decrease in water-levels and hence mining of the resource, as the water is several thousand years old and can only be replaced by recharge occurring in the Tibesti Mountains several hundred kilometres away. This situation can only be acceptable when the water is required for a definite time period, such as until another source of supply (e.g. desalination) is available. The amount of precipitation also has an effect on groundwater quality, water often being of poorer quality in desert areas. This is due to high evaporation rates tending to concentrate soluble salts at the ground surface, ready to be dissolved and carried down into the aquifer by any water that does succeed in infiltrating during a rare rainfall event. Desert sediments also often contain evaporite deposits that are highly soluble. Additionally, where recharge is limited, water movement will be slower and therefore the residence time of water in the aquifer will be longer, allowing more time for the aquifer to interact with it.

On a global scale, areas with high precipitation are relatively small. However, even in these areas the rainfall is not necessarily distributed evenly throughout the year, and dry periods can occur for which water must be stored for domestic and agricultural uses. Generally in these areas water supplies come from streams, springs and rain-water collection systems rather than from wells and boreholes. The quality of this water is generally good but turbid. The abundant amounts of decaying organic matter present in these, usually heavily forested, areas tend to decrease the pH and increase the concentrations of iron and manganese, making groundwater development a viable alternative where these factors plus biological purity and the lack of suspended solid particles are important.

Temperature, solar radiation, humidity and wind speed affect water resources by controlling the amount of evaporation that takes place. In hot climates sufficient surface water often cannot be stored through the dry season and it has to be supplemented or replaced by groundwater. At the other climatic extreme, ground temperatures never rise above freezing and permafrost occurs. It is not practicable, in these areas, to melt enough water for large settlements, due to the amount of energy that would require. Therefore springs, deep lakes or wells are the alternatives. Large springs are rare in the arctic, most only flowing during late summer and early winter, when they are fed by meltwater from the active zone; lakes are susceptible to contamination. This leaves groundwater development which depends on boreholes penetrating beneath the permafrost or locating areas where it is locally absent (e.g., close to deep lakes or major rivers). Groundwater in these regions is generally of good quality, although high iron, manganese and organic matter contents have been reported.

Climate also affects the rate at which weathering occurs, with thicker sequences forming in the tropics. Similarly it affects erosion and deposition. For example, glacial sediments, which can form good aquifers, are only deposited in cold climates.

Geology

Geology affects groundwater and surface water resources in many ways—different rock types have different hydrogeological characteristics, geological structure controls drainage and groundwater flow systems and age has some effect on the amount of induration of sediments and the precipitation of secondary minerals in volcanic rocks.

Typical porosities and permeabilities of various rock types are given in Table II. The most common sedimentary rocks are fine-grained silts, clays and shales; these have relatively high porosities (decreasing with both age and depth of burial) but very low permeabilities and do not generally form aquifers. However, they act as confining beds for water in underlying aquifers and in situations where differential heads are created, induced drainage of

Table II Typical ranges in values for the porosity and permeability of common rock types

Rock type	*Porosity* (%)	*Permeability (m/day)* (*min.*)	(*max.*)
Clay	45–60	5×10^{-7}	– 10^{-4}
Loess	40–50	0.01	– 1.0
Sand	25–50	0.1	– 500
Gravel	25–40	50	– 50 000
Sand and gravel	20–35	5	– 100
Till	20–40	10^{-7}	– 0.5
Salt	⩽1	5×10^{-6}	– 5×10^{-3}
Limestone, dolomite	0.1–20	5×10^{-6}	– 0.5
Karst limestone	5–50	0.1	– 1 000
Chalk	5–40	up to 5	
Sandstone	5–30	5×10^{-5}	– 20
Shale	1–25	5×10^{-8}	– 10^{-4}
Lignite	⩽1	0.1	– 10
Friable tuff	35–60	0.02	– 2
Welded tuff	10–40	5×10^{-5}	– 0.2
Dense basalt	0.1–1	10^{-6}	– 10^{-3}
Fractured basalt	5–50	10^{-4}	– 1
Vesicular lava	10–50	10^{-4}	– 10^{-3}
Pumice	up to 85		
Slate	0.001–1	5×10^{-9}	– 5×10^{-6}
Dense crystalline rock	0.0001–3	5×10^{-8}	– 10^{-5}
Fractured crystalline rock	2–10	10^{-3}	– 10

the large volumes of water stored in them occurs. Due to their low permeability these rocks form good dam sites and catchment areas for reservoirs.

Sands are highly permeable as well as being porous and make good aquifers if wells and boreholes are properly designed with suitable sand screens and gravel packs. Runoff from unconsolidated sands is generally limited to very heavy rainfall events. Sandstones often have permeabilities several orders of magnitude lower than the corresponding unconsolidated sediments—this is due primarily to the presence of cement, reducing their porosity, but also to compaction. Highly indurated well-cemented sandstones and quartzites do not, therefore, form good aquifers unless their permeability is increased by fracturing.

Carbonate rocks have high porosities when deposited, but this is often later reduced by compaction, solution of aragonite and calcite and the redeposition of calcite cement. Their permeability is generally low. Carbonates may be greatly affected by processes occurring even later after deposition, which alter their hydraulic characteristics again. Dolomitization of a lithified limestone reduces its volume by 13% producing a similar proportion of voids. However, dolomite crystals are so small that the permeability generally remains low. Also, rainwater percolating through a carbonate aquifer slowly dissolves it, producing solution channels and caverns through which water can move very rapidly. Carbonate rocks are commonly well-jointed and fissured, and water can additionally move through cavities other than solution features. The yields from boreholes into these rocks, therefore, vary considerably depending on whether water-bearing fissures are encountered or not. Water quality is also variable, being hard but of good quality under low flow conditions, but similar to surface water with high suspended solids after heavy rain. Surface water courses often disappear completely when they reach outcrops of carbonate rocks, not to reappear again until they reach other strata.

Because most sedimentary rocks are stratified (layered), they tend to be anisotropic, with greater permeabilities generally occurring in the horizontal rather than the vertical direction.

Fresh metamorphic and intrusive igneous rocks have minimal porosities and permeabilities, but groundwater may be obtained from fractured and weathered zones and its quality is normally good. Where not extensively fractured these low porosity and permeability rocks provide excellent sites for dams and reservoirs. Volcanic rocks are very variable in character, and typical porosities vary from less than 1% for basaltic lavas to more than 85% for pumice. In lava flows, permeability is a function of the structures within the rocks: cooling joints, lava tubes, intersecting vesicles, fracturing caused by the buckling of partially solidified lava and voids left between flows. Therefore, the development of groundwater from lavas depends on intersecting a localized water-bearing system where yields will generally be high, whereas elsewhere they will be minimal. Fragmental volcanic rocks such as tuffs and pumice commonly form good layered aquifers, similar to coarse-grained sediments, with high intergranular permeabilities. Mesozoic or older volcanic rocks, however, tend to form poor aquifers due to the crystallization of secondary minerals and compaction. Groundwater quality in volcanic rocks is generally good to excellent, containing predominantly calcium, magnesium and bicarbonate or sodium and bicarbonate ions. However, near hot springs and fumaroles, the water may have a high sodium and chloride content, and a low pH and high sulphur and fluoride concentrations may cause problems. Where extremely high permeabilities exist, biological contamination is a potential danger. However volcanic rocks tend to weather rapidly and the resulting thick soils filter out most pathogens.

The structural geology of an area controls both its permeability and often the direction of flow. Rocks that have undergone considerable structural deformation will generally become more permeable during this, with fracture zones in crystalline rocks becoming areas of higher permeability. However fractures decrease with depth due to increasing overburden pressures. On the other hand, secondary mineralization of joints and faults can both reduce the permeability and prohibit flow moving along and across the discontinuity. In sedimentary or volcanic rocks where some degree of layering occurs, faults may, depending on the size of throw, either provide hydraulic continuity between two different formations or prohibit movement downdip where an aquifer abuts against an aquiclude. Where rocks are folded, higher permeabilities will be encountered in anticlines, and lower values in synclines as the former are tensional features with the fractures open nearer the surface (Fig. 2). Conversely, therefore, synclines form better reservoir sites. The structural geology of an area is linked to its past and present positions on a tectonic plate. Sites with high seismicity such as those on or near plate boundaries are not suitable for the construction of impounding reservoirs. However, the pressing need for water storage may outweigh this risk. The weight of water behind a dam may trigger off earthquakes in a previously stable area (e.g., the Kariba dam on the Zambezi River).

Geomorphology

The geomorphology of a catchment affects the water resources in many ways. Firstly topographic position is

(a)

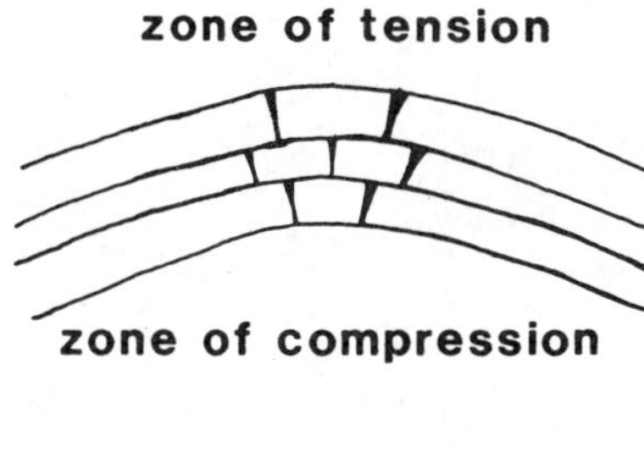

(b)

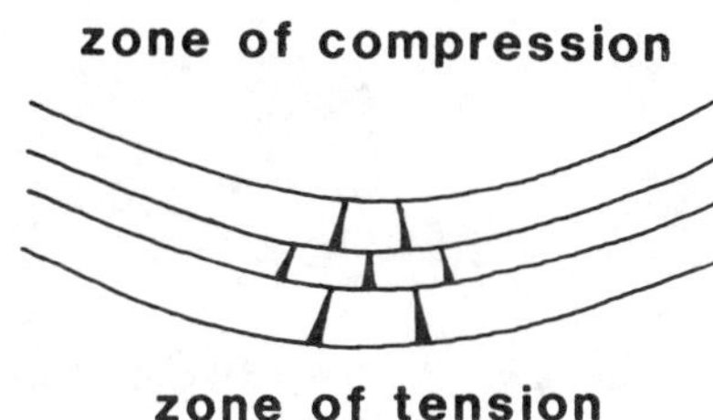

Fig. 2. Location of open joints in (a) anticlines, and (b) synclines.

important, in general the yields of boreholes will be higher in valleys than on interfluves. There are several reasons for this. The smaller a catchment area is, the less water there is available, therefore sites further up slope have smaller resources than those lower down. Secondly, water moves downhill due to gravity and, therefore, flow paths converge towards lower ground. Lastly, valleys are often structurally controlled with higher yields being obtained from the discontinuities beneath them than from the surrounding rocks. However there are some exceptions. In certain areas in the tropics, clay minerals tend to become concentrated under the lower lying ground and hence better yields are obtained on the relatively flat interfluves.

Rivers flowing through flat areas are more likely to deposit sediments; these are often of a higher permeability than the surrounding rock and hence will provide improved supplies. The deposits will also provide bank storage from which to feed the river in times of low flow.

The age of an erosion surface or the length of time since uplift last occurred, has some effect on water resources, as this represents the time that weathering, erosion and deposition have been operational, and hence is related to the thickness of any soil or superficial deposits present. Whether erosion is occurring also needs to be taken into consideration when siting a reservoir, as sediment will be deposited prematurely if the river is slowed down by an artificial feature such as a new lake and dam. In the case of the Aswan High Dam in Egypt, not only is sediment filling up the reservoir and hence reducing its capacity, but previously the Nile was relied upon to provide fresh sediment and nutrients annually to the farmers downstream of the dam, while now they are dependent on artificial fertilizers. As the amount of sediment a river carries increases considerably in times of flood, dams are now often constructed in such a manner that during these periods sediment-laden water bypasses the reservoir.

THE POWER, USES AND PROBLEMS OF WATER

Water is a very powerful medium and is the main mechanism whereby material is transported over the earth's surface. It is also often a cause of major catastrophes. This section looks at the ability of water to weather, transport and redeposit material, the uses it is put to and the problems it causes.

Weathering

Rocks at or near the surface are continually being attacked by various atmospheric and biological processes causing them to weather. This occurs through both mechanical and chemical means and the products vary depending on climate, topography and geological setting. In mechanical weathering, rocks are disintegrated by temperature changes, frost action and organisms. In a very dry climate the stresses created by diurnal variations in temperature causing a thin outer shell of rock to expand and contract are insufficient to rupture it, but if it is wet repeated cycles of heating and cooling will. In cold climates, any water present in cracks, pores and crevices will freeze when the temperature drops below 0°C, this leads to an increase in volume of 9% which exerts pressure rupturing the rock. Waves can have a similar effect to ice. Water in the form of high-pressure spray is forced into cracks and crevices, compressing any air already in them. When the waves retreat, the compressed air expands suddenly loosening whole blocks of rock. Earthworms and other burrowing organisms, such as termites, help to break up the soil and the roots of shrubs and trees force rock crevices apart as they grow.

Chemical weathering causes the minerals of a rock to decompose, dissolve or be loosened by the water, oxygen and carbon dioxide in the atmosphere, by soil water or by organisms and the products of their decay. Carbon dioxide dissociates water into hydrogen and bicarbonate ions. These react with limestone to form calcium bicarbonate which is readily soluble in water, causing the formation of both typical karst landscapes and large underground solution channels and caverns. Silicates decompose and become hydrated to form clay minerals and silicic acid when dissolved oxygen and carbon dioxide are present in water. Rocks weather at different rates, the higher the anion to cation ratio (the ratio of the number of negatively charged ions to the number of positively charged ions in the chemical formula of a mineral) of a mineral is, the more resistant it is to weathering. Therefore, basic rocks particularly those containing olivine (ratio 1:33) are more rapidly weathered than granites and granophyres. Organisms decay to produce carbon dioxide, organic acids and traces of ammonia and nitric acid. The presence of these substances in soil water increases its ability to weather rocks. All types of plants, including fungi and lichens, also contribute to chemical weathering as they abstract certain elements from rocks and liberate others.

The physical, chemical and biological agents of weathering work in conjunction with each other. The stresses required to shatter a rock purely by physical means are significantly reduced by prior decomposition, and the formation of cracks provide opportunities for air and water to attack fresh rock at greater depths.

However, depending on the temperature and rainfall, one or other mechanism will predominate. In humid climates with low relief, decomposition and dissolution take precedence, while in areas with steep slopes and cold or dry climates disintegration is the dominant mechanism.

Erosion and deposition

Once the weathering products have been formed they are transported away by water (in the form of rain, rivers or the sea), ice (as glaciers, frost or melting snow) or wind. As rain falls on dry ground the splashes mobilize grains of soil which, if the rain is of sufficient intensity, flow down slope as sheet wash or flash floods. Once flatter ground is reached, the flood water spreads out and slows down, the fine-grained material is deposited and the water evaporates or infiltrates.

When rivers flow down slope, they carry with them any particles of weathered material that they can move. Their ability to move material is related to their velocity and turbulence. The maximum load of debris of mixed shapes and sizes that a river can carry is proportional to its velocity to the power of 3 or 4, but the size of the largest fragments of a given shape that can be transported is proportional to its velocity to the power of 6, provided an adequate depth of water is present. Under normal conditions flow is generally laminar and little erosion takes place. However during floods water movement becomes turbulent and eddies occur. These eddies cause changes in pressure, create vacuums and rocks are sucked into streams and carried away. They also bounce loose material off the bottom which is then carried along by the current until it is no longer fast enough and they sink. Larger material is rolled along the bottom and the finest material is carried in suspension. Erosion occurs most rapidly near the sources of rivers where the gradients are steepest, here rivers are continually cutting deeper valleys which also eventually become wider due to lateral erosion. Where the velocity of a river decreases, such as away from the main channel or in bodies of standing water, it can deposit a considerable proportion of its load. Rivers, however, still carry over 8000 million tons of sediment into the seas each year.

The sea operates as an agent of erosion in several different ways. Currents, tides and, in particular, waves pick up loose material and transport it to be redeposited elsewhere. Loose fragments of rock are worn away by attrition and destroy the coastline by corrasion. Cliffs are undercut by waves.

Although ice is a crystalline solid, it deforms and flows when subjected to sustained pressure. As glaciers move down slope under the force of gravity they pluck away loose rocks in their path, and these form an abrasive basal surface which then grinds down other rocks on the glacier's route. As a glacier melts and retreats, it leaves behind this debris to form deposits known as moraines.

Water quality

The oceans have a dissolved solids content of 3.5% but the water evaporating from them is distilled. However, the atmospheric precipitation is slightly acidic because the rain or snow picks up dissolved carbon dioxide from the atmosphere and has a low dissolved solids content (typically between 10 and 20 mg/l) from the material forming the nuclei of water droplets. Once it reaches the ground surface, the quality of water is influenced by the natural materials and pollutants with which it comes

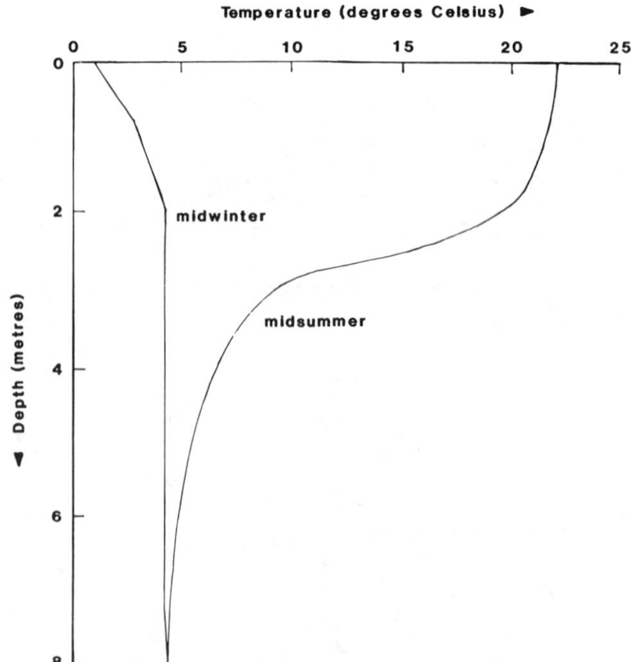

Fig. 3. Temperature gradients of a lake at different times of the year.

into contact. Where water is continually moving, the constant dilution gives it a remarkable resistance to pollution. This is assisted by its temperature-density curve, which ensures that sub-arctic and temperate lakes are overturned bi-annually (Fig. 3). As groundwater is normally of better quality than surface water, it usually only receives chlorination treatment before being put into supply, while surface water undergoes coagulation, sedimentation, filtration and chlorination.

Water as a source of energy

The downhill movement of surface water due to gravity has been utilized to turn waterwheels and produce mechanical power for centuries, and the process has been used to generate electricity since the early nineteenth century by the use of water-driven turbines. Hydro-electric power has many advantages over other forms of power generation; there are no problems of transport or increasing costs of fuel, no waste products to be disposed of and it can be switched on or off within minutes without wasting energy. Its disadvantage is the uncertainty of a reliable water supply, but often this can be solved by the creation of a large storage reservoir to smooth out seasonal and, in the case of the Kariba Dam on the Zambezi River, annual variations in rainfall.

Water can also be used in electrical storage schemes. In pumped storage schemes, excess base output of electricity is stored as potential energy to meet peak demands. At Ffestiniog in North Wales, a natural lake is filled at night using off-peak electricity from the Trawsfynydd nuclear power station to pump water uphill. This water is then available to drive turbines to generate hydro-electric power to meet peak demands the following day.

Tidal and wave power can also be harnessed to produce electricity. The first tidal power station was completed in 1966 on the Rance estuary between St Malo and Dinard, on the north coast of France. A major disadvantage of tidal power is that the timing of the output is dependent on the ebb and flow of the tide and therefore, even if the generators are designed to function at both low and high tides, the output is cyclical. Thus, a tidal plant can neither provide a steady base output nor possesses the ability to meet peak load requirements as and when they arise. Wave power is even less useful as it has no reliability.

Soil moisture occurring within a few metres of the earth's surface undergoes diurnal and seasonal changes in temperature, related to the ambient air temperature. These decrease with depth and in temperate regions die out within a few metres of the ground surface. Below this depth, the water is at a constant temperature, similar to the mean annual air temperature, throughout the year. Groundwater can therefore be used as a source of water at a constant temperature, and groundwater heat pumps that pass the water through heat exchangers are at the experimental state in the UK but are used for space heating in West Germany and the USA.

The earth is continuously generating heat through the decay of naturally occurring radioactive minerals and, as a result, groundwater temperatures increase with depth. This increase is called the geothermal gradient. The average gradient for the world is 25°C/km; however, it varies significantly depending on the geological history of the area and the natural radioactivity of the bedrock. Geothermal gradients are therefore lowest in old stable shields and highest in areas of recent tectonism. Thermal groundwater near the surface in areas with neither recent earth movements nor higher than normal radioactivity indicate deep circulation of groundwater (e.g., Bath, Avon).

The earth's natural heat can be exploited as geothermal energy. Naturally occurring steam or hot water from convective hydrothermal systems can be used to generate electricity. The first such field to be developed was at Lardarello in Italy. This used natural steam to drive a turbine. However, vapour-dominated systems are rare, the only other ones known being at Geysers in California and Metsukawa, Japan. Most geothermal fields, which are associated with rifting and volcanism along plate boundaries, yield hot water (e.g., Iceland, New Zealand, Kenya).

In areas with lower geothermal gradients, it is necessary to drill to considerable depths to encounter the required temperatures and as permeabilities decrease with depth, well yields are often too low for the scheme to be economic. However geothermal energy is being used in the Paris Basin, and in the UK it has recently come on-stream for space and water heating in Southampton. Water can also be used as a carrier of energy. The geothermal gradient in granites is high, but their permeability is low. This can be increased artificially by drilling two boreholes and using hydrofracing to interconnect them. Water can then be circulating through the system and the heat it collects recovered back at the surface by a heat exchanger. An experimental scheme is currently in operation in the Carnmenellis granite in Cornwall.

Flooding

Flooding is a problem caused by both surface and groundwater. Surface flooding occurs when the available channels are unable to cope with the volume of water running off. Once the channel is full, a stream will overflow its banks and may cause widespread flooding with related damage. It is possible to widen and deepen river channels by dredging to increase their storage capacity, however as well as being expensive, this also has the disadvantage of decreasing the velocity of the river under normal conditions leading to premature deposition of its load and silting up of the enlarged channel. Therefore continued maintenance of artificial channels is generally required. An alternative is to increase the height of the banks by artificial levees. However, where this has been done the rivers often flow locally at levels higher than the adjoining land, inviting a major disaster if flooding occurs. A preferable method of flood control is to create a new channel, down which any dangerous excess of water can be diverted to an alternative discharge point or to bypass an area where flooding would cause a particularly severe problem.

Work has been carried out on predicting events likely to lead to flooding so that any damage incurred is minimized. The Flood Studies Report (Natural Environment Research Council, 1975) relates the mean annual instantaneous peak discharge to various catchment characteristics plus the lithology of the soil and the antecedent soil moisture. The statistical methods used can be applied throughout the world, although the actual equations devised are only valid for the United Kingdom.

Under natural conditions, groundwater will flow into any excavation in the saturated zone of an aquifer. The size of the problem depends on the quantity of water and, hence, the permeability of the surrounding material. The inflow of groundwater can be prevented or reduced by one or more of the following techniques: sheet piling, diaphragm walling, grouting, freezing, compressed air, or dewatering. None of these methods will totally exclude the inflow of groundwater into an excavation, therefore some method of drainage is normally required in addition. This is done by pumping water from sumps constructed at low points in an excavation. In low permeability rocks, such as those encountered in deep metalliferous mines, and also in some coal mines, and for shallow road and rail cuttings that only just go below the water-table, this is often the only method of water control required. The existence of a water-table above the floor of a gravel pit places severe constraints on its reclamation and restoration by landfilling (see Chapter 5).

Flooding has also been recorded as occurring in cities which are sinking due to depletion of the underlying confined aquifers (e.g., Venice, Gambolati *et al.*, 1974; Bangkok, Rau, 1988). Another interesting case of flooding is recorded by George (1988)—certain cities, towns and developments in arid and semi-arid environments are being endangered by rising groundwater-levels (mostly due to man-made causes, such as irrigation and the decline in usage of local wells with the introduction of new piped supplies from distant reservoirs). Both the above illustrate the hazard of water-levels rising relative to the land surface.

Ground stability

The extent to which the water present in any excavation that extends below the water-table causes instability is dependent on the pore-water pressure. This bears no relation to the quantity of water flowing in or the permeability of the surrounding material. At any point in a confined aquifer the total stress is equivalent to the weight of the overlying beds per unit area. This stress is balanced by upward pressures comprising the grain-to-grain contact pressure or effective stress and the pore-water pressure relative to atmospheric pressure.

In most situations the total stress remains constant (as the weight of the overlying material does not alter); however, the relative proportions of this stress taken by the framework of the aquifer and the water may change. If the pore-water pressure is reduced, due to water being abstracted, then the effective stress must increase and subsidence may occur due to compaction. If the overlying material is progressively removed, the total stress is reduced and the effective stress gradually decreases until it is zero and all the weight of the overlying material is taken by the pore-water.

In consolidated rocks, the grains are held together by cement. However in unconsolidated rocks, the effective stress is provided by the resistance to shearing of the friction between grains. If the effective stress is reduced to zero, there is still some tendency for individual grains to stick together, particularly when clay is present, leaving some shear strength or cohesion. However, clean sand has little or no cohesion and once its effective stress is reduced to zero, the sand loses its shear stress and behaves in a similar manner to a liquid. Even if the effective stress is decreased slightly, the reduction in shear strength can lead to material slumping into excavations or to landslips; most landslips occur during or at the end of rainfall events.

Movements along fractures in consolidated rocks are caused by changes in pore pressure. Stresses, associated with movement the crust, build up until they exceed the strength of the rock which then fractures, forming a fault and releasing energy in the form of an earthquake. It should therefore be possible to predict movement and hence earthquakes by monitoring pore pressures along fault zones. A future possibility is that by injecting water into these zones, frictional resistance can be kept low, stresses are relieved a little at a time and major earthquakes can be prevented.

Although impounding reservoirs are sited on low permeability rocks, due to the considerable head differences that build up between the upstream and downstream sides, water flows under the dam leading to increased pore pressures in the rock beneath it. This reduces the effective stress and hence the frictional forces stopping the pressure of the impounded water forcing the dam downstream. The problem can be alleviated by the injection of grout to strengthen the rock and reduce its porosity and permeability and by draining water into a cavity within the dam to relieve the build-up of pore pressure.

WATER AND LAND USE

Although water is a pre-requisite for life, too much can also be a nuisance. Thus, the presence or absence of water is a major constraint on land use. In the United Kingdom, where rainfall is plentiful and water resources are not a major problem, planners tend to take the presence of water for granted, and are therefore often more concerned about the effects of flooding. However, this is only a very minor problem compared with the floods that occur periodically in low-lying countries like Bangladesh, where extensive areas of land are submerged with loss of people, livestock and houses, together with the destruction of vital crops. Elsewhere in the world, too little water is a problem, though. The site for the new capital city of Nigeria was chosen due to its central location, but limited water resources may now constrain its development.

Pollution of water may be accidental (due to a spillage of oil or chemicals as the result of an accident), incidental (the effects of fertilizers and biocides getting leached into the hydrological cycle) or intentional (the disposal of waste materials either onto the surface or into below surface excavations); however, it is nearly always related to the past or present use of land, and affects the potential for future land use.

The two main types of water pollution are biological and chemical. The major cause of biological pollution is sewage containing bacteria and viruses. Chemical pollution is caused either by natural compounds present in the catchment area or aquifer, or by the addition of man-made effluents. Physical pollution, resulting in a change of temperature, colour or cloudiness, can also be a problem.

Urban development

When the first settlers chose locations for their villages on the banks of perennial rivers, water was available for transport as well as supply. Today urban areas still require water for supply and usually for waste disposal as well. The possibility of flooding and ground stability must be taken into account when designing buildings if the water-table is shallower than the foundations or likely to become so during the life of the building. Water-levels in the Chalk in London originally overflowed above ground level, but due to abstraction exceeding recharge in the first half of this century the aquifer became unconfined. Levels are now rising again and could, for example, eventually reach the lowest basement levels of the proposed new British Library building. This has required expensive modifications to be made to the design of the piles. Surface water flooding may be a problem in towns, either due to rivers overtopping their banks or because storm sewers cannot cope with the large volume of runoff created, due to roofs and tarmac being impermeable. The storage of salt for the de-icing of roads and its concentration in runoff can also be a cause of water pollution.

A water-borne sewage disposal system relies on the principles of dilution and bacteriological activity. If an insufficient quantity of water is available to dilute the waste or the bacteriological activity is limited or curtailed the system breaks down. However, water treatment has generally stopped pollution from sewage reaching sources of public supply. Areas off mains drainage often utilize cesspits to dispose of waste. These allow the waste to leak away slowly and if situated over a minimum of 2 m of unsaturated aquifer (except in highly permeable aquifers), aerobic decomposition and attenuation should be sufficient to protect the underlying groundwater body.

Industrial development

Certain industries require water with specific properties to enable a particular process to be carried out. For example, the clothing industry requires soft water with a low iron content, hence its traditional home in the Pennines, using soft gritstone water. Water used in processes where it will be boiled must have a low temporary hardness otherwise the system will fur up with scale. The food processing industry requires water low in iron, calcium and magnesium; however, moderately hard water is required for baking. China clay processing is sensitive to the presence of humic acids in the water, requires electrical conductivity to be within the range 250–400 microsiemens per centimetre, and also that the calcium and chloride ion concentrations are low. Electricity generating requires huge volumes of water for cooling; therefore, power stations are generally situated on rivers or the coast, although some use groundwater. There is also the problem that even after passing through cooling towers the water may still be several degrees higher than its source when it is discharged. This can increase algal growth to the extent that fish and other water life have insufficient oxygen and die. Some work has been carried out on the possibility of passing waste heat from power stations through heat exchangers to increase its temperature, and then injecting it into the ground to form a thermal anomaly that decays over a period of time, producing sufficient heat at the surface to heat the soil and enable crops to be grown out of season. Water resources used to generate heat need to be located near to the area of supply, otherwise transport is expensive and heat is lost in transit.

Certain industrial processes need to be located well away from sources of supply due to the possibility of pollution. Storage tanks for chemicals or fuel can leak or spill endangering both surface and groundwater supplies. During the industrial revolution, unregulated discharges of phenols, cyanides and compounds of chromium and lead were responsible for the chemical poisoning of many rivers. Legislation has now stopped unauthorized discharges in most of the world. However, accidents can still occur and in the UK there are several incidents of industrial wastes causing serious harm to river systems each year. Similar unauthorized discharges also still take place into the Great Lakes of Canada and the United States.

Downwind of industrial areas sulphur and nitrogen compounds are increased in the atmosphere and hence in precipitation. This has led to the problem of acid rain over parts of North America and Europe. There is some controversy over the amount of damage this causes, and it is more of a problem where the soils already have a low pH. In Britain this tends to be upland, often forested, catchments. In some areas (e.g., Canada and Scandinavia), surface waters, particularly lakes, have become so acidified that fish can no longer live in them. The problem has been solved locally by the addition of lime to the water to raise its pH.

Waste disposal

To minimize the pollution risk, it is important that hazardous or potentially hazardous wastes are disposed of in such a way that they do not contaminate either surface water sources or aquifers. In general, surface water dilutes any pollutant or unwanted waste, but it still remains in the system. Exceptions are wastes with a high biochemical oxygen demand; this decreases naturally in flowing rivers due to aeration. Groundwater is generally free of pathogenic micro-organisms, suspended solids and certain chemical pollutants due to the natural filtering effect of the unsaturated zone on any percolating water, plus its comparatively long residence time. Other chemicals, however, are more persistent. This is true only if the water-table is not at the surface, the aquifer does not contain large fissures or comprise coarse gravels, and groundwater sources are properly sealed to prohibit the entry of contaminated surface water.

Many landfill waste disposal sites therefore operate on the principle of dilute and disperse, the leachate produced being allowed to enter a surface water course or the unsaturated zone of an aquifer, on the assumption that concentrations will be harmless by the time they reach a source of water supply. However, some waste disposal sites produce such toxic leachates (generally those with high metal concentrations) that they have to be contained. In this case rainfall onto, and runoff from, the waste site must be minimized to reduce the volume of leachate requiring collection and disposal.

The disposal of radioactive waste is a major problem. Some uranium-derived materials will remain radioactive for hundreds of thousands of years, and thus it is important that they are isolated from all sources of water. Over this period of time formations normally considered to be aquicludes will allow significant amounts of groundwater movement to occur. The current trend is towards deep burial of the wastes, in extremely low permeability formations, encapsulated in borosilicate glass within metal canisters. However, over long periods of time groundwater could corrode the metal and leach the wastes out from the glass. In addition, radioactive waste gives out heat when it decays. This could alter the characteristics of potential storage formations such as salt and clays. Research is being carried out in several countries to find solutions to these problems.

Quarrying and mining

Quarries and open-cast mines can only operate by normal methods if the water-table is below the base of the workings or can be artificially lowered to this level; otherwise problems of flooding and stability occur. In areas with both high rainfall and permeable rocks, it may not be possible to lower the water-table below a certain level and working will then have to cease, unless special extraction methods such as drag-lines or dredgers are adopted. In deep mines, dewatering is again required but because of the general decrease in permeability with depth this is often not too much of a problem. However the initial sinking of the access shafts through permeable strata, which will eventually be lined out, can cause problems. Waters from coal mines are acidic and also often have high iron and sulphate contents. Mine drainage into surface water courses can therefore have a major effect on water quality. Large volumes of water are generally needed for the processing of minerals and in some cases the location of valuable deposits in arid areas with little groundwater may preclude exploitation. In the Kalgoorlie Goldfields of Western Australia, for example, water had to be piped about 600 km from near the coast before large-scale exploitation became feasible.

Agriculture, fish farming and forestry

The presence of water is often a crucial factor as to whether agriculture is viable or not. Plants need sufficient water to grow, but in arid or semi-arid areas where potential evapotranspiration is considerably higher than rainfall, the presence of evaporating water on the land surface or at a depth where capillary action can take place leads to a concentration of salts in the soil. In these circumstances irrigation may exacerbate the situation as soluble salts are washed down into the groundwater by either natural recharge or excess irrigation water, and over a period of time the salinity of the groundwater increases. In the Indus Valley in Pakistan, irrigation over a long period of time has caused water-levels to rise and salinities to increase so much that large areas of land are now infertile.

The salinity tolerance of different crops varies and some crops that are tolerant to saline water will also grow in fresh water while others will not. Two elements to which plants and soils are particularly sensitive are boron and sodium. If an area is likely to become flooded or has a high seasonal water-table it is more suitable for grassland than growing arable crops.

Pollution from agriculture is currently a major cause for concern. Nitrate is present in the soil due to both the fixation of atmospheric nitrogen by leguminous plants and the oxidation of organic material by bacteria. However, the levels leached out from these sources into either surface or groundwater are relatively low except beneath areas of intensive livestock raising where nitrates can reach several hundred mg/l. Nitrate is also used widely as a fertilizer to increase yields. Unless applied at exactly the right time and quantity in relation to the crop, weather and soil conditions, any excess will run off into surface water courses or infiltrate down through the unsaturated zone to the water-table. The EEC maximum admissible concentration of nitrate in public water supplies is 11.3 mg/l (as N). Already there are several areas on the outcrops of the major aquifers, in the UK, where levels of nitrate in groundwater exceed this limit. Concentrations decrease once the aquifer becomes confined due to the existence of reducing conditions.

Agriculture also causes pollution from biocides, used to control weeds, pests and fungus. These organic compounds can be highly persistent and, even in extremely small quantities, render water unfit for human consumption. Pesticide concentrations in surface runoff from arable fields are generally tens to hundreds of parts in 10^{12}, but they may exceed this locally. Nevertheless, serious pesticide contamination of groundwater is rare, as those compounds with a high molecular weight generally have a low solubility in water and are adsorbed onto clay minerals and organic matter in the soil and those with low molecular weights are volatile and therefore often returned directly to the atmosphere. Many herbicides are decomposed by anaerobic bacteria and thus continue to degrade at depths where oxidizing conditions no longer exist. Groundwater contamination only occurs where soils are permeable, cracked or thin and underlain by a thin unsaturated zone of fractured or karst bedrock.

The third main cause of agricultural pollution is silage. This has an extremely high biochemical oxygen demand and if it reaches a surface water course, due to an accident or a storage tank leaking, it will kill fish and other biota.

Fish farming requires the presence of large quantities of water at a constant temperature throughout the year—this can only readily be supplied by groundwater. There are few locations off major aquifers capable of providing sufficiently high yields. Thick superficial deposits along river valleys may not be a suitable source as the water is eventually recycled from the river and, unless there is a long time-lag, it will not be at a constant temperature all year round. Fish are very sensitive to the presence of zinc in water, levels in excess of 0.02 mg/1 being toxic to them.

Forestry has a considerable effect on water resources. Trees obviously require water to grow. As a result they decrease the amount of surface runoff available for groundwater recharge. Afforestation of a catchment prone to downstream flooding is a good idea, except in the arid tropics where water resources are limited. Trees also affect the transport of sediment within a catchment, as they bind the soil together with their roots. However, more sediment runs off from an afforested catchment than a grassland one.

Land use and water resources

It is important that the land use of a catchment is considered in relation to its water resources. Will the proposed use affect the amounts of erosion, runoff and infiltration that occur or have a deleterious effect on the quality of the water? Both surface water courses and their surrounding catchment areas, plus groundwater recharge areas must be protected from sources of pollution if they will be required at any time for supply. As most human activities are concentrated in the top 10 m of the earth's surface, water above this level is the most vulnerable. In general, it is easier to clear up surface water sources than groundwater, due in the latter case to the aquifer as well as the resource becoming polluted and therefore problems potentially persisting for thousands of years.

WATER AND HEALTH

Water is required for life, and to remain healthy everyone needs access to a clean and reliable supply of water. Consuming insufficient quantities of liquids leads to dehydration and ultimately to death. Water may be of unsuitable quality for drinking on biological, chemical or physical grounds, and standards have been laid down by the World Health Organization (Table III) as well as the EEC (Commission of the European Communities, 1980) and many individual countries.

Coliform bacteria occur in soils and in the intestines of warm-blooded animals. Soil bacteria cannot normally live in water, but, if there is sufficient organic matter present, it may support heavy growths of highly toxic amoebae. Water contaminated by faecal bacteria and viruses can transmit severe, and in some cases fatal, diseases such as cholera, typhoid, bacillary dysentery and infectious hepatitis.

Most chemical compounds or elements present naturally in water are present in sufficiently low concentrations that it remains suitable for public supply. However, levels of fluoride in drinking water in excess of 2.5 mg/l cause dental fluorosis or mottling, though levels of 1–1.5 mg/l are beneficial to health as they prevent tooth decay. These values are for temperate climates and in the tropics the recommended concentrations are about half those cited above, due to a higher average daily consumption of water. Waters with concentrations of sodium in excess of 22 mg/l are unsuitable for people with heart disease, although the recommended limit for supply is considerably higher than this. High levels of nitrate in drinking water can lead to methaemoglobinaemia in babies and there is a possible link with cancer of the stomach.

A correlation has been shown to exist between the concentration of aluminium in water supplies and the occurrence of Alzheimer's Disease, but as aluminium is only soluble in soft waters it can be removed by adding lime or caustic soda to raise the pH. Similar measures are required where aluminium is added in the form of alum in treatment plants, as a coagulant to remove colour and iron.

Most metals are toxic if ingested in sufficient quantities, as they accumulate in certain organs of the body.

Table III Drinking water quality (from World Health Organization, 1984)

		Guideline value
(1)	*Microbiological quality*	
	Faecal coliforms	0 per 100 ml
	Coliform organisms	0 per 100 ml
(2)	*Inorganic constituents of health significance*	
	Arsenic	0.05 mg/l
	Cadmium	0.005 mg/l
	Chromium	0.05 mg/l
	Cyanide	0.1 mg/l
	Fluoride	1.5 mg/l
	Lead	0.05 mg/l
	Mercury	0.001 mg/l
	Nitrate	10 mg/l (as N)
	Selenium	0.01 mg/l
(3)	*Organic constituents of health significance*	
	Aldrin and dieldrin	0.03 μg/l
	Benzene	10 μg/l
	Benzopyrene	0.01 μg/l
	Carbon tetrachloride	3 μg/l*
	Chlordane	0.3 μg/l
	Chloroform	30 μg/l
	2,4-D	100 μg/l
	DDT	1 μg/l
	1,2—dichloroethane	10 μg/l
	1,1—dichloroethene	0.3 μg/l
	Heptachlor and heptachlor epoxide	0.1 μg/l
	Hexachlorobenzene	0.01 μg/l
	Gamma—HCH (lindane)	3 μg/l
	Methoxychlor	30 μg/l
	Pentachlorophenol	10 μg/l
	Tetrachloroethene	10 μg/l*
	Trichloroethene	30 μg/l*
	2,4,6—trichlorophenol	10 μg/l
(4)	*Aesthetic quality*	
	Aluminium	0.2 mg/l
	Chloride	250 mg/l
	Colour	15 TCU†(true colour units)
	Copper	1.0 mg/l
	Hardness	500 mg/1 (as $CaCO_3$)
	Hydrogen sulphide	not detectable by consumers
	Iron	0.3 mg/l
	Manganese	0.1 mg/l
	pH	6.5–8.5
	Sodium	200 mg/l
	Sulphate	400 mg/l
	Taste and odour	inoffensive to consumers
	Total dissolved solids	1 000 mg/l
	Turbidity	5 NTU†(nephelometric turbidity units)
	Zinc	5.0 mg/l
(5)	*Radioactive constituents*	
	Gross alpha activity	0.1 Bq/l
	Gross beta activity	1 Bq/l

* tentative value
† standard units for measuring colour and turbidity
mg/l = milligrams/litre
μg/l = thousandths of milligrams/litre (or micrograms/litre)
Bq/l = bequerels/litre

Soft waters will dissolve lead from pipes and solder, so that levels increase as the water travels through the distribution network. Lead affects the functioning of the red blood cells and can cause serious damage to the kidneys, liver, brain, reproductive and central nervous systems, as well as mental retardation in children. Other metals will be present in water if the catchment area or aquifer is mineralized or polluted. Cadmium can cause hypertension, nausea and softening of the bones. Other poisonous metals that may be present in water are arsenic, copper, chromium and mercury. Ions that are not harmful to health but may cause problems with water being used for domestic purposes are iron and manganese which discolour laundry and food, and coat pipework with black slime at low concentrations.

Man-made organic chemicals can also give rise to problems. Phenol is not thought to be toxic to humans until it reaches levels considerably higher than those at which problems of odour and taste occur when the water is chlorinated. However, other organic chemicals (e.g. pesticides) could be carcinogenic or mutagenic; but insufficient information is at present available to determine which are present in water, and whether they are harmful or not.

The concentration of radionuclides in drinking water should be as low as possible, because any unnecessary exposure to radiation should be avoided. Naturally occurring radioactive isotopes are not normally a problem, as their levels are low; however, consumption of water contaminated by reactor or uranium processing wastes would present a major health problem.

The connection between water chemistry and health is part of the wider problem of the effects of environmental geochemistry on health—various review papers in Bowie and Webb (1980) deal with the problems caused by trace elements.

Irrigation water containing high concentrations of trace metals or polluted by sewage effluent can contaminate crops creating an indirect danger to health.

WATER-RELATED MAPS

Maps indicating water resources or showing water-related parameters take many forms, from simple, single parameter maps comprehensible to a large range of users to complex multi-parameter maps aimed at specialists. Unlike geological maps, most of these maps are time-dependent, constantly requiring updating to portray changes in both the physical location of the resource and the chemical nature of the water. With the recent arrival of geographic information systems (GIS), these maps are on the verge of a major transformation. Currently traditional maps are often used as the starting point for solving a hydrological problem, however in the future the starting point will be a data base which after manipulation could produce the results in several formats, one of which would be a map. However, at present digitization of the large volumes of data is expensive and can be justified only if they require frequent revision or a considerable amount of manipulation. The availability of satellite data already digitized providing information on geological structures, snow cover, flooded areas, the position of submarine springs and water quality (to name just a few hydrologically useful parameters) is already influencing the method of map compilation for areas with limited ground data.

Hydrological maps

Hydrological maps include meteorological maps indicating the areal distribution of factors such as precipitation, temperature and evaporation and the location of rain gauges and climate stations. Generally they are relatively simple and are often incorporated as small-scale insets on other more complex maps.

True hydrological maps indicate surface water features and are therefore generally related to catchment areas. They vary from factual data (e.g., drainage pattern) to derived or predictive information (e.g., estimated flooding frequency, rate of erosion of river banks). A surface water drainage map could include the location of swamps, glaciers, artificial features such as dams with reservoir capacity and gauging stations with flow data and catchment areas, in addition to rivers and standing water.

Flow maps have been produced using lines of different width for differing flows. Water quality data may also be indicated, either one particular parameter or an index of quality taking into account several factors. This can be superimposed on the flow data by using differently coloured lines for differing qualities. The position of weirs (aeration) and inflows (dilution) are required to define class boundaries accurately. Figure 4 indicates the flows and water quality (based on an index of several parameters) for the River Wear and its tributaries. Maps have also been produced indicating the difference between the actual and the objective quality of a river.

Drainage system maps are readily computerized as they show only topographic contours and the course of the surface water features. Figure 5 indicates part of a 1:50 000 topographic map of Northern Ireland and a slope or line of steepest descent map produced by computer from the topographic data at a 50 m grid interval. Each point on the grid has an elevation and drainage direction associated with it and this data can then be used to define catchment boundaries, drainage patterns and calculate the area draining to any point in a catchment in a more objective manner than the traditional 'by eye' method (Morris and Heerdegen, 1988).

Other useful information that can be produced from such data is the location of saturated areas; these are related to the catchment area and the angle of slope. For given weather conditions they give an indication of the ability of the ground to support vehicles and machinery and, therefore, have agricultural, engineering and military applications. This information is also useful when considering restoration of land. With the addition in the future of gridded thematic data (e.g., land use, soil, lithology, precipitation and evaporation data) to the digital terrain model, it should be possible to obtain a wide range of catchment characteristics for any point of interest. These could then be used to assess the risks of flood or drought, or to aid in the selection of sites with sufficient energy potential for small-scale hydroelectric power installations.

The base flow index is a measure of the characteristic

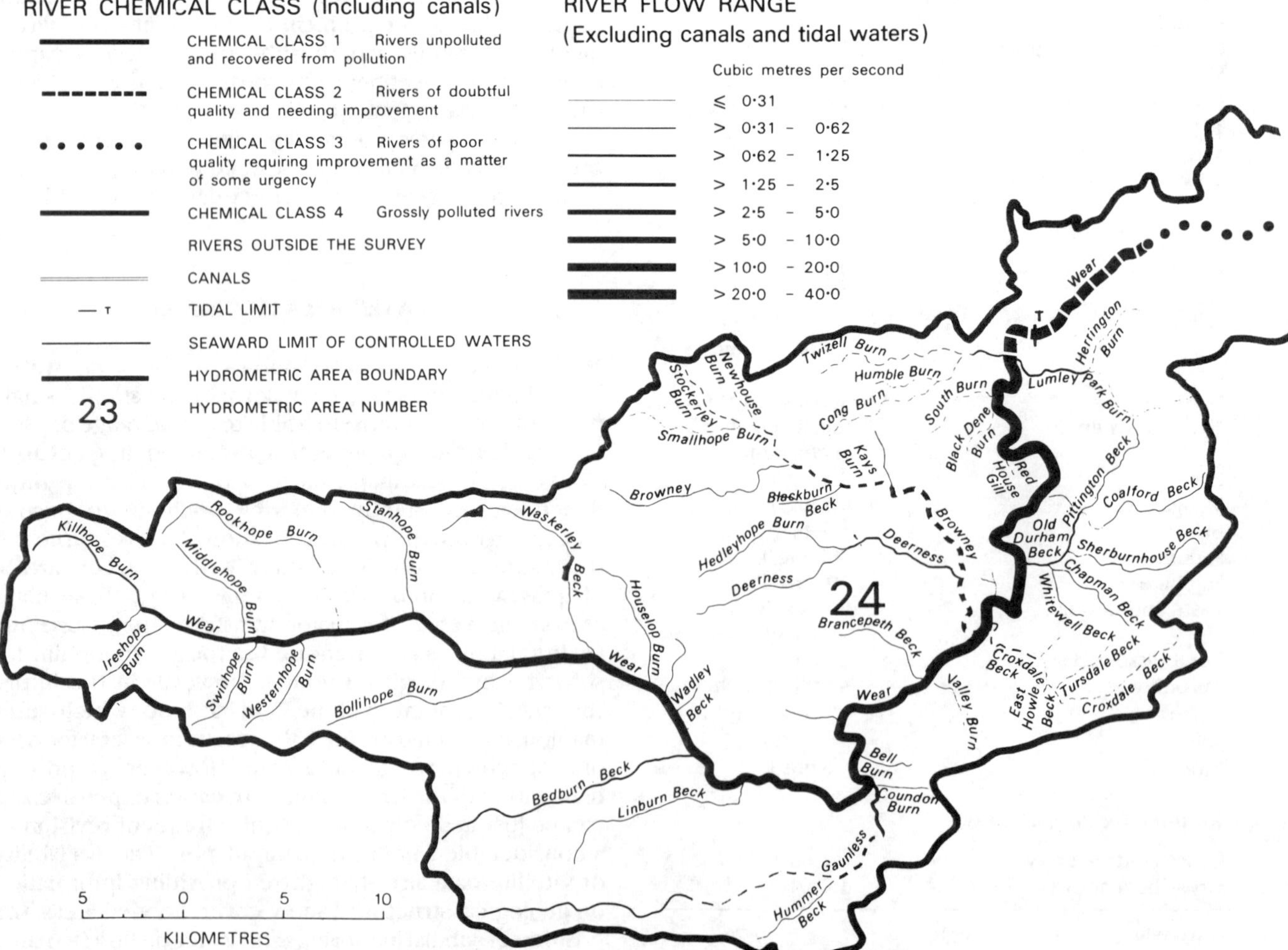

Fig. 4. River flow and chemical class for the Wear Catchment, England (from Department of the Environment, 1976) (black and white illustration derived from a coloured original).

(a)

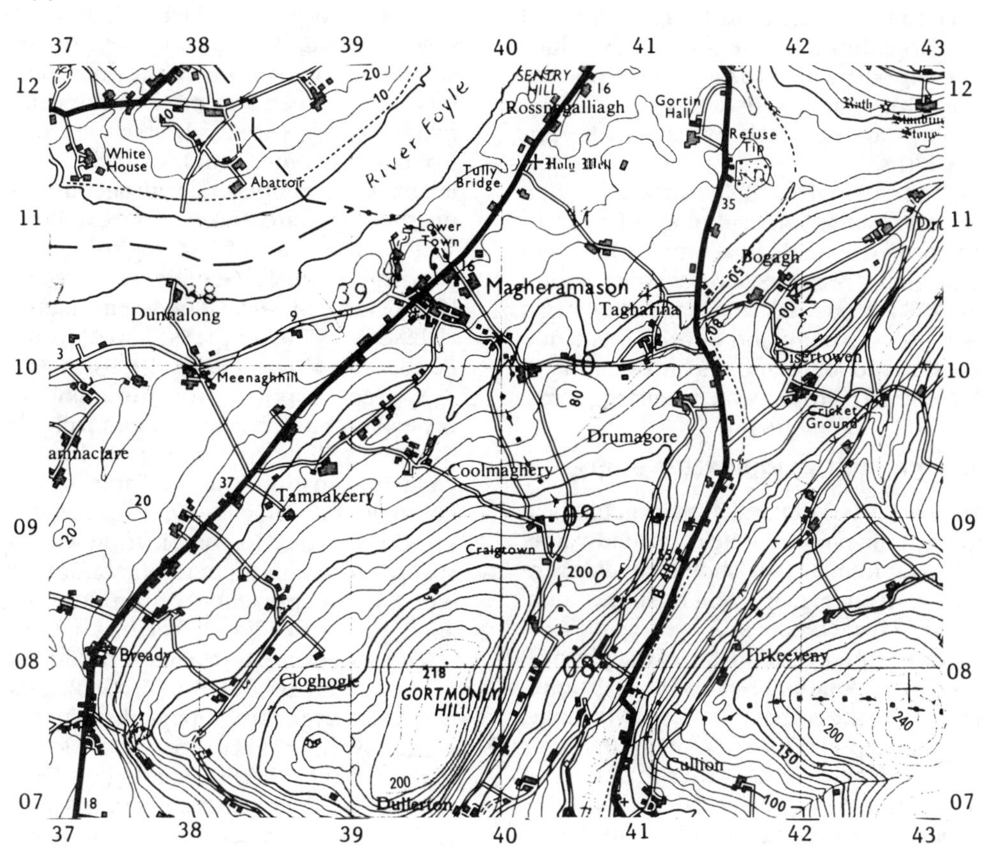

(b)

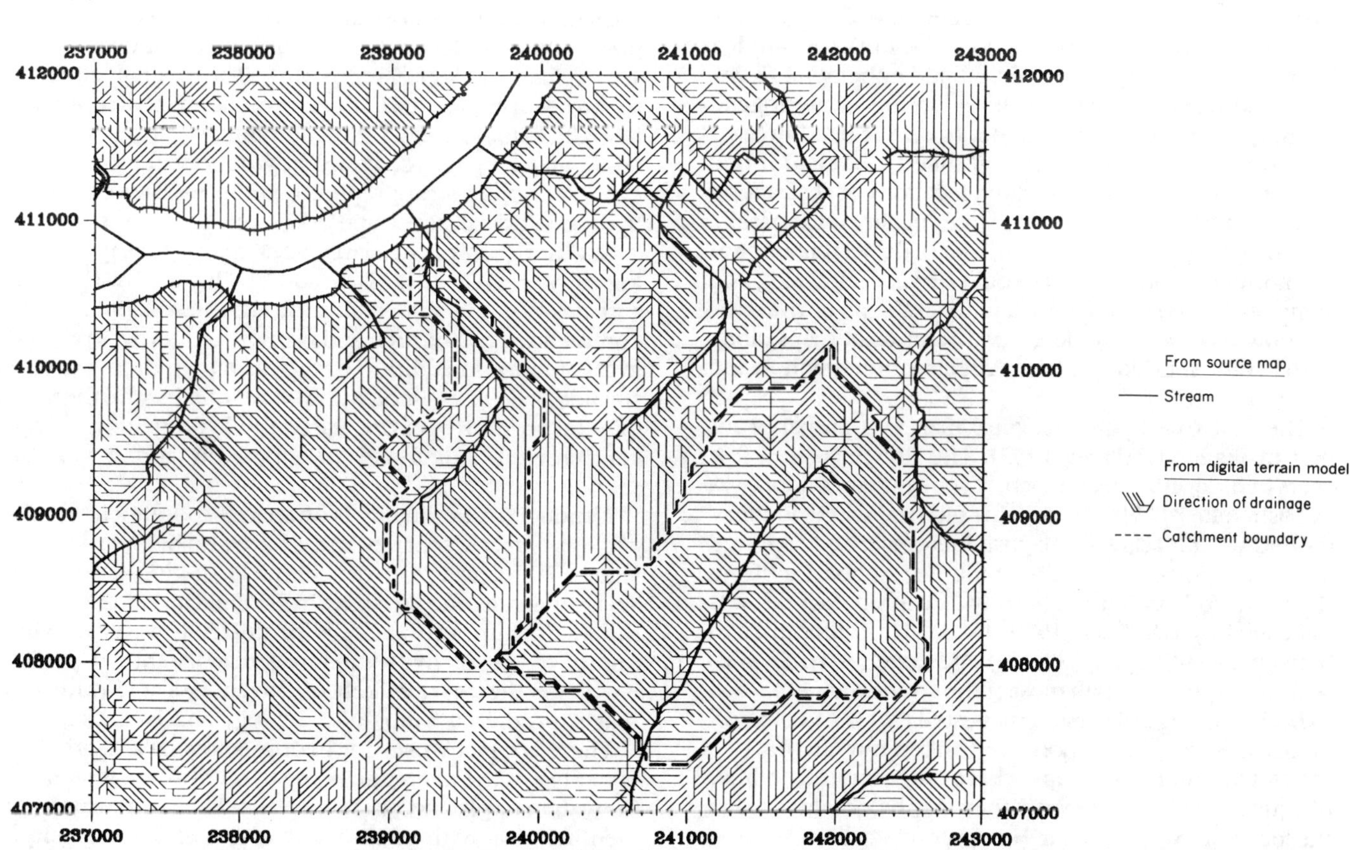

Fig. 5.(a) Portion of the 1:50 000 topographical map of Northern Ireland (Sheet 7) (Ordnance Survey reproduction). **(b)** Computer generated catchment boundaries and directions of drainage derived from a digital terrain model by the Institute of Hydrology (© NERC copyright).

response of a catchment to rainfall and is very useful in both low flow and flood estimation. Gustard *et al.* (1987) have produced a map indicating base flow index values for Scottish rivers using both gauged data and interpolated values from areas with similar soils and geology. The different index values are depicted by using different colours and line widths.

Dickerson (1974) predicted areas likely to be flooded by mapping the geology and geomorphology of a catchment which were formed in geological time, rather than using flow measurements recorded over a short period of time. Svatos (1979) produced land capability maps identifying areas with constraints on development due to soil losses, erosion, flooding and the protection of surface water catchment or groundwater recharge areas.

Hydrogeological and groundwater maps

Ambroggi and Margat (1960) defined a hydrogeological map as 'the mapping of the physical state of groundwaters within their geological frame'. This should be differentiated from 'a groundwater map which shows only physical and chemical characteristics'. A hydrogeological map therefore shows the inter-relationship of the physical and chemical parameters of the aquifers and their groundwater within an area plus the relevant surface hydrology, while a groundwater map just looks at one parameter such as water points, specific yield, aquifer thickness or chemical quality and should be considered as a source of additional information to a hydrogeological map. A prerequisite for any hydrogeological map is therefore a geological base, preferably one that subdivides the formations on the basis of rock type rather than age. The hydrogeological data is then superimposed on this, so that the final map portrays the solid and drift geology (subdivided into aquifers and non-aquifers), the surface hydrology (perennial, intermittent and ephemeral rivers, lakes and reservoirs, karst features, springs, waterlogged areas, glaciers), structural contours on various horizons (top, base or isopachytes), potentiometric levels (for one or more aquifers or seasons), the locations of abstraction and monitoring boreholes and water quality information. For complex areas or those with a lot of detail; this amount of information on one map would be very confusing and some of it is separated onto additional sheets or overlays, or shown as smaller-scale groundwater inset maps along with parameters like rainfall or effective precipitation and relief.

The first true hydrogeological map was prepared for part of France (Delesse, 1862). The British were a few years behind and a map of part of south-west London was published in 1877 by Lucas. This map indicated aquifers, groundwater levels and the positions of wells and springs. Maps of various parts of the world were published from then on and an internationally accepted legend was published by UNESCO (1970, 1983). This differentiated between small ($\geqslant$1:1 000 000) and large ($\leqslant$1:250 000) scale maps and although most of the symbols are common to both, the original version suggested that the geological information could be portrayed differently at the two scales. On larger-scale maps, where only a limited number of aquifers will be dealt with, they may be ornamented in the locally accepted colour. However on small-scale maps, which are often of whole countries, the legend requires that the geological formations are subdivided into aquifers in which intergranular flow predominates, those in which fissure flow predominates and those with only local potential and non-aquifers. These are indicated by blue, green and brown respectively overprinted with a relevant lithological ornament. Each colour may be used in several shades, with darker blues and greens indicating the better aquifers and darker brown the poorest aquifers. This latter method has been accepted world-wide and ensures that hydrogeologists are easily able to pick up salient features from maps of other countries, including those in unfamiliar languages. Examples of hydrogeological maps prepared on this basis are those of Great Britain (England and Wales (Institute of Geological Sciences, 1977) and Scotland (British Geological Survey, 1988)), Europe (Bundesanstalt für Geowissenschaften und Rohstoffe UNESCO, 1974—in press) and Mozambique (Republica Popular de Moçambique, 1989).

A large proportion of Mozambique is underlain by crystalline intrusive and metamorphic rocks. As a result, the map will be predominantly brown. As only limited yield information was available for these rocks, they could not be subdivided on this basis. Therefore remote sensing was used in combination with available borehole records to assign them to one of three categories on the basis of the presence of a thick weathered horizon and/or fractures which might contain water.

The series of regional maps produced at 1:100 000 for Great Britain by the British Geological Survey are examples of larger-scale maps. A portion of the one for the area between Cambridge and Maidenhead is shown in Fig. 6. On this, the drift geology is indicated by a colour wash (apart from sands and gravels that are ornamented) and the aquifers are ornamented with bricks for the limestones (Chalk) and dots for the sandstones (Lower Greensand), and the non-aquifers (e.g. Gault) are left blank. Onto this geological base are superimposed surface water features (rivers, principal gauging stations), the sites of licensed abstractions, structural contours and potentiometric levels for the principal aquifers in the area (Chalk, Lower Greensand). The overall effect is relatively complex and, therefore, the map was printed in eight colours. Insets include rainfall and relief, a location map, well hydrographs, groundwater quality, a cross-section, a stratigraphic column with explanatory hydrogeological notes and a key to the main map.

Another example of a large-scale map is the Hydrogeological Map of Croatia (Sarin and Urbiha, 1977). This is at a scale of 1:200 000 and, although based on the recommended blue, green and brown colours, was printed in 20 colours. In addition to the main map there are also smaller-scale insets indicating water quality, the thermomineral properties of the water, the hydrogeology of the karst region, relief and precipitation and a vertical section.

Although water quality is often incorporated in hydrogeological maps, separate maps have also been produced. Figure 7 indicates a hydrochemical section across the Triassic Sandstone in south-west England, subdividing the aquifer up into zones with differing water quality. Another type of hydrochemical map is shown in Fig. 8 which indicates the likely occurrence of acid waters in Wales, predicted from soils, geology and land use.

There are many uses for hydrogeological and groundwater maps, the main one being to identify areas suitable for water supply and waste disposal. They can be used to identify areas with strata likely to landslip and groundwater circulation patterns. They have also been used to solve engineering problems. Chilton (1979) used a hydrogeological map to identify the area draining to a proposed motorway tunnel, and Coe (1981) used one to assess the impact of, and the area to be dewatered for, a new coal

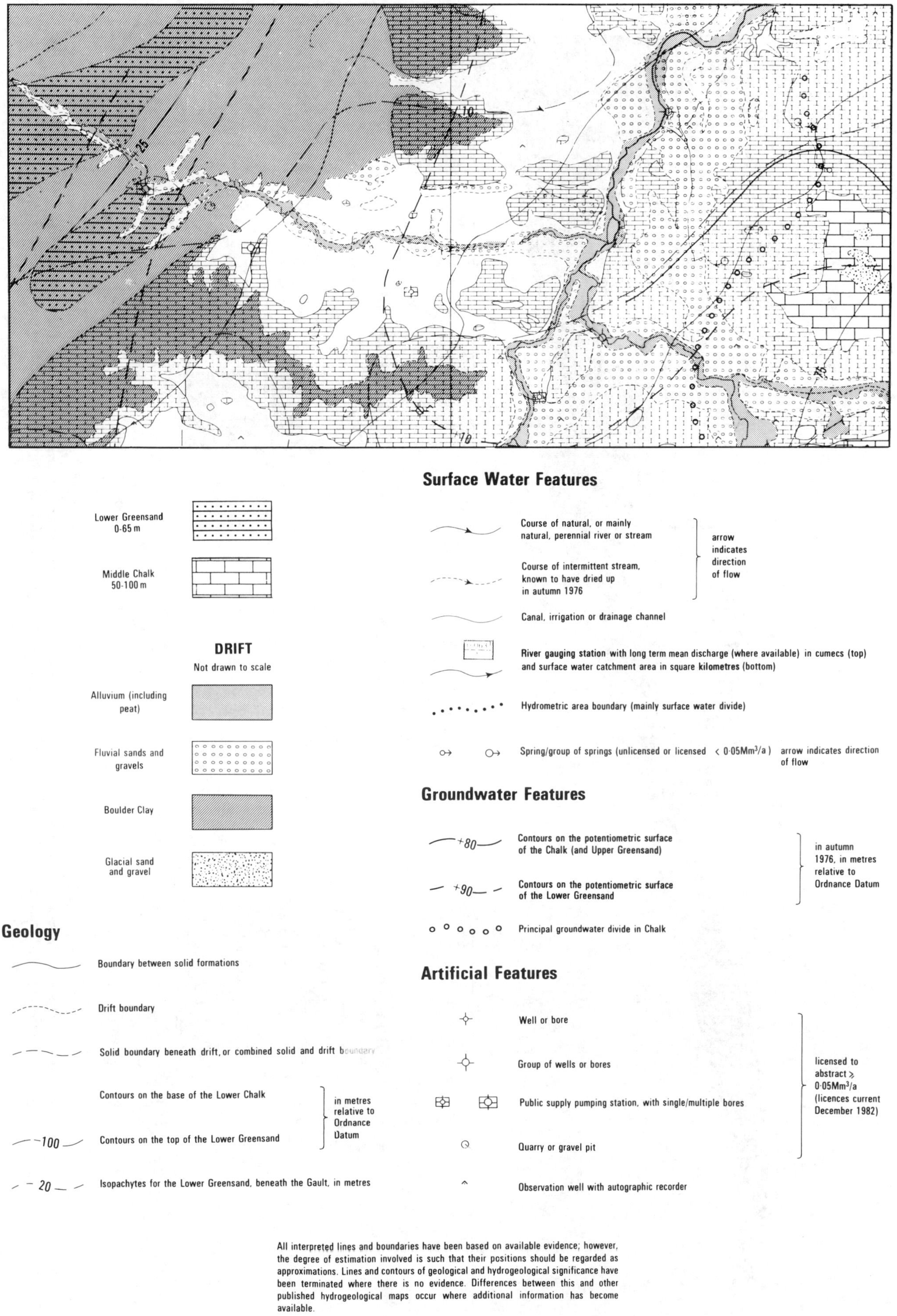

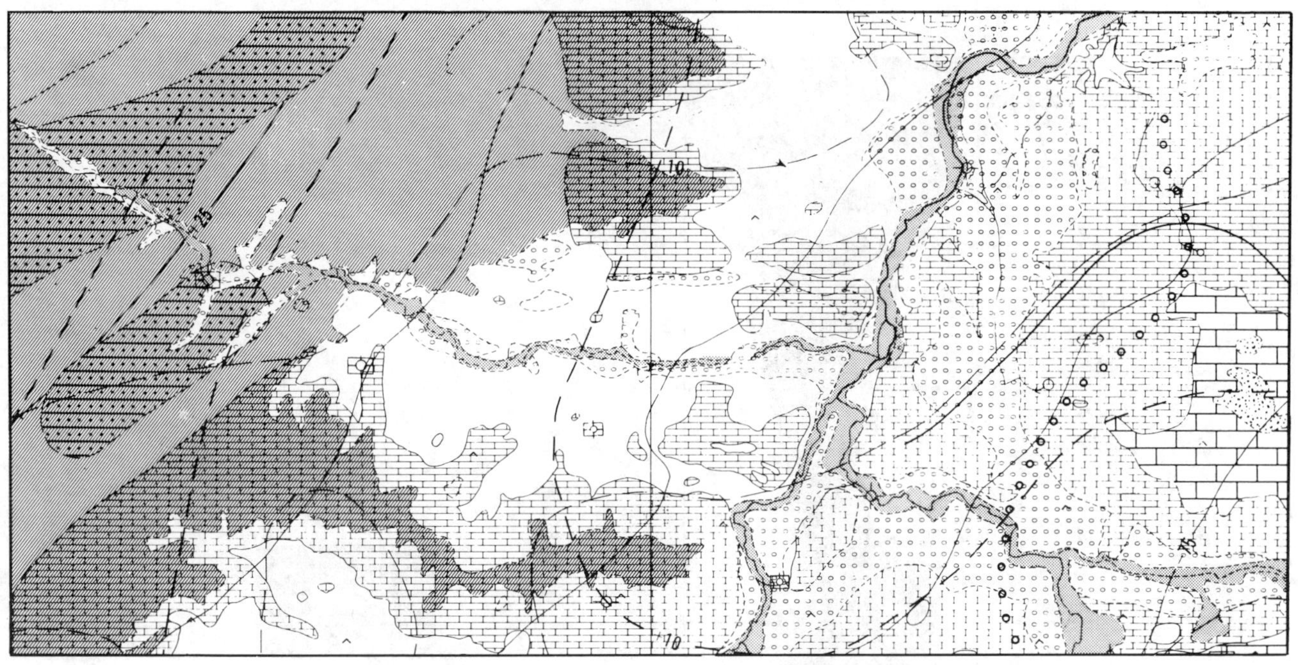

Lower Greensand 0-65 m

Middle Chalk 50-100 m

DRIFT

Not drawn to scale

Alluvium (including peat)

Fluvial sands and gravels

Boulder Clay

Glacial sand and gravel

Geology

Boundary between solid formations

Drift boundary

Solid boundary beneath drift, or combined solid and drift boundary

Contours on the base of the Lower Chalk } in metres relative to Ordnance Datum

—100— Contours on the top of the Lower Greensand } in metres relative to Ordnance Datum

— 20 — Isopachytes for the Lower Greensand, beneath the Gault, in metres

Surface Water Features

Course of natural, or mainly natural, perennial river or stream } arrow indicates direction of flow

Course of intermittent stream, known to have dried up in autumn 1976 } arrow indicates direction of flow

Canal, irrigation or drainage channel

River gauging station with long term mean discharge (where available) in cumecs (top) and surface water catchment area in square kilometres (bottom)

Hydrometric area boundary (mainly surface water divide)

Spring/group of springs (unlicensed or licensed < 0·05Mm³/a) arrow indicates direction of flow

Groundwater Features

+80 Contours on the potentiometric surface of the Chalk (and Upper Greensand) } in autumn 1976, in metres relative to Ordnance Datum

+90 Contours on the potentiometric surface of the Lower Greensand } in autumn 1976, in metres relative to Ordnance Datum

Principal groundwater divide in Chalk

Artificial Features

Well or bore

Group of wells or bores

Public supply pumping station, with single/multiple bores

Quarry or gravel pit

} licensed to abstract ⩾ 0·05Mm³/a (licences current December 1982)

Observation well with autographic recorder

All interpreted lines and boundaries have been based on available evidence; however, the degree of estimation involved is such that their positions should be regarded as approximations. Lines and contours of geological and hydrogeological significance have been terminated where there is no evidence. Differences between this and other published hydrogeological maps occur where additional information has become available.

Fig. 6. Hydrogeological map of the area between Cambridge and Maidenhead (BGS 1984, NERC copyright) (black and white illustration derived from a coloured original).

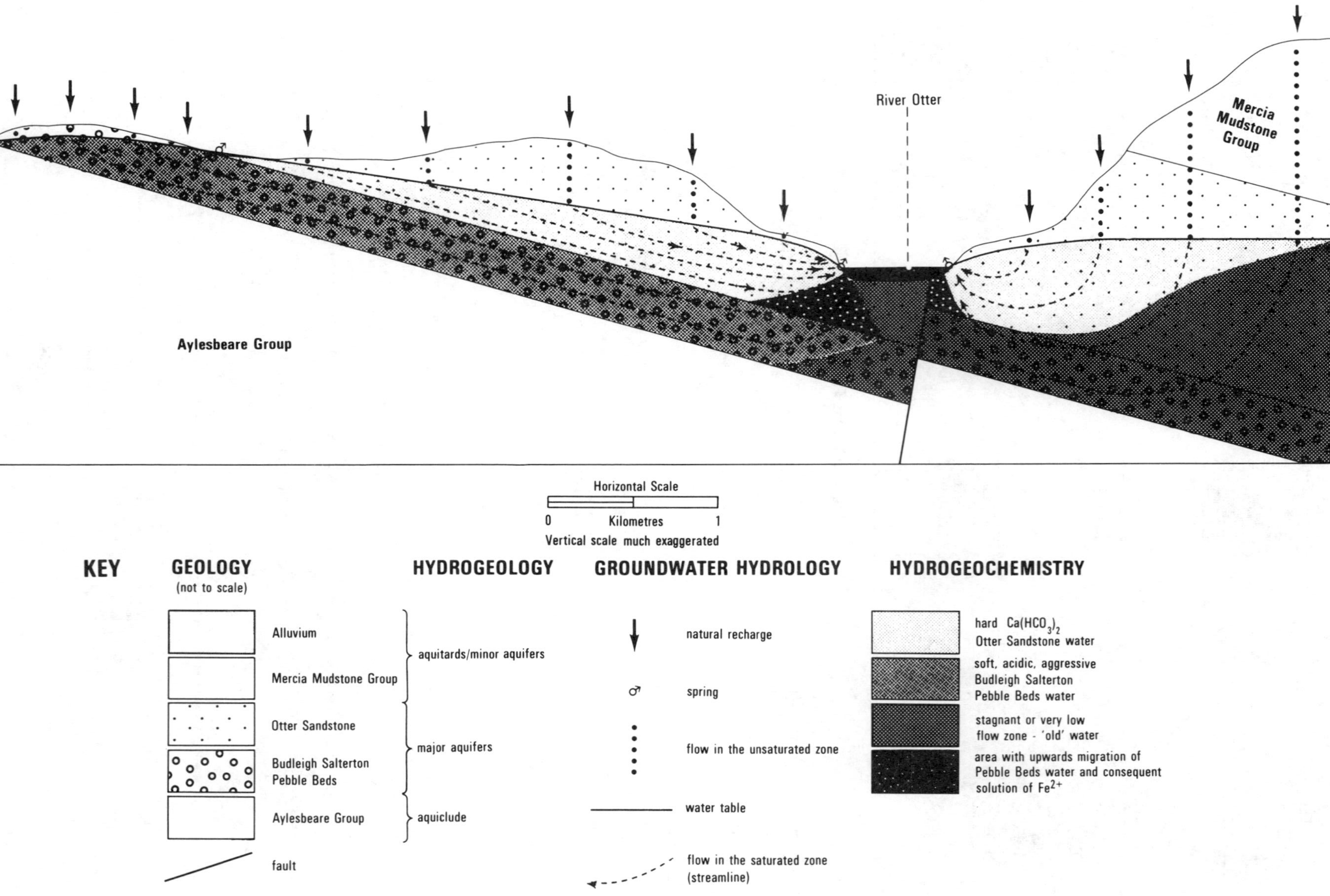

Fig. 7. Diagrammatic representation of flow processes and groundwater chemistry of the Triassic aquifer in the Lower Otter Valley, South West England (Institute of Geological Sciences, 1982, NERC copyright) (black and white illustration derived from a coloured original).

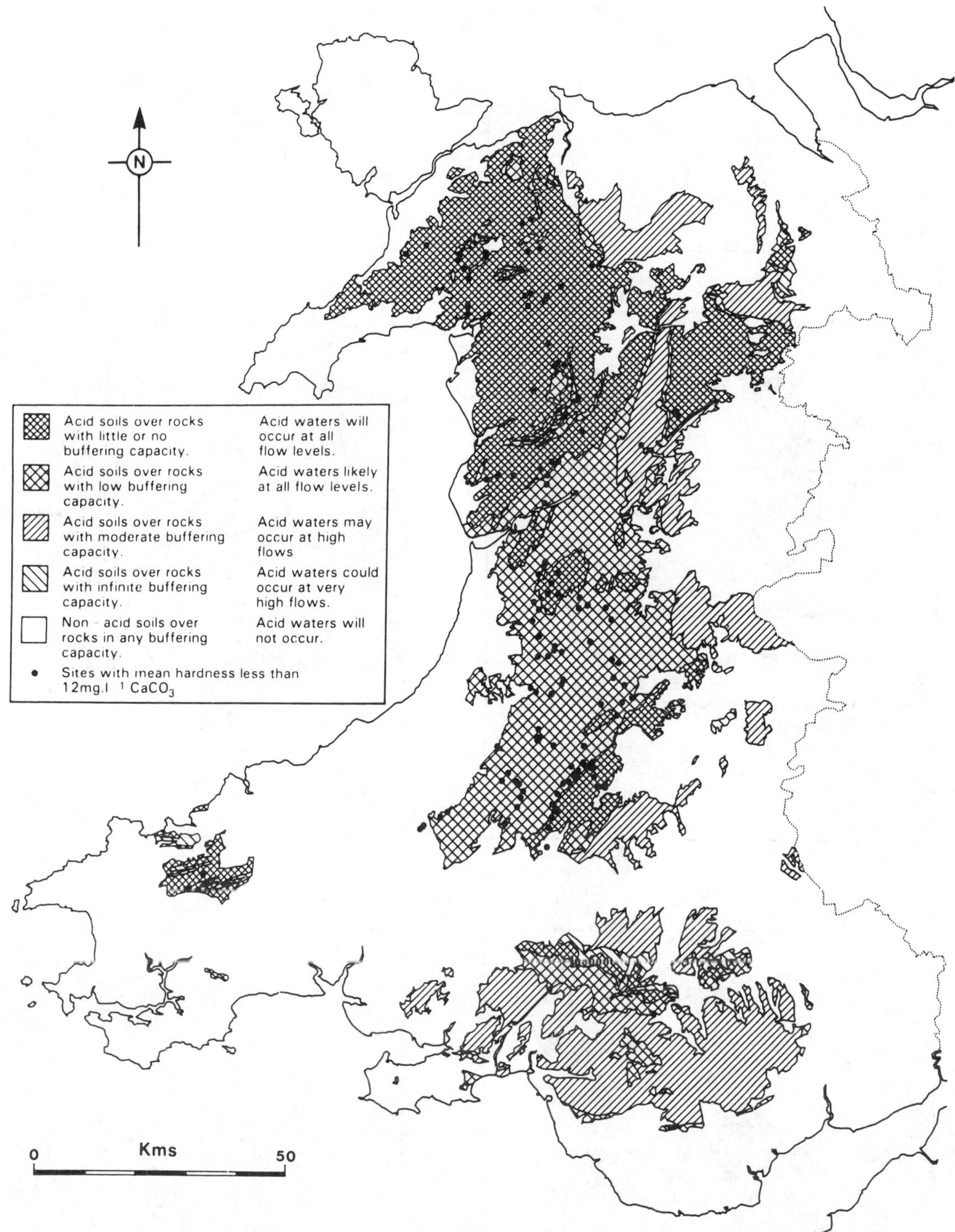

Fig. 8. The likely occurrence of acid waters in Wales predicted from soils, geology and land use (Department of the Environment copyright, 1989).

mine. The large number of parameters portrayed on hydrogeological maps means that most existing maps are complicated and consequently aimed at specialists.

Water resources maps

These maps indicate the potential water resources available in an area. They can range from a diagrammatic sketch indicating the abstractions from both surface and groundwater sources within an area to complex multi-coloured maps. Figure 9 shows a portion of south-east England from a series of maps indicating groundwater resources for the European Community (Commission of European Communities, 1982). The first map indicates the areal extent of the aquifers, including their useable lateral extent beneath overlying formations. On the original, the aquifers were colour-coded in accordance with the international legend to indicate whether inter-granular or fissure flow predominates. The second map indicates potentiometric contours for the major aquifers and the direction of groundwater flow. The third one subdivides the aquifer outcrop areas into units and indicates groundwater abstractions (including springs and mine drainage discharges—the latter, not occurring in this sample area), the total volume of water pumped from each unit and the abstraction per unit area. The

(a)

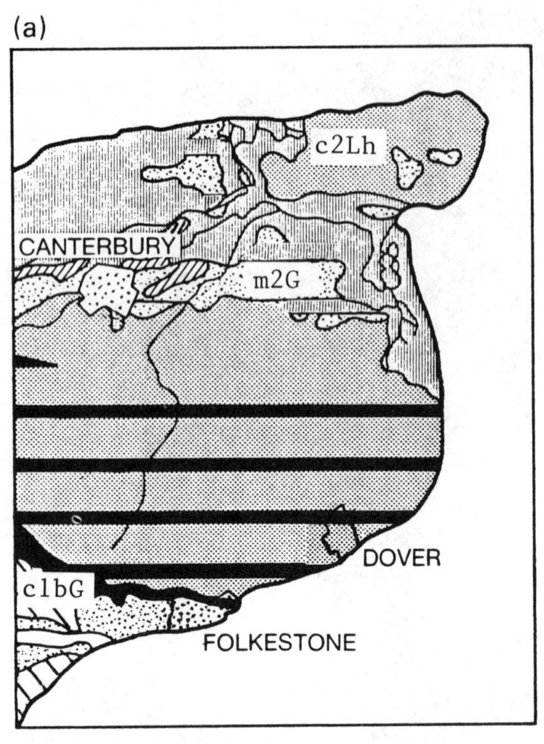
c2Lh
CANTERBURY
m2G
DOVER
c1bG
FOLKESTONE

(b)

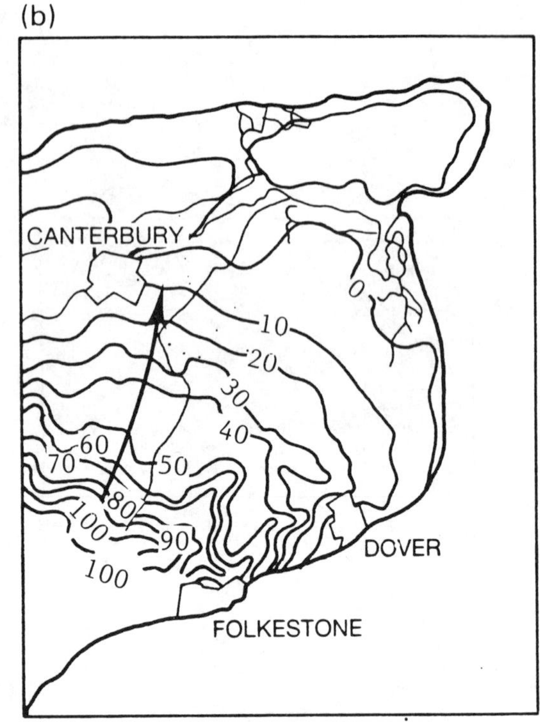
CANTERBURY
0
10
20
30
40
50
60
70
80
90
100
100
100
DOVER
FOLKESTONE

(c)

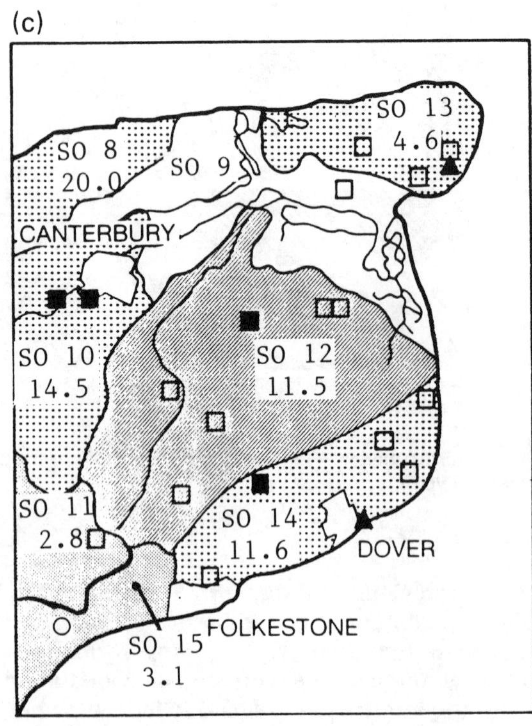
SO 13
4.6
SO 8
20.0
SO 9
CANTERBURY
SO 10
14.5
SO 12
11.5
SO 11
2.8
SO 14
11.6
DOVER
SO 15
3.1
FOLKESTONE

(d)

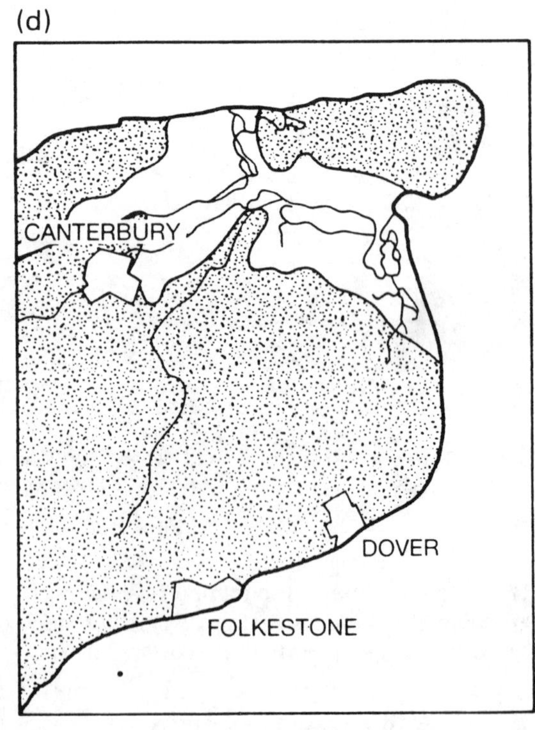
CANTERBURY
DOVER
FOLKESTONE

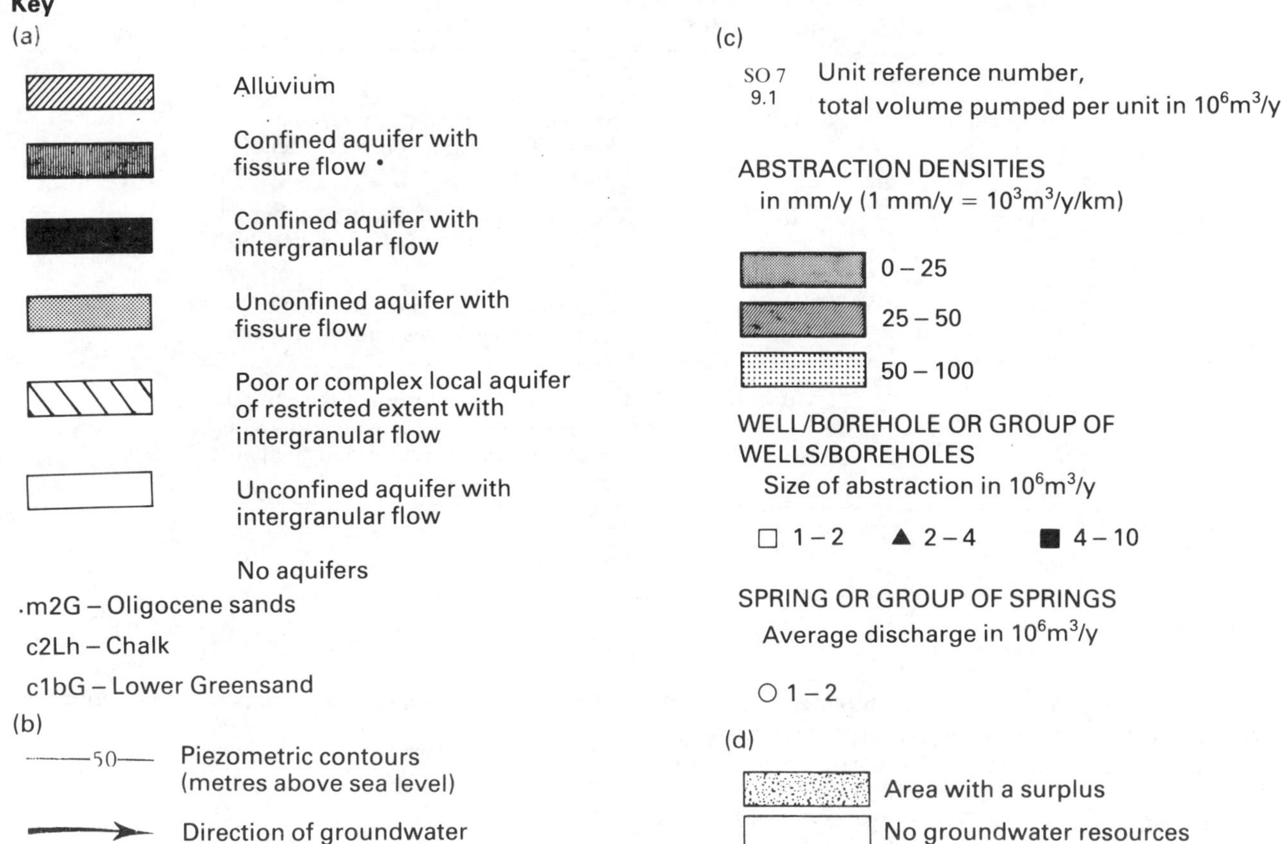

Where an aquifer continues to be of importance beneath another aquifer, the ornament of the lower aquifer is continued as thin stripes across the outcrop of the upper one.

Fig. 9. Groundwater resources of Southeast England (a) aquifers (b) groundwater hydrology (c) groundwater abstraction (d) balance of resources (after Commission of European Communities, 1982, Sheet 13).

final map indicates whether the groundwater resources for each unit are overdeveloped or if a surplus is available.

The French have produced a map of the potential groundwater resources of Western and Central Africa (Bureau de Recherches Géologiques et Minières/Géohydraulique, 1986). The final map is a product of three intermediate maps. These intermediate maps were prepared by combining available borehole and other data in pairs to produce accessibility, exploitability and reliability maps. The accessibility map is a combination of the probability of success (obtaining more than 0.7 m³/hr) and the well depth required to obtain this yield. The exploitability map is a combination of median or mean yields and the depth the water is abstracted from or pumped water-levels. The reliability map indicates the mean annual effective precipitation and the local volume of the aquifer. Each of the parameters was given a score from 1–4 (e.g., Fig. 10), and from this four classes were defined (*poor, moderate, good and very good*) for each descriptor. An intermediate map was then produced indicating the extent of each of the four classes. The points from all three intermediate maps were then added together (Table IV) to produce a final map of water

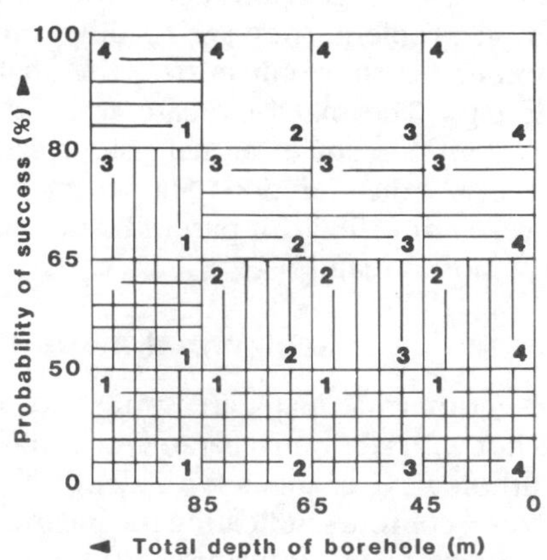

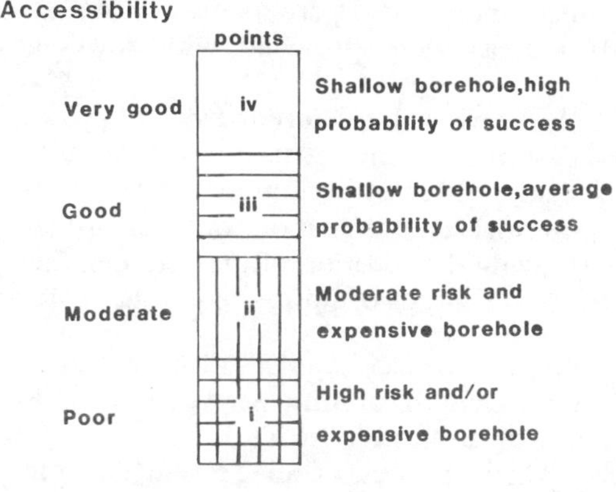

Fig. 10. Classification matrix for the accessibility descriptor (after BRGM/Geohydraulique, 1986).

Table IV Classes of water resource potential (after Bureau de Recherches Géologiques et Minières/Géohydraulique, 1986)

Number of points	*Category*	*Potential*
iii–iv	I	Unfavourable location, limited to potable water exploration only on identified structures
v–vii	II	Not very favourable formation, rather uneconomic and poor possibilities, high failure risk
viii–x	III	Favourable formation, possibilities of advantageous development, at a small risk
xi–xii	IV	Very favourable formation, excellent possibilities of development presenting good economic return and minimum risk

resource potential. Figure 11 is a simplified version of part of this, the original being overprinted with the relevant lithological ornament. The starting point for the whole project was a data base, which could be readily corrected and updated. The scores for the six parameters were calculated by the computer—it classified each area, defined its boundaries and produced all the maps (intermediate and final). It would also have been possible to specify a range for each of the criteria and obtain a map indicating the areas where all six were met. The final map produced assumed that the weight given to each parameter was of equal value. However, with a computerized data base the map could easily be redrawn weighting one or more of them, or even changing the parameters used from the original data base.

Vulnerability to pollution maps

Up to 20 years ago, the major problem for hydrogeologists was providing sufficient groundwater to satisfy demand. More recently the quality of groundwater has become important and in most of the developed world aquifers are now being assessed for their vulnerability to pollution. One of the first maps to consider the protection of aquifers was the Hydrogeological Map of the Rhine Valley between Bonn and Dusseldorf (Dieler and Diesel, 1964). This map indicated hydrochemical data and whether urban areas were provided with sewerage systems.

Vrána (1977) reviewed the different types of protection maps and described a series of maps being produced for Czechoslovakia which superimpose land use and sources of potential contamination onto hydrogeological maps showing the location of the aquifers, their transmissivities and sources of supply requiring protection.

A series of maps that place aquifers in various zones depending on the length of time it would take for a pollutant with similar characteristics to water to reach the water-table is in the process of being produced for the European Community. These maps use the lithology and minimum thickness of the unsaturated zone to indicate whether the aquifer was vulnerable to pollution in:

- less than a week;
- a week to a year;
- 1–20 years;
- more than 20 years.

Other pollution maps of areas where detailed information is available portray vulnerability as a measure of the permeability. De Smedt *et al.* (1987) describe pollution time maps of Belgian aquifers based on the permeability and lithology of the overburden and underlying aquifer and the thickness of the unsaturated zone. Work in Denmark by Villumsen *et al.* (1983) produced maps of groundwater vulnerability based on an index calculated from the thickness of the unsaturated zone, its permeability and the potential for chemical reactions to take place in it.

In Great Britain, the Soil Survey and Land Research Centre and the British Geological Survey have produced a series of maps indicating the vulnerability of the major aquifers in the Midlands to nitrate leaching. These maps take into account the thickness and lithology of any drift deposits overlying the aquifer and the overall permeability of the soil to classify the aquifer outcrops into zones of low, moderate, high and extreme risk. Figure 12 shows part of the map for the Mansfield area. Although the maps represent a significant advance in the study of the nitrate pollution problem, they are readily comprehended by non-experts as they only portray the final interpretation of the data. The maps indicate areas of potential risk, rather than current or actual risk, as land use was not taken into consideration. Similar maps, prepared by Desprez *et al.* (1982) for part of France, also incorporate information on land use.

Geothermal maps

Some geothermal maps are similar to hydrogeological maps but with additional parameters such as temperature and lithological changes with depth shown as well as structural contours indicating the location of the aquifer, piezometric levels, transmissivities and salinities. They

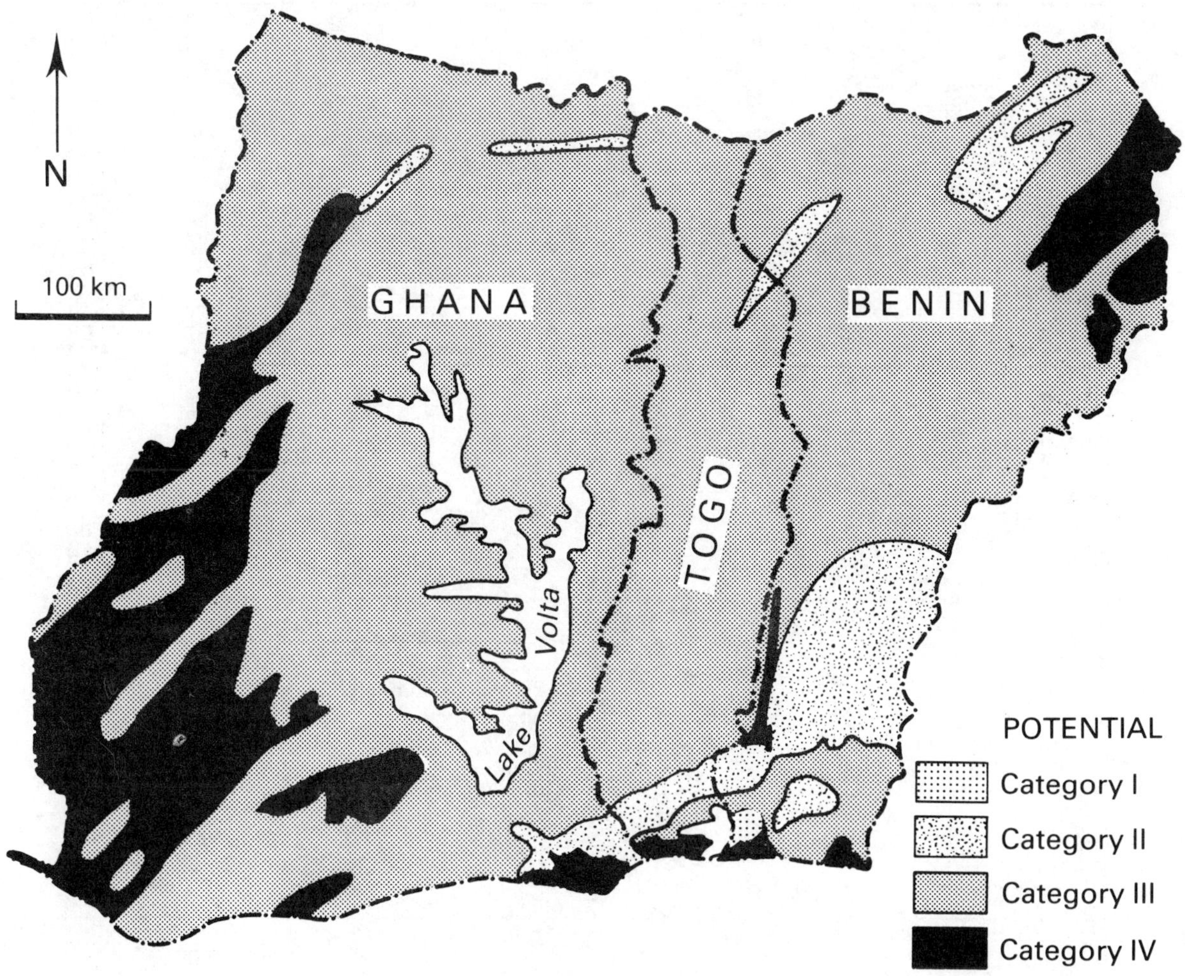

Categorical synthesis combining three descriptors: accessibility, exploitability, and safety

Category I: Unfavourable terrain, limited to exploration for drinking water in identified structures

Category II: Relatively unfavourable terrain typified by barely economic potential and a high risk of failure

Category III: Favourable terrain, with advantageous development potential and a moderate degree of risk

Category IV: Very favourable terrain typified by extensive development potential with good economic return and minimum risk

Fig. 11. Water resource potential map (after BRGM/Geohydraulique, 1986).

are usually less detailed than hydrogeological maps of shallow aquifers, due to the smaller amount of data available.

Figure 13 shows part of the Geothermal Map of the United Kingdom (British Geological Survey, 1985). This is generalized indicating the locations of the potential geothermal fields, the form of the Mesozoic basins, the location of the radiothermal granites and thermal springs and heat flow measurements and contours. Insets include estimated temperatures at a depth of 2 km and the estimated depth of the 100°C isotherm. Each potential source of geothermal energy is annotated.

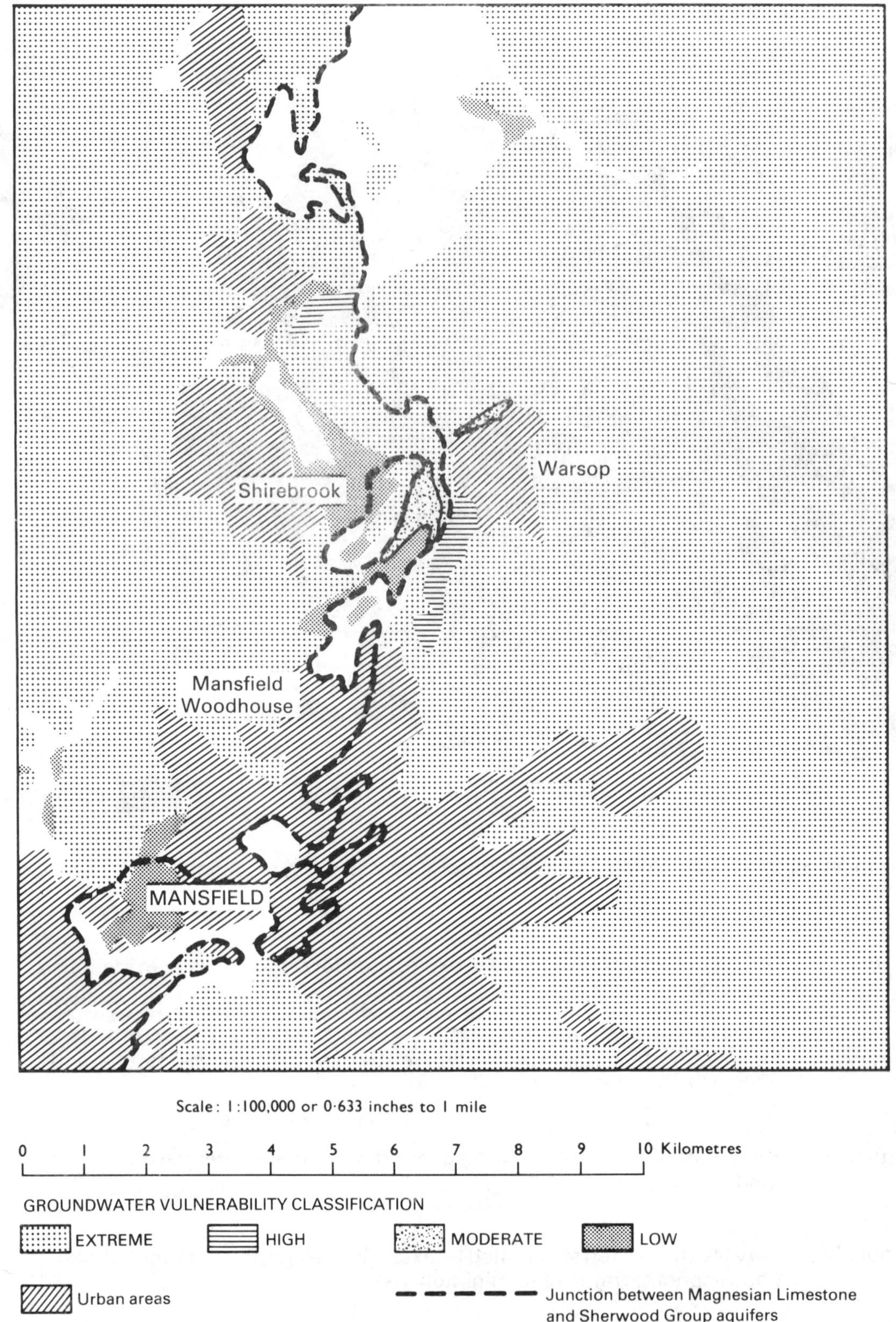

Fig. 12. Groundwater vulnerability in the Mansfield area (adapted from Palmer, 1987) (simplified drawing derived from a coloured original).

Water maps for land use planning

The various parameters that can be mapped and are relevant to water and land use planning are summarized in Table V. This table should be treated as a list of examples rather than as a comprehensive guide, and it should be noted that several factors (e.g., water quality and recharge areas requiring protection) fall into more than one category. As water is required for virtually every aspect of life from human consumption, agriculture, mineral processing, manufacturing power generation to waste disposal, and even though an excess of water can cause problems, it is essential that information on the presence or absence of water is readily available to planners. The large variety of maps, either available or that can be produced if required, and

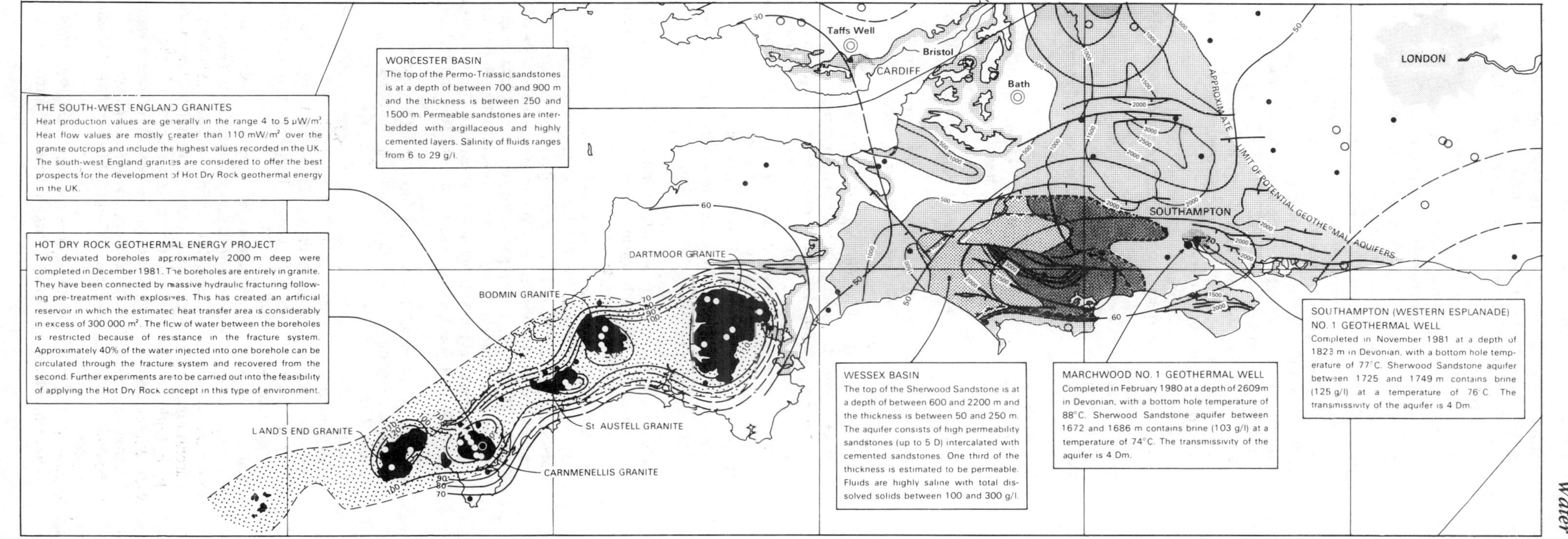

	Buxton	Matlock	Taffs Well	Bristol	Bath
Temperature (°C)	28	20	20	24	47
Mean discharge (l/s)	10	1.3	1.0	2.0	15
Total dissolved solids (mg/l)	410	450	290	910	2150

Fig. 13. Geothermal map of Southern England (reproduced from British Geological Survey, 1985, Crown copyright) (black and white illustration derived from a coloured original). Scale: 1:1 500 000

Table V Mappable features of relevance to water resources and land use planning

Meteorological	*Geological*
Precipitation	Soil type
Evapotranspiration	Lithology
Effective precipitation	Structure
Quality of precipitation	Structural contours
Temperature	Landslips
Climate stations	
	Hydrogeological
Geographical	Potentiometric levels
Relief	Direction of groundwater flow
Angle of slope	Abstraction/Monitoring boreholes
River gradient	Transmissivity
Distribution of vegetation	Permeability
Rate of erosion	Porosity
Land use	Specific yield
Quality of agricultural land	Specific capacity
Areas under irrigation	Water quality
Population density	
	Water Resources
Hydrological	Capacity of reservoirs
Drainage network	Borehole success rate
Drainage density	Mean/median yields
Perennial, ephemeral, intermittent rivers	Water use
Artificial drainage features	Resources in relation to abstraction
Glaciers	
Snow cover	*Pollution*
Areas of permafrost	Potential pollutants
Swamps and waterlogged areas	Thickness, unsaturated zone
Drained areas	Permeability, unsaturated zone
Flow rates	Travel time through unsaturated zone
Minimum desirable flows	Protection areas
Variations in flow	
Gauging stations	*Geothermal*
Water quality	Thermal springs
Springs	Isotherms
Karst features	Heat flow
Flood risk	Radiothermal granites
Drought risk	Deep sedimentary basins
Base flow index	Potential geothermal fields
Potential hydroelectric sites	Lithological changes with depth

which cover the complex details of water availability, quantity and quality, etc., are therefore important tools for assessing the suitability of land for various purposes and for pin-pointing various problems before they occur.

REFERENCES

Ambroggi, R. and Margat, J. (1960). *Légende générale des Cartes Hydrogéologiques du Maroc (General legend for hydrogeological maps of Morocco)*. Publication No 50, IASH. 32 pp (in English and French). Also published as *Notes et Mémoires No. 148 du Service Géologique du Maroc*. 32 p. Rabat (in English and French).

Bowie, S.H.U. and Webb, J.S. (1980). *Environmental Geochemistry and Health*. Phil. Trans. Roy. Soc. London, Ser. B., 216 pp.

British Geological Survey (1984). *Hydrogeological Map of the area between Cambridge and Maidenhead, 1:100 000.*

British Geological Survey (1985). *Geothermal Map of the United Kingdom. 1:1 500 000.*

British Geological Survey (1988). *Hydrogeological Map of Scotland. 1:625 000.*

Bundesanstalt für Geowissenschaften und Rohstaffe/UNESCO (1974). International Hydrogeological Map of Europe. 1:1 500 000, 30 sheets, *in press.*

Bureau de Recherches Géologiques et Minières Géohydraulique (1986). *Carte de Potentialité des Ressources en Eau Souterraine de l'Afrique Occidentale et Centrale. 1:5 000 000.*

Chilton, P.J. (1979). Hydrogeological mapping for an investigation into the design of a motorway tunnel. *Bull. Inter. Assoc. Eng. Geol.*, **19**, 242–247.

Coe, C.J. (1981). The use of geologic, hydrologic and geochemical mapping techniques in environmental assessment. *Groundwater*, **19**(6), 626–634.

Commission of the European Communities (1980). Council directive on the quality of water for human consumption. *Official Journal 23, No. 80/778 EECL. 229*

Commission of European Communities (1982). *Groundwater Resources.* 38 sheets.

Delesse, A. (1862). *Une carte hydrologique du Département de la Seine avec légende, 1:25 000.*

Department of the Environment (1976). *River Pollution Survey England and Wales, 1975*, Chemical classification of rivers for the Northumbrian Water Authority.

Department of the Environment (1989). *Acidity in United Kingdom Fresh Waters—United Kingdom Acid Waters Review Group 2nd Report*, HMSO.

De Smedt, P., de Breuck, W., Loy, W., van Autenboer, T. and van Dijck, E. (1987). Groundwater vulnerability maps. *Aqua*, **5**, 264–267.

Desprez, N., Landreau, A. and Vogt, D. (1982). Mineralisation of groundwater by agricultural practices: mapping of sensitive areas—application to the Indre-et-Loire District (France). In *International Symposium. Int. Assoc. Hydrogeologists, Prague, Czechoslovakia, 'Impact of Agricultural Activities on Groundwater'.* Mem. Vol. XVI, Part 1, 195–206.

Dickerson, E.J. (1974). Environmental geologic mapping of flood-prone areas: an alternative to engineering methods. Approaches Environ. Geol. (Austin) Conference, 13–14 May.

Dieler, H. and Diesel, E. (1964). Die Grundwasserbeschaffenheit im Rheintal zwischen Bonn und Dusseldorf. *Geol. Mitt.* Band 3, Heft 4, 339–396, Aachen.

Downing, R.A. and Gray, D.A. (eds.) (1986). *Geothermal Energy—the Potential in the United Kingdom*, HMSO.

Gambolati, G., Gatto, P. and Freeze, R.A. (1974). Predictive simulation of the subsidence of Venice. *Science*, **183**, 849–851.

George, D. (1988). Rising groundwater—a problem of development. Abstract Volume 'Geoscience in Development'. Conference held at Nottingham University, 26–30 September (AGID, Geol. Soc. London, Inst. Geol., Inst. Min. and Met.), *in press.*

Gustard, A., Marshall, D.C.W. and Sutcliffe, M.F. (1987). Low flow estimation in Scotland. Institute of Hydrology, *Report no. 101.*

Institute of Geological Sciences (1977). *Hydrogeological Map of England and Wales, 1:625 000.*

Institute of Geological Sciences (1982). *Hydrogeological Map of the Permo-Trias and other Aquifers of South-west England, 1:100 000.*

Lucas, J. (1877). *Hydrogeological Survey, Sheet 1 and accompanying explanation.* Edward Standford, London.

Morris, D.G. and Heerdegen, R.G. (1988). Automatically derived catchment boundaries and channel networks and their hydrological applications. *Geomorphology*, **1**, 131–141.

Nace, R.L. (ed.) (1971). Scientific framework of world water balance; *UNESCO Tech. Papers Hydrol.*, **7**, 27 pp.

Natural Environment Research Council (1975). *Flood Studies Report.* 5 vols.

Palmer, R.C. (1987). *Groundwater vulnerability: Map 6 Mansfield.* Soil Survey and Land Research Centre.

Rau, J.L. (1988). Geology for urban development in Asia. 'Geosciences in Development'. Conference held at Nottingham University, 26–30 September (AGID, Geol. Soc. London, Inst. Geol., Inst. Min. and Met.), Proceedings, *in press*.

Republica Popular de Moçambique (1989). 'Hydrogeological Map of Mozambique, 1:1 000 000', *in press*.

Sarin, A. and Uřbiha, M. (1977). Cartographic novelty in a Hydrogeologic Map of Croatia. *Mem. Int. Assoc. Hydrogeol.*, **13**, Part 1, Papers, B29–B38.

Svatos, V.C. (1979). A case study of the analysis of the capability of undeveloped lands to contribute to water quality degradation, Computer Mapping in Natural Resources and the Environment, including applications of satellite-derived data. *Harvard Univ. Lab. Comput. Graphics and Spat. Anal.*, 115–133.

Thornton, I. and Webb, J.S. (1979). Geochemistry and health in the United Kingdom. In *Environmental Geochemistry & Health* (Bowie, S.H.U., and Webb, J.S., eds.), Phil. Trans. Roy. Soc. London, Ser. B., 151–168.

UNESCO (1970). *International legend for hydrogeological maps*. Paris.

UNESCO (1983). *International legend for hydrogeological maps (revised version)*.

Villumsen, A., Jacobsen, O.S. and Sønderskov, C. (1983). Mapping the vulnerability of groundwater reservoirs with regard to surface pollution. *Aarbog-Danmarks Geologiske Undersoegelse*, 1982, 17–38.

Vrána, M. (1977). Development of methods for preparation of groundwater protection maps. *Mem. Int. Assoc. Hydrogeol.*, **13**, part 1, Papers, B22–B28.

Watson, D.C. (1988). *UK Acid Waters Review Group—Second Report*.

World Health Organisation (1984). *Guidelines for drinking water quality. Volume 1 (of 3), recommendations.*

Chapter 5

Mineral Resources

Joseph McCall

GENERAL DISCUSSION

Introduction

Mineral resources are relevant to land use planning from both the resource potential and the resource constraint aspect. The community requires these resources to maintain its existence, yet their extraction must conflict with alternative demands for land use, and may severely affect the environment.

Definition

Mineral resources are defined as those useful materials that man extracts from the Earth, in order to carry out his industrial and other activities. Mineral resources are utilized, and so consumed by us, in virtually every human activity in the modern, industrialized society.

It is essential to realise that 'Earth is a closed system where matter cannot be created. There are finite limits of such resources as minerals and fossil fuels . . . when mined out they cannot be replaced. Thus mineral resources, as contrasted with other products of the Earth, are non-renewable, must be mined in place, have extraordinarily high discovery costs, and development costs may increase with time' (Coates, 1985). Any mine, quarry or hydrocarbon field has a finite life, and must eventually become 'depleted', i.e. either exhausted or no longer economically viable.

The requirement for minerals

The following issues are considered in this chapter:

(1) the need for minerals and the scale of the industry;
(2) their value to the economy and society;
(3) the consequent need to discover minerals and to safeguard them from sterilization.

There is an ever increasing demand for minerals and energy (Lüttig, 1987). With increasing world population, increased industrialization and the concomitant increase in world wide living standards, it cannot be otherwise. Lüttig (1987) provided two useful diagrams (Figures 1(a), 1(b)) showing the most important minerals in terms of: (1) quantity produced, and (2) value. Figure 1(c) illustrates the annual *per capita* consumption of non-energy minerals in the USA (after US Dept. of the Interior, Bureau of Mines).

The requirement for minerals is well illustrated by Lüttig's estimate of the consumption of mineral resources *per capita* in an average lifespan of 70 years by a citizen of the Federal Republic of Germany (Table I). It is at once apparent that, far from being a luxury, a continuous flow of mineral resources into the industrialized community is a necessity of life.

The magnitude of the task of extracting minerals from the rock in which they are harboured is well illustrated by Table II. World production figures are given in Table III.

As minerals are used, there is a constant need to replenish reserves. This, in turn, requires development of more sophisticated methods of deposit detection and utilization of low-grade material which had previously been regarded as marginal or non-viable (Fig. 2). Also required is the development of improved extraction techniques (for example, finding ways of removing thick overburden cheaply and so making extraction of deeper buried deposits cost effective). Even the bringing into

Earth Science Mapping for Planning, Development and Conservation

(a)

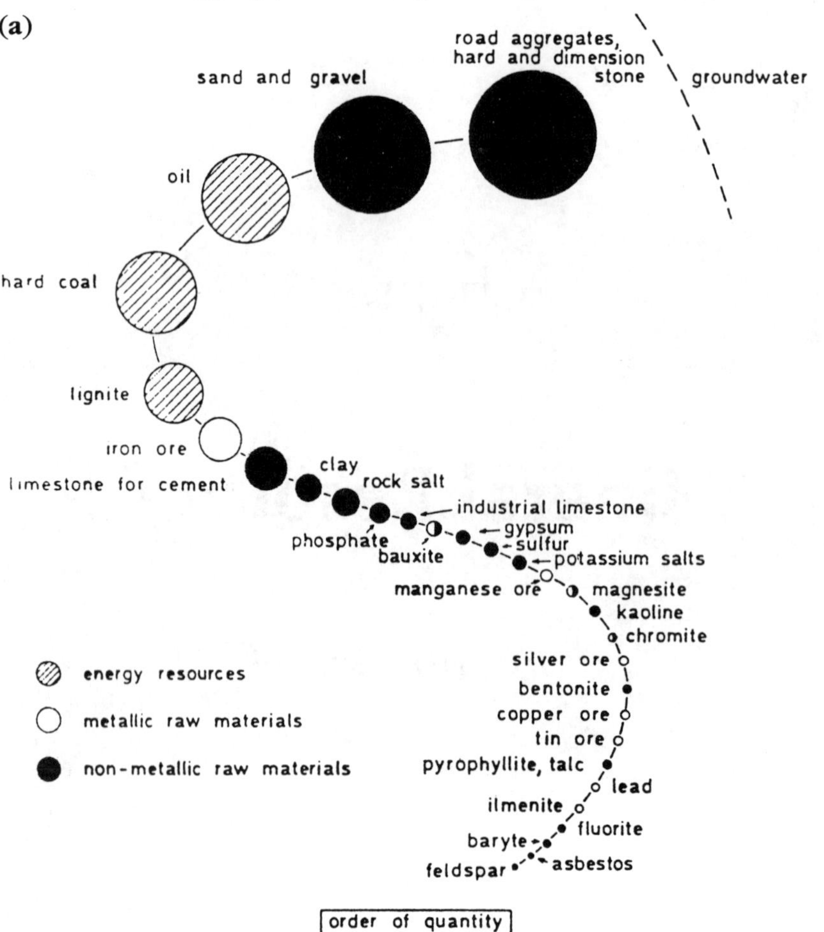
road aggregates, hard and dimension stone
sand and gravel
groundwater
oil
hard coal
lignite
iron ore
limestone for cement
clay
rock salt
industrial limestone
gypsum
sulfur
phosphate
bauxite
potassium salts
manganese ore
magnesite
kaoline
chromite
silver ore
bentonite
copper ore
tin ore
pyrophyllite, talc
lead
ilmenite
fluorite
baryte
asbestos
feldspar
energy resources
metallic raw materials
non-metallic raw materials
order of quantity

(b)

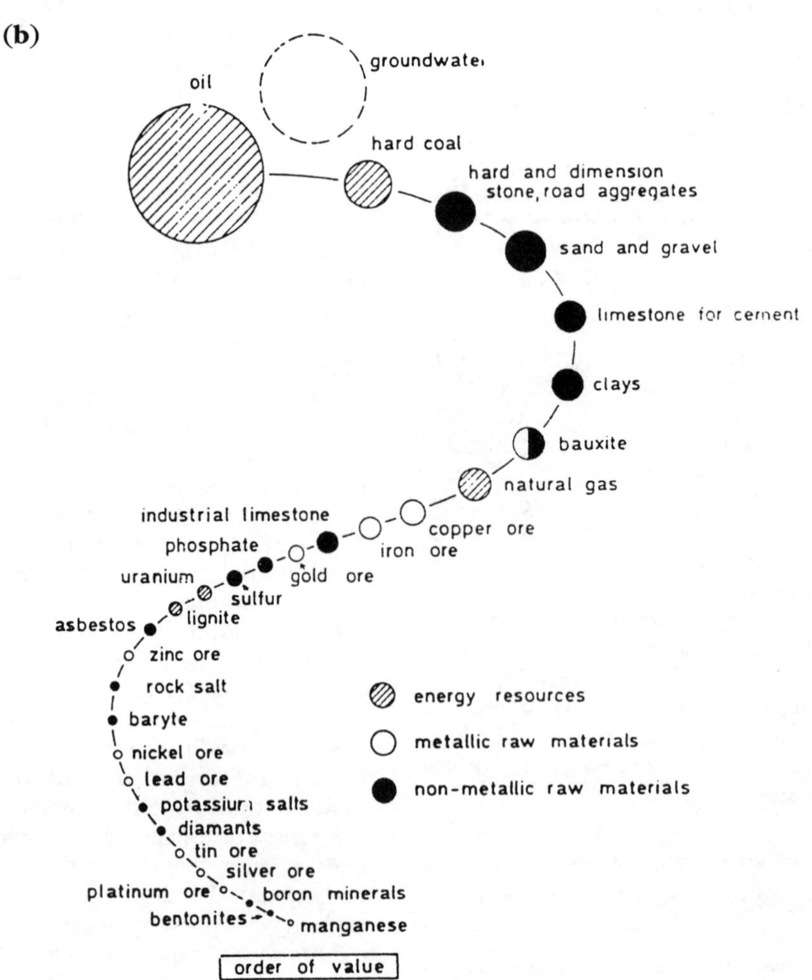
oil
groundwate
hard coal
hard and dimension stone, road aggregates
sand and gravel
limestone for cement
clays
bauxite
natural gas
industrial limestone
copper ore
iron ore
phosphate
gold ore
uranium
sulfur
lignite
asbestos
zinc ore
rock salt
baryte
nickel ore
lead ore
potassium salts
diamants
tin ore
silver ore
platinum ore
boron minerals
bentonites
manganese
energy resources
metallic raw materials
non-metallic raw materials
order of value

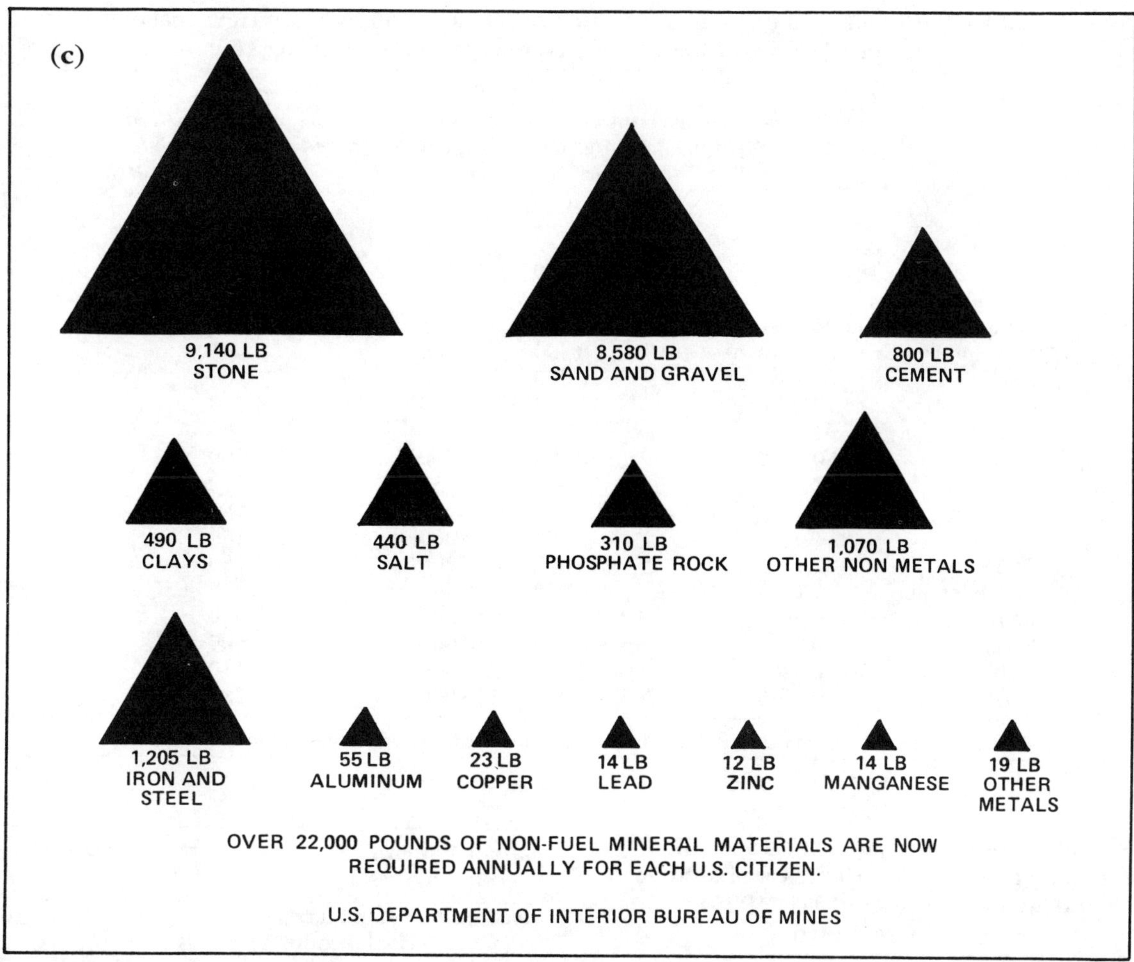

Fig. 1. Order of (**a**) quantity, (**b**) value of the thirty most important raw materials (after Lüttig, 1987), (**c**) annual *per capita* consumption of non-energy minerals in the USA (after US Department of the Interior, Bureau of Mines).

Table I Consumption of mineral resources and some products by a single citizen of the Federal Republic of Germany during a lifespan of 70 years (after Lüttig, 1987)

Mineral resources/ products	*Tonnes*	*Mineral resources/ products*	*Tonnes*
Sand and gravel	460	Dolomite	3
Petroleum	166	Phosphate	3
Crushed rock	146	Sulphur	2
Lignite	145	Peat	1.8
Limestone	99	Dimension stone	1.8
Steel	39	Potassium salt	1.6
Cement	36	Aluminium	1.4
Clays	29	Kaolin	1.2
Industrial sand	23	Steel alloys	1
Rock salt	13	Copper	1
Gypsum, anhydrite	6		

play of substitute resources (for example, crushable hard rock deposits in place of river gravels for aggregate use) is a recovery method that needs development.

Equally it becomes increasingly important with the spread of urban fringe development, the relocation of industry, and the introduction of new industry, to carry out proper resource potential studies to prevent sterilization of remaining resources by alternative uses of mineralized land.

The constraints on location and quantity imposed by past geological processes

The location and availability in quantity of any mineral deposit is entirely dependent on the geological processes which produced it or subsequently modified it.

Whereas certain resources such as sand and gravel or limestone, do permit a little latitude in selecting the least environmentally damaging extraction sites, because they are related to common geological processes and are thus of widespread occurrence, others are less common and allow little site selectivity. These must be mined where they occur, however environmentally damaging, or not at all. The extreme cases are exemplified by platinum, which can be mined in only a handful of small areas of the Earth's land surface, or diamond, which occurs in small volcanic explosion 'pipes', of extreme rarity and individually only measuring a few metres to tens of metres across. Coal was formed in ancient swamps, and though a false sense of dimension may be gained from seeing thick coal bearing sequences and thick coal seams, the actual coal basins occupy very restricted areas compared with the total land surface. Hydrocarbons, discovered onshore in the UK, occupy village-sized 'fields', which are sites where favourable geological trap structures coincide with permeable reservoir rocks, nearby source rocks containing fossil vegetable or animal matter, and impervious cap-rock seals—every oilfield depends on a coincidence of circumstances (Fig. 3).

Table II Comparison of minimum-grade ore and common rock tonnages that must be mined to yield a ton of selected mineral commodity materials

Metal	*(1) Crustal abundance (parts per million)*	*(2) Tons of average continental rock required*	*(3) Tons of minimum-grade ore required*	*(4) Concentration above background required (2÷3)*
Aluminum	83 000	12	4	3
Iron	48 000	20	3	6
Titanium	5 300	190	100	2
Manganese	1 000	1 000	3	330
Vanadium	120	8 300	100	83
Zinc	81	12 000	40	300
Chromium	77	13 000	3	4 300
Nickel	61	16 000	100	160
Copper	50	20 000	200	100
Lead	13	77 000	30	2 570
Uranium	2.2	455 000	670	680
Tin	1.6	625 000	6 000	104
Tungsten	1.2	830 000	200	4 150
Molybdenum	1.1	910 000	400	2 275
Mercury	0.08	12 500 000	200	62 500
Silver	0.065	15 000 000	10 000	1 500
Platinum	0.028	36 000 000	330 000	110
Gold	0.0035	285 000 000	125 000	2 280

Source: Mining and minerals policy, 1977, Annual Report of the Secretary of the Interior.

Table III World production figures for the major mineral commodities. Figures are based on World Mineral Statistics (1981–1985), except where shown

Major mineral commodities	*Tonnes (unless stated otherwise)*
Aggregate minerals	12×10^{9}*
Aluminium (bauxite)	543×10^{5}
Antimony (metal content)	533×10^{2}
Asbestos (all types)	42×10^{5}
Barytes	62×10^{5}
Bentonite	642×10^{4}
Beryl	3×10^{3}
Borates	$3\,015 \times 10^{3}$
Chromium (chromite)	102×10^{5}
Coal	$4\,422 \times 10^{6}$
Cobalt (metal content)	273×10^{2}
Copper (metal content)	85×10^{5}
Diamond	66×10^{6} carats
Diatomite	183×10^{4}
Feldspar	43×10^{5}
Fluorspar	49×10^{5}
Fuller's Earth	445×10^{4}
Gold (metal)	15×10^{5} kg
Graphite	609×10^{3}
Gypsum	74×10^{6}
Iron Ore	847×10^{6}
Kaolin	2×10^{7}
Lead (metal content)	35×10^{5}
Limestone	*c.*1×10^{9}†
Lithium (metal content)	75×10^{2}
Magnesite	121×10^{5}
Manganese ore	245×10^{5}
Mercury	63×10^{5} kg
Mica	238×10^{3}
Molybdenum (metal content)	98×10^{3}
Nepheline syenite	467×10^{3}
Nickel (metal content)	797×10^{3}
Perlite	17×10^{5}
Petroleum (crude)	272×10^{7}
(natural gas)	18×10^{11} m^3
Phosphate rock	145×10^{6}
Platinum group minerals (metal content)	251×10^{3}
Potash (K_2O content)	295×10^{5}
Rare earths (concentrates)	536×10^{2}
(as REO)	32×10^{3}
Salt	$1\,708 \times 10^{5}$
Sillimanite (+andalusite, kyanite)	51×10^{4}
Silver (metal)	$13\,469 \times 10^{2}$ kg
Sulphur (+pyrites) (S content)	543×10^{5}
Talc	75×10^{2}
Tantalum/Niobium (concentrate)	38×10^{3}
(Nb metal)	16×10^{3}
(Ta metal)	3×10^{2}
Tin (metal content)	185×10^{3}
Titanium minerals Ilmenite	59×10^{5}
Rutile	387×10^{3}
All forms (TiO_2)	33×10^{5}
Tungsten (metal content)	45×10^{3}
Uranium (metal content)	*c.* 1×10^{5}
Vanadium (metal content)	317×10^{2}
Vermiculite	*c.* 5×10^{5}
Zinc (metal content)	9×10^{5}
Zirconium minerals	814×10^{3}

* Statistic based on various sources (rough estimate only).
† Statistic based on Wolfe, 1984.

The restrictions on occurrence of workable mineral deposits, imposed by geological factors, thus impose severe constraints on land use planning for minerals.

Classification of mineral varieties and their settings

There are several acceptable classifications of mineral resources. Bates (1969) presents a useful and very broad classification of mineral resources:

(1) metals and metallic ores;
(2) mineral fuels;

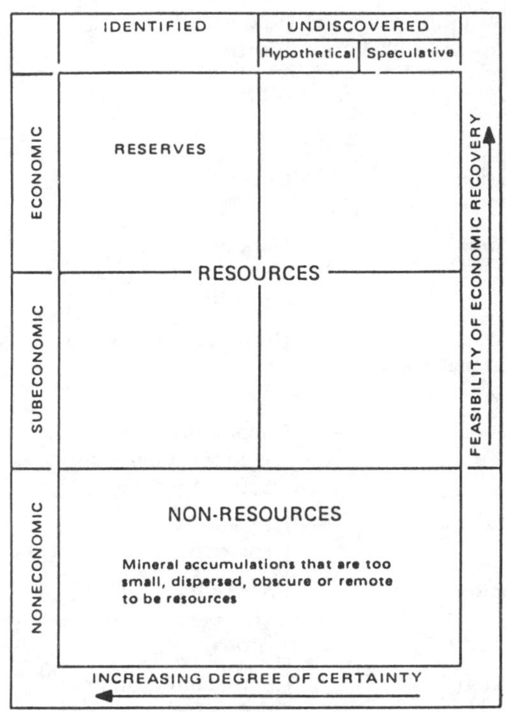

Fig. 2. Classification system for mineral resources and reserves (after US Geological Survey).

(3) groundwater;
(4) a large number of substances that do not fit into the above classes ('Industrial Minerals' *sensu lato*)

Groundwater is covered in Chapter 4 of this book. The remainder concern us here. These have been classified slightly more fully, separating the industrial minerals (*sensu stricto*) from the constructional minerals:

(1) precious minerals and base metals (gold, silver, diamond, platinum, copper, tin, lead, zinc, iron, etc.);
(2) industrial minerals* (limestone, gypsum, talc, asbestos, etc.);
(3) constructional minerals (building stone, aggregate minerals, road metal, etc.);
(4) energy minerals (coal, hydrocarbons, radioactive minerals, etc.).

A single deposit of any of these minerals may, depending on grade and market demand, be used in several ways, and may thus fall within more than one of the above categories. Nevertheless, these broad categories are useful.

In Tables IV and V below the variety of known minerals and the settings in which they occur are summarized.

Respective needs for information on mineral resources by industry and planning authorities

The land use planner must have full information concerning mineral resource potential and constraints. This allows rational choices to be made between various land uses in a general way. The level of detail required is, therefore, limited—'broad-brush' studies serve the purpose. This contrasts with the detailed, site specific data on mineral reserves required by the extractive industry. Studies of reserves focus on quality and commercial viability.

In the case of such resources as precious minerals and base metals, public authorities usually rely on broad assessments based on information supplied by the explorer or operator; the same is true for hydrocarbons. Company confidentiality may be a barrier to obtaining such information, but broad estimates are usually possible because of certain statutory requirements for data supply built into licences to explore and operate.

The more widespread industrial minerals and constructional minerals are more closely related to the activities of the public authority, since they account for most land used for mineral extraction. Thus it is not uncommon for the public authority to commission detailed resource evaluations of these minerals. In the United Kingdom most of such evaluations have been commissioned by the Department of the Environment, though explorers and operators may commission such studies. It is also not uncommon for coal resource studies to be commissioned by the Public Sector, through the agency of the same department.

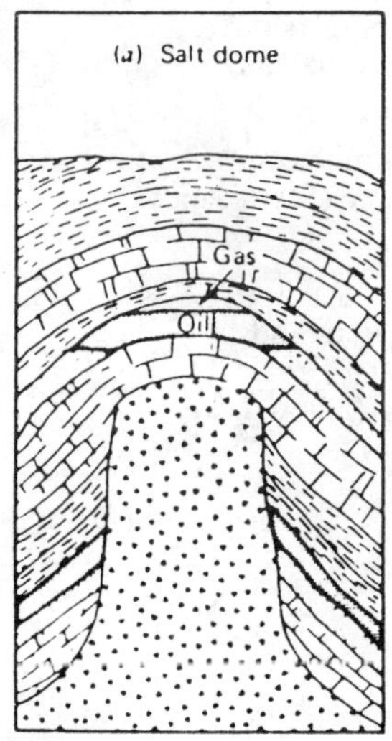

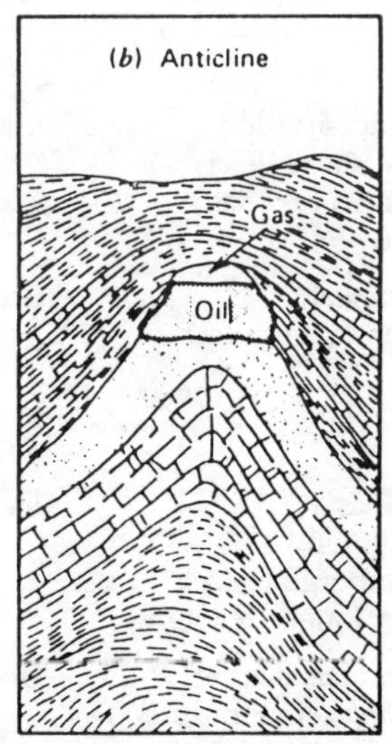

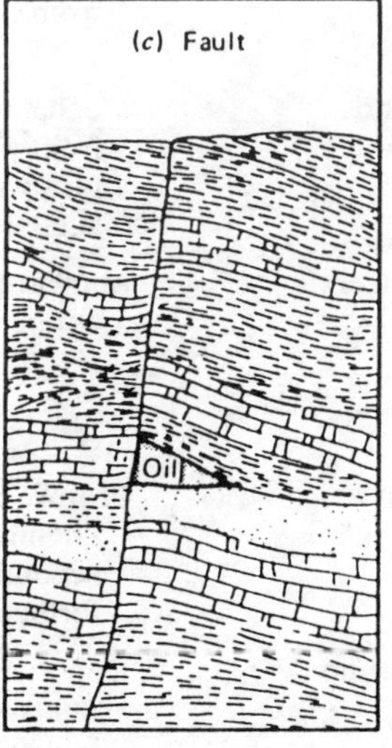

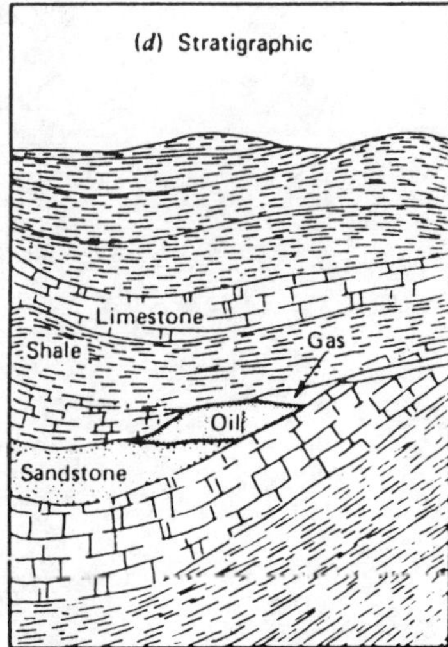

Fig. 3. Types of oil traps (after US Geological Survey).

**sensu stricto.*

Table IV Classification and occurrence of minerals (precious minerals, base metals, etc.), relationships between ore types, genetic processes and commodity

Association	*Ore type*	*Main process of formation*	*Mineral commodity*
(i) *Igneous*			
Ultramafic–mafic rocks	Magmatic	Magmatic segregation (oxides)	Chromite
	Magmatic	Magmatic segregation (sulphides)	Platinum group, Nickel-copper (gold, platinum)
Ultramafic rocks	Magmatic (explosion pipes)	Magmatic crystallization from gas charged magmas (exact process obscure)	Diamond (kimberlite)
Breccia	Magmatic related?	Exact process obscure	Copper, uranium, gold (Olympic Dam, South Australia)
Carbonatites	Magmatic	Magmatic crystallization	Rare earths (pyrochlore)
Carbonatites	Magmatic	Magmatic segregation	Copper, gold, nickel (Palabora, South Africa)
High-level granitic rocks (calc-alkaline)	Magmatic, hydrothermal	Subsequent alteration by fluids of mixed origin	Copper, molybdenum, gold, silver (Porphyry Coppers), Tin (Cornwall)
Pegmatites (late-phase granitic igneous rocks)	Late magmatic	Direct crystallization from gas-rich magmatic fluids	Tantalite, columbite, beryl, lithium minerals, etc.
Mafic volcanics	Magmatic, hydrothermal (volcanogenic)	Deposition from hydrothermal fluids related to submarine lava effusions	Copper, lead (Exhalative Massive Sulphide Deposits)
(ii) *Hydrothermal*			
	Vein mineralization	Deposition from hot (+200°C) epigenetic hydrothermal fluids	Gold, silver, tellurides (vein)
	Sediment hosted	Deposition from hydrothermal fluids at varying temperatures (50–260°C)	Copper, lead, zinc, silver (in argillaceous, sandy and calcareous sediments)
(iii) Sedimentary			
	Alluvial and eluvial	Mechanical concentration from water	Tin, gold, diamond, columbite–tantalite (Placer Deposits)
	Beach sands	Mechanical concentration from water at shore lines plus wind blown (aeolian) accumulation	Ilmenite, rutile, zircon, monazite, xenotime (Black Sands)
	Palaeoplacers	Ancient mechanical concentrations from water	Gold (Witwatersrand, South Africa) Uranium (Elliott Lake, Canada)
	Chemical	Protore formed as a chemical sediment; secondary (supergene) ore formed by prolonged leaching	Iron ores
	Unconformity	Deposition from circulating fluids at varying temperatures focussed on an unconformity between two stratigraphic sequences	Uranium
	Sandstone hosted	Migration of oxygenated meteoric (ground) water introduced the minerals	Uranium, vanadium

Table IV (*cont.*)

Association	*Ore type*	*Main process of formation*	*Mineral commodity*
(iv) *Weathering*	Residual	Intense surficial weathering, mainly under moist tropical conditions	Iron ores, bauxite (alumina), nickel, cobalt (Lateritic Deposits)
(v) *Metamorphic*	Metamorphic	Crystallization mainly as disseminations in metamorphic rocks; contact metamorphic processes due to heat from igneous bodies; pervasive regional metamorphism associated with mountain building	Copper, zinc, tin, tungsten, molybdenum; ruby, sapphire, etc.

Table V Classification and occurrence of minerals

INDUSTRIAL MINERALS (*sensu lato*, includes CONSTRUCTIONAL MINERALS)	
Igneous rocks	*Igneous minerals*
Granite	Feldspar*
Nepheline syenite	Mica*
Basalt and diabase	Lithium minerals*
Pumice and pumicite	Beryl*
Perlite	Diamond (pipes)
	Vein and replacement minerals
	Quartz (crystal, massive)
	Fluorspar
	Barite
	Magnesite
	Kaolinite
Metamorphic rocks	*Metamorphic minerals*
Slate	Graphite
Marble	Asbestos
	Talc
	Vermiculite
	Kyanite, Andalusite, Sillimanite
	Wollastonite
Sedimentary rocks	*Sedimentary minerals*
Sand and gravel	Diatomite (biogenic)
Sandstone	Potash and soda minerals
Clay (including kaolinite, bentonite)	Borates
	Nitrates
Limestone, dolomite	Sulphur (bedded)
Phosphate rock (biogenic)	Diamond (alluvial, eluvial)
Gypsum	
Salt	
	Volcanic minerals
	Sulphur (fumarolic)
ENERGY MINERALS	
	Igneous minerals
	Uranium (some deposits)
	Vein and replacement minerals
	Uranium (some deposits)
Sedimentary rocks (biogenic)	*Sedimentary minerals*
Coal, lignite, peat	Uranium (some deposits)
Tar sands	
Oil shales	
	Biogenic minerals
	Oil (liquid, displaced from oil shale etc. sources to secondary reservoirs and traps, mainly animal origin)
	Natural gas (gaseous, displaced from sources in coal seams etc., to secondary reservoirs and traps, mainly vegetable origin)
	Geothermal (mostly igneous/volcanic related)
	Geothermal steam
	Hot rock

*Mainly in pegmatites (coarse late-stage igneous intrusions)

Marker (1984) listed the uses of resource studies commissioned by the Department of the Environment, London:

(1) formulation and evaluation of development plan policies;
(2) considering planning applications for extraction of minerals;
(3) safeguarding resources from sterilization by other development;
(4) quantifying resources in areas of large-scale development proposals;
(5) selection of deposits for detailed evaluation.

He was concerned specifically with aggregate minerals, but the principles are of general application. He noted four levels of investigation:

(1) desk studies evaluating existing geological maps and collating relevant published and archival data;
(2) rapid assessment studies incorporating limited borehole investigation and testing of samples, geophysics or other indirect methods;
(3) full-scale assessment including extensive borehole, sampling and testing programmes;
(4) regional collations using computerized data banks of results from full-scale assessment programmes.

Mineral resource aspects are an important component of the multi-thematic ('Environmental Geology Mapping') mapping studies covered in Chapter 9. Areas of former mining activity, those with mineral resource potential, and those with conflicts between mining and quarrying and other land uses provide ideal subjects for the Environmental Geology Mapping (EGM) approach.

It is essential to mount mineral resource evaluations early wherever development is proposed, irrespective of whether the mineral resource is likely to be mined now, within decades or in the long term. It is unwise to sterilize mineralized land by allowing other development over it, and it is also imprudent to allow new housing or industrial development over hazardous areas, the legacy of former mining or the result of current mining operations.

Types of investigation

There are a large number of methods of exploration for minerals. The most remote methods are essentially reconnaissance techniques. There is the remote sensing approach, utilizing satellite imagery (Landsat etc.) and various enhancement techniques (infra red radiation etc.). Remote sensing is most suitable for large area regional coverage from a resource potential viewpoint. However, from the constraint viewpoint, remote sensing can be very useful in delineation of old workings, existing workings, present land use, problem areas, and areas of sterilization.

Air photography, supported by air photo interpretation, is a major aid in delineating superficial and bedrock geology, unravelling the geomorphology and geological structure, and the resource boundaries. It is particularly useful in glaciated terrains or river terrace settings. It is invaluable, again, in detecting former workings and outlining problem areas and areas of sterilization. It can also be used simply as a very rapid method of erecting a base map.

There are also the indirect methods—geophysics (magnetometry, scintillometry, electromagnetic, induced polarization, resistivity, etc.), which can be applied from the air or on the ground—and the geochemical methods (blanket study of soil or heavy mineral samples collected on the ground, and specialized searches for either pilot elements (for example, arsenic for gold) or the mineral itself (directly for gold, for example)). The geochemical methods require extensive back-up laboratory services, and may be coupled with moderately accurate assays (for example, atomic absorption) or very accurate procedures (for example, fire assay for gold). A very useful review of such methods is provided by Plant *et al.* (1988).

Geophysics is widely employed in the search for precious minerals and base metals, and also hydrocarbons, as a precursor to direct exploration of the promising anomalies revealed. However, it is also useful in the industrial and constructional mineral context, especially in the regional rather than site specific context. In the latter there really is no substitute for direct methods. Ambrose *et al.* (1987) reviewed the applications of geophysical methods in the English Midlands and Welsh Borderlands. While accepting that direct drilling for aggregate minerals is the main technique, they found that resistivity, seismic, electromagnetic, induced polarization and ground radar methods had some potential for evaluating scattered drift deposits. Coupling them with direct drilling to test their efficiency, it was found that the resistivity method (Offset Wenner) was most cost effective and rapid, predicting bedrock contact (base of drift) to 5% accuracy, and being particularly effective for terrace deposits such as those of the Rivers Avon and Severn. The EM31 conductivity meter was also useful in detecting channelled bases of terraces. These authors stressed that a back-up programme of boreholes is essential, and that sole reliance on geophysical methods was dangerous. Hamilton (1984) cited geophysics, coupled with vegetation mapping, photogrammetry, and remote sensing, as viable back-up methods for aggregate mineral mapping. He noted that both resistivity and seismic refraction will provide depths to interfaces, but gravity and magnetic methods are of limited use, though they can pick up the heavy minerals contained in sands and gravels.

Direct exploration by drilling, pitting, trenching, bulk sampling is obviously the most accurate way of establishing reserves, but it is expensive. However, the accuracy may outweigh cost. Thus care should always be taken before employing indirect methods such as geophysics, which are quick and effective over large areas, but are also expensive and only give approximate indications. Very often it is advisable to utilize these indirect methods as a first stage in resource evaluation, but in other cases it is better to proceed straight into direct evaluation. The indirect methods are never adequate on their own to establish ore reserves and quality.

Mappable information on resources

Some types of information that can be included in a mineral resource map are summarized in Table VI.

Types of extraction

Minerals are extracted from the ground by one of four methods:

(1) opencasting, quarrying;
(2) deep mining;
(3) solution mining;
(4) wells.

Table VI The type of information that can be included in a mineral resource map

(1) QUANTITY	Extent Thickness	Volume and tonnage
(2) QUALITY	Physical properties	Hardness, porosity, abrasion effects, polishing value, reflectivity, grain size, thixotropy viscosity, colour, brightness
	Chemical properties Grade Impurities	 (a) disseminated (b) interbeds (c) channels, infilled cavities
(3) ACCESSIBILITY (*re* ease and type of extraction)	Type of overburden Thickness of overburden Structure of deposit Depth to groundwater/ direction and amount of flow Nature of strata beneath mineral	
(4) WORKABILITY (*re* ease of extraction and stability on excavation, blasting hazards)	Internal structures (jointing, faulting, veining, dip) Weathering Underground cavities	

Opencasting and quarrying are only viable at shallow depth and with relatively thin overburden. Solution mining is rare and only applicable to soluble minerals such as potash salts.

Hydrocarbons (oil and natural gas) are liquids or gases which are released up wells with or without pumping. Tar sands and oil shales are opencasted, however.

The method of extraction adopted will depend on a number of factors:

(1) depth, shape, size and structure of the mineral deposit;
(2) physical properties and internal structure of the mineral deposit;
(3) thickness, nature and physical properties of the overburden;
(4) nature of the underlying material;
(5) hydrogeological conditions.

Opencasting and quarrying are predominantly used for extracting near-surface, bulk, low price, industrial and constructional minerals—near-surface iron ores, bauxite (aluminium) and coal seams, and the black (titanium) beach sands used for pigment. Quite shallow rock deposits such as limestone or gypsum may, in some cases, be more conveniently worked from shallow adits, underground. Deep underground mining is invariably very costly and frequently requires backfilling to support the cavity after mining; shaft haulage costs are also very heavy. It is thus utilized mainly for valuable metalliferous minerals (gold, copper, lead, zinc, tin, platinum) or diamonds in pipes. Large bodies of these minerals can also be effectively worked in great open pits several hundred metres deep, and very commonly the upper levels of large deposits are so worked, even if eventually deep mining has to be employed. Deep coal seams and potash deposits are worked by deep mining methods.

Environmental effects of extraction

Any past, present or proposed extraction of minerals must affect land use because it must raise some conflict with present or proposed land use, or may harmfully affect the environment by virtue of extraction, processing, disposal of waste, transport, and contamination of water and air.

Examples of such harmful effects are:

(1) pollution of surface or groundwater by emissions or waste products;
(2) chemical contamination from mine and quarry waters into surface and groundwaters;
(3) blighting of the landscape by mine workings, quarries or spoil dumps;
(4) risks imposed by large unstable spoil dumps (the Aberfan disaster in South Wales epitomizes this);
(5) destruction of topsoil—overburden must be removed in quarrying or opencasting, and once dissipated cannot be replaced;
(6) sterilization of future development by mines and quarries; ancient abandoned workings will present a major environmental hazard;
(7) environmental risks to miners themselves must not be overlooked—they are epitomized by disasters at the Belmoral Gold Mine, Quebec, and a gold mine in the Rand, Transvaal, South Africa, where mines have been engulfed by mining upwards into overlying swamps or water-charged dolomites.

Very specialized mapping programmes, backed by computer analysis, are nowadays undertaken with the aim of minimizing the visual impact of surface workings (Fig. 4). This represents a whole new field of mineral related mapping.

Of the energy minerals, coal presents much the same

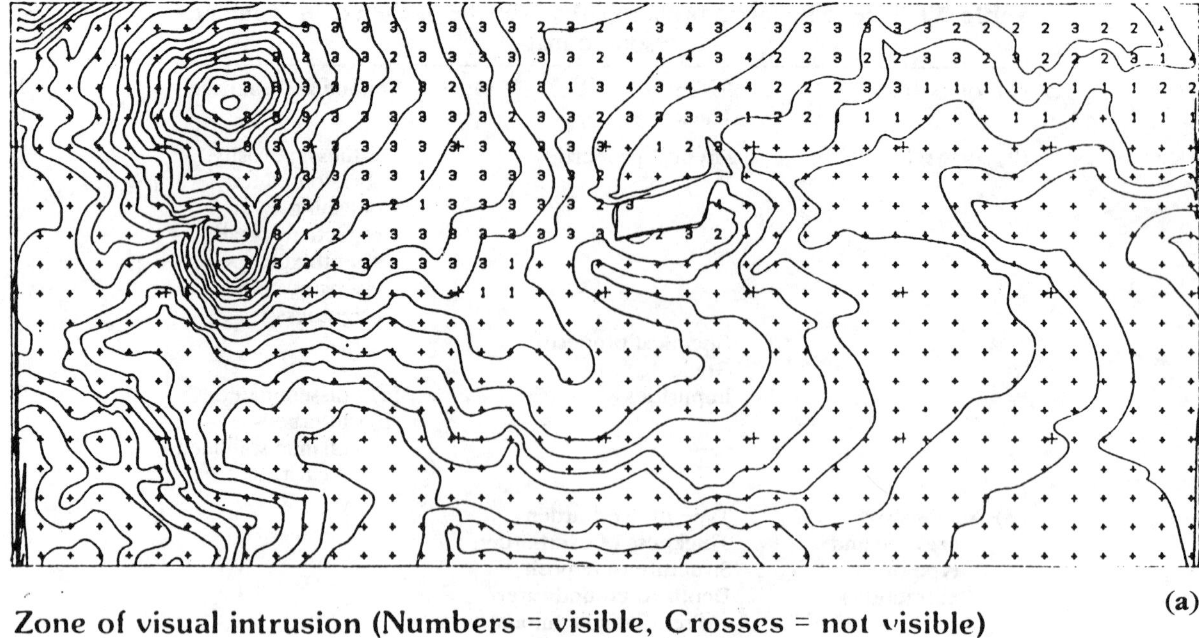

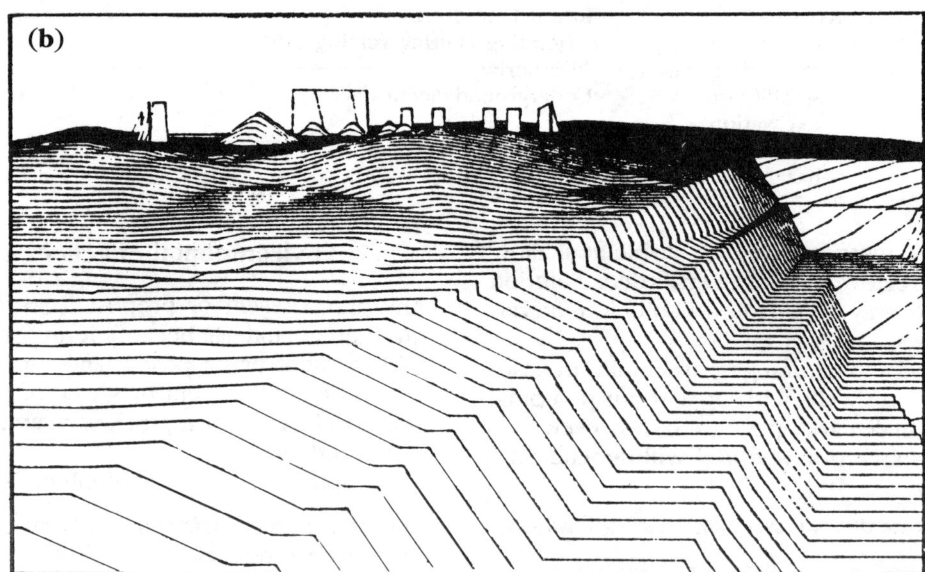

Fig. 4. Examples of computer diagrams utilized for minimizing landscape impact of opencast workings (from Jarvis, 1987).

problem as other minerals, but with the additional environmental problems of dust pollution, fire and explosion and immense waste tips. Oil and gas extraction onshore presents very special problems of land use conflict and environmental damage, there being a requirement for large installations and pipelines, and there are risks of fire and air pollution by waste gases, and of pollution of both surface and groundwater. These problems have been largely minimized in the United Kingdom through excellent co-operation between these modern industries and the public authorities. Oil shales in the western United States and tar sands in Alberta, Canada, opencasted on an immense scale, scar the land and present risks of surface and groundwater pollution. Strip opencast coal mining has likewise scarred a number of areas of the United States.

Radioactive minerals (uranium) present very special problems. There are critical problems of potential air and water contamination, of harmful radiation affecting people, livestock and vegetation, grazing land, surface and groundwater bodies. There are considerable problems of disposal of radioactive wastes, and accidental plant water discharges, and there is, in the processing plants, the risk of seismic hazards, landslides, subsidence and flooding.

Constraints on extraction

The land use planner aims to optimize the mineral resource usage, while at the same time minimizing the environmental impact. Only in very clear cases of severe environmental impact or damage can extraction be opposed. One of the planner's most important functions is to delineate areas of potential sterilization and problem. These may be incorporated in the maps of mineral consultation areas (areas in which the industry needs to be consulted on non-mineral development proposals). The main constraints are shown on Table VII.

Table VII The main mappable constraints on mineral working

Planning constraints	*Economic constraints*
Agricultural land quality	Market location (related to transport costs)
Landscape value	
Recreational use (National Parks, etc.)	
Urban Development and allocations	
Communications	
Service infrastructure	
Conservation areas (sites of special scientific interest, archaeological sites)	

Another economic constraint may be the lack of market demand at a realistic price related to mining or quarrying, processing and transport costs. It is important to realize that the economic constraints may vary with time and with new policies. Likewise, the planning constraints may, with the passage of time, lose their force. As other mineral supplies of a viable grade and mineable at a reasonable price, become exhausted, less economically viable, lower grade deposits, formerly 'marginal' or 'non-viable' may become an economic proposition.

Mines close up and reopen with changing economic conditions. An example is the gold mines in Canada and Australia that have lately reopened as a result of high gold prices, after years of closure due to low prices. New technology also may allow once 'worked out' mines to reopen, and may indeed introduce new deposits that would formerly have been regarded as sub-economic. The land use planner must be wary of regarding all 'old mines' as irrevocably worked out, and the existence of any mines, active or closed must be fully taken into account in any planning for other land use. Even so there must come a time when a mineral deposit becomes irrevocably depleted and worked out.

Restoration after land use for mineral extraction

The restoration of mineral workings after completion of extraction to their former condition and use, or other beneficial use, is now a major issue in heavily populated countries, where there is strong competition for use and reuse of land in agricultural areas and areas of natural beauty. Restoration is a statutory requirement of most mineral extraction projects nowadays and rightly so. Complete restoration of the land to its pre-mining state may not be possible because of drainage problems or high water-table, but we see many excellent examples of land restored to agricultural, housing or industrial, and recreational use.

Opencast and quarry sites are widely subjected to landfill, which may be domestic or industrial waste, and thus there is a close relationship between waste disposal and opencasting/quarrying. Any such waste disposal requires very careful investigation and planning, to prevent toxic waste contaminating surface and groundwater; even domestic waste can cause dangerous build-ups of methane gas. Another common use of old quarry sites is recreational. This is a very ancient practice (the Norfolk Broads are old peat diggings), and a more recent example is the Cotswold Water Park in Gloucestershire and Wiltshire, England, where recent gravel diggings have been converted to recreational lakes.

Techniques of restoration vary according to the type of mineral extraction, the type of rehabilitation and the nature of the after-use.

When minerals are extracted from open pits and quarries, the working may be:

(1) relatively shallow, laterally extensive, and in minerals that are largely saleable (e.g., extraction of river terrace gravels);
(2) deep and yielding minerals which are largely saleable (e.g., brick clays, limestone, igneous rock);
(3) shallow, but worked for minerals which comprise only a small proportion of the material extracted (e.g., nodular celestite, gypsum or ironstone);
(4) deep, but with the saleable mineral comprising only a small proportion of the material extracted (e.g., open cast coal, fullers earth).

The last two give rise to workings which can be backfilled with waste rock as working progresses. The other types of workings must be left as voids or infilled with materials from elsewhere.

Open pits may be located in areas where the water-table is deep and where they therefore remain dry on abondonment. Such pits can be infilled with waste materials where this does not, for example, give a potential for pollution of the groundwater—common infill materials are demolition rubble or waste rock from elsewhere (e.g., remote disposal of colliery spoil). The resulting infilled site will be landscaped, topsoil usually being replaced (having been stored during exploitation). If the infilling materials include putrescible wastes, precautions must be taken against migration of gases such as methane, and of leachates.

Alternatively sites may be put to use without infilling, though this will commonly be preceded by trimming or grading of the face, which will need to be properly designed to avoid risk of subsequent slippage. Rock faces left behind after working may be extremely large and at several levels. Commonly this precludes restoration of the face, though the floor may be returned to agriculture or used for built development. Steep rock faces are unstable for long periods and must be regularly monitored and repaired at intervals (e.g., the many limestone quarry faces enclosing industrial sites in the city of Plymouth—McCall *et al.*, 1985). Where unsightly, particularly in areas of high landscape value, such faces are often regraded. Conversely, such regrading may be opposed by those interested in conserving them as a resource for geological research and education. Attempts have been made recently to utilize controlled blasting of large faces to produce forms which mimic natural limestone landforms (e.g., the Peak District, Gagen and Gunn, 1988). This technique would both blend these obtrusive faces into the landscape and preserve geological exposures.

Extraction from pits which extend below the groundwater table can be achieved by dewatering or working the pit wet (e.g., dredging, or use of a drag-line or bucket wheel excavator). Restoration can be undertaken to such uses as farming at a lower level, but this would require permanent pumping out of the site. Such sites are more commonly left wet as reservoirs or areas for water-based recreation, or are infilled. The filling of such pits does require special precautions to prevent contamination of groundwater, and is usually limited to the use of 'inert fill' (chemically non-reactive, non-putrescible materials).

Underground mining gives rise to surface effects

which may require restoration. These include shafts and adits, and waste tips (pits receiving the fines separated out during processing). Shafts and adits are considered in detail in Chapter 7. Waste tips may be reworked as a resource (e.g., to extract small quantities of valuable residues such as gold or uranium), transported for void filling elsewhere, or simply, as is more common, landscaped after tipping ceases.

The principle factors in any of the above procedures that can be assisted by earth science mapping are:

(1) geomorphological mapping to identify the restoration landforms which will best reduce landscape damage (this approach has been little used so far, with the result that old workings tend to be obtrusive);
(2) hydrogeological mapping (see Chapter 4) to establish the effects of the restoration system on the general groundwater regime;
(3) mapping of agricultural soils and potential (see Chapter 6), since it is commonly a requirement that land returned to agriculture should be of a grade as close as possible to that of the original land;
(4) mapping of factors which influence rock face and slope stability in sites that are abandoned but not infilled (see Chapter 7). This might include the general deterioration over a period of time of trial sites to give prior indications of the likely maintenance requirements of future residual rock faces resulting from proposed development of workings nearby.

The legacy of former mineral extraction

Whereas relatively shallow surface pits can be restored to their former condition or to some other use, sites of deep mining remain as major hazards, though they can be tidied up at the surface, and large spoil dumps can be levelled with release of their former sites for other use.

The accurate delineation of the sites of former deep mining is a very important field of mineral related mapping (Holt and Marker, 1987). Since 1976, more than a dozen such studies have been commissioned by the Department of the Environment, London (e.g. Freeman, Fox Ltd, 1988 (a), (b); Ove Arup and Partners, 1976, 1983, 1985, 1986). Another example is the Polish study, briefly cited as a case study below (Dragowski *et al.*, 1983). Such studies are mainly aimed at returning mined out land to other development, while at the same time excluding zones of hazards. This activity is further considered in Chapter 7.

The value of mapping

The basic geological map (solid, solid and drift, or drift only) is an invaluable aid to mineral resource evaluation. It shows the extent on the surface of the formations which are likely to harbour minerals and allows projections in depth to be made from the sections that normally accompany it. From it the presence of favourable structures can also be deduced. The economic significance of stratigraphy unravelled in such maps should not be overlooked—Prentice (1988), for example, demonstrated the direct relevance of stratigraphy to brick-clay reserves in part of southeast England.

There are also numerous types of specialized maps that map be used both to illustrate the resource and the constraint aspect. Of great importance is the selection of an appropriate scale to the particular objective of the mapping. These specialized maps may comprise a Thematic Set, some being single theme ('Element') maps; some being composite theme ('Derived') maps: and some being 'Potential' maps.

The map is the most convenient way of presenting data for a wide variety of users. Those with no technical knowledge of geology and the related earth sciences find they can readily assimilate the technical information if it is clearly and simply presented by map and section. However, three-dimensional representations such as block and fence diagrams, and clearly written supporting text will greatly enhance the value of the map presentation.

CASE STUDIES

Introduction

There is an immense body of literature on this subject, so it is only possible to select here a number of cases that illustrate the variety of approaches adopted in mapping for planning and development purposes, the variety of subjects, the contrast between resource-oriented and constraint-oriented studies and the studies which combine both aspects, and also the contrast between regional reconnaissance mapping and detailed site-specific studies of small areas.

Precious minerals and base metals

Three case studies are very briefly cited below illustrating:

(1) *Tucson, Arizona*: the conflict between mineral and water resources and urban development.
(2) *Walbrzych, Poland*: the hazards imposed by underground mine workings beneath urban areas.
(3) *Hemerdon, Southwest England*: the establishment of the resource potential and environmental impact of development of a large-scale minerals deposit on the fringes of a city (Plymouth) and within agricultural land.

Tucson, Arizona—controlling hillside development and protecting and preserving mineral and water resources (Davidson *et al.*, 1978)

Eastern Pima County has yielded millions of tons of copper metal since mining commenced in 1858, and copper mining is, and will continue to be, a major economic factor in the region. The huge embankments and dried up ponds that form the waste products of the copper mining are encroaching on the satellite developments of Tucson, a city of half a million people. These developments commenced before mining activity in their immediate vicinity, but housing is now spreading onto land that may hold copper and also onto extensive mine installations and their tailings. Mining here is by opencast methods, a form of mining that is completely incompatible with housing development. Mineral rights are reserved by the Federal Government, so privately owned land, including housing developments, can be opened up to mining claims. This study was mounted on account of this obvious conflict.

The study was essentially a multithematic study (see Chapter 9). It embraced hillside development, mineral and water resource conservation, mining waste control and the hazards of old mine workings. It emphasized both the mineral potential and the constraints aspects, and illustrates the way differences in law from one country to another can affect the need for and design of such mineral resource related projects.

A resource potential map derived from this study is illustrated in Fig. 5.

The influence of an underground vein ore mine on engineering geology considerations in urbanized areas (Dragowski *et al.*, 1983)

This study relates to environmental problems created by deep mining of barite, silver and galena in a system of steeply inclined veins, mined out and backfilled with waste rock, and extending beneath the town of Walbrzych. The effects of variation in vein dip are complex and little understood. The effects of differing vein attitudes and distances from the surface on building foundations were evaluated.

The basis for the study was cartographical and geological mapping on a scale of 1:2000, data being added on soil groupings to a depth of 5 m encountered in building foundations. This mapping was followed by registration of the present state of buildings and symptoms of ground surface deformation; studies of rocks around the deposits, their types and physico-mechanical properties; and registration of the whole range of ancient mine cavities, with projection of the mine galleries onto the land surface. Data were also recorded from a number of boreholes and research pits.

This is an example of a mineral related hazard study of a small area, site specific and quantified in detail, and with strong engineering geology emphasis. The results enabled land use planners to avoid founding buildings on hazardous sites. There is an obvious overlap here with Chapter 6 (Soils), Chapter 7 (Hazards), and Chapter 8 (Construction), well illustrating the close relationship of these subdisciplines with minerals.

Hemerdon, an evaluation case history (Christoffersen and King, 1988)/Environmental planning and mineral development—the case of Hemerdon Mine (Jones, 1987)

This study related to a proposal to extract tungsten and tin from Hemerdon Mine on the outskirts of the city of Plymouth in southwest England, a proposal which would have involved one of the largest open pit mines in Britain, with a projected life of 21 years. Tungsten is a relatively rare element used in steel manufacture to produce extremely hard, wear resistent cutting tools, armour piercing ammunition and electric lamp filaments. It is used in other alloys also—pigments, lasers and catalysts. The main economic minerals are wolframite and scheelite.

The site is on the edge of the kaolinized Dartmoor granite on its contact with metamorphosed siltstones, mudstones and basic volcanics, and it adjoins the village of Hemerdon, amid agricultural land 11 km to the northeast of the centre of Plymouth and only 3 km from the edge of Plympton, now a suburb of the city.

Tungsten was discovered in 1867 and open working commenced in 1917. Prospecting shafts were sunk from 1936–38 and deep mining occurred for a brief period from 1943–1944. The resource potential was established recently by 25 000 m of large diameter core drilling and percussion drilling, underground bulk sampling, and metallurgical test work (including pilot plant scale testing). Very sophisticated computerized methods of ore reserve evaluation were utilized. By 1977, some 59 million tons of low grade ore had been established.

Whereas this is not primarily a mapping study, it is of relevance as it illustrates the considerations involved in planning a major open pit metal mining operation in an area of land use conflict. The operators emphasized the resource aspect, tungsten being a strategic mineral and the mine, if commissioned, being capable of supplying the entire UK requirements. The employment opportunities were also stressed. The main environmental problems are visual impact, noise, dust, archaeology, ecology, the effect on the existing agricultural economy, and waste disposal. The operation would produce: 2.4 million tons of broken waste rock per annum; 1.5 million tons of coarse washed granite from the processing plant; and 1 million tons of finely ground waste granite and clay fines residues. There would also be a significant discharge of waste pit water into the surface water system via the Smallhanger brook. Though close to the National Park Boundary, the case concerning this amenity was weakened by the existence of immense china clay workings in the form of an open pit and spoil dumps at Lee Moor, between Hemerdon and the Park Boundary.

A major undertaking of this sort requires a number of very specialized investigations involving mapping, carried out both by the operator and the planning authority before a balanced judgement can be achieved on the planning application. The initial application to mine was rejected, but was resubmitted. The project would probably only be commercially viable at a time of high wolfram and tin prices, and these commodities are at the present time subject to very wide price fluctuations.

Industrial minerals

Two cases are cited below, illustrating:

(1) *Limestones of the Peak District Central England*: a study of the resource potential of a mineral resource ranging from high to low grade and used both as an industrial and as a constructional mineral, situated in an area of extreme environmental sensitivity (National Park).

(2) *Magnesian Limestone, Central and Northern England*: a study of a similar industrial mineral resource but with both potential and constraints aspects covered.

Limestones of the Peak: a guide to limestone resources of the Peak District, Derbyshire and Staffordshire (Harrison and Adlam, 1985)

Commissioned by the Department of the Environment, London, this study covers a resource which supplies both high chemical quality limestone for industry (45%) and low grade material for aggregate and roadstone. The area of 540 km^2 contains some 28 active quarries.

Striking the proper balance between the needs to protect the landscape in a scenic and recreational area and the resource needs of nation and local community is the task of the land use planners in this sensitive area,

POTENTIAL FOR COPPER DEPOSITS NEAR TUCSON (at a scale of 1:250 000)

AREAS — A = more potential
B = intermediate potential
C = less potential

The Tucson region contains many large deposits of copper minerals that have been commercially mined, and more are still to be discovered. The relative copper bearing potential of the rocks has been appraised, based on the location and characteristics of known ore bodies and prospects; on the age, type and alteration of rocks; on the intensity of rock fracturing; and on geophysical measurements. The relative potential for silver, gold, uranium, lead and zinc is similar to that for copper.

Because depth of cover may significantly deter ore discovery and mining, the maps shows where blanketing deposits are more or less than 1000 feet thick. The boundaries will surely change as knowledge and experience accumulate. The map is a reduced and simplified excerpt from Preliminary Map Showing Potential for Copper Deposits in the Tucson 2 degree quadrangle, Arizona, OF 76-458.

DEPOSITS & OCCURRENCES

● Major copper deposit producing in 1974

● Minor productive copper deposit

○ Copper occurrences

OVERBURDEN

More than 1000 ft

----- 1000 -----

Less than 1000 ft

ORDINANCE NUMBER ________

AMENDING THE PIMA COUNTY ZONING ORDINANCE NUMBER 1952–III BY ADDING ARTICLE 44 MINERAL EXTRACTION ZONES

BE IT ORDAINED BY THE BOARD OF SUPERVISORS of Pima County, Arizona:

Section 1. That the Pima County Zoning Ordinance Number 1952–III be amended by adding Article 44 Mineral Extraction Zones to read as follows:

ARTICLE 44

MINERAL EXTRACTION ZONES

Sec. 4401 PURPOSE

The purpose of this article is to protect economic mineral resources and to insure the orderly and systematic development of these mineral resources while protecting the public health, safety, peace, comfort, convenience, and general welfare of the people of Pima County and to secure for them the social and economic advantages of an orderly efficient use of land and also to inform potential land owners of the possibility of mining in the area in the future as authorized by the Arizona Revised Statutes (A.R.S. 11–806).

Sec. 4403 AREAS WHERE RESIDENTIAL DEVELOPMENT SHALL BE REGULATED

A. Active Metallic Mining Areas

B. Metallic Mineral Occurrence Areas

C. Metallic Mining Residual Hazardous Areas

D. Patented Claims and Active Metallic Mines within Withdrawn Areas

Sec. 4405 REVIEW

The Active Metallic Mining Area Zones and the Metallic Mineral Occurrence Zones will be reviewed no less than once every five years. Areas no longer designated Class A on U.S.G.S. Geology, Production and Mineral Occurrence Open File Maps (74–143) shall be considered for rezoning in accordance with the adopted area and neighborhood plans of Pima County.

A. Active Metallic Mining Areas

a. Major Areas

Contiguous land areas 100 acres or greater in size owned or leased by mining companies or persons which contain or are utilized for metallic mining currently in production including but not limited to open pit mining, underground mining, subsidence areas, shops, dumps, tailings ponds, shafts, mills, smelters, haul roads, and all other adjacent and ancillary structures associated with metallic mining operations.

1. A buffer zone will extend no less than one mile from the perimeter of the Major Active Metallic Mining Area.

2. Residential densities shall not exceed one dwelling unit per thirty-six acres (1 D.U./36 acs.) within the Major Active Metallic Mining Buffer Zone.

3. All other uses permitted by Section 801 of the General Rural Zone shall be permitted within the Major Active Metallic Mining Buffer Zone.

b. Minor Areas

Contiguous land areas less than 100 acres in size owned or leased by mining companies or persons, which contain or are utilized for metallic mining currently in production including but not limited to open pit mining, underground mining, subsidence areas, shops, dumps, tailings ponds, shafts, mills, smelters, haul roads, and all other ancillary structures associated with metallic mining operations.

1. A buffer zone extending no less than six hundred and sixty feet (660 feet) shall extend from the perimeter of all sides of the Minor Active Metallic Mining area.

2. No residential or commercial uses shall be allowed within the minor active metallic mining area buffer zone.

3. All other uses permitted by Section 801 of the General Rural Zone shall be allowed within the Minor Active Metallic Mining Area Buffer Zone.

B. Metallic Mineral Occurrence Areas

Metallic Mineral Occurrence Areas are defined as lands classified as "A" and underlain by less than 1,000 feet alluvium on U.S.G.S. Map I–844–G. "(Preliminary) Map showing potential for copper deposits in the eastern three-quarters of the Nogales 2° quadrangle, Tucson area, Arizona," and open-file map (76–458) "(Preliminary) Map showing (potential for) copper deposits in the Tucson 2° quadrangle, Arizona."

a. Residential development shall be restricted to one dwelling unit per 10 acres (1 D.U./10 acs) or lower densities.

b. All other uses permitted by Section 801 of the General Rural Zone shall be permitted within the Metallic Mineral Occurrence Areas.

C. Metallic Mining Residual Hazardous Areas

Metallic Mining Residual Hazardous Areas are defined herein as, but not limited to: Open pits, shafts, tunnels, tailings ponds, mining subsidence area, and dumps as a result of metallic mining activities.

1. No residential, commercial, nor industrial development with the exception of mine uses as defined in Section 4402, paragraph 1 of this Article shall be permitted on residual hazardous areas until these lands are certified non-hazardous by a geologist and soils engineer registered in the State of Arizona. The geologic and soils reports shall be submitted to, approved and certified by the Pima County Highway Department.

2. If applicant has shown that the residual hazardous area is non-hazardous and non-mineral as certified by the Pima County Highway Department, these areas shall be considered for rezoning in accordance with adopted areas and neighborhood plans for Pima County.

[D. omitted]

Fig. 5. Potential for copper deposits—Tucson, Arizona. Extract from map legend (after Davidson *et al.*, 1978) and the related ordnance relating to mineral extraction zones (from Davidson *et al.*, 1978).

EXPLANATION OF COLOURS*

CATEGORIES OF LIMESTONES

	Composition (% $CaCO_3$)
Very high purity	≥ 98.5
High purity	≥ 97.0-98.5
Medium and low purity	≥ 85.0-97.0
Impure	< 85.0
Undivided Categories 1 and 2	≥ 97.0
Undivided Categories 2 and 3	≥ 85.0-98.5
Undivided Categories 3 and 4	< 85.0-97.0
Very high purity limestone likely to be partially dolomitised	

DOLOMITES

High purity (≥ 97.0% carbonate)

Low purity (≥ 85.0-97.0% carbonate)

MUDSTONES

Mudstones with subordinate limestone

Mudstones with calcareous sandstone and sandstone

IGNEOUS ROCKS

Undivided (basalt, dolerite, tuff)

POCKET DEPOSITS

Sands and clays

INSET MAPS

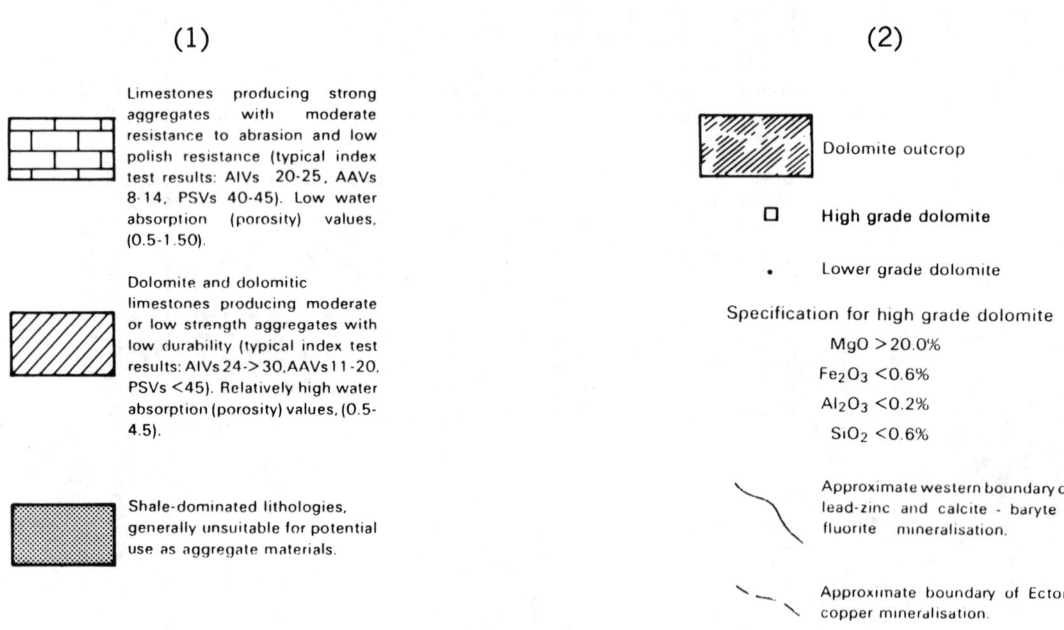

*Ornamentation has been used here in place of the colouring used on the original.

EXPLANATION OF GEOLOGICAL SYMBOLS

CARBONIFEROUS LIMESTONE (DINANTIAN)

Derbyshire Shelf Province

LsM	Longstone Mudstones
EyL	Eyam Limestones
K	Knoll-reef limestones
Mo	Monsal Dale Limestones
Mo/dk	Monsal Dale Limestones (dark facies)
BLL	Bee Low Limestones
Rap	Apron-reef limestones
WDL	Woo Dale Limestones

Off-Shelf Province and Staffordshire Shelf Province

OnS	Onecote Sandstones
MX	Mixon Limestone-Shales
WdF	Widmerpool Formation
L(WdF)	Limestone in Widmerpool Formation
EcL	Ecton Limestones
Hp	Hopedale Limestones
K	Knoll-reef limestones
Kv	Kevin Limestones
Mi	Milldale Limestones
Mi/dk	Milldale Limestones (dark facies)

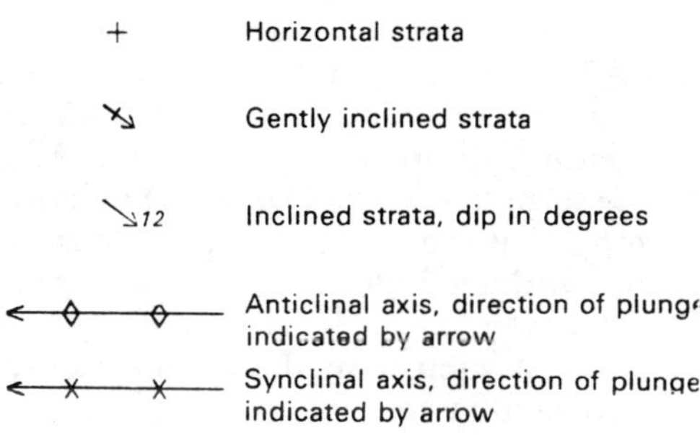

Faults and Mineral veins

Geological boundary, solid

Geological boundary, drift

Boundary between categories (for boundaries not coincident with geological boundaries)

Inferred boundary between categories of limestones

Landslip

Made ground

Drift, undifferentiated

OUTLINE OF DINANTIAN GEOLOGY AND SUMMARY OF RESOURCES

FORMATION	LITHOLOGY	THICKNESS	RESOURCES
Derbyshire Shelf Province (Buxton - Castleton - Bakewell - Wirksworth area)			
Longstone Mudstones	Mainly mudstones with minor argillaceous limestones.	40m (Little Longstone) [191 716], 14m (Matlock) [300 600], 6-9m (Crich) [353 544].	Mudstones and impure limestones. Unlikely to be potentially useful for aggregate.
Eyam Limestones	Thin-bedded, dark grey cherty limestones and knoll-reefs of massive and well-bedded, pale grey fossiliferous limestone.	Marked local variation largely dependent on presence or absence of reef facies. Bedded facies normally 10-25m thick, reef facies up to 40m thick.	Knoll-reefs of high purity or very high purity limestones. Bedded-facies limestones of medium to low purity, or are impure (dependent on chert content). Suitable for use as aggregate materials.
Monsal Dale Limestones	Lithologically variable limestones. Mainly pale to medium grey, thin to thick-bedded with sporadic cherts. A dark facies, which is principally developed in the Miller's Dale - Ashford area, consists of dark, thin - bedded cherty limestones with mudstone partings. The formation contains several volcanic members consisting of basaltic lava and tuff. They make up a large proportion of the total thickness in the eastern half of the Province. Dolomitised in the Monyash - Matlock - Brassington area	143m (Cressbrook Dale) [173 731], increasing to 200m around Ashford [193 695] and to about 375m at Longstone Edge [205 731]. Around 100m thick in the Buxton area [066 735] and 50m thick in the Cromford - Wirksworth area [294 569 - 285 542].	Variable chemical purity. Mainly high or medium/low purity limestones, although some beds are classified as very high purity limestone. The dark facies limestones are mostly classified as impure. The dolomites and dolomitised limestones are generally of low purity. Limestone aggregates suitable for most constructional purposes (except road-surfacing); dolomite aggregates are generally less strong and durable.

Fig. 6. Extract from the resources map key, limestones of the Peak, Derbyshire and Staffordshire, UK. Inset: Keys to inset maps: (1) aggregate resources of the carboniferous limestone, and (2) distribution of high-grade dolomite, and limits of mineralization (after Harrison and Adlam, 1985).

and the aim was to evaluate potential, providing information on both quality and quantity.

The main product is a 1:50 000 coloured resource map. The limestones are classified on their $CaCO_3$ content—this parameter relates closely to potential end use. Blue colour-coding is used to differentiate four grades (Fig. 6) of purity, and blue colour-banding is used for areas of combined grade; blue- and green-banding denotes areas of partly dolomitized limestones. Green colour-coding denotes two purity grades of dolomite. Several smaller maps support the main map in the text, denoting geological structure, distribution of high grade limestone and dolomite, and of aggregate quality material. The various limestone formations are differentiated on the resource map by stratigraphic (letter) symbols, and likewise on the supporting maps. Detailed description and analyses are given in the supporting text. There is a set of small maps which show the distribution of significant chemical contaminants (SiO_2, Al_2O_3, Fe, MgO, MnO, K_2O, Na_2O, P_2O_5, F, Pb, Zn). A short section of the text covers waste rocks in the limestone and their effect on resource potential. Hydrology and hydrogeology are not extensively covered, though there is a small map of groundwater levels in the limestone. There is glossary of terms and a bibliography.

This is an excellent resource potential study, and it illustrates how complex the parameters may be that are required for defining a quite simple rock mineral resource. It also illustrates how careful map design allows a large number of parameters to be shown on a single map sheet. Again, a separate study of the constraints aspects would be required for land use planners to perform their task, but the resource potential study is so large, separation of the two aspects into two studies is a practical necessity.

Planning constraints and their influence on quarrying of magnesian limestone (Clark *et al.*, 1987)

Magnesian limestone is a commodity of national importance, quarried for centuries for a range of uses. Exploitation began in Roman times, and in the 18th century industrial use (agricultural, iron, steel, paper) expanded, while in this century constructional use has dominated and it has supplied 10% of the nation's limestone requirements. The uses are given in Table VIII.

This study was commissioned by the Department of the Environment and carried out by Rendel, Palmer and Tritton. It reflects the greater degree of planning control being exerted to preserve the environment and safeguard the resource. Quarries are developed along the 230 km outcrop from Newcastle to Nottingham, and up to now their development has been controlled by presence of the limestone resource, its quality and market requirements, and constraints on quarrying due to physical and planning restrictions. The aim of the study was to review existing information on geology and mineral resources, and to assess planning and physical factors that have determined development along the outcrop.

Topics covered throughout the outcrop length were geology, mineral resources and market use, past and present mineral exploitation. Constraints on mineral extraction were covered by a case study of a small part of the outcrop length in County Durham.

The physical constraints are summarized in Table IX and the planning constraints in Table X. A summary of the nine sets of maps is given in Table XI. Figure 7 shows the legend for the constraints map.

This study is very factual and states the existing situation. Resource potential and constraints aspects are evenly balanced. The point is made at the end that though the many constraints add up to a rather bleak picture regarding future exploitation, the Magnesian Limestone Escarpment Local Plan for Durham of 1982 does not imply that the constraints present an absolute bar to future working. Indeed, Thrislington Quarry was permitted to extend operations into a Site of Special Scientific Interest subject to the Nature Conservancy Council transplanting the unique flora to another site over five years.

A valuable function of this study was to define high grade areas and the local plan protects such areas for future working.

The report lists 400 disused quarries which present

Table VIII Current uses of Magnesian Limestone (after Clark *et al.*, 1987)

Market use	*Location of main active workings*
(1) LOW-GRADE MATERIAL (LIMESTONE AND DOLOMITE)	
Construction materials	
Roadstone	
Sub base	Throughout outcrop
Coated wearing course	Durham (Raisby Quarry) S. Yorkshire (Holme Hall)
Concrete aggregate	Throughout outcrop
Hardcore	Throughout outcrop
Building stone	Mainly N. Yorkshire, Notts
Armourstone/rockfill	Mainly N. Yorkshire, Notts
Agricultural lime	Throughout outcrop
(2) HIGH-GRADE MATERIAL (DOLOMITE)	
Ferrous metallurgical flux	Durham (Thrislington Quarry) Derbyshire (Whitwell Quarry)
Sinter	Durham (Thrislington Quarry)
Refractories	Durham (Raisby Quarry)
Glass Manufacture	S. Yorkshire (Warmsworth Quarry)
Magnesia production	Durham (Thrislington Quarry)

Table IX Physical constraints on Magnesian Limestone working (after Clark *et al.*, 1987)

Constraint	*Location*	*Potential problem*
Physical		
Presence of thick overburden		
Soil (mainly glacial drift)	Mainly east and central County Durham	Costly stripping; contamination of top of limestone if fractured
Rock	Within and east of outcrop	Costly and difficult stripping
Presence of solution features	Local occurrence throughout outcrop	Affects safety of operations and quality of worked material
Heavily fractured and faulted zones	Predominantly Yorkshire counties	Problems of face stability and design. Unsuitable for building stone or armourstone/rockfill use
Presence of deep weathered zones	Areas to be confirmed by study	Costly stripping
Presence of clay or marl layers	Local occurrence throughout outcrop	Slope stability problems. Contamination by mixing
Areas of high groundwater level	Areas to be confirmed by study	Costly dewatering. Problems of groundwater contamination
Variable material test properties	Throughout outcrop	Failure to pass current material specifications
Economic		
Uneconomic transport distances to markets	Areas to be confirmed by study	High haulage/transport costs

Table X Planning constraints on Magnesian Limestone Extraction (after Clark *et al.*, 1987)

Areas of sterilization (areas already worked)

Formal constraints

Buffer zones (protecting existing urban development etc.)
High-grade agricultural land
Special landscape areas
Sites of special scientific interest
Recreational areas
Aquifer protection zones

Informal Constraints

Areas occupied by non-mineral development (such as residential areas of towns and villages, roads, railways, rivers, etc.)
Refusals of planning permission
Protection of areas of identified high-grade mineral reserves from use as a source of low-grade materials for construction purposes

Table XI Magnesian Limestone study—summary of the maps

The outcrop of 230 km was split into seven areas based on convenient map size. The maps, at a scale of 1 : 50 000, show:

(1) Detrital geology
(2) Isopachytes of limestone (total limestone or Upper and Lower Divisions of Magnesian Limestone, as appropriate to the northern and southern parts of the outcrop respectively)
(3) Isopachytes of overburden (3 areas only)
(4) Contours of the base of the Permian (4 areas only)
(5) Groundwater contours
(6) Quarry locations and status
(7) Positive and intermediate planning status areas, consultation areas, and outstanding planning permissions. Key shows:
  (a) planning permissions for extraction and existing mineral workings
  (b) known resources indicated in structure plans or minerals local plans
  (c) consultation areas (mineral planning authority to be notified of supplies)
(8) Negative planning status areas and informal constraints (see Table X for the list of these). The key shows:
  (a) areas covered by one constraint
  (b) areas covered by two constraints
  (c) areas covered by three constraints
  (d) Grade 1 and Grade 2 agricultural land
(9) Areas of sterilization of resources

both a hazard and waste disposal resource (though waste disposal in permeable rocks such as limestone requires very careful monitoring for groundwater pollution).

Constructional minerals

Constructional minerals are among the most important minerals from a land use planner's viewpoint. They are all subject to near-surface extraction, and they are extracted in immense tonnages. Their extraction raises numerous conflicts with other land uses, both existing agricultural activity and settlement and other planned developments. Particular problems arise with thin deposits such as many sands and gravels, since large takes of land are required to produce adequate tonnages. Large tonnages of these minerals are an essential requirement of any major urban, transportational or energy development project.

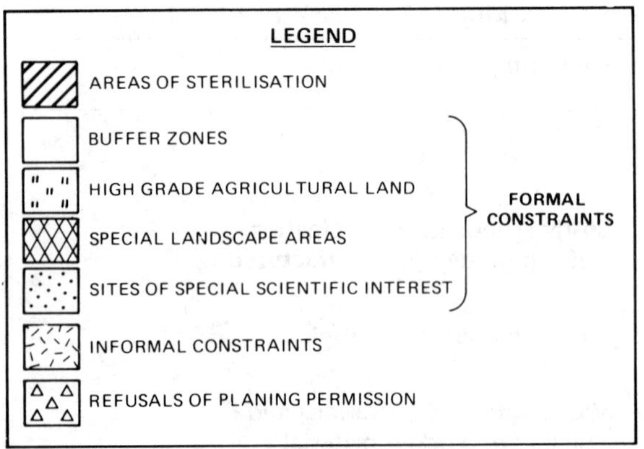

Fig. 7. Planning constraints affecting magnesian limestone extraction in county Durham, UK. Extract from map key. Reproduced by permission of the Geological Society from 'Planning constraints and their influence on quarrying of Magnesian limestone' by Clark *et al* in *Planning and Engineering Geology*, 1987.

The seven case studies cited below illustrate the following:

(1) *Northeast Illinois*: a large area, regional (reconnaissance) resource potential study embracing a wide variety of potential mineral sources in glaciated terrain.
(2) *Cricklade, Wiltshire, England*: a moderately large area regional resource potential study of river terrace gravel deposits.
(3) *Wreake Valley, Leicestershire, England*: a small area, detailed study of river terrace gravel deposits, covering both resource potential and constraints aspects.
(4) *Shrewsbury, Shropshire, England*: a small area, detailed study of glacial gravel deposits, covering both resource potential and constraints aspects.
(5) *River Arroux, France*: a site specific study dealing entirely with environmental constraints.
(6) *Norway (various areas)*: a regional appraisal of sand and gravel resource potential closely integrated with considerations of concrete performance.
(7) *Caerphilly, Wales*: a regional mapping study of the potential of hard rock resources.

Sand and gravel and peat resources in NE Illinois (Masters, 1978)

The Chicago Metropolitan Area makes heavy demands on sand and gravel, and the aim was to describe the deposits to aid the extractive industry, land owners and land use planning agencies.

Published and unpublished reports and maps, records of engineering test wells and other borings, modern soil maps, geological survey field notes and pit descriptions, and deductions based on the geological history of the area were all incorporated. The main product was a 1 : 125 000 resource map utilizing stipple and hatching to denote the varied glaciofluvial and glaciolacustrine deposits (bouldery, cobbly and pebbly outwash plains, fans and deltas, kames and eskers, beach ridges, spits, bars and deltas, sand dunes and peat bogs).

The classification into divisions numbered I–VII (Fig. 8) progresses from the thickest, coarsest, and most widespread to the finest and most restricted. Supporting text describes each subdivision, gives production figures and location of past and present extraction, and specifications for various uses.

This is an example of a regional study covering a large area (about 1500 km^2). Although it is necessarily generalized, it does relate geology to quality, but beyond general predictions of thickness and sizing relationships, lacks detail and quantification. It is valuable as a regional resource map, but would need to be followed up by site specific resource potential studies. It lacks any coverage of constraints.

Sand and gravel resources of the Thames Valley, the country around Cricklade, Wiltshire (Robson, 1975)

This example represents over 100 studies commissioned by the Department of the Environment to this format. The case study selected here deals with river terrace sand and gravel deposits of the Upper Thames and its tributaries. The area covered was quite large—220 km^2. The main product was a resource sheet on a scale of 1 : 125 000, in which various shades of pink colour were used to denote mineral deposits.

On the Cricklade map (Fig. 9) four river terraces of Pleistocene age were denoted by symbols, the solid colour being used to denote mineral (overburden less than 1 m thick), and colour stipple being used to denote areas known to contain minerals but not assessed.

Borehole sites are marked on the map and there is a panel with stipple ornamentation set against each hole to denote the mean particle size distribution (fines, sand, gravel). Boundaries of resource blocks erected in the course of this survey are shown, and these blocks are described in the supporting text with inferred resource estimates, thickness of mineral and overburden, and grading details. Borehole logs are given in full at the end of the text. Appendices include the statistical basis for the resource estimates.

This is a regional resource potential evaluation, intended as a basis for site specific investigation of potential extraction sites. It covered a much larger area than either the recent Wreake Valley study or the recent Shrewsbury study, and much less concentration of boreholes was possible. However, it did not include any consideration of constraints on extraction of the mineral.

Research programme to assess the potentially workable sand and gravel resources in the Wreake Valley, Leicestershire (Anon., 1985b)

This study for the Department of the Environment relied on 221 previous boring records, 650 new shallow auger holes, geomorphological study and stereoscopic air photography to aid in the delineation of resource boundaries and areas of sterilization. The area covered was only 39 km^2. The study was fairly intensive and fourteen resource blocks were delineated and evaluated within Pleistocene to Recent glacial sands and river terraces. Block by block descriptions and bound-in maps of the blocks were accompanied by particle size distribution diagrams, and details of lithology of clasts, together with details of grading. Maps on a scale of 1 : 25 000 cover location and extent of mineral resources, contours of mineral thickness, overburden ratios and classification of barren areas, and four schematic geological cross-sections. Stipple denoted areas of primary sterilization.

Type	Subtype	Description
		Type I—Bouldery outwash plains
I		Areas known or believed to be underlain by thick (more than 20 feet [6.6 meters] and often greater than 60 feet [19.7 meters]) sheetlike deposits containing abundant very coarse sand, gravel, and boulders with less than 10 feet (3.3 meters) of overburden. Deposition occurred as outwash plains and deltas.
	I_1, I_2	Areas the same as type I but much thinner (less than 20 feet [6.6 meters]) and with less gravel and fewer boulders are designated subtype I_1. Subtype I_2 areas are similar to those of I but have more than 10 feet (3.3 meters) of overburden and have more variable textures.
		Type II—Cobbly outwash plains
II		Areas known or believed to be underlain by sheetlike deposits containing abundant sand, gravel, and cobbles. Deposits are thinner (20 to 50 feet [6.6 to 16.4 meters]) and finer than type I.
	II_1, II_2	Very thin (less than 20 feet [6.6 meters]) deposits of this type are designated subtype II_1. Deposition occurred as outwash plains and deltas. Subtype II_2 areas are similar to those of II but have more than 10 feet (3.3 meters) of overburden and usually have more variable textures.
		Type III—Pebbly outwash plains, fans, deltas
III		Areas known or believed to be underlain by sheetlike deposits of silt, sand and some gravel more than 20 feet (6.6 meters) thick.
	III_1	Some areas may contain largely silt or sand-size material. They are designated subtype III_1 where they are very thin (less than 20 feet [6.6 meters]). Deposition occurred as outwash fans, plains, and deltas, sometimes grading into valley trains.
		Type IV—Kames, eskers
IV		Areas of hills and ridges (kames and eskers) known or believed to be composed wholly or partly of sand and gravel, probably greater than 20 feet (6.6 meters) thick.
	IV_1	Where they become very thin (less than 20 feet [6.6 meters]), they are designated subtype IV_1. Deposition occured as ice contact deposits.
		Type V—Valley trains
V		Areas known or believed to be underlain by elongate, well sorted, evenly bedded deposits of silt, sand and gravel, probably more than 20 feet (6.6 meters) thick. Different areas may have different size material.
	V_1	Very thin (less than 20 feet [6.6 meters]) deposits of this type are designated subtype V_1. Deposition occurred as valley-trains (sometimes gradational from outwash plains) and occasionally from recent streams.
		Type VI—Beach ridges, bars, spits, deltas
VI		Areas known or believed to be underlain by lacustrine deposits of fine to coarse-grained sand, often in the form of elongate beach ridges and bars, or sometimes of sheetlike deltas probably greater than 20 feet (6.6 meters) thick.
	VI_1	Very thin (less than 20 feet [6.6 meters] thick) deposits of this type are designated subtype VI_1.
		Type VII—Sand dunes
VII		Areas known or believed to be underlain by wind-blown deposits of mainly fine-grained sand greater than 20 feet (6.6 meters) thick.
	VII_1	Very thin (less than 20 feet [6.6 meters]) deposits of this type are designated subtype VII_1.
		Peat
		Areas that are known or believed to contain peat. They extend over more than 40 acres and are usually 2 to 10 feet (0.7 to 3.3 meters) thick; a few might contain more than 20 feet (6.6 meters) of peat.
		Areas known or believed to be devoid of large sand and gravel or peat deposits, although localized accumulations may be present.
SM / M		Areas of past or present excavations (SM) plus areas of fill (M) that usually involve sand and gravel or bedrock resource areas.

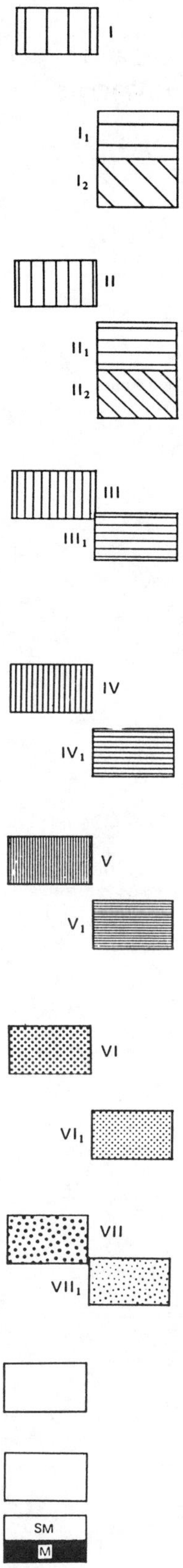

Note: Where two types of deposits are given, separated by a dash (I-I_2), both occur in that area.

Note: Where two types of deposits are given, above and below a line (Pt/II), the deposit changes with depth from the upper to the lower type.

Fig. 8. Extract from the key to the resource map: sand and gravel and peat resources in Northeast Illinois. Figure adapted from plate 1, Masters, 1978.

EXPLANATION OF SYMBOLS AND ABBREVIATIONS

DRIFT

RECENT AND PLEISTOCENE

~	**Alluvium** -sands silts, peat and clay, with lenses of gravel
1A / 1B	**First Terrace (Northmoor)** -gravels and sands of the Thames and its tributaries
	Second Terrace (Summertown-Radley) -gravels and sands of the Thames and its tributaries
	Third Terrace (Wolvercote) -gravels and sands of the Thames and its tributaries
	Fourth Terrace (Hanborough) -gravels and sands of the Thames and its tributaries
	Head -pebbly clay, solifluction material
	Glacial Drift, undifferentiated -pebbly clay

SOLID

JURASSIC

Cr	**Corallian Beds** -oolitic limestone, sandstone, clay and rubbly coral rock
OxC	**Oxford Clay** -grey or bluish-grey clay
KlB	**Kellaways Beds** -grey and bluish-grey clay with an impersistant bed of sand near the top
Cb	**Cornbrash** -shelly oolitic limestones and sandy mudstones
Fmb	**Forest Marble** -oolitic limestones, mudstones and sandstones
GO	**Great Oolite** -white or greyish-cream oolitic limestone
AO	**Athelstan Oolite** -cream or white oolitic limestone

BOUNDARY LINES

- - - - - Geological boundary, Drift

— — — Geological boundary, Solid

Fault at surface, crossmark indicates downthrow side

Inferred boundary between recognised categories of deposits

Resource Block boundary

Broken lines denote uncertainty

Worked-out Ground (sand and gravel)

BORHOLE DATA

SITE LOCATIONS

⊙ Mineral Assessment Unit (M.A.U.) boreholes

⊙ Other boreholes

M.A.U. BOREHOLES

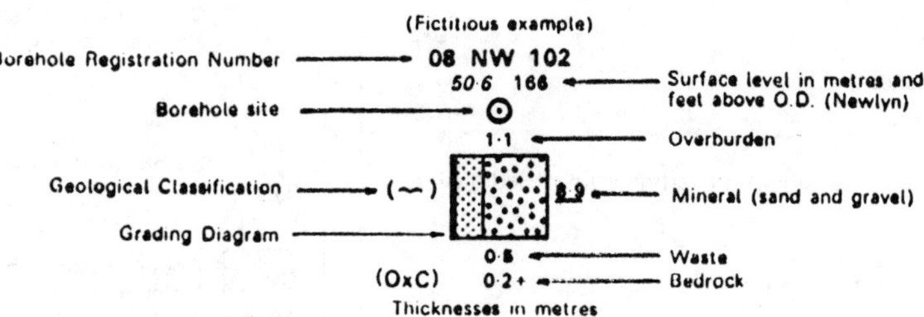

Note.
(i) Figures underlined denote thickness used in the assessment of resources
(ii) The + sign indicates that the base of the deposit was not reached
(iii) The figures in italics are the metric conversions of measurements recorded in feet
(iv) The Geological Classification is given only for mineral and bedrock

Borehole Registration Number

Each M.A.U. borehole is identified by a Registration Number e.g. 19 NE 14. The first number and letters refer to the quarter sheet, and the second number to the I.G.S. serial number for that quarter. The unique designation for borehole 19 NE 14 is SU 19 NE 14

Grading Diagrams

Each grading diagram shows the mean particle size distribution of a distinct deposit of mineral

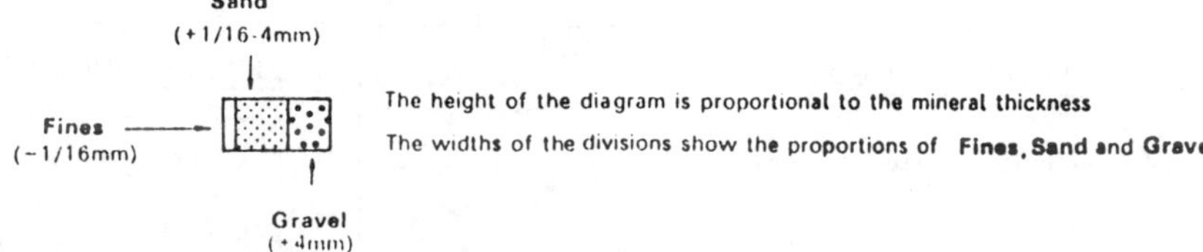

OTHER BOREHOLES

The layout of information is the same as for M.A.U. boreholes, although data available may not be as comprehensive. They are registered in the same series, except for records in the Hydrogeological Department, for example 252/53 signifies Hydrogeological Department borehole 53 on New Series One-Inch Geological Sheet 252. The final depth of deep boreholes is quoted in metres above or below O.D.

EXPOSURES AND HAND AUGER DATA

Information from the inspection of quarries, stream sections and ditch sections, and information obtained by hand auger is shown in the same way as for boreholes, but it is located by an asterisk thus *

CATEGORIES OF DEPOSITS

Exposed mineral. (Overburden generally less than 1m)

Sand and gravel not assessed

Sand and gravel either not potentially workable, or absent

RESOURCE BLOCKS

For the purpose of assessment the mineral-bearing land is divided into Resource Blocks (see Report). Each is designated by a letter

Fig. 9. Extract from the key to the resource map, sand and gravel resources of the Thames Valley, Cricklade area, UK (after Robson, 1975).

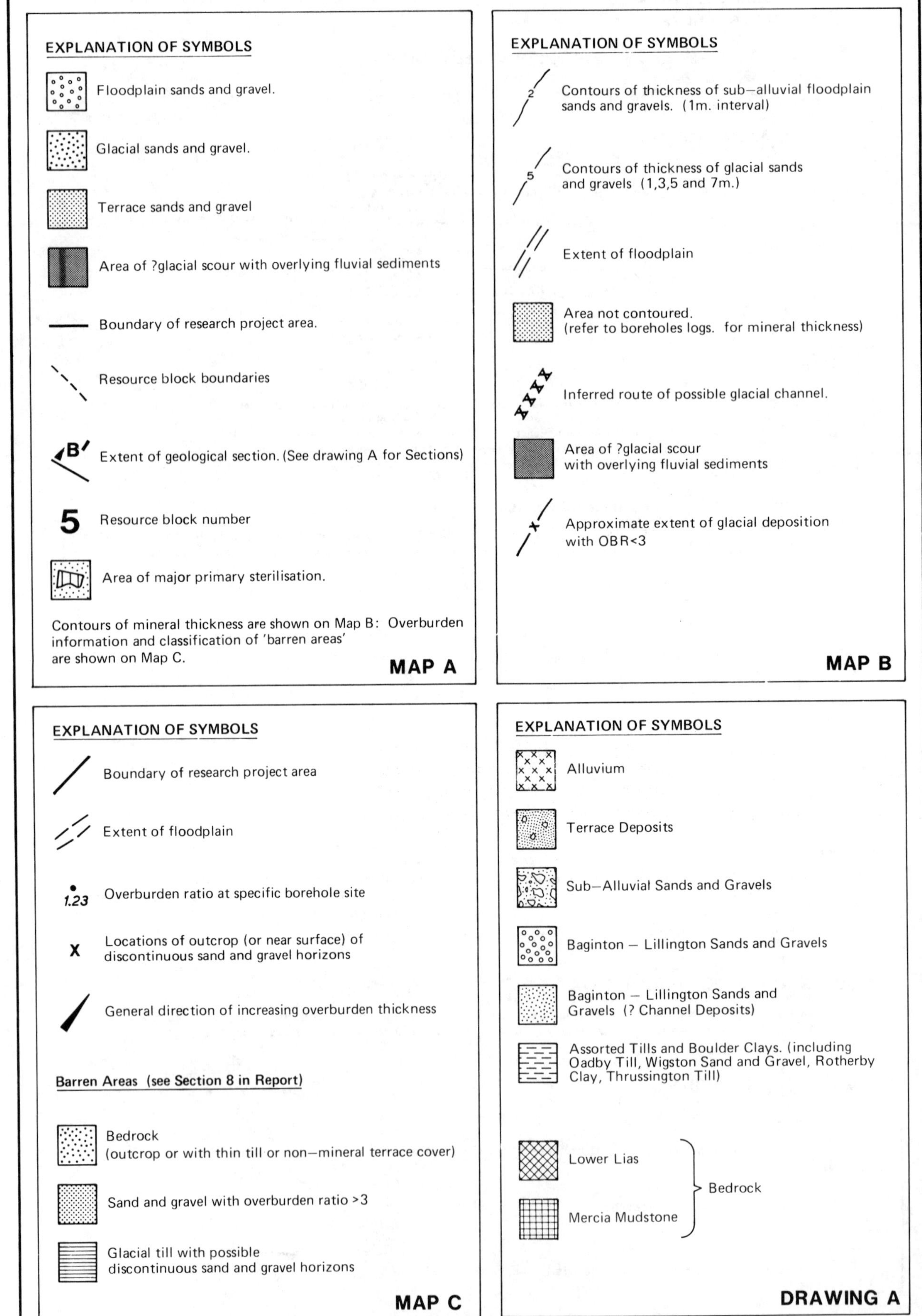

Fig. 10. Illustration of the key design of three maps 1:25 000 scale showing **(a)** location and extent of mineral resource; **(b)** contours of mineral thickness; and **(c)** overburden ratios and classification of barren areas: and **(d)** four schematic geological cross-sections (after Anon, 1985b).

In this report, primary sterilization is taken to cover areas where it is more or less irreversible: commercial and residential development; rivers, roads and railways; and areas worked out previously. Such areas were deducted from resource block estimates. Secondary sterilization was taken to cover areas that might conceivably be exploited in the future: dense woodland, allotments, recreational areas. Air photography used dated from 1969, but updating of the sterilized areas was effected by field survey. The maps bound into the text show the areas of primary and secondary sterilization in each resource block (Fig. 10 illustrates the map key design used in this report).

Appendices cover field procedures, statistical procedures and methods, and borehole logs accompanied by particle size distribution diagrams.

The simple map design is very effective. There is limited cover of the constraints aspect in the supporting text, and archaeological sites, SSSIs, dedicated woodlands and nature reserves are not delineated as such. Hydrology and hydrogeology appear to be underemphasized. Any sand and gravel working would be likely to influence ground and surface water, and will also require large volumes of water for processing, so water aspects should normally be fully covered.

Assesment of sand and gravel resources to the south of Shrewsbury (Crimes, 1985)

Like the Wreake Valley study, this study was focussed on a quite small area—in this case an area of only 36 km^2. The Pleistocene and Recent geology is dominated by glacial and fluvioglacial deposits (sand, gravel, boulder clay, stony morainic drift). Geological remapping was required and a geomorphological survey was also carried out, the landforms being described and their likely interpretation and resource potential evaluated for each grid square.

Forty eight shell and auger holes were directed at selected sites of promise, indicated by field mapping and geomorphology. They were thus neither random nor systematically arranged on a grid. Sampling (25 kg sand, 50 kg gravel) was at 1 m intervals downhole and at significant lithological changes. A full laboratory testing programme used the descriptive categories shown in Fig. 11.

Six resource blocks were outlined and summarized in text, with resource estimates and graphic representation of particle size distribution. The study involved a thorough data search and hydrology/hydrogeology were considered in some detail; existing gravel extraction had affected public borehole supplies and riparian and fishing rights.

The planning constraints, such as aquifer protection zones, sites of special scientific interest, nature conservancy prime sites, free preservation orders, scheduled areas of outstanding natural beauty, and archaeological sites were delineated on an accompanying map, though no separation into primary and secondary, or informal and formal sterilization constraints was attempted.

Deleterious minerals were searched for and, as coal was present in the gravels, both derived locally from coal measures underlying them and brought in with the ice movement from the north, this included coal particles.

Appendices cover the glaciation history in some detail; resistivity and seismic surveys (pilot surveys to test the methods which met with mixed results), and statistical methods. Calculation of reserves was based on isopachytes and on averaging borehole results within a block. Resource blocks were designed with ease of working in mind.

Maps on 1 : 10 000 scale covered: location of boreholes, auger holes, quarry sections and trenches; distribution of sands, gravels and fines as revealed in the boreholes; drift; boreholes, wells and pipelines with related hydro-

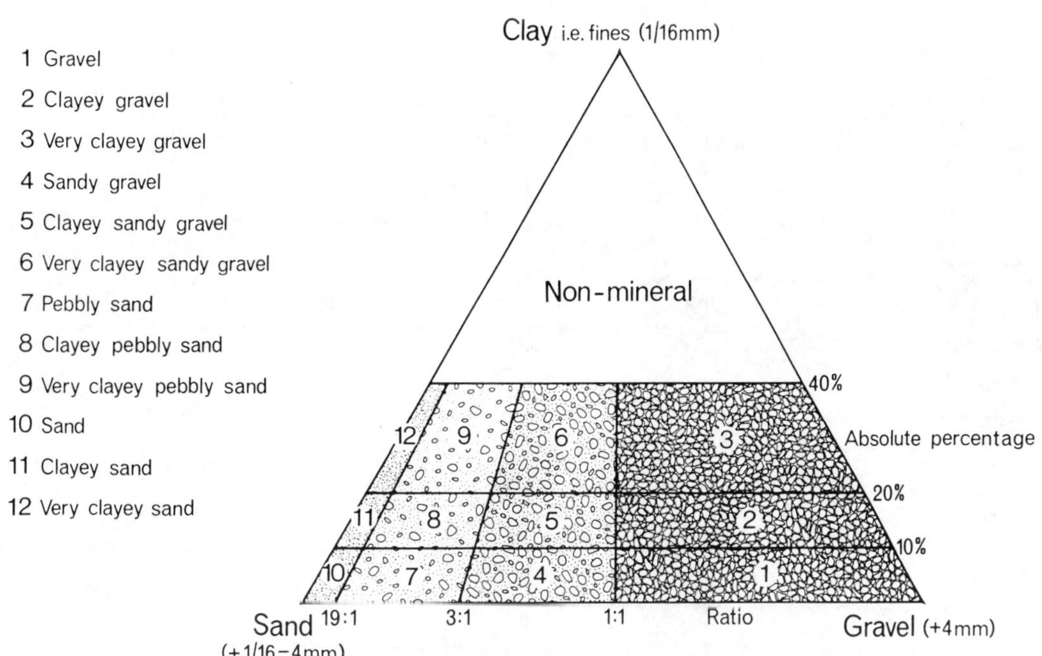

Fig. 11. Diagram showing the descriptive categories used in the classification of sand and gravel in the Shrewsbury area, UK (after Cannell, 1982; and Crimes *et al.*, 1985).

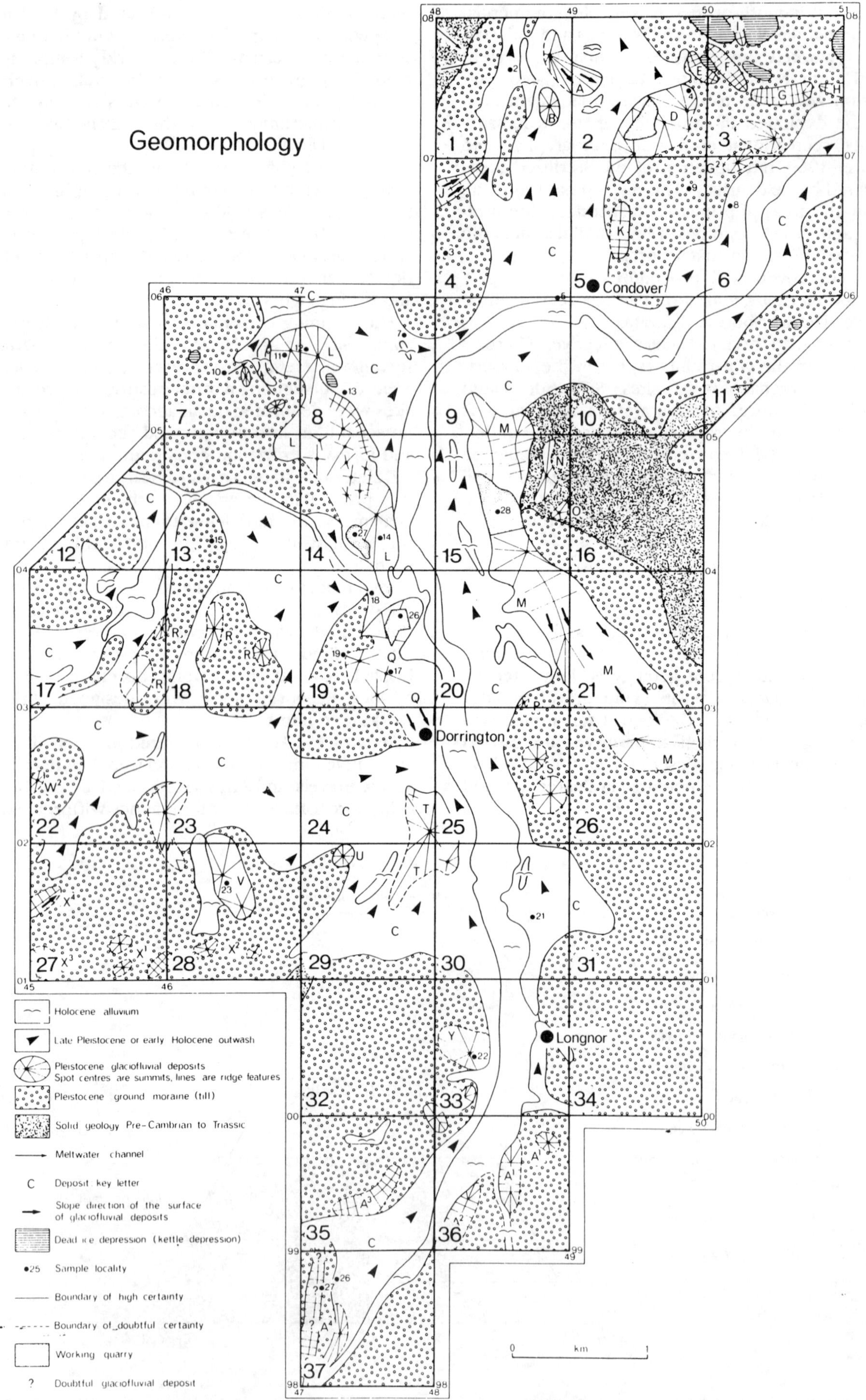

Fig. 12. Glacial geology map, Shrewsbury area (from Crimes *et al.*, 1985).

geological data; and hydrogeological data related to the geology of boreholes and wells. A geomorphological map was presented on a scale of 1 : 16 500 (Fig. 12). Two versions of the report were presented—a full technical report and a simplified report for non-specialists.

Much attention was paid to the design of maps and figures. Fence diagrams supported the text and a block model was also prepared (Fig. 13). The areal distribution of sand and gravel and fines as revealed in assessment boreholes is shown in Fig. 14.

This study epitomizes the modern, detailed, small area approach in which both resource potential and constraints are adequately investigated.

Example d'étude d'impact pour l'exploitation d'un gisement alluvionaire (Laviron *et al.*, 1985)

This is an environmental impact study, apparently commissioned by a mineral extraction company in the valley of the River Arroux in the Morvan Region, France. Terrace gravels are the principal aggregate resource near Gueugnon and Toulon-sur-Arroux, and were covered by a previous potential study. However, quantities are limited and they are likely to be superseded in time by crushed granites of the Morvan. The alluvials have long been exploited by the Department of Saone and Loire for constructional uses.

There is a conflict between groundwater supplies to the local community and resource extraction because these supplies are within the terrace gravels. The possible harmful effects of working were evaluated, as were the effects on the countryside, fauna, flora, human economy, agriculture and vegetation. The effects of noise, dust, vibrations and transport were also assessed. The reasons for selecting the particular area and the measures necessary to reduce or compensate for environmental damage were stated.

This site study includes a lengthy section on restoration of the land and water-table after extraction, former pits being converted to lakes or ponds bordered by tree planted bunds. The proposed stages in working are illustrated (Fig. 15).

It is an example of a site specific environmental impact study related to a single extraction project.

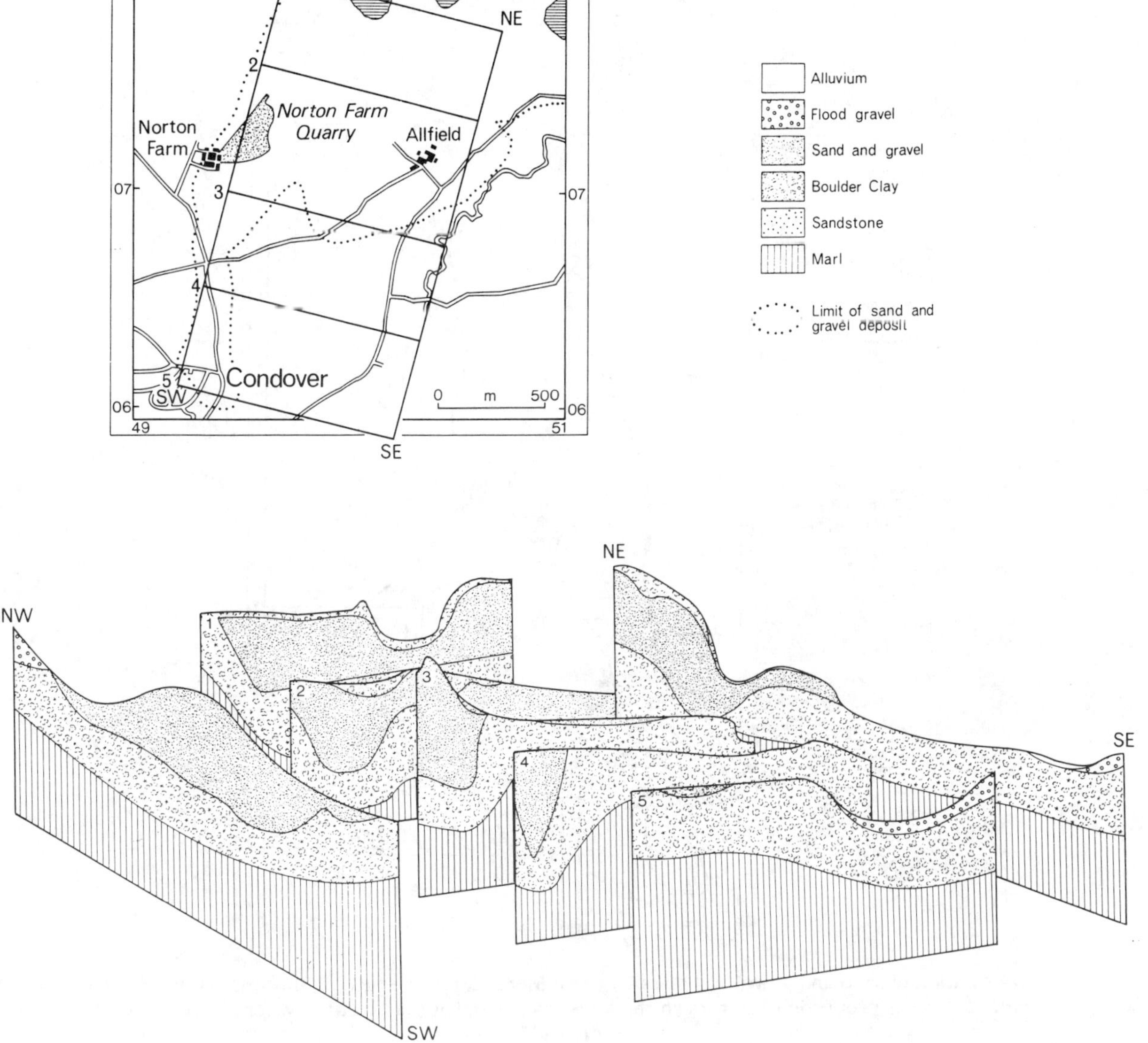

Fig. 13. Use of a three-dimensional diagram to illustrate sand and gravel occurrences south of Shrewsbury (from Crimes *et al.*, 1985).

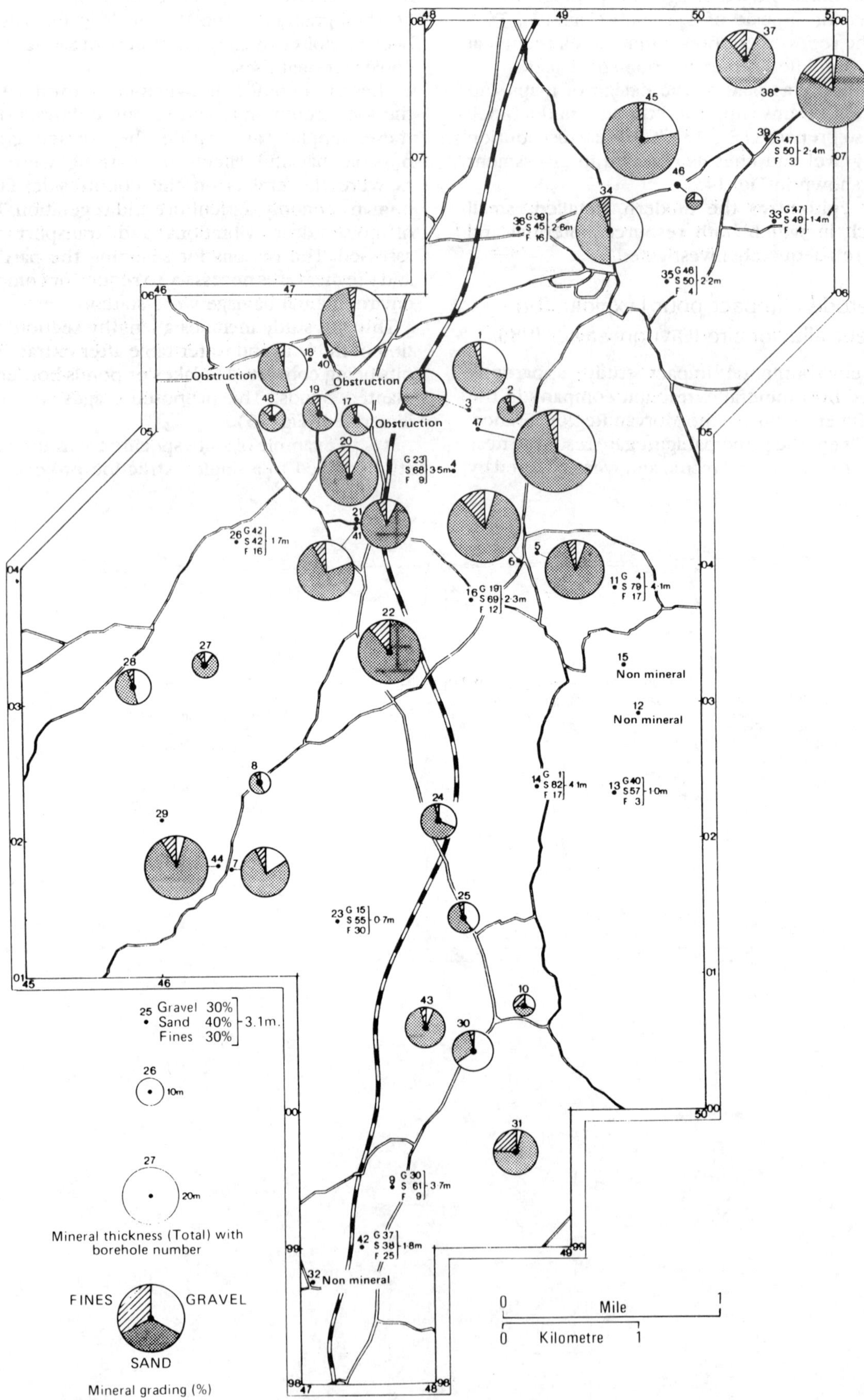

Fig. 14. Areal distribution of sand and gravel and fines, in the mineral deposits south of Shrewsbury as revealed in assessment boreholes (diameter of circle is proportional to proven thickness of mineral at that location where it exceeds 5 m) (from Crimes *et al.*, 1985).

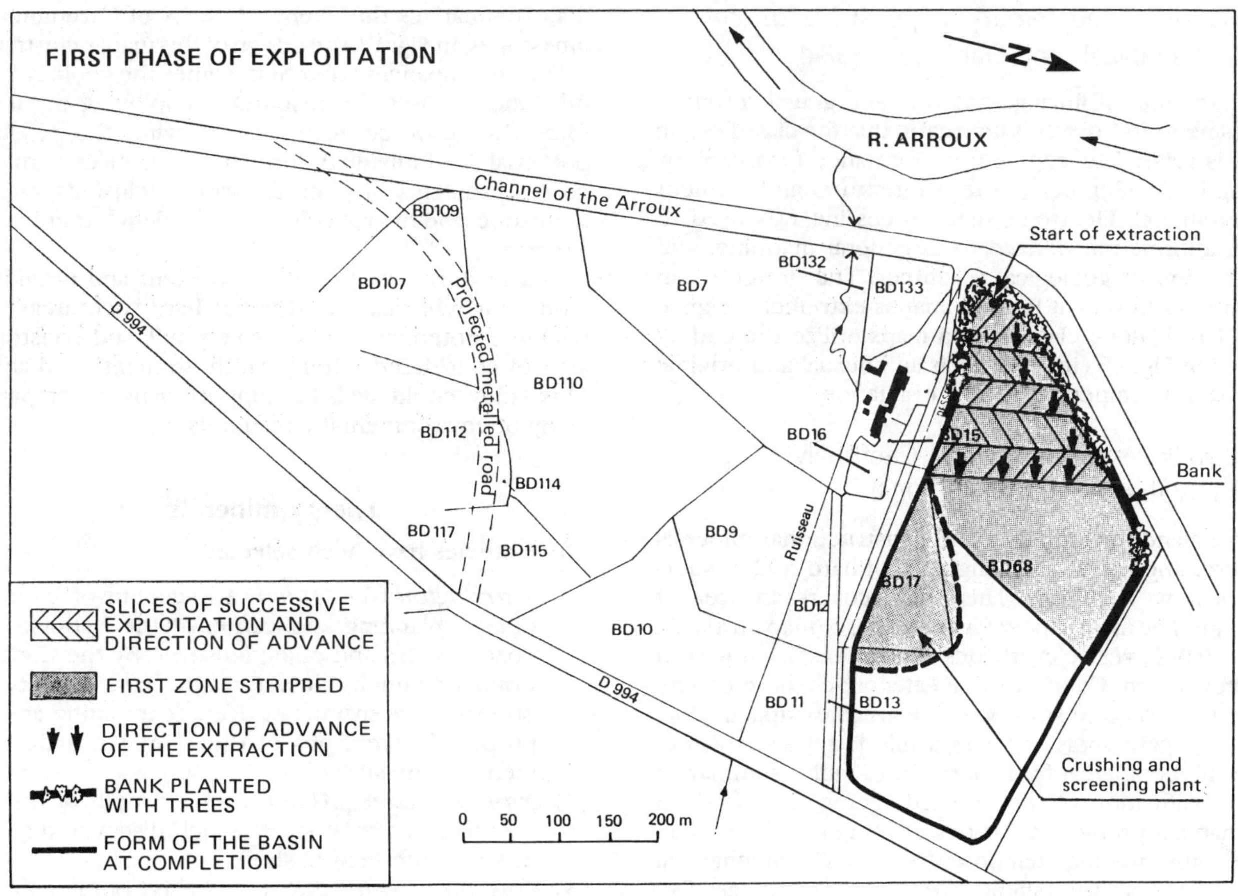

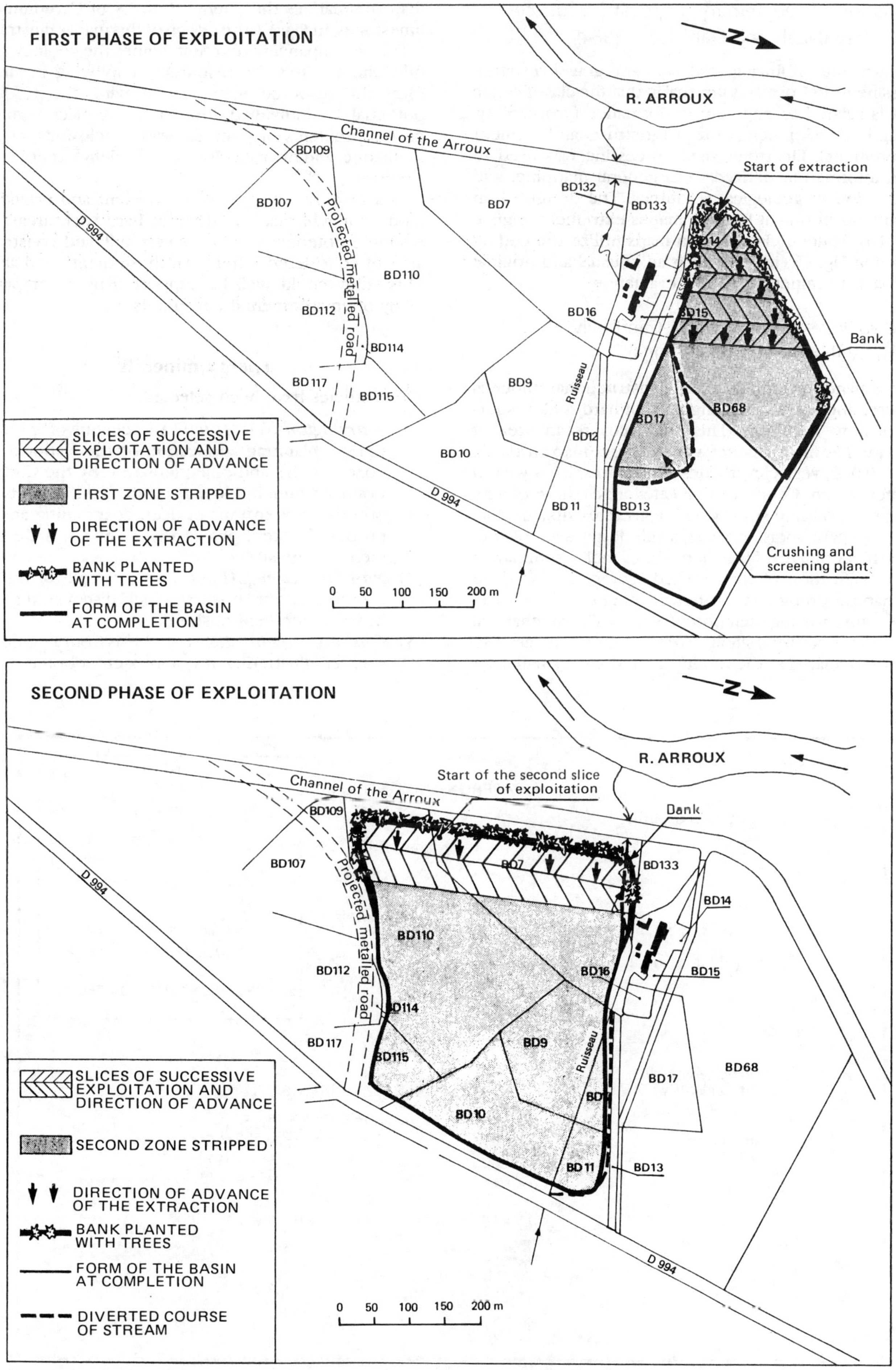

Fig. 15. Maps illustrating two phases of extraction proposed for an aggregate quarry, River Arroux, France (after Laviron *et al.*, 1985).

Evaluation of Quaternary aggregate resources in Norway (Neeb and Danielson, 1984)

A programme of fluvioglacial sand and gravel resource assessments in Norway is unusual in that the classification used is related to concrete performance (workability, strength development, internal texture and cement requirement). Electronic data processing was used to store information derived from general mapping, and registration of geological resources. The design of an 'Areal Distribution at Present' map is extremely original (Fig. 16). Mineral classification maps utilize the coding shown in Fig. 16 (inset). This is an unusual and original approach to mineral resource evaluation.

Hard rock resources around Caerphilly, South Wales (Adlam *et al.*, 1984)

The demand for industrial and constructional minerals is increasing and stocktaking even of hard rock resources is now necessary. This study covers an area of 400 km^2. The main product was a resource map on a scale of 1 : 50 000, very cleverly designed to display a wealth of information. Colour-coding categorizes three quality groups of polished stone value, aggregate impact value, and aggregate abrasion value, while letter symbols denote the geological formations. Inset tables summarize the classification of the crushed aggregates based on mechanical properties, further defining the three categories and showing their potential uses. The mechanical test data for fresh and slightly weathered samples are also given and related to the various formations. A small inset map summarizes the chemical purity of Carboniferous limestones. In Fig. 17, the design of this map is illustrated.

The accompanying text first defines the geology quite fully and assesses the resources, amplifying in tabular form the resource map and showing the resource potential of individual formations. A brief summary covers problems of jointing, waste inclusions such as mudstone, and hydrogeology (very briefly); there is also a glossary.

As a resource study this is excellent and provides a model for what can be done in hard rock areas. The resource potential study was very full, and constraints are not considered other than those mentioned above. This study could well be supported by a companion study of environmental constraints.

Energy minerals

Three studies have been selected:

(1) *Warrington New Town*: An account of investigations, a planning application by the National Coal Board (NCB), and objections to it by the Cheshire County Council, illustrating the conflict between proposed extensions of deep coal mining and the proposed extensions of the New Town development on the surface.
(2) *Rainbow Lake, Alberta*: An example of a major resource potential study which discovered a giant hydrocarbon field onshore.
(3) *Onshore oilfields, UK*: A synthesis from a number of sources illustrating the approach adopted to the

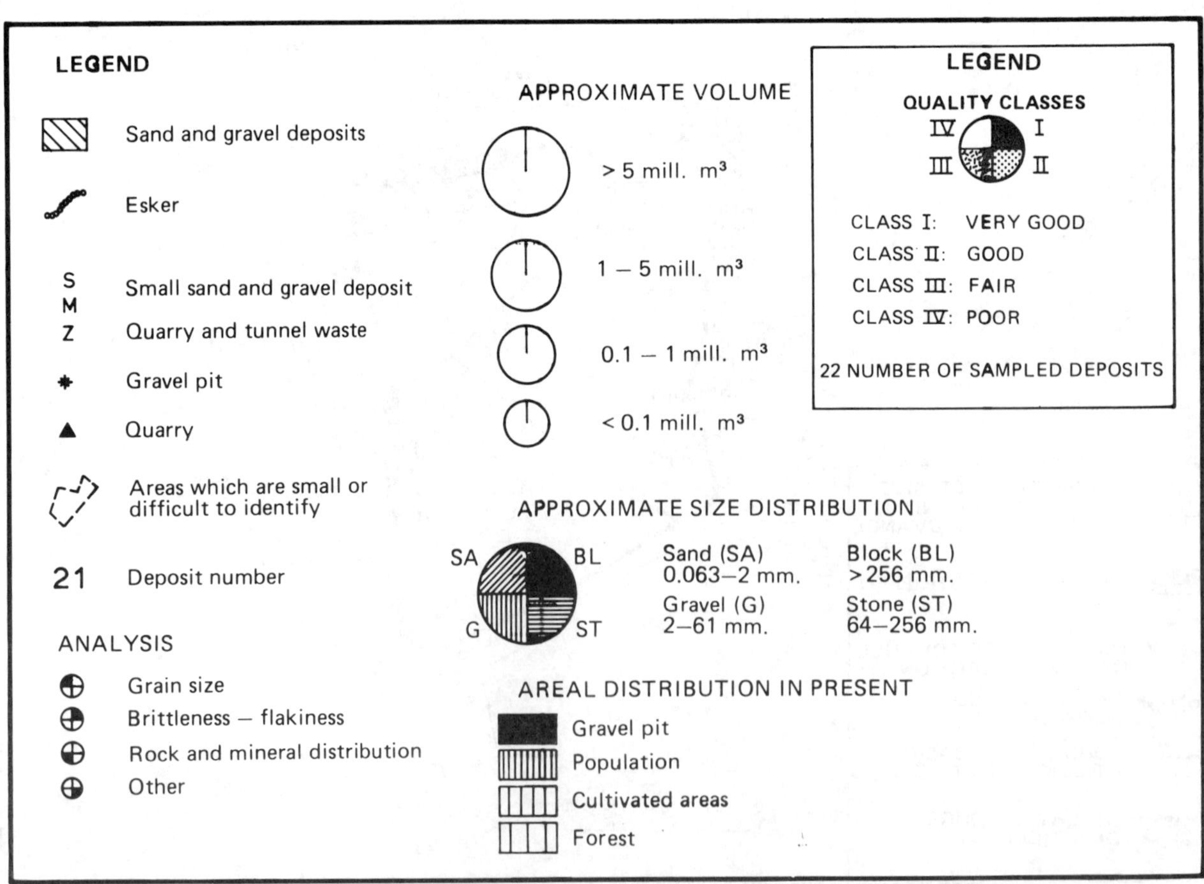

Fig. 16. Extract from the key to ADP (Areal Distribution at Present) map of sand and gravel resources in Norway (inset: key to map showing quality classes at different locations) (after Neeb and Danielson, 1984).

environmental constraints confronting hydrocarbon exploration and development.

In addition, the very important Alaska pipeline study, related to the environmental impact of an immense oil pipeline construction across wilderness terrain in Alaska and Canada is included in Chapter 9 of this volume, being essentially an environmental geology mapping (multi-thematic) study.

The impact of proposed underground coal mining on the planning of Warrington New Town/Planning consent for the extension of underground coal mining beneath Warrington New Town (Atkin, 1978)

The example of Warrington New Town is the largest illustration of a relatively short-term multi-purpose urban development, within an active subsidence area related to proposed extension of ongoing coal mining beneath it. Two collieries are situated close to the scheduled New Town development area: Bold Colliery to the northeast (sunk in 1880 and geologically well known) and Parkside Colliery to the north (opened in 1964 and then at a stage of initial development, so geologically not so well defined in its possible southward extension). The proposed extension of the New Town at Great Sankey and on the disused Burtonwood Airfield (obsolete because of mining related subsidence) was situated directly above the proposed extension of the Parkside workings. Coal production at that colliery employed 1600 people and production was 740 000 tons per annum, with reserves (gross) estimated at 60 million tons. The coal was needed to supply Fiddler's Ferry power station nearby, by direct rail link. This extension was the subject of the NCB planning application.

The NCB had devised and used successfully elsewhere in built-up areas a system of partial extraction leaving pillars of coal separating extraction panels. Provided pillars were of adequate size not to yield, and were placed vertically above one another where more than one seam was to be worked, lateral strain was reduced so much that surface damage, if any, was slight. The system is calculated to leave no build-up of strain over the pillar edges and to produce a flat-bottomed subsidence profile.

The County Council, responsible for planning the New Town of 170 000 population, had to consider the risk to the new housing developments, to large pile-founded buildings (of which several already existed), to sewers and gas installations. They engaged consultants to weigh the national interest in energy sources against the physical constraints on the existing and planned development above ground.

The extent and general effects of mining were predicted as shown in Fig. 18, and ultimately they agreed to mining under the restrictions outlined in Fig. 19.

Another paper by Wilde in the same symposium volume (Wilde, 1978) covers the design and subsidence precautions in the buildings and infrastructure to counter, as far as possible, the effects of coal mining below.

Rainbow Lake, Alberta, Canada (McQuillin *et al.*, 1980)

The Rainbow Lake investigation, covering five years from 1965 onwards, was a resource potential evaluation for hydrocarbons onshore. A then relatively new concept of 'common depth-point seismic reflection' contributed to the discovery and development of reservoirs in a series of stratigraphic traps (see Fig. 3), which yielded recoverable reserves estimated at 2.2 billion barrels of oil and 1.5 trillion cubic feet of gas.

The location of the area is shown in Fig. 20 and the geology in Fig. 21. A sequence of sediments about 7300 ft thick overlies a Precambrian basement of sediments and basic rocks. The hydrocarbon targets were in Devonian sediments about 5700 ft thick, which are unconformably succeeded by 700 ft of Mississippian (Carboniferous) marls, and they in turn are overlain unconformably by 1200 ft of Lower Cretaceous shales. Glacial deposits and muskeg cover make the area difficult to explore and the best time for exploration is during the winter freeze.

The main targets were Devonian reef carbonates of the Rainbow Member of the Keg River Formation. The laterally impersistent reefs grew up to 800 ft above the general platform of the Lower Keg River Formation. Anhydrite deposits above provided an impervious seal.

The investigation involved methodical seismic mapping and tying-in of well data to construct maps of the Keg River Formation/Rainbow Lake Member. The seismic surveying (Figs. 21 and 22) was based on the recognition on a number of reflectors, and from delineation of these stratigraphic horizons isopachyte maps were constructed showing variations in the thickness of the sequence and, hence, the subsurface geological structure.

The investigation hinged on two main factors:

(1) initial successful interpretation of old data, including the data that led to the initial discovery of oil and gas in the area by Banff-Aquitaine;
(2) the coincidence in time of this investigation and a major technical advance in seismic methods.

In five years, all the reef reservoirs shown in Fig. 22 were discovered.

This case history illustrates the main procedures involved in a major onshore investigation for hydrocarbons. Although the principal tools were geophysics and drilling (though seismic is the dominant tool, other geophysical methods are important in hydrocarbon exploration, particularly gravity and magnetics), basic geological mapping and the compilation of all existing data are also essential components of such an investigation.

Onshore oil, UK

Constraints studies (from a variety of published papers) predominate among mapping studies related to land use planning for hydrocarbon exploitation in the onshore UK. The fundamental approach to planning control was outlined in Department of the Environment Circular 2/85 (1985), entitled 'Planning Control over Oil and Gas Operation'. The philosophy stated therein is to encourage exploration and development subject to proper environmental safeguards.

Licences to explore are issued by the Department of Energy. From the environmental protection viewpoint, certain safeguards are built into these licences covering exploratory surveys (geophysical etc.) and drilling. The approach to development at county authority level is outlined in *Landward Oil and Gas* (Anon., 1986(a)). The county, in this case Dorset (wich contains the largest onshore oilfield in Britain, Wych Farm) set out its policies in this paper. Following the initial discovery,

EXPLANATION OF STRATIGRAPHICAL SYMBOLS

DRIFT

		Approximate thickness (m)		
(alluvium pattern)	Alluvium, river and estuarine			RECENT AND PLEISTOCENE
(glacial pattern)	Glacial deposits, undifferentiated			RECENT AND PLEISTOCENE

SOLID

		Approximate thickness (m)		
BLi	Blue Lias - limestones and mudstones	up to 20		JURASSIC
PnG	Penarth Group - mainly mudstones	7		TRIASSIC
MMG	Mercia Mudstone Group - mainly mudstones	113		TRIASSIC
UCM	Upper Coal (Pennant) Measures - mainly sandstones (S) with some mudstones (M)	up to 850	COAL MEASURES	CARBONIFEROUS
MCM	Middle Coal Measures - mainly mudstones and coals; some sandstones (S)	107-213	COAL MEASURES	CARBONIFEROUS
LCM	Lower Coal Measures - mainly mudstones and coals; some sandstones (S)	82-183	COAL MEASURES	CARBONIFEROUS
MG	Millstone Grit - mainly mudstones; some sandstones (S) at base	37-55	MILLSTONE GRIT SERIES	CARBONIFEROUS
ML	Main Limestone - mainly dolomite (Do); some limestone (L) at top	12-540	CARBONIFEROUS LIMESTONE SERIES	CARBONIFEROUS
LSh	Lower Limestone Shale - upper mudstone division, lower limestone/dolomite division (L)	36-107	CARBONIFEROUS LIMESTONE SERIES	CARBONIFEROUS
UORS	Upper Old Red Sandstone - mudstones; some quartz conglomerate/sandstone (QC) at base	76-122	OLD RED SANDSTONE (UPPER)	DEVONIAN
BrG	Brownstone Group-mainly sandstone	122-183	OLD RED SANDSTONE (LOWER)	DEVONIAN
SMG	St. Maughans Group-mainly mudstones; some sandstones (S), and limestones (L)	457-610	OLD RED SANDSTONE (LOWER)	DEVONIAN
RG	Raglan Marl-mainly mudstones; some sandstones (S) and limestones(L)	335-610	OLD RED SANDSTONE (LOWER)	SILURIAN
Lu	Ludlow Series-mudstones, siltstones, sandstones and limestones	c.50		SILURIAN

EXPLANATION OF ASSESSMENT COLOURS

Colours are used to indicate the aggregate potential of the rocks at surface.

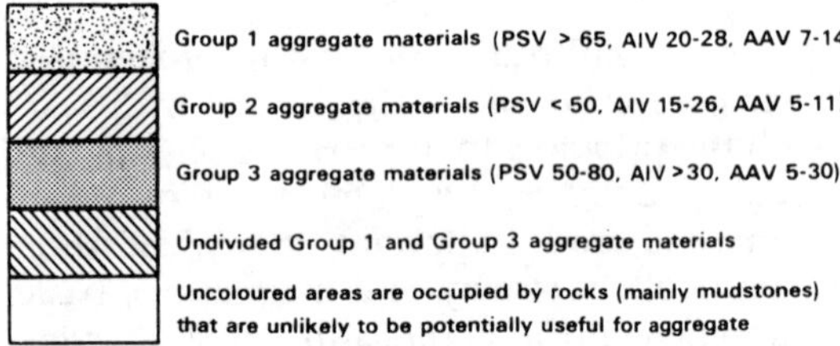

BOUNDARY LINES

Geological boundary, Solid

Geological boundary, Drift

STRUCTURAL SYMBOLS

Fault (crossmark indicates downthrow side)

Axial trace of anticline

Axial trace of syncline

15 Inclined strata, dip in degrees

SITE LOCATIONS

o Industrial Minerals Assessment Unit (IMAU) borehole

\+ Collected section

LANDSLIP

SUMMARY OF THE CHEMICAL PURITY OF THE CARBONIFEROUS LIMESTONE

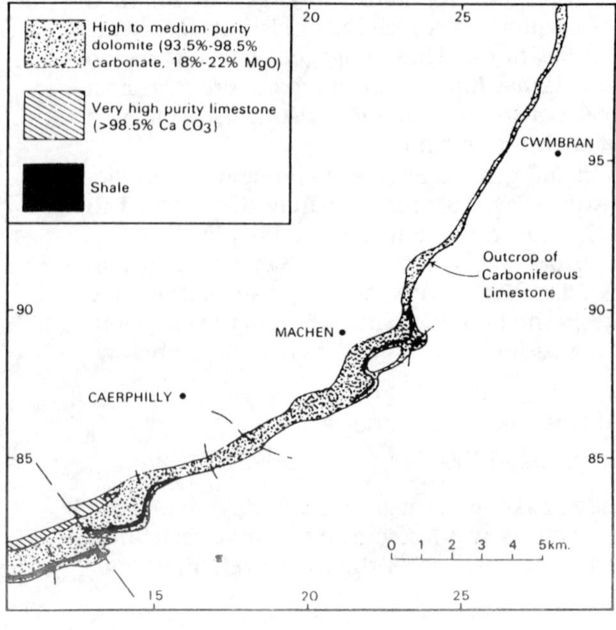

CLASSIFICATION OF CRUSHED ROCK AGGREGATES BASED ON THEIR MECHANICAL PROPERTIES

	DESCRIPTION	TYPICAL INDEX TEST RESULTS	POSSIBLE USES
GROUP 1 aggregate materials	Polish-resistant aggregate, moderate or good strength and abrasion resistance	PSV>65 AIV 20-28 AAV 7-14	Road surfacing, most other construction purposes
GROUP 2 aggregate materials	Strong aggregate, good abrasion resistance, low polish-resistance	PSV<50 AIV 15-26 AAV 5-11	Roadstone (except surfacing materials), concrete aggregate, railway ballast, armour stone etc.
GROUP 3 aggregate materials	Weak aggregate, variable resistance to abrasion and polishing	PSV 50-80 AIV>30 AAV 5-30	Unsuitable for many construction purposes, except rock fill

SUMMARY OF MECHANICAL TEST DATA OBTAINED FROM FRESH AND SLIGHTLY WEATHERED ROCK SAMPLES

ROCK TYPE	AGGREGATE IMPACT VALUE (AIV)			AGGREGATE CRUSHING VALUE (ACV)			AGGREGATE ABRASION VALUE (AAV)			POLISHED STONE VALUE (PSV)		
	Range	Mean	S.D.*	Range	Mean	S.D.	Range	Mean	S.D.	Range	Mean	S.D.
PENNANT SANDSTONE	47-16	25	4.7	36-13	21	3.9	37.5-2.7	9.6	4.5	79-64	71	3.5
LOWER AND MIDDLE COAL MEASURES SANDSTONE	44-21	30	8.1	38-16	23	6.5	12.8-3.3	7.0	2.4	80-63	69	7.7
CARBONIFEROUS LIMESTONE – LIMESTONE	26-20	22	1.8	-	-	-	10.8-7.8	9.3	1.0	-	48	-
CARBONIFEROUS LIMESTONE – DOLOMITE	32-15	20	4.3	27-13	19	3.6	16.0-4.8	7.5	2.4	52-42	47	4.3
OLD RED SANDSTONE SANDSTONE	47-21	33	7.8	34-21	27	4.9	43.7-2.6	13.0	11.8	83-55	68	14.1

*S.D. Standard Deviation

GEOLOGICAL MAP OF THE PENNANT MEASURES

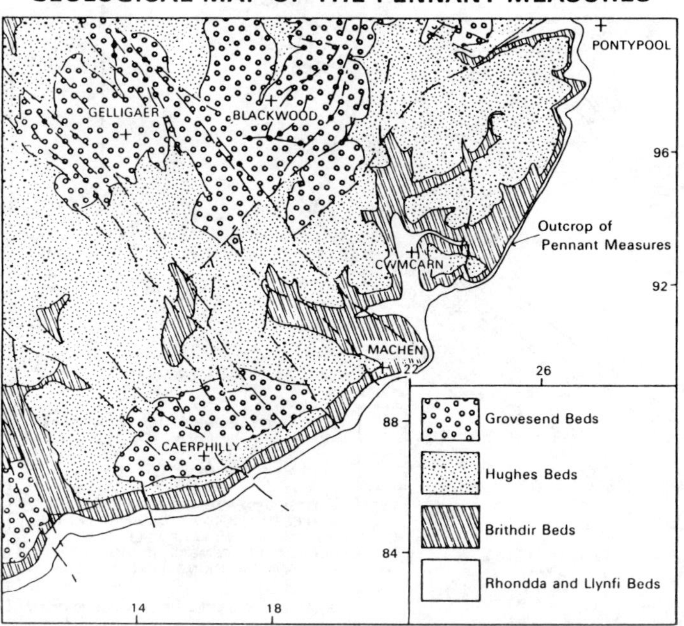

Fig. 17. Extract from resource map, Crushed Hard Rock Resources around Caerphilly, South Wales (this explanatory material was placed alongside a normal geological column showing the stratigraphical units) (after Adlam *et al.*, 1984).

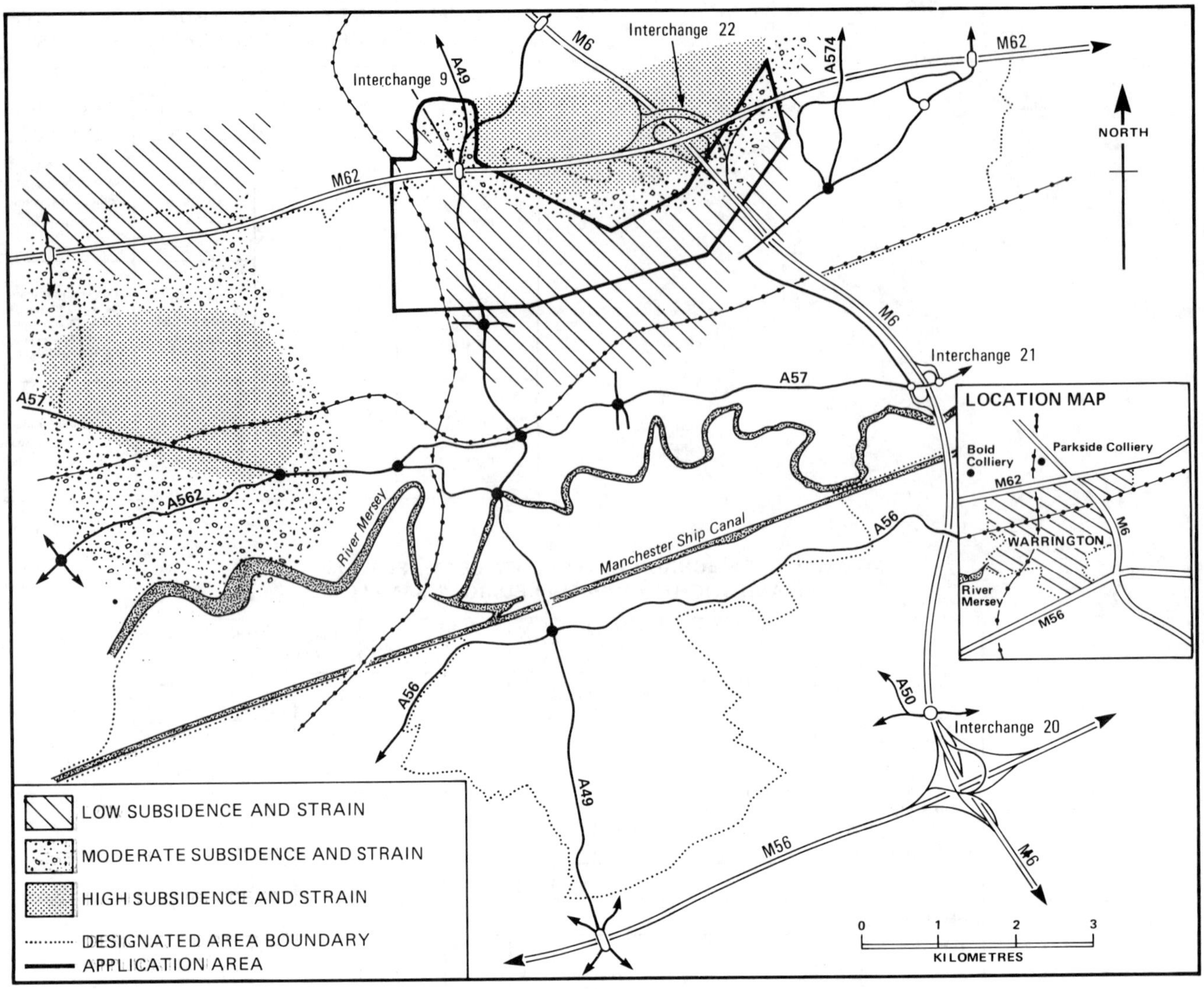

LOW SUBSIDENCE AND STRAIN
MODERATE SUBSIDENCE AND STRAIN
HIGH SUBSIDENCE AND STRAIN
DESIGNATED AREA BOUNDARY
APPLICATION AREA

Fig. 18. Extent and general effects of mining under Warrington New Town (after Walsh, 1978; Atkin 1978).

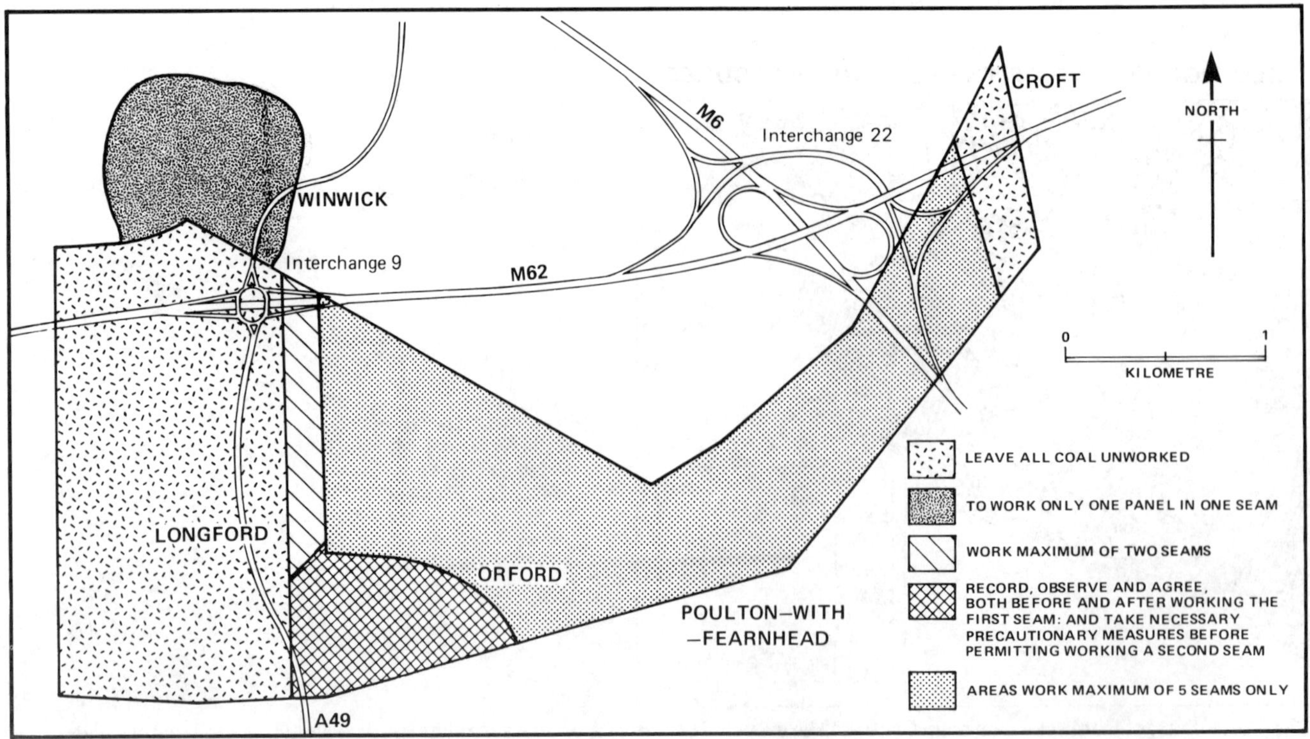

Fig. 19. Restrictions imposed on coal mining under Warrington New Town (after Atkin, 1978).

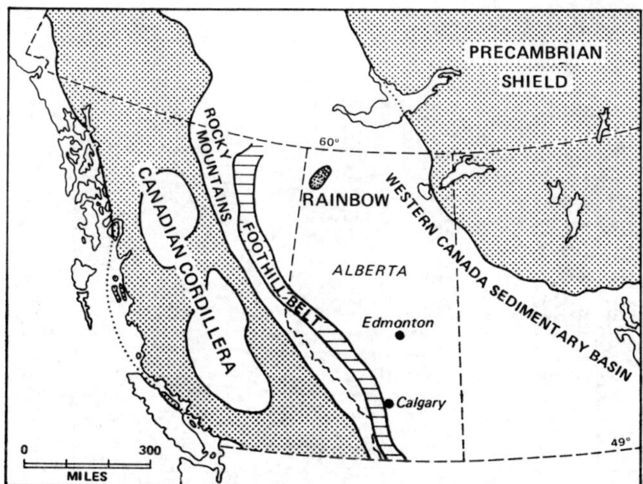

Fig. 20. Rainbow Lake Field, Canada, location diagram.

the developer must submit an overall scheme for appraisal and delineation of the field. In these early stages the strictest confidentiality is maintained. Planning applications for further drilling are processed in the light of this submission.

Major developments in open countryside must be supported by detailed justification of the local or national need for the resource, including possible alternative locations and a statement of the environmental impact. The authority will require designs for temporary drilling sites and long-term installations, minimizing environmental impact by landscaping etc. Plans for restoration on decommissioning are also a common requirement. Schemes for eliminating pollution, especially of water resources, and for minimizing effects on vegetation and wildlife will also be necessary.

It is clear from the above that the onus is on the developer to commission a number of environmental impact studies, which will involve both earth science and other scientific disciplines. So far oilfields discovered in the UK have not been areally extensive and, in general, they have required a number of areally concentrated environmental impact studies. In addition to those commissioned by the developer, there are likely to be a number of such studies commissioned by the county or national authority to aid it in formulating its plans.

The Nature Conservancy Council (NCC) has formulated guidelines (Anon., 1986(b), 1987) covering hydrocarbon exploration onshore. There are six major guidelines:

(1) thorough consultation between all involved and affected parties is essential;
(2) sensitive areas should be identified and avoided where at all possible;
(3) if this is not possible, environmental assessment procedures are recommended for such areas;
(4) site management should aim at all times to reduce to a minimum the environmental impact;
(5) sites of significant nature conservation value should be avoided;
(6) opportunities to enhance nature conservation should not be overlooked.

The NCC emphasizes that after initial consultation with various interests, environmental constraints mapping should identify potential problem areas and guide the location of production, processing and transportation facilities.

They take the view that sites of special scientific interest should and can be kept free from any development installations, for such scheduled sites number only 400 in the entire country and are rarely extensive. However, the Department of the Environment maintains that there are no absolute 'no go' areas for mineral resource exploration provided adequate safeguards are adopted.

An example of the sort of installation that is required for deep drilling for hydrocarbons (often as much as 3000 m or more) is shown in Fig. 24(a), and the same site is shown in Fig. 24(b) six months later, after restoration. This illustrates well the attention to environmental protection paid by oil explorers in the UK.

There is clearly a major field of earth science mapping related to planning and development for onshore hydrocarbon exploration. Of course, in addition, offshore discoveries require large onshore terminals and pipeline installations, and planning these also requires a substantial earth science input.

CONCLUSION

From the general discussion above and the case studies briefly cited to illustrate the relationships between mineral resources and land use planning, it is apparent that there is nowadays a very extensive earth science mapping and investigation activity directly related to land use planning and development.

The following important conclusions may be drawn.

(1) The fundamental resource is land, which may or may not contain beneath it valuable physical resources of a non-renewable nature.

(2) There will be inevitable conflicts between mineral resource extraction and other potential uses for the land, and even between two or more resources contained in the same geological formation (for example, the case of the River Arroux, Laviron *et al.*, 1985, cited above).

(3) It is essential that weight be given to resource potential and physical constraints aspects, such as hydrogeology, either by including both in a single investigation, or by commissioning coupled studies.

(4) The geology and geomorphology of a subject area should always be comprehensively summarized in maps, sections and supporting text, in a form sufficiently simple (aided by glossaries) to be understood by non-technical readers. Many important mineral resources are closely related to the superficial geology, though ore minerals etc. are more often related to bedrock geology. Both should be described, and particular attention should be paid to very full and careful analysis of glacial superficial cover, which has very complex origins.

(5) Indirect aids such as geophysics and remote sensing can only be used with discretion, and only when it can be shown that they are sufficiently accurate for the project requirements. These methods are expensive and their use should always be weighed against spending more on direct methods (drilling, sampling and testing), which generally have to be coupled with them. Remote sensing

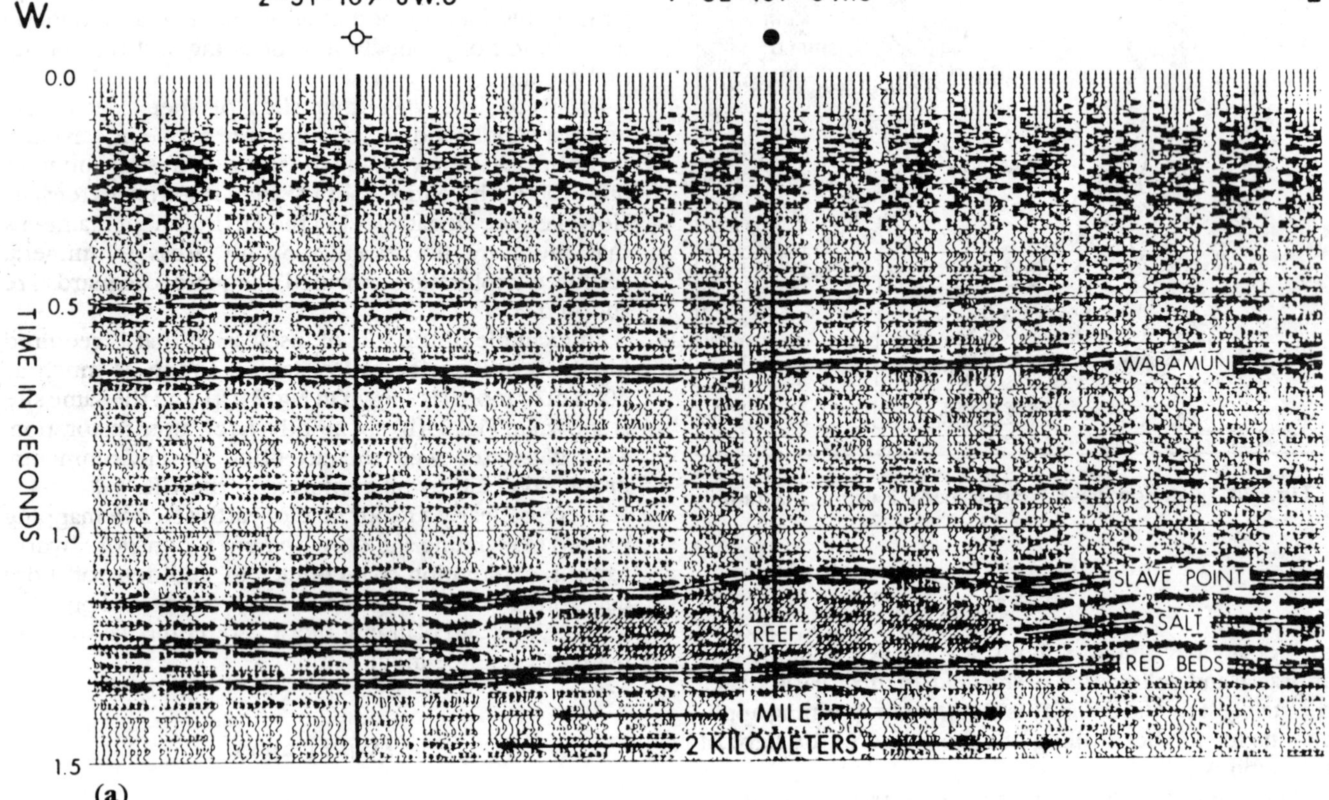

(a)

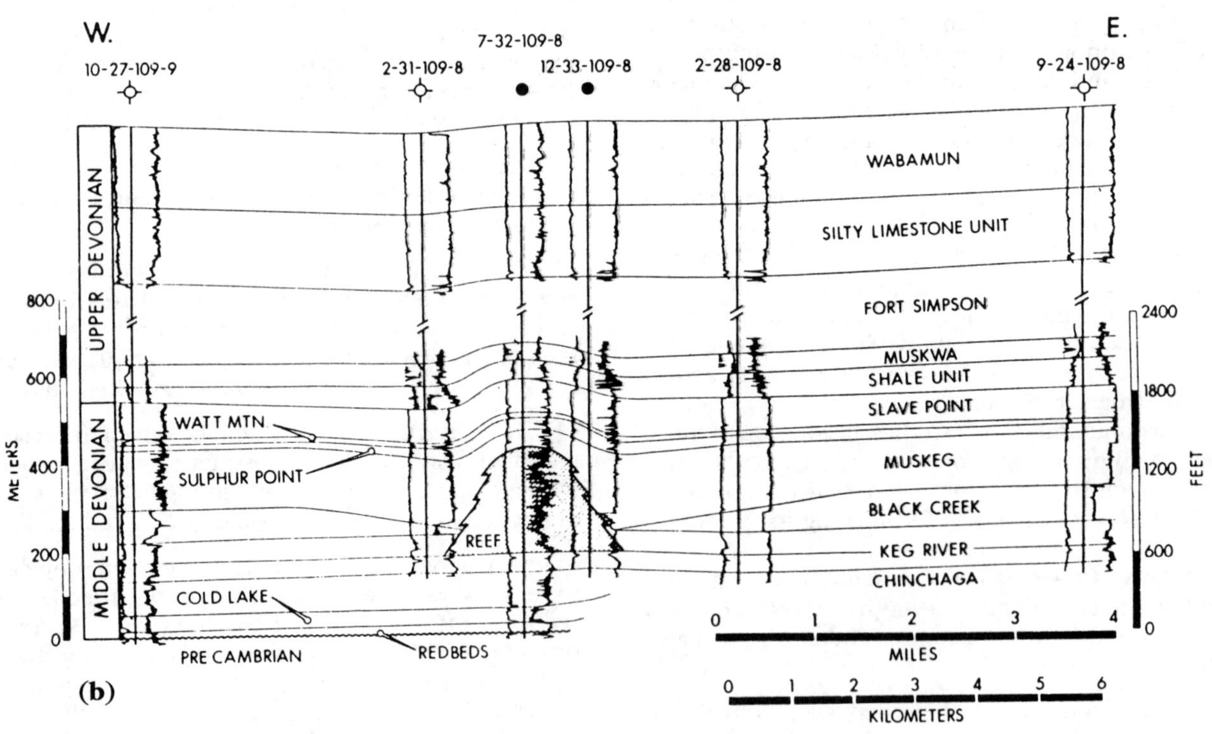

(b)

Fig. 21. Rainbow Lake Field, Canada. Seismic section across discovery pool and geological cross-section across the discovery pool (from McQuillin *et al.*, 1980).

has important applications in defining constraints and air photography is an invaluable tool both for defining the basic geology and the constraints.

(6) Intelligent and original map design has been illustrated in several of these case studies. Maps need to be coupled with sectional, fence and block diagrams to provide a three-dimensional view, and the computer is an invaluable aid to modelling in three dimensions.

(7) Computerized data processing is also of wide application to mineral resource studies of this kind.

(8) Hydrology and hydrogeology have been neglected

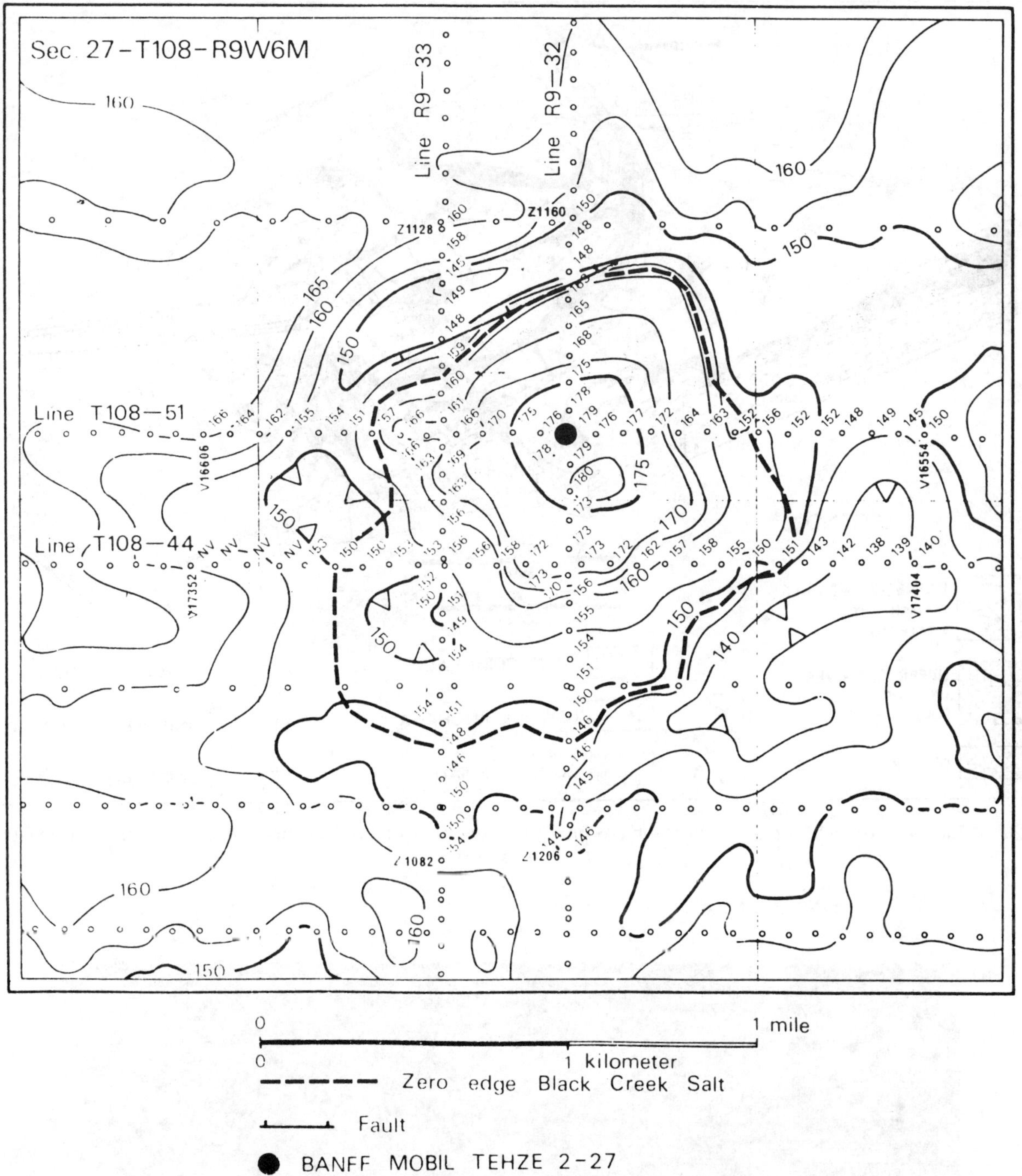

Fig. 22. Rainbow Lake Field, Canada. Map of two-way reflection time interval between Slave Point and Base Cold Lake Salt Horizons (from McQuillin *et al.*, 1980).

in some of the case studies cited above. It seems fair to comment that any mineral resource related study commissioned for land use planning purposes should fully explore the surface and groundwater systems, as any mineral development cannot fail to affect them.

(9) Data base collection is also sometimes neglected. This is of paramount importance. Every item of existing data on the subject, both published and accessible on open file, should be ferreted out and used, and archived for future easy reference.

(10) In the world at the end of the twentieth century, the population pressures, the conflicting pressures for land use, the need to protect valuable resources from sterilization and the need to protect the fragile environment, demand that any mineral resource development is subjected to rigorous planning and any other development takes into account the mineral resource constraints, as well as the legacy of previous mineral resource exploitation.

(11) Geologists rightly give great regard to the traditional regional mapping practise which is fundamental to extending their knowledge. Yet they should not be blinded to the fact that in the changing world at the end of the twentieth century, applied geology mapping and investigation is going to require more and more of the available public funds, and a major part of this will be mineral resource related mapping for land use planning purposes.

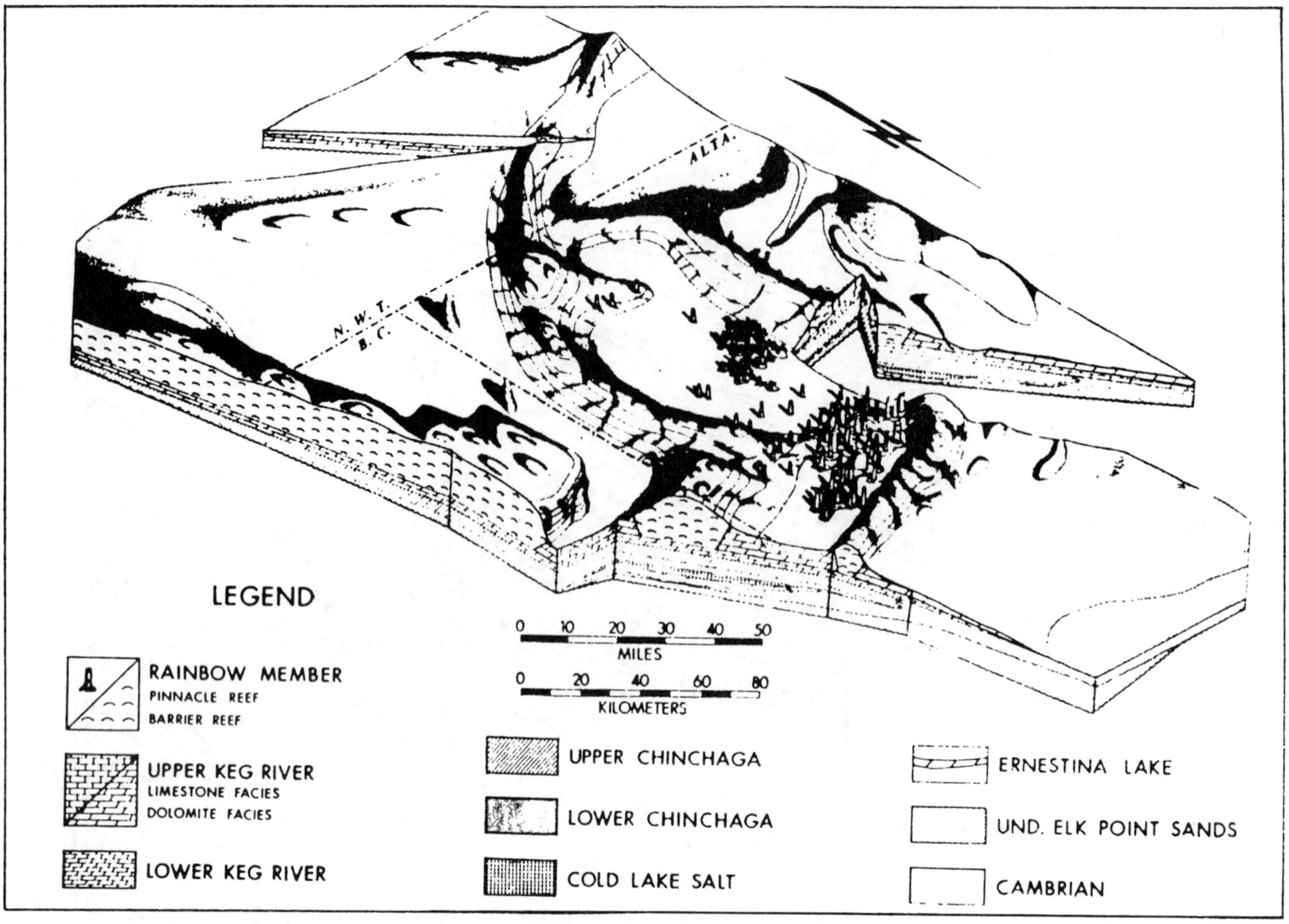

N.W.T.
B.C.
ALTA.

LEGEND

0 10 20 30 40 50
MILES
0 20 40 60 80
KILOMETERS

RAINBOW MEMBER
PINNACLE REEF
BARRIER REEF

UPPER KEG RIVER
LIMESTONE FACIES
DOLOMITE FACIES

LOWER KEG RIVER

UPPER CHINCHAGA

LOWER CHINCHAGA

COLD LAKE SALT

ERNESTINA LAKE

UND. ELK POINT SANDS

CAMBRIAN

Fig. 23. Rainbow Lake Field, Canada. Three-dimensional representation of the Keg River Reef Formation (from McQuillin *et al.*, 1980).

(a)

(b)

Fig. 24. Heath Farm onshore drilling site, Staffordshire, in the UK. (**a**) Aerial view of site with Kenting rig 31 in position. (**b**) Aerial view after the rig has been removed and the field reinstated (Shell Photographs WW55747-8af).

REFERENCES

Adlam, K.A.McL., Harrison, D.J. and Wild, J.B.L. (1984). The hard rock resources around Caerphilly, South Wales. *British Geol. Surv. Mineral Assessment Report No. 141.*

Ambrose, K., Cannell, B. and Moorlock, B.S.P. (1987). The mapping and assessment of aggregate resources in the south Midlands and Welsh Borderland. In *Planning and Engineering Geology* (Culshaw, M.G. *et al.*, eds.), Geol. Soc. London, Eng. Geol. Spec. Publ. No. 4, 347–353.

Anon. (1977). Annual Report US Secretary for the Interior.

Anon. (1982). Magnesian limestone escarpment—minerals and landscape restoration, local plan (written statement). Durham County Council, Durham.

Anon. (1985a). Research programme to assess the potentially workable sand and gravel resources in the Soar Valley, Leicestershire. *Report No. 261/UK/*0883/3, Mar 1985, Engineering Geology Ltd/Geomorphological Services Ltd, for the Department of the Environment, unpublished, 297 pp.

Anon. (1985b). Research programme to assess the potentially workable sand and gravel resources in the Wreake Valley, Leicestershire. *Report No. 300/UK/*0984/1985, Engineering Geology Ltd/Geomorphological Services Ltd, for the Department of the Environment, London, unpublished, 169 pp.

Anon. (1986a). *Landward Oil & Gas,* **1**(2), 27.

Anon. (1986b). Keep real hedges and flatten the symbolic fences between industry and conservation. *Landward Oil & Gas,* 3(1), 35–37.

Anon. (1987). *Nature Conservancy Council Guidelines for Onshore Oil and Gas Development.* Nature Conservancy Council, London (unpublished: 23.6.1987).

Atkin, D.K. (1978). Planning consent for the extension of underground coal mining beneath Warrington New Town. In *Urban Regeneration and the Inner Cities.* Royal Institution of Chartered Surveyors, Symposium, Annual Conference, Harrogate, September 26–29, Session 1.

Bates, R.L. (1969). *Geology of Industrial Minerals and Rocks.* Dover Publications, New York.

Christofferson, J.E. and King, P.M. (1988). Hemerdon, an evaluation case history. In *Extractive Industry Geology, 1985* (Scott, P.W., ed.), Geol. Soc. London. Misc. Paper No. 18, 27–39.

Clark, A.R., Morey, A.J., Gilbert, S.T. and Bacciarelli, R.E. (1987). Planning constraints and their influence on quarrying of Magnesian Limestone. In *Planning and Engineering Geology* (Culshaw, M.G. *et al.*, eds.), Geol. Soc. London, Eng. Geol. Spec. Publ. No. 4, 379–387.

Coates, D.R. (1985). *Geology and Society.* Chapman and Hall, New York and London.

Crimes, T.P., Marshall, P., Harris, A.L., McCall, G.J.H., Chester, D.K., Mussett, A.E., Wilson, C.D.V., Brenchley, P.J., Flinn. D. and Hunt, C. (1985). *Assessment of Sand and Gravel Resources to the South of Shrewsbury.* Dept. of Geol. Sci. Liverpool University, Report for the Department of the Environment, London, unpublished.

Davidson, E.S., Feth, J.H. and Spieker, A.M. (1978). Tucson, Arizona: controlling hillside development and protecting and preserving mineral and water resources. In *Nature to be Commanded* (Robinson, G.D. and Spieker, A.M., eds.), USGS Prof. Paper 950, 22–27.

Department of the Environment (1988). *Outline Study of the Magnesian Limestone in the United Kingdom.* (Rendel, Palmer and Tritton, and the University of Hull), Open File Report, DOE, London.

Dragowski, A., Kaczynski, R. and Pininska, J. (1983). The influence of an underground vein ore mine on engineering geology considerations in urbanised areas. *Bull. Int. Assoc. Eng. Geol.*, **28**, 122–128.

Freeman, Fox Ltd (1988a). *Investigations of Groundwater Characteristics in the Area Around Chacewater and St. Day in the County of Cornwall.* 4 vols, Department of the Environment, London, Open File Report.

Freeman, Fox Ltd (1988b). *Methods of Compilation, Storage and Retrieval of Data on Disused Mine Openings.* HMSO, London.

Gagen, P.J. and Gunn, J. (1988). A geomorphological approach to limestone quarry restoration. In *Geomorphology in Environmental Planning* (Hooke, J.M., ed.), John Wiley and Sons, New York.

Hamilton, C.G. (1984). Exploration methods for aggregates. *Bull. Int. Assoc. Eng. Geol.*, **29**, 89–93.

Harrison, D.J. and Adlam, K.A.McL. (1985). *Limestones of the Peak: a Guide to Limestone Resources of the Peak District, Derbyshire and Staffordshire.* British Geological Survey Mineral Assessment Report No. 144.

Holt, D.N. and Marker, B.R. (1987). Benefits of engineering geology for land use planning in areas of post-metalliferous mining. In *Planning and Engineering Geology* (Culshaw, M.G. *et al.*, eds.), Geol. Soc. London, Eng. Geol. Spec. Publ. No. 4, 75–80.

Jarvis, D. (1987). The use of computers in quarry planning & design. *M and Q Environment*, **1**(1), 10–11.

Jones, A.R. (1987). Environmental planning and mineral development: the case of the Hemerdon Mine. In *Planning and Engineering Geology* (Culshaw, M.G. *et al.*, eds.), Geol. Soc. London, Eng. Geol. Spec. Publ. No. 4, 389–395.

Laviron, F., Gand, G., Longere, P. and Cógnard, E. (1985). Example d'étude d'impact pour l'exploitation d'un gisement alluvionaire. *Bull. Liaison des Laboratoires des Ponts et Chaussees*, **12**, 271–285.

Lüttig, G. (1987). Keynote address: Geology for Environmental Planning. Proc. of the International Symposium. *Geological Mapping in the Service of Environmental Planning* (Wolff, F.C., ed.), Norges Geologiske Undersøkelse, Trondheim, 5–10.

Marker, B.R. (1984). A technical basis for regional aggregate policies. *Bull. Int. Assoc. Eng. Geol.*, **29**, 121–124.

Masters, J.M. (1978). *Sand and Gravel and Peat Resources in Northeast Illinois.* Illinois State Geol. Circular No. 503, 11 pp.

McCall, G.J.H., Chandler, P., Coote, V.J., Crimes, T.P. and Roxburgh, I.S. (1985). Environmental geology study in the Plymouth-Plymstock Area. Report prepared for the Department of the Environment, London, unpublished (open file).

McQuillin, P., Bacon, M. and Barclay, W. (1980). *An Introduction to Seismic Interpretation.* Graham and Trotman, London, 147–160.

Neeb, P.R. and Danielson, S.W. (1984). Evaluation of Quaternary Aggregate Resources in Norway. *Bull. Int. Assoc. Eng. Geol.*, **29**, 129–135.

Ove Arup and Partners (1976). *Reclamation of Derelict Land: Procedures for Locating Abandoned Mine Shafts.* DOE Report, London.

Ove Arup and Partners (1983). *A Study of Limestone Mines in the West Midlands: the Legacy of Mines Long Abandoned.* DOE Summary Report (unpublished).

Ove Arup and Partners (1985). *Study of Mining Subsidence, Ebbw Vale.* DOE Report unpublished).

Ove Arup and Partners (1986). *Colliery Spoil Disposal Options.* DOE Report, HMSO (London).

Plant, J.A., Halc, M., and Ridgway, J. (1988). Developments in regional geochemistry for mineral exploration. *Trans. Inst. Min. Met.* (Sect. B), **97**, 116–121.

Prentice, J.E. (1988). Evaluation of brick-clay reserves. *Trans. Inst. Min. Met.*, **97**, B9–B14.

Robson, P. (1975). *The Sand and Gravel Resources of the Thames Valley, the Country around Cricklade, Wiltshire.* Inst. Geol. Sci. Mineral Assessment Report No. 18, 89pp.

Walsh, T. (1978). The impact of proposed underground coal mining on the planning of Warrington New Town. In *Urban Regeneration and the Inner Cities*, Royal Institution of Chartered Surveyors, Symposium, Annual Conference, Harrogate, September 26–29, Session 1.

Wilde, P.M. (1978). The design and subsidence precautions in the buildings and infra-structure of Warrington New Town. In *Urban regeneration and the Inner Cities*, Royal Institution of Chartered Surveyors, Symposium, Annual Conference, Harrogate, Sept. 26–29, Session 1.

Wolfe, J.A. (1984). *Mineral Resources—a world review.* Chapman and Hall, New York and London.

Chapter 6

Soils

Malcolm J. Reeve

INTRODUCTION

Soil is defined (*Oxford English Dictionary*, 1976) as the 'upper layer of earth in which plants grow, consisting of disintegrated rock, usually with admixture of organic remains'. It is one of the fundamental components of terrestrial ecosystems. Soil not only supports plant growth but, through it, supports the world-wide diversity of habitat and wildlife so important to man. With the underlying rocks, soil supports our roads, railways and buildings and provides some of the constructional materials for them. It forms the surface on which most recreational activities on land take place. It absorbs excess rainfall to reduce the risk of flooding and provides a buffer zone between the potentially polluting efforts of man at the land surface and vital water supplies at depth. It is thus an important factor in planning and resource development.

Although soil can be conceived as having an upper surface delimited by the atmosphere and lateral boundaries at deep water or bare rock outcrops, its lower limit is more difficult to define. Though there is often a sharp change at shallow depth from soil to hard rock, in many areas of the world, such as in alluvium or in the humid tropics, unconsolidated soil-like materials extend to a depth of several metres and there is thus an overlap between the deeper horizons of interest to geologists and the shallower horizons which are studied by soil scientists. However, in this chapter, the concept of soil is taken to be that used currently by soil scientists (pedologists), namely that the lower limit is normally the lower limit of biological activity (Soil Survey Staff, 1975). This generally coincides with the deepest rooting of native perennial plants, commonly 1–2 m, and usually is the zone in which the original parent material has undergone the greatest change. It is important to distinguish this concept of soil from those of the layman or soils engineer. The former might only consider soil as the surface 30 cm or so that is ploughed by farmers or dug by gardeners, while the latter considers soil to be any unconsolidated material regardless of depth or mode of formation.

Earth Science Mapping for Planning, Development and Conservation
© G.J.H. McCall & B.R. Marker (Graham & Trotman, 1989), pp. 119–156.

SOIL FORMATION

The study of soils as a science (pedology) evolved from the Russian school of science led by Dokuchaiev (1846–1903). They conceived that soils were independent natural bodies each with a unique morphology expressed by a succession of distinctive layers roughly parallel to the land surface. A vertical section through the different layers of soil reflects the particular set of complex interacting processes responsible for its development.

Soil processes

In humid climates where there is a cover of vegetation, plant remains decompose and become incorporated with the mineral matter. Decomposition is rapid in base-rich, well-aerated soils and mixing is aided by faunal activity, but in strongly acid soils, where there is little mixing by fauna, decomposed plant litter accumulates on the soil surface. Decomposition is also inhibited by waterlogging or low temperatures, causing organic matter to accumulate at the surface.

Soils develop further through the accumulation of unconsolidated mineral grains and rock fragments by weathering. This physical breakdown produces progressively finer particles which are then more accessible to chemical weathering by percolating rain-water containing dissolved oxygen, carbon dioxide and acids derived from decomposing organic residues. Carbonates are gradually dissolved, micas and other minerals in sedimentary rock become hydrated, ferrous and sulphide ions are oxidized and primary silicate minerals, such as feldspar, are subject to hydrolysis, whereby basic cations (calcium, magnesium, sodium and potassium) and some silica pass into solution. 'Leaching' (Fig. 1), the extent of which is related to annual rainfall and soil texture, tends to move these soluble constituents downwards beyond the root zone or laterally through the soil by seepage, leaving relatively insoluble weathering residues such as clay minerals, quartz and hydrated ferric oxides. Some soluble constituents may be deposited lower down in the soil as coatings or nodules, occasionally cementing layers to form calcretes, gypcretes or silcretes.

In many soils dispersed clay is carried in suspension by water percolating to lower levels. The clay is redeposited as skins or coats on the surface of structural aggregates or in pores and around stones when films of water dry out.

Intense leaching and translocation can lead to 'podzolization'. In this process an acid humus layer accumulates at the surface and releases organic acids. These form complexes with iron and aluminium compounds that readily migrate downwards in percolating water. A depleted, bleached subsurface zone is thus left above layers that have been correspondingly enriched by precipitation of humus and/or iron and aluminium.

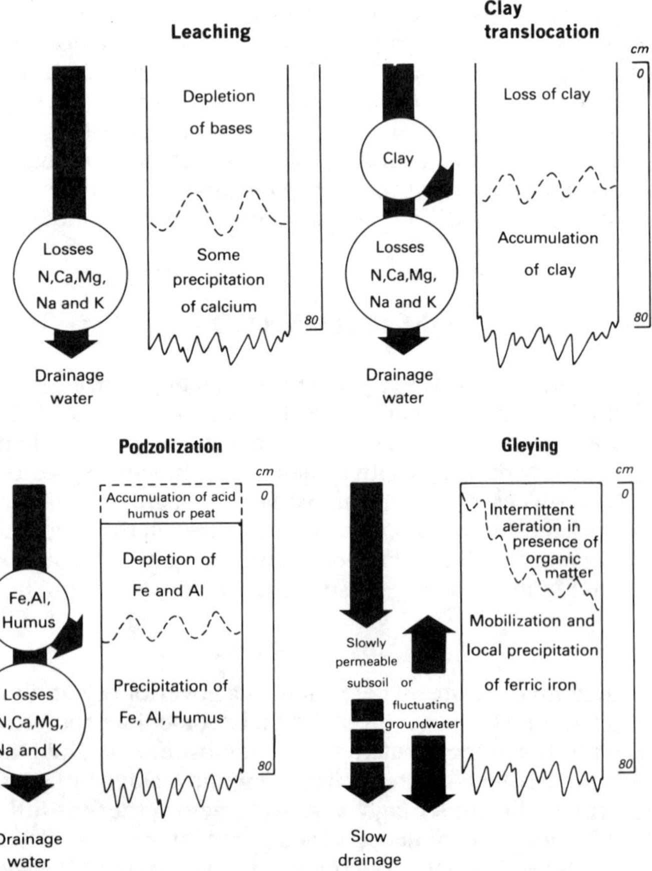

Fig. 1. Soil forming processes and the movement of water (from Ragg *et al.*, 1984).

Waterlogging of soils leads to the reduction, mobilization and removal or redeposition of iron compounds. The reduction of ferric iron to more mobile, colourless or grey ferrous iron complexes, by micro-organisms or products of decomposing organic matter, is known as 'gleying'. This results in grey layers or in distinctive grey and orange mottles where poorly and better aerated parts of a soil layer show the differential effects of gleying. Gleying might result from an impermeable subsoil impeding surface water drainage or may occur in more permeable subsoils affected by groundwater.

In hot climates soil processes are further modified. Dehydration during dry seasons results in the transformation of hydrated ferric oxides into haematite giving the soils a bright red colour ('rubification'). Groundwater at shallow depth can result in 'salinization' of soils as dissolved salts are precipitated out from water moving towards the surface in response to evaporation. In hot humid conditions podzolization effects are modified and 'laterization' produces thick red or yellow subsoil layers high in sesquioxides and low in silica.

In arid climates, chemical weathering by rain-water is an infrequent occurrence. Accumulation of organic matter and leaching are both very slow processes and the related soil features are poorly developed.

Soil horizons

The layers developed under the influence of the various soil forming processes are known as soil 'horizons'. They are distinguished from each other by variations in such properties as colour, texture, structure, chemical status and amount of organic matter. A summary of the main horizons recognized and the symbols used to represent them is given in Table I (fuller details are in Avery, 1980, and FAO-Unesco, 1974).

The recognition of different sequences of soil horizons and the understanding of how these influence utilization of soils as a world resource has formed the basis of classifications to distinguish and order the soils, and surveys to map and delineate the areal extent of soils of similar characteristics differing significantly from the characteristics of adjacent soils.

SOIL CLASSIFICATION

Soils can be classified in many ways (e.g., for their suitability for a specific crop or for their capacity to bear off-road vehicles). These are special purpose classifications that will be discussed later along with examples of applied soil maps. This section deals with pedological soil classifications, most of which are based on the presence or absence of horizons or combinations of horizons having certain characteristics which affect the general use of the soils as a resource for primary production. Though the detail of such classifications is beyond the scope of this book, it is important to understand the meaning of some of the terms used as they often feature on the legends of soil maps produced by 'general-purpose' surveys (see below).

The early pedological classifications of the Russians identified zonal, intrazonal and azonal soils (Sibirtsiev, 1901). 'Zonal' soils reflected the influence of climate and vegetation and thus could be related to major climatic regions of the world; 'intrazonal' soils occurred where local physiographic factors, such as poor drainage or specific parent materials (such as chalk), overrode the

Table I Horizon notation

Main horizon type	*Definition*	*Most common subdivisions*	*Definition*
L	Fresh plant litter with original plant structures little altered.		
F	Partly decomposed or comminuted litter remaining from earlier years; some plant structures remain visible.		
H	Well decomposed litter, often mixed with mineral matter in which original plant structures cannot be seen.		
O	Peaty horizons accumulated under wet conditions and saturated for at least 30 consecutive days in most years (or artificially drained).	Of	fibrous peat;
		Om	semi-fibrous peat;
		Oh	uncultivated humified peat;
		Op	peaty surface horizon mixed by cultivation.
A	Mineral soil horizon formed at or near the surface, and characterized by incorporation of humified organic matter intimately associated with the mineral fraction. Incorporation of organic matter is presumed to result from biological activity or artificial mixing during tillage.	Ah	uncultivated A horizon;
		Ap	cultivated A horizon;
		Ahg, Apg	gleyed A horizon.
E	Subsurface mineral horizon underlying the A horizon that is lighter in colour and contains less organic matter, sesquioxides of iron and/or clay than the horizon beneath.	Ea	pale E horizon depleted of sesquioxides;
		Eb	brown E horizon depleted of clay;
		Eg and Eag	gleyed E horizon
B	Mineral soil horizon differentiated from adjacent horizons by colour and structure. It usually underlies an A or E horizon and is characterized by illuvial concentration of silicate clay, iron, aluminium or humus. Other forms of B horizon result from alteration of the parent material by removal of carbonates; formation, liberation or residual accumulation of silicate clays or oxides.	Bf	sharply defined thin ironpan;
		Bs	ochreous coloured sesquioxide enriched B horizon;
		Bt	clay enriched B horizon;
		Bw	an altered and/or structurally reorganized B horizon not qualifying as a Bf, Bs, Bt, Bw, Bg, Bh;
		Bg	gleyed B horizon;
		Bh	B horizon containing translocated organic matter.
C	Unconsolidated or weakly consolidated mineral horizon which retains evidence of rock structure and lacks pedological properties.	Cu	an unconsolidated C horizon;
		Cr	weakly consolidated C horizon dense and/or coherent enough to prevent penetration by roots;
		Ck	a C horizon containing secondary $CaCO_3$;
		Cg	gleyed C horizon;
		Cgy	a Cg horizon contaning secondary $CaSO_4$;
		CG	intensely gleyed C horizon;
		Cm	continuously cemented material;
		Cx	compact dense fragipan.
R	Hard or very hard bedrock.		

influence of regional climate; and 'azonal' soils occurred in freshly deposited or exposed parent materials. The zonal type of soil classification persisted well into this century but has gradually been replaced by new classifications.

The basic unit employed and classified in modern soil studies is the soil 'profile'. This comprises a sequence of soil horizons, roughly parallel to the ground surface. As a soil profile is only two-dimensional, various three-dimensional units have been identified for purposes of classification and mapping. The 'pedon' (Soil Survey Staff, 1975) has been defined as a basic three-dimensional sampling unit, its area ranging from 1–10 m^2 depending on the variability of the soil. For the purposes of classification, a larger unit or 'polypedon' (Fig. 2) has been defined. This is a group of contiguous pedons that differ from the surrounding pedons to such a degree that there are different responses to management for growing

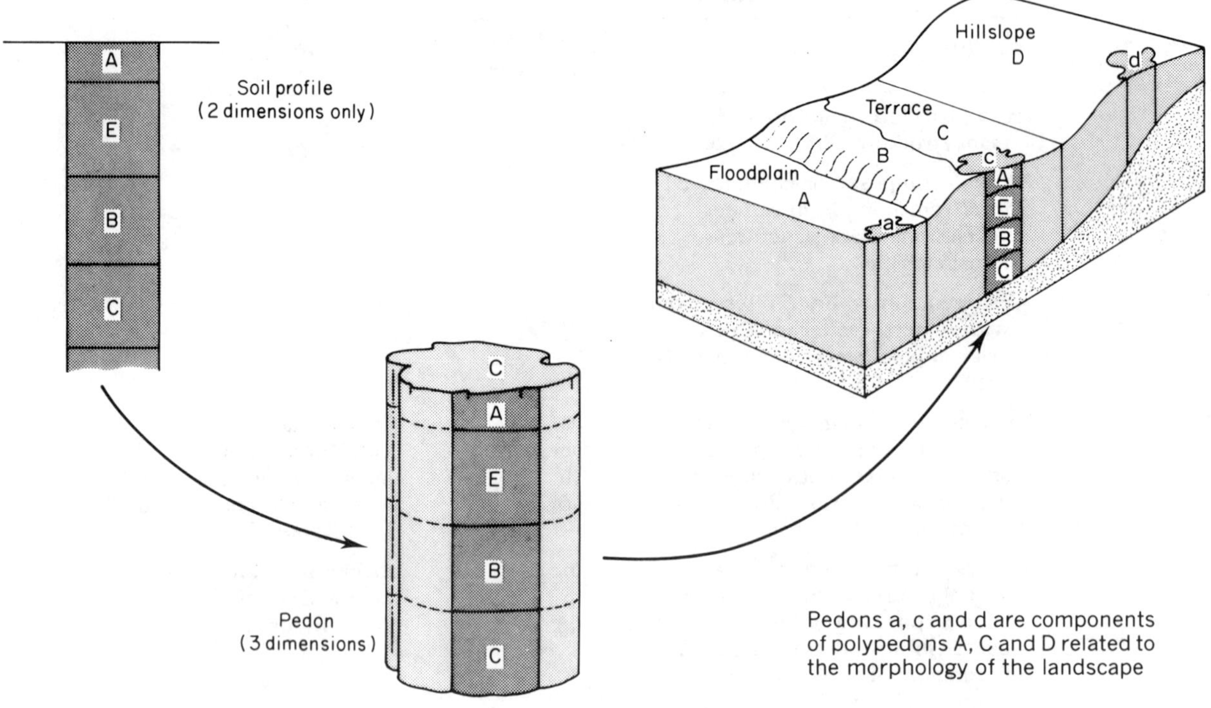

Fig. 2. The soil profile, pedon and polypedon (modified after Bridges, 1982).

plants, to engineering manipulations or other uses that require appreciable investment (Soil Survey Staff, 1975).

Many classifications have evolved to order pedons, polypedons or similar three-dimensional units into a logical system. Most advanced countries have soil research or survey organizations that have developed their own pedological classifications tailored to the soils of their own country (e.g., Avery, 1980; Commission de Pedologie et de Cartographie des Sols, 1967; de Bakker and Schelling, 1966). There are also two classifications that have achieved world-wide use: the classification developed for the 1:5 000 000 FAO-Unesco soil map of the world (FAO-Unesco, 1974), and the US Taxonomy (Soil Survey Staff, 1975). The FAO classification is based on the geographic and evolutionary background of soils and is mainly suited to small-scale soil maps. The US Taxonomy is a hierarchical taxonomy which gives major emphasis to precisely defined diagnostic subsurface horizons and can be applied to soil maps at a range of scales. The level of greatest generalization, the 'order', is subdivided through the levels of 'suborders', 'great groups' and 'subgroups' to the 'family', which is the level of greatest detail.

A list of some of the more common terms and their meaning is given in Table II. It is not meant to be exhaustive or definitive but to give the means for the non-specialist user of soil maps to derive more information from map legends couched mainly in soil classification terms.

MAPPING SOIL DISTRIBUTION

Soil survey is the field, and subsequent laboratory, study of soils, carried out to determine and delineate on maps their characteristics, properties and/or evolution. These are elements which define the uniqueness of individual soils and their possibilities for utilization and development in the light of competing demands. In the same way that there are different concepts of soil, there are different types of soil survey. Brink *et al.* (1982) describe engineering soil surveys which aim to characterize all unconsolidated material down to bedrock. However, the use of the term 'soil survey' in this chapter is taken to refer to surveys of the upper 1–2 m of land, sometimes called 'pedological' or 'agricultural' soil surveys.

Beckett and Burrough (1971) state that a soil survey of an area may be expected to answer any or all of the following questions:

1. (a) What kinds of soil (i.e. what classes of soil) are present?
   (b) In what proportions do they occur?
   (c) What are their properties?
   (d) What proportions of the area are occupied by soils with particular properties, or particular ranges of one or more properties?
2. (a) What is the soil class at any site of interest in the area?
   (b) What are the properties of the soil at any site of interest in the area?
3. (a) Where can soils of a particular class be found in the area?
   (b) Where can soils of particular properties be found in the area?

A soil map is not essential to answer all of questions 1(a) to 1(d) but is particularly useful for preparing answers to questions 2 and 3.

Types of survey

It is common (Bridges, 1982; Avery, 1987) to describe soil maps as the products of 'general purpose' or 'special purpose' soil surveys.

A general purpose survey is intended to serve as

a systematic inventory of the soil resources of an area, providing information that can be used or interpreted for a wide range of purposes, some of which may be unforeseen when the survey is initiated (Avery, 1987). The information is customarily displayed as a soil map showing the distribution of specified classes of soil, either singly or in combination (Fig. 3), the differentiating and behavioural characteristics of which are described either in the map legend or in an accompanying text. Surveys of this type have been commissioned by governments from national soil survey organizations in many countries and commissioned by overseas aid organizations for under-developed countries. Although they may be interpreted for a range of developmental uses, their primary aim is usually for making predictions about land use potential and management of the soil for growing crops in areas which may range in size from a single field to a district, region or country.

Special purpose surveys have a single purpose in mind, such as determining the suitability of land for, or planning the layout of, an irrigation scheme. As such, they may only record the limited range of soil properties directly relevant to the intended use. However, in practice, many special purpose surveys are conducted in a similar way to general purpose surveys recording many of the same parameters but including extra field sampling or analysis relevant to the specified objective. Likewise many general purpose surveys are applied at a later date, via interpretive models and/or additional sampling or analysis, to specific objectives. The main difference in survey types and their usefulness is between the comprehensive (pedological) approach in which at least a standard range of soil properties is recorded and the objective approach in which the only data recorded are single properties or a limited range of properties perceived as relevant to the objective.

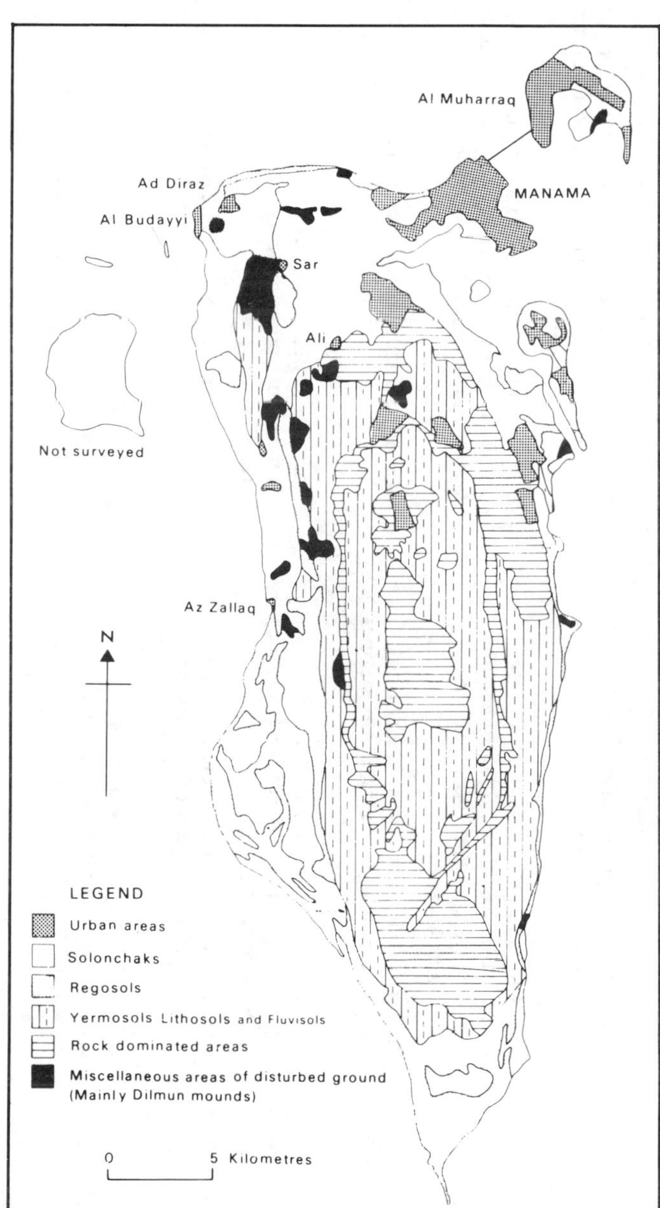

Fig. 3. Soils of Bahrein (from Bridges and Burnham, 1980).

Survey methods

Both approaches to mapping soil resources can be greatly assisted by remote sensing, but both require some input of field observation of soil properties. This is achieved by three main kinds of mapping procedure (Beckett, 1968; Avery, 1987).

Grid survey

Soil and associated observations are made at regularly spaced, pre-determined sampling points. This provides a reasonably unbiased systematic sample of the project area and the information gained is particularly suitable for computer-assisted manipulation. Reliable estimates of proportions and of mean values and variance of particular properties can be obtained. However, location of, and access to, sites can cause difficulty and, in the absence of other observations, interpolated boundaries can be of limited validity.

Physiographic survey

Mapped boundaries are based on external features of the soil and landscape perceived by systematic air-photo interpretation or other remote sensing techniques. Field observations are made, not to locate boundaries but to identify or describe the soil within each map unit. A physiographic survey is fast and cheap but soil units can only be identified if they possess consistent and mutually exclusive external expression.

Free survey

By observing the soils in a survey area and relating their position to specific segments of the landscape, a model of how the soils were formed is developed. Observations spaced along random transects or clustered within small 'sample' areas can then be used to predict with considerable accuracy the kinds of soil at a specific location and the boundaries between them. The advantages of free survey are in using the skill of an experienced field operative to reduce observation intensity in terrain with a simple soil pattern so that effort can be concentrated in the mapping of complex areas.

Although distinct in principle, these three procedures can be combined in different ways with the aim of achieving maximum utility for a given sampling effort (Avery, 1987).

In order to understand what can be represented on soil maps and how that information can be interpreted and applied in a planning context, it is important to know what is recorded in the field or measured in the laboratory as a result of the sampling effort, and what

Table II A selection of international soil classification terms, their meanings and example interpretations for development

Soil characteristics	*Example interpretations*	*Keyword/ stem*	*Example use*
Soils with a strongly bleached horizon (leached of free iron oxides) and/or associated horizons of accumulated organic matter, iron or aluminium.	Dependent on high fertilizer inputs for agriculture, corrosive to buried pipes, limited absorbance of pollutants.	Podzol, Spod-	Podzols, Spodosols
Soils of humid climates with brown weathered subsoils but no layers of significant depletion or accumulation.	Agriculturally fertile and productive, reasonable drainage status, year-round accessibility, versatile for most uses.	Brown, Cambi-, Brun-	Brown earth, Cambisols, Brunisols
Soils with an excess of water as indicated by grey and/or orange mottle colours, typical of low-lying ground or over impermeable rocks.	Dependent on artificial drainage for agricultural use, deep excavations unstable, poorly suited for houses with basements, weak bearing strength, potential for wetland habitat, corrosive to buried pipes, poorly suited for soakaways.	Plano-, Stagno Gley, Aqu-, Hydro-,	Planosols, Stagnogley Aquents, Sols hydromorphes
Soils with hard rock at shallow depth, typical of hill country.	Difficulty with deep excavation, droughty for agricultural crops in many areas, stable construction site but susceptible to erosion.	Litho-	Lithomorphic soils, Lithosols
Soils with an horizon in which silicate clay has accumulated to a significant extent. Acrisols and Ultisols are also deficient in bases.	Moderate to low fertility, stable for construction, can be corrosive, subject to compaction and disintegration, restricted permeability for soakaways.	Argillic, Luvi-, Alfi-, Ulti-, Acri-	Luvisols, Argillic brown earths, Alfisols, Ultisols, Acrisols
Base-rich soils, dark coloured and rich in organic matter, typical of the steppes.	Fertile for agriculture, resistant to erosion.	Molli-, Cherno-, Phaeo-	Mollisols, Chernozens, Phaeozems
Clayey soils with a high shrink-swell potential, resulting in seasonal cracking and soil mixing.	Vulnerable to heave. Special design required for foundations and pavements, difficult to cultivate but high fertility.	Verti-, Pelo-	Vertisols, Pelosols
Low density soils rich in volcanic glass and allophane, typical of volcanic areas.	Large phosphorus inputs required for agriculture, high compressibility, low density and low strength for construction.	And-	Andosols, Andepts
Calcareous soils shallow over highly calcareous parent materials such as chalk or marl.	Agriculturally fertile, deep excavation often difficult, local subsidence problems associated with sinkholes, not corrosive, mining potential for cement or lime, habitat value.	Rend-	Rendzinas, Rendolls

Soils in alluvial deposits, often with weakly developed soil profiles.	Fertile, compressible under load, ill-suited to urban development where prone to flooding, mining potential for sand and gravel.	Fluvi-, Alluvial	Brown alluvial soils Fluvisols
Soils in thick organic materials (peat).	Habitat uniqueness, weak bearing strength (high compressibility), extractive potential for horticultural uses, corrosive to buried pipes.	Histo-, Peat	Histosols, Peat soils
Soils with horizons having high concentrations of soluble salts, especially of sodium; common in coastal or low-lying hot arid areas.	Problematic for irrigated agriculture (structural instability, drainage problems) potentially corrosive to iron pipes, potentially damaging to pervious bituminous coatings, vulnerable to erosion and underground 'piping'.	Sal-, Sol-, Natr-, Sod-	Solonetz, Solonochaks Sols salins, Sols sodiques
Soils characterized by red colours resulting from weathering of iron compounds in a warm climate.	Stable for construction, moderate fertility and permeability.	Red, Rossa	Terra rossa, red brown earths
Soils of deserts or semi-deserts, often with little darkening of surface horizons.	Water shortage restricts most uses unless irrigated, vulnerable to erosion.	Aridi-, Xer-, Yerm-	Aridisols, Xerosols, Yermosols
Deeply weathered soils of tropical and subtropical climates, having high contents of hydrated oxides of iron or aluminium.	Low fertility without fertilizer, permeable, resistant to erosion, cemented layers provide constructon material.	Lato-, Ferral-, Oxi-	Ferralsols, Oxisols, Lateritic soils
Weakly developed soils consisting dominantly of stones or sand as in materials formed by erosion.	Consolidation required for construction, droughty for agriculture.	Rego-, skeletal	Regosols, skeletal soils
Weakly developed coarse-textured (sandy) soils.	Liable to erosion, limited absorbance of pollutants, good drainage and accessibility, suitable for recreational use, droughty for agriculture.	Areno-, Psamm-,	Arenosols, Psamments
Very acid soils in wet alluvium or peat.	Weak bearing strength, corrosive, agricultural use required drainage and very high fertilizer inputs.	Sulph-	Acid sulphate soils, Sulfaquepts

additional environmental information is collected to complement it.

Table III lists the site and soil information which may be collected during the field and laboratory work associated with a pedological soil survey (though additional laboratory analyses might be carried out for studies into soil genesis). Only on rare occasions is every feature in the list measured or recorded, probably only for 'benchmark' experimental sites on which a large amount of interpretative effort is being expended. On other occasions the description, sampling and analysis is tailored to the interpretation. For example, emphasis is placed on topographic detail, salinity, water retention and permeability properties in surveys for irrigation schemes, on vegetation and soil flora and fauna in surveys for ecological interpretation, on heavy metal content (or other contaminants) in investigations of soil pollution and on the grading analysis, consistency limits and potential shrink/swell in soil surveys for development involving earthmoving or construction. For most detailed soil surveys involving intensive sampling, only selected sites are described and analysed in detail in order to characterize the soil map units present. Other sampling sites may be merely soundings (recording a restricted list of properties) in order to locate boundaries between adjoining soils.

What must be emphasized is that despite the basic requirements as stated by Beckett and Burrough (1971) above, most soil surveys involve site and vegetation description and incorporate other environmental features. The information that results can be presented as separate maps, for example of land use and vegetation (Obihara *et al.*, 1964), or in accompanying reports. It is clear that, far from solely being inventories of soil properties, most soil mapping exercises can be considered land inventories.

Survey intensity levels

Whatever the type of survey, the detail of information shown varies with the intensity level of survey input and the scale of the map. Both are determined by the objectives of the project (Table IV). The main intensity levels are described below, illustrated mainly by examples of general purpose soil surveys.

Compilations

Soil maps produced from existing information can be categorized as either generalized or schematic. The former type generalize delineations on soil maps made by field surveys at a larger scale, thus displaying geographic soil relationships over larger areas and for broader purposes than on the maps from which they are derived. Schematic soil maps differ in that they are compiled mainly from information other than existing soil maps. Both types can be presented as atlases (e.g. Ganssen and Hadrich, 1965) or as a map accompanied by an explan-

Table III Description and sampling of soils—features assessed or measured in the field or measured in the laboratory on field samples

*Type of information feature**		
Site	*Relief, slope, aspect* *Rock outcrops* *Land use and vegetation* Soil surface condition	Elevation Soil erosion and deposition Flood hazard
Soil profile description	*Depth and thickness of horizons* *Recognition of horizons (see text)*	
Soil horizon description	*Colour (and mottling)* *Organic matter content* *Particle size (texture) class* *Stone type and content* *Carbonate content* *Presence of mineral accumulations* *Boundary to next horizon*	Soil structure Pores and fissures Roots and other soil flora Soil fauna
Quantitative field measurements on profiles	Infiltration rate Water-table height and amplitude Penetration resistance	
Quantitative field measurements on horizons	Soil reaction (pH) Salinity Saturated hydraulic conductivity (permeability)	
Quantitative laboratory measurements on horizon samples	<u>Chemical</u> Soil reaction (pH) Plant nutrient content Carbonate Cation exchange capacity Exchangeable cations and base saturation Organic matter content Clay mineralogy Heavy metal content Electrical conductivity	<u>Physical</u> Particle size distribution Atterberg limits Potential shrink/swell Bulk density Pore size distribution (water retention properties) Hydraulic conductivity (saturated and unsaturated) Structural stability

*Features in italic are the minimum that are normally recorded during 'soundings' to establish the position of soil boundaries once the soil map units have been properly defined by fuller descriptions and analyses.

Table IV Soil map scales and survey intensity levels (after Avery, 1987; Dent and Young, 1981; FAO, 1969)

Kinds of survey and intensity level	*Scale*	*Inspection density*	*Kind of mapping unit*	*Typical objectives*
Compilations and syntheses	1:1 000 000 and smaller		Taxonomic classes—suborders, great soil groups and phases of great soil groups	Promoting awareness (atlases, teaching, etc.), global planning
Exploratory (very low intensity)	1:250 000–1:1 000 000	1/km^2	Land system or landform-soil units enclosing great soil groups	Broad regional or national land use planning or development
Reconnaissance (low intensity)	1:100 000–1:250 000	1–5/km^2	Landform-soil units based on associations of subgroups or series	Regional soil resource inventories, feasibility studies
Semi-detailed (medium intensity)	1:100 000–1:25 000	5–20/km^2	Soil and landform-soil units based on series and associations of series	Appraisal for use of land at farm level, district land use planning
Detailed (high intensity)	1:10 000–1:25 000	20–100/km^2	Soil series, phases of series, some associations and complexes	Appraisal or use and management of farmland at field level (e.g., irrigation works) or for urban development
Very detailed (very high intensity)	1:10 000 and larger	100/km^2	Phases of soil series, single properties of soils	Special purpose surveys of experimental plots, development sites, etc. Problem solving

atory legend (e.g. FAO, 1974). Mapping units are usually related to the higher orders of the relevant pedological classification system, such as in the FAO/Unesco 1:5 000 000 Soil Map of the World (for an extract, see Bridges, 1982). Many national soil maps (e.g. Roeschmann, 1986) are compilations.

Exploratory surveys

These surveys provide the necessary background material for development plans at the regional or national level and are commonly the first type of mapping exercise carried out in completely unknown terrain. They rely heavily on remote sensing and most boundaries are drawn in the laboratory. Field inspection is limited to one observation per kilometre and is carried out to check that the identified boundaries are valid and to identify the major class of soil at great group or order level or in more general terms. The 1 000 000 soil map of Kenya (Sombroek *et al.*, 1982) is an example of an exploratory soil map designed so that planners can select areas to be developed for one or another specific type of land use. It is a refinement of an earlier 1:3 000 000 soil map using information from detailed surveys and additional purposive sampling. Stocking (1983) used a more rapid approach to produce computer-generated exploratory soil maps of an area in Tanzania (Fig. 4).

Often a 'land system' or 'integrated' approach is taken to group areas of land with a certain homogeneity of landform, soil, vegetation, groundwater conditions and climate. The 1:500 000 land resource study of NE Zambia (Mansfield *et al.*, 1975) used a land system approach to make a preliminary assessment of the potential of the area for agricultural development. While that study relied mainly on maps to convey the information, the land systems of Queensland (Galloway *et al.*, 1970) are described by block diagrams and tabular material (Fig. 5).

Terrain classification (Grant 1976) is another form of land-system mapping carried out primarily for soil engineering purposes, such as military trafficability (Brink *et al.*, 1966) or road building (Brink *et al.*, 1982). Though based largely on landforms with their associated geology and regolith cover, a description of soil (in non-pedological terms) forms an important component of the analysis.

Reconnaissance surveys

Reconnaissance surveys differ from exploratory surveys in that they involve a greater level of field observation and allow map units to be defined more precisely in terms of soil classes. Often, they include areas of land for which more detailed soil information is available (Ragg *et al.*, 1984). The map units are often termed soil 'associations'. These are a geographic grouping of soil types (polypedons) that occur in a certain topographical juxtaposition but differ due to variations in parent material, relief and hydrology (similar to the concept of 'catenas', Milne 1935). Figure 6 is an example of part of a soil association map of an area in Canada. The legend is more comprehensive than on smaller-scale maps and gives information on soil, slope, moisture regime, parent material and vegetation in addition to soil classification names. Also, an accompanying cross-section shows the relationship of soil associations to geology and topography. The map units of the 1:250 000 regional and county soil maps of England and Wales are also associations but, in the accompanying explanatory bulletins (e.g. Jarvis *et al.*, 1979) block diagrams show the juxtaposition of individual soils within each association (Fig. 7). Surveys such as these are useful to identify the forms of development that are possible within large regions of a country and the level and nature of national effort required to implement such development.

Semi-detailed and detailed surveys

Many of the more populated or agriculturally productive parts of industrialized countries of the world have been covered by general purpose semi-detailed or detailed

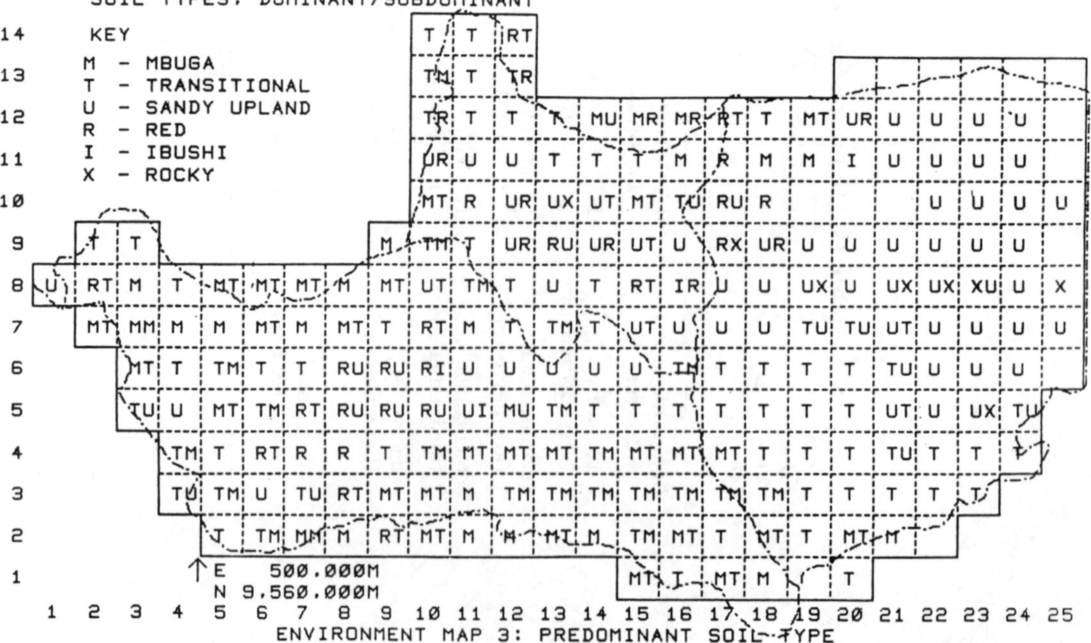

Fig. 4. Computer-generated soil map of Shinyanga and Maswa districts, based on 100 km² interpretation units (from Stocking, 1983).

(13) LEINSTER LAND SYSTEM* (4330 SQ MILES)

Extensive uniform old alluvial plains; leached grey and brown massive earths with hardpan; paperbark or bloodwood-stringybark woodland.

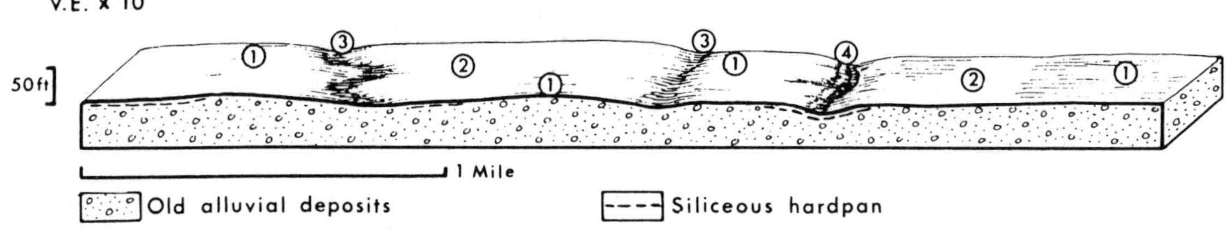

Land Unit	Area and Distribution	Land Forms	Soils	Vegetation
1 7 obs.	30% Throughout	Plains; level to very gently undulating; often on slight rises and old levees; slopes mainly 0–1%, locally up to 2%; mainly medium-textured old alluvium; fairly deeply weathered; siliceous hardpan common below 3 ft	Mainly leached grey and brown massive earths with hardpan horizons, Crowbar (Gn2.94, 2.85); minor loamy yellow earths, Kalinga (Gn2.85) and Welcome (Gn2.86), and sandy yellow earths, Clark (Gn2.25)	Bloodwood–stringybark, box, or paperbark woodland (*M. viridiflora*, *M. nervosa*), some evergreen mixed scrub; ground cover *Schizachyrium*, *Sorghum*, *Chrysopogon fallax*, *Drosera*, Cyperaceae
2 10 obs.	60% Throughout	As for unit 1 but some tendency to be on slightly lower sites	Mainly leached grey and brown massive earths with hardpan, generally mottled, Crowbar (Gn2.94); extensive brown acid to neutral mottled structured loams, Wyaaba (Gn3.05, 3.25); minor sandy yellow earths, Clark (Gn2.45), and alkaline loamy yellow earths, Lorraine (Gn2.93)	Paperbark woodland (*M. viridiflora*) with scattered bloodwood and mixed grasses and Cyperaceae (*Sorghum*, *Schizachyrium*, *Eriachne*, *Alloteropsis*, *Chrysopogon fallax*, *Heteropogon contortus*)
3 3 obs.	5% Throughout	Old valley floors forming broad shallow depressions with no active stream incision; sand to clay; sometimes siliceous hardpan below 3 ft	Mainly leached grey and brown massive earths, Crowbar (Gn2.95, 2.85); some texture-contrast soils with thick sandy surface and neutral subsoils, Gamboola (Dy3.22); often containing concretionary ironstone	Paperbark woodland (*M. viridiflora*) with some box and bloodwood over *Schizachyrium*, Cyperaceae, *Eriachne*, *Alloteropsis*, *Aristida*
4 3 obs.	5% Sporadic	Shallow incised valleys 30–100 ft deep, ¼–2 miles wide; scalded and dissected areas common on weathered silts and siliceous hardpan outcrops; streams incised 5–20 ft into hardpan with frequent water-holes	Alkaline leached grey and brown massive earths, Lorraine (Gn2.83), texture-contrast soils, Stewart (Dy3.23), and uniform fine-textured soils, Mimosa (Uf6.41)	Scalds paperbark scrub (*M. foliolosa*) and some box over *Schizachyrium* and *Aristida;* elsewhere box woodland (*E. microneura*) with bloodwood and paperbark over *Schizachyrium*, Cyperaceae, and *Sorghum*

* Similar land forms and vegetation to those of Mayvale land system of the Leichhardt–Gilbert area.

Fig. 5. A description of a land system in Queensland, Australia (from Galloway *et al.*, 1970).

surveys (e.g. Olding *et al.*, 1956; Macvicar and Loxton, 1967).

In semi-detailed surveys, ground observation becomes increasingly important and soil classes can be defined more precisely. Surveys are usually carried out using free survey or a combination of free and grid survey techniques. A map unit on semi-detailed soil maps at a scale of 1:50 000 or 1:63 360 is commonly defined by the name of the dominant soil type or 'soil series' (the 'series' can be equated with the classification concept of the polypedon) although more complex areas might be indicated. In addition, important soil-related properties such as groundwater depth and fluctuation (Stichting voor Bodenkartering, 1976) might be overlain on the soil information. The information from semi-detailed soil maps might be sufficient to assess the economic feasibility or even permit the implementation of less intensive forms of development but more detailed information will usually be required before investment in intensive development can proceed.

At a scale of 1:25 000–1:15 000, individual field enclosures can be cartographically represented or recognized on air photographs and soils can be related directly to the parcels of land which enclose them. 'Phases' of soil series might be identified where certain minor properties relevant to the purpose of the survey (e.g. topsoil texture or slope) vary between fields (Fig. 8). At this scale of information, detailed plans for fairly intensive development of land can be formulated. For example, the soil map of Tuscola County, Michigan, is contained in a report which gives interpretations for a wide range of land uses including recreational development, building site development, agriculture, woodland and wildlife reserves.

Very detailed surveys and site investigations

Very detailed surveys are usually carried out on land receiving or destined to receive a large financial investment. In agriculture or forestry this will usually be experimental plots; otherwise it will be land being taken out of primary production for urban development, road construction or similar use. Grid survey is often used and boundaries between soil series, phases of series or single soil properties are then interpolated between grid points. The survey of Stackyard field on Woburn Experimental Farm, Bedfordshire, England (Catt *et al.*, 1979) is an example of a very detailed survey. The field had been used for agricultural experiments for more than 100 years but there were clearly variations in crop yields not related to weather or experimental treatment. Detailed sampling using a 20 m grid enabled a series of isoline maps to be drawn (Fig. 9) showing a significant variation in soil texture and related moisture retention

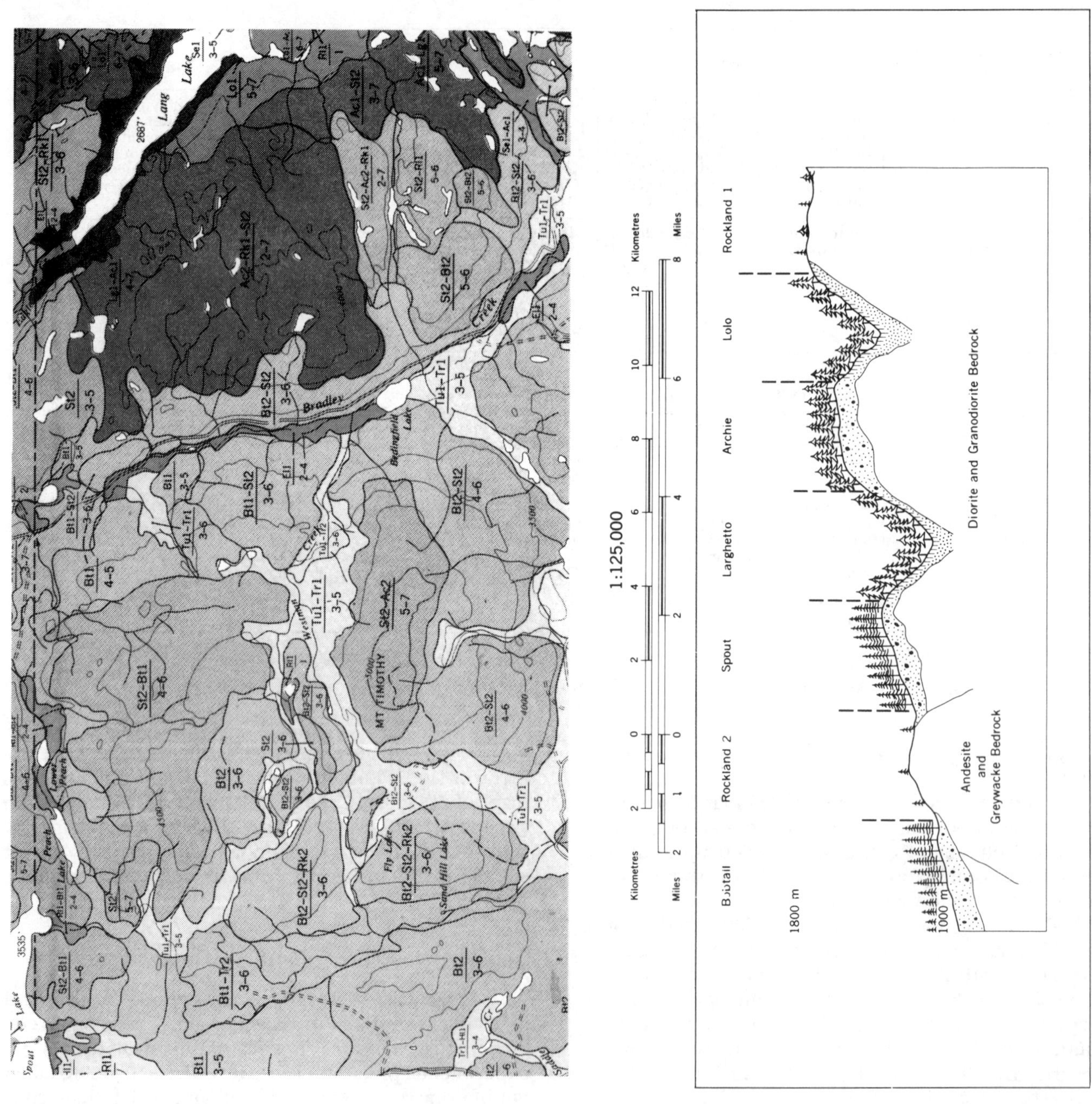
1:125,000
Kilometres
Miles
Bjutall
Rockland 2
Spout
Larghetto
Archie
Lolo
Rockland 1
1800 m
1000 m
Andesite and Greywacke Bedrock
Diorite and Granodiorite Bedrock
MT TIMOTHY
Lang Lake
Bradley Creek
Fly Lake
Sand Hill Lake

SOIL ASSOCIATION LEGEND

COLOR AND SYMBOL	SOIL ASSOCIATION[1] (SOIL CLASSIFICATION OF THE DOMINANT SERIES[2])	PARENT MATERIAL AND SURFACE EXPRESSION	SOIL MOISTURE REGIME[3] AND DRAINAGE CLASS[4]	BIOGEOCLIMATIC ZONES[5] AND VEGETATION COMMUNITY	MAP UNITS[6] AND MINOR SOILS OR LAND TYPES
THE SOIL ASSOCIATIONS OF THE QUESNEL HIGHLANDS The elevation of this physiographic area ranges from 1,000 to 1,800 m asl and the bedrock is composed principally of diorite and granodiorite, with some augite andesite in the west. The topography is rolling to mountainous and the soil parent materials are glacial till or colluvium. Soils are often shallow with some exposures of bedrock especially in the northeast.					
Ac	**ARCHIE (Podzolic Gray Luvisol)**	Medium textured[7], gravelly, strongly to medium acid[8] glacial till on strongly to very steeply sloping topography.[9]	Humid and moderately well drained	Engelmann Spruce – Subalpine Fir zone	Ac1 Some shallow to bedrock soils and Orthic Humo-Ferric Podzols Ac2 Inclusions of imperfectly drained soils and some shallow to bedrock soils
Bt	**BOBTAIL (Orthic Gray Luvisol)**	Medium textured, gravelly, slightly acid to mildly alkaline glacial till on moderately to steeply sloping topography.	Humid and moderately well drained	Interior Douglas fir zone. Rocky Mountain Douglas fir – pine grass community	Bt1 Minor inclusions of imperfectly drained soils Bt2 Some shallow to bedrock soils and imperfectly drained soils
Lg	**LARGHETTO (Degraded Eutric Brunisol)**	Moderately coarse textured, gravelly, slightly acid to neutral colluvium on steeply to very steeply sloping topography.	Humid and well drained	Engelmann Spruce – Subalpine Fir zone	Lg1 Some shallow to bedrock soils
Lo	**LOLO (Degraded Dystric Brunisol)**	Moderately coarse textured, gravelly, strongly to medium acid colluvium on steeply to very steeply sloping topography.	Humid and well drained	Engelman Spruce – Subalpine Fir zone	Lo1 Some shallow to bedrock soils
Rk	**Rock: land type**	Granitic rock outcrops and talus slopes. Andesite and Greywacke rock outcrops and talus slopes.		Little or no surface vegetation	Rk1 Rk2
St	**SPOUT (Podzolic Gray Luvisol)**	Medium textured, gravelly, slightly acid to neutral glacial till on strongly to very steeply sloping topography.	Humid and well drained	**Interior Douglas fir zone (wet subzone)**	St1 Minor inclusions of imperfectly drained soils St2 Some shallow to bedrock soils and imperfectly drained soils

Fig. 6. Extract from the 1:125 000 soil map of the Lac La Hache-Clinton area, British Columbia, showing the extended legend. (from Valentine and Schori, 1980).

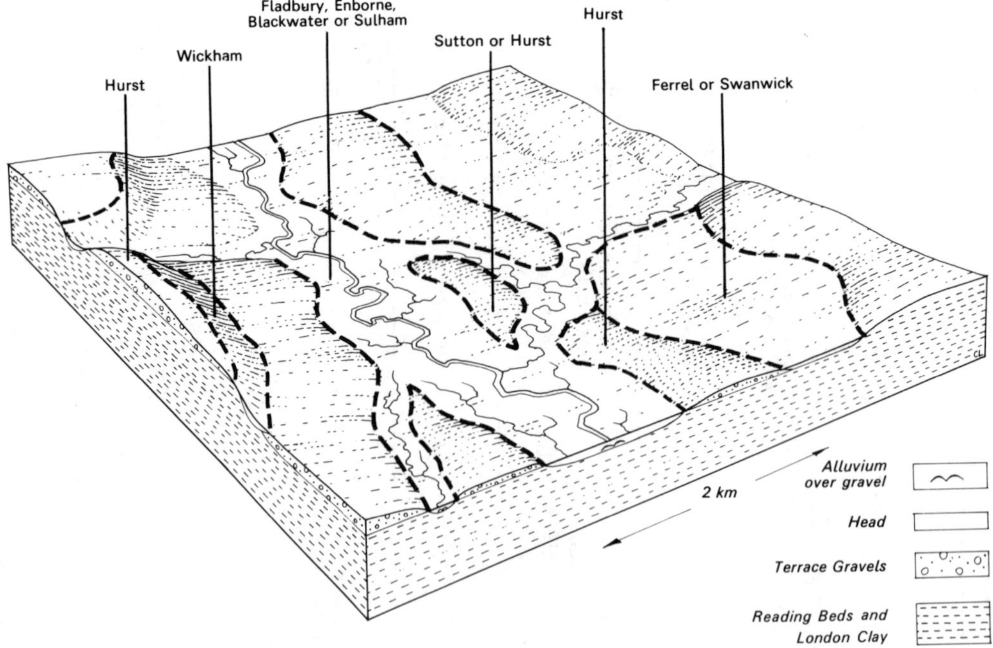

Fig. 7. Distribution of soils in map unit 13 (from Jarvis *et al.*, 1979).

characteristics. Future experiments can now be planned accordingly.

INTERPRETIVE AND SPECIALIST SOIL MAPS AND THEIR USE

It has been stated that most soil surveys are not solely inventories of soils but are concerned with land in the broader sense. In many parts of the world land is a scarce or fragile resource.

Careful planning is required to make the best use of land for the future. The planner looking at broad land use options wants answers to questions such as (Eswaran, 1978):

(1) What can I do with the area?
(2) What are the alternatives?

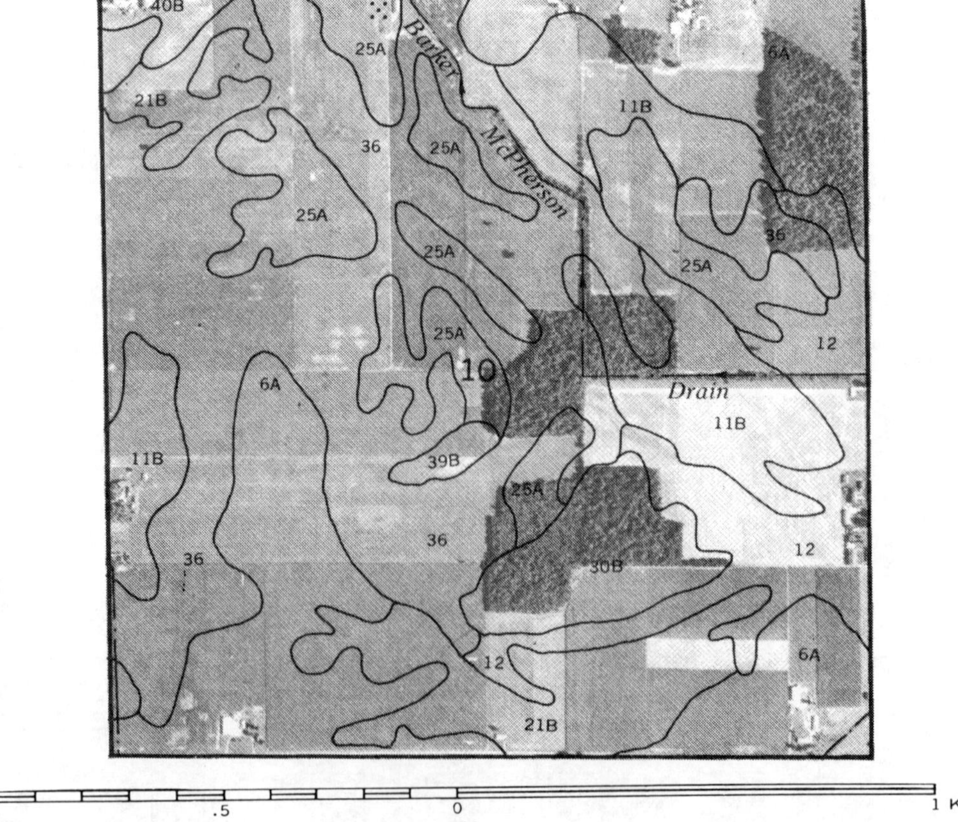

Fig. 8. Extract from Sheet 97 of the 1:15 840 soil survey of Tuscola county, Michigan (from Mettert, 1986).

6A Tappan-Avoca complex, 0–3% slopes
11B Metamora sandy loam, 0–4% slopes
12 Corunna sandy loam
21B Wixom loamy fine sand, 0–4% slopes
25A Londo loam, 0–3% slopes
30B Spinks loamy fine sand, 0–6% slopes
36 Tappan loam
39B Ottokee loamy fine sand, 0–6% slopes
40B Chelsea fine sand, 0–6% slopes

Fig. 8 (*Cont.*). Key.

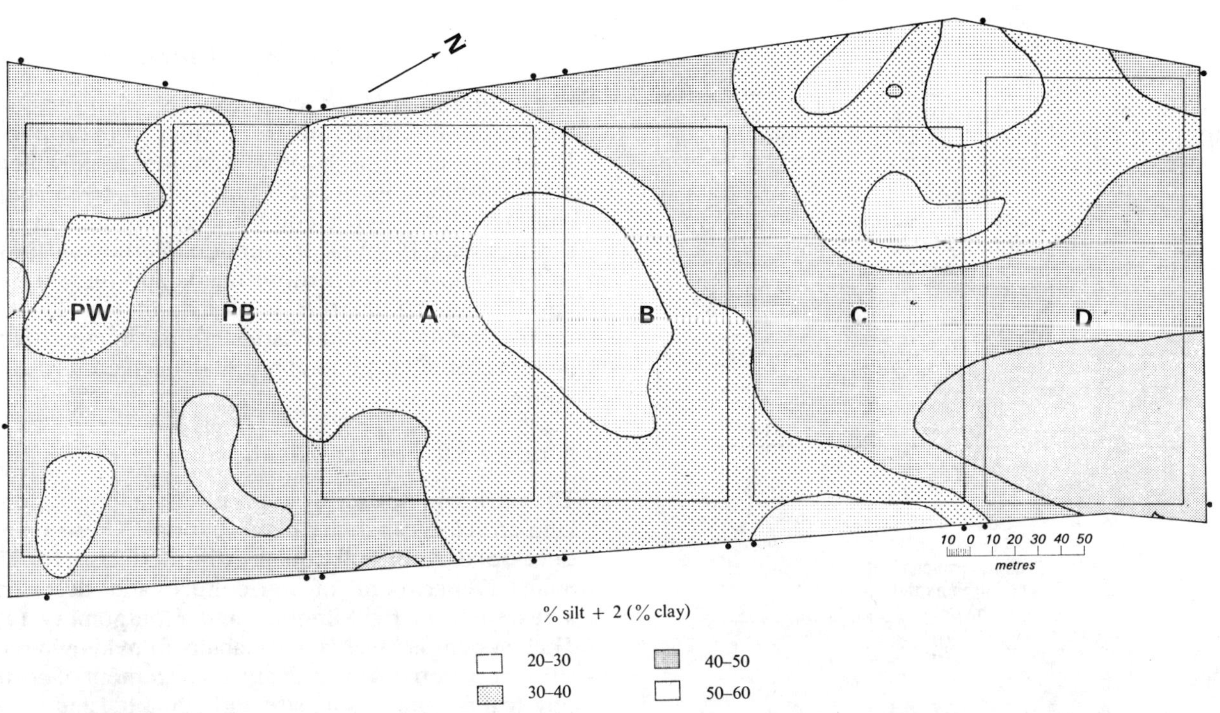

Fig. 9. Variations in surface (0–25 cm) texture (% silt + 2 (% clay)) on Stackyard Field and boundaries of main experimental sites (Permanent Wheat (PW), Permanent Barley (PB) and Series A, B, C and D) (from Catt *et al.*, 1979).

(3) What are the risks?
(4) How much flexibility do I have?

Mapping soil and site characteristics helps to answer these questions by establishing what are the land qualities which influence the suitability of land for a specific use (e.g. growing onions) or capability for a broad use (e.g. recreation). The distinction between land qualities and land suitabilities is shown in Table V. By combining suitabilities with economic assessments and studies of environmental impact (Fig. 10) land use decisions can be made in a considered fashion. However, often the planner will be faced with a determined developer, wanting the land for a specific use. The questions the planner will want answering about the proposed use will be:

(1) Will it detrimentally affect surrounding land?

Table V Land qualities and suitabilities relevant to agriculture, forestry, hydrology, mineral extraction, engineering, leisure and conservation

Use	*Land qualities*	*Land suitabilities (for)*
Agriculture and forestry	Droughtiness, wetness, erodibility, workability, trafficability, plantability, windthrow risk	Specific crops Specific techniques (e.g., irrigation)
Hydrology	Run-off acceptance Effect on river flow hydrograph	Surface water gathering, groundwater abstraction
Mineral extraction	Workable reserves present	Access to workings, restoration of waste material
Engineering and construction	Corrosivity, stability, compactibility, wetness, shrinkage, susceptibility to frost heave, permeability	Roads, soakaways, pipe-lines, landscaping, buildings
Leisure/amenity	Walkability, drainage characteristics, trafficability, landscape quality	Rural car parks, picnic areas, camp sites, foot-paths, playing fields
Conservation	Ecological richness Landscape appeal	Maintaining *status quo*, reclaiming damaged land

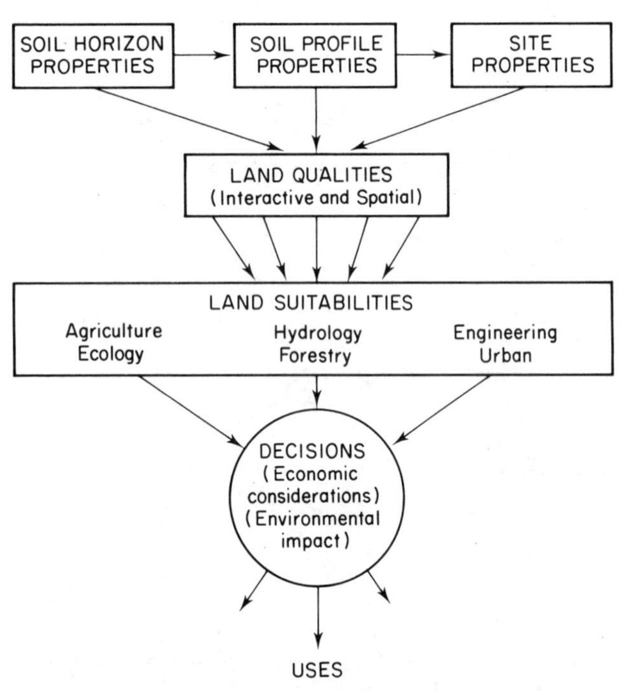

Fig. 10. Processes of evaluating land.

(2) Does it have an environmental impact elsewhere?
(3) Does it have a short- or long-term reclamation need?

The examples that follow give examples of ways in which soil maps have been used in both types of situation. For convenience, the examples have been grouped into three sections, agriculture and forestry, urban and industrial development, and environmental protection.

Agriculture and forestry

The requirements for crop growth, whether horticultural crops or coniferous trees, differ from crop to crop. However, all crops have a common requirement for a reasonable moisture and nutrient supply, a sufficient rooting depth and a good energy regime for photosynthesis and biomass production (Verheye, 1986). The relationship between these needs and measurable environmental parameters is shown in Fig. 11. All the environmental characteristics are parameters that may be represented in general purpose soil maps as described above.

Land capability classifications

The most widely used systems for evaluating agricultural land are land capability classifications. These were first developed by the Soil Conservation Service of the United States Department of Agriculture and described in a tested form by Klingbiel and Montgomery (1961). Their system assesses land capability from known relationships between the growth and management of crops and physical factors of soil, site and climate. Land is graded according to its potential and the severity of limitations for crop growth into eight classes, class 1 having few limitations and class 8 having very severe limitations. The land capability system has been applied in many forms around the world (Murthy *et al.*, 1968; Prickett, 1966, Canadian Land Inventory, 1965). Typical is the seven class system introduced into Britain (Bibby and Mackney, 1977) and summarized below:

Class 1. Land with very minor or no physical limitations to use.
Class 2. Land with minor limitations that reduce the choice of crops and interfere with cultivations.
Class 3. Land with moderate limitations that restrict the choice of crops and/or demand careful management.

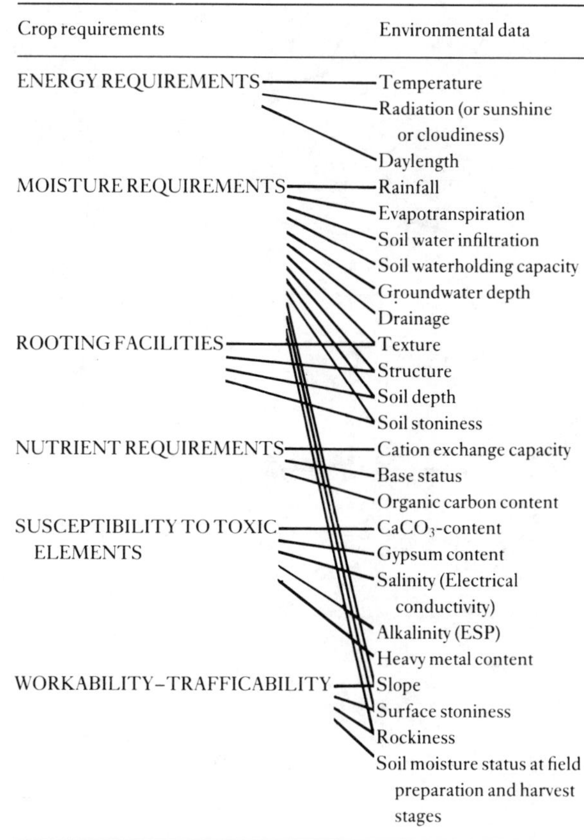

Fig. 11. Relationship between crop growth requirements and climatic, soil and physiographic properties of the land (from Verheye, 1986).

Class 4. Land with moderately severe limitations that restrict the choice of crops and/or require very careful management.

Class 5. Land with severe limitations that restrict its use to pasture, forestry and recreation.

Class 6. Land with very severe limitations that restrict use to rough grazing, forestry and recreation.

Class 7. Land with extremely severe limitations that cannot be rectified.

Classes are subdivided into subclasses based on the kinds of limitation affecting land use. Subclasses described by Bibby and Mackney (1977) include:

w wetness;
s soil limitations;
g gradient and soil pattern limitations;
e liability to erosion;
c climatic limitations.

Many soil maps produced in Britain have been accompanied by land capability maps using this system and, on maps at scales of 1:25 000 and larger (e.g. Fig. 12), land capability units have been employed to group land capability subclasses that respond in a similar way to management and improvement practices. Despite the widespread use of land capability classifications, their subjectivity can result in debates such as whether a particular tract of land is a poor grade 3 or a good grade 4. Attempts to use yield data to refine capability classifications have largely been unsuccessful but various parametric approaches have been used to define agricultural land capability or value (e.g. Leamy, 1974). In them, scores are assigned according to individual features of the soil and environment within an area which might be a field, a soil type, or an agricultural holding. An additive or multiplicative system is then applied to arrive at a points total which reflects the land quality. By dividing points totals into ranges, land of different agricultural quality can be represented cartographically. McRae and Burnham (1981) give a comprehensive review of parametric agricultural land capability systems. One of their important uses has been in assessing agricultural land for taxation as in West Germany (Weiers and Reid, 1974), though the New York State land taxation system relies solely on estimated yields and input costs (Olson, 1982).

Land suitability classifications

Whereas agricultural land capability classifications only give vague appraisals, any detailed soil maps can be interpreted to give a suitability appraisal for a specific crop or for specific management practices, such as irrigation or mole draining. Less detailed soil maps can be used to give suitability ratings for specified kinds of cropping use.

Suitability for broad land uses

The FAO Framework for Land Evaluation (FAO, 1976) classifies the fitness of a given type of land for a specified use through four levels of decreasing generalization (in a similar manner to land capability classifications):

(1) ***Land suitability orders***: these indicate kinds of suitability where S = suitable and N = not suitable.
(2) ***Land suitability classes***: these reflect degrees of suitability within orders where S1 = highly suitable, S2 = moderately suitable, S3 = marginally suitable, N1 = currently not suitable, N2 = permanently not suitable.
(3) ***Land suitability subclasses***: these indicate the kinds of limitation or the main kinds of improvements required within classes (e.g. S3e, N1m); for example e = poor erosion resistance and m = moisture deficiency, though there is not a formal list of subclasses.
(4) ***Land suitability units***: these indicate minor differences in the management required within subclasses (e.g. S3e1, S3e2).

The FAO Framework can be applied to detailed surveys at the precision of unit level, but reconnaissance or exploratory surveys are not normally interpreted beyond class level. Figure 13 is an example of its use applied to a land system map already completed for the Guhiya catchment in the arid zone of Western Rajasthan. Shankarnarayan *et al.* (1983) describe how the requirements and limitations of the three kinds of land use were compiled from field survey experience and observations on general land use patterns, cropping patterns, crop yield and alternative land use systems for drylands. The requirements of each land use were then compared with the attributes of each land mapping unit to arrive at the suitability classifications. Their approach gives most emphasis to suitability for dryland arable farming. They take into consideration the non-suitability of certain lands for arable farming when assessing suitability for pasture growth and livestock production. Likewise, land is only classified for woodland forestry if it is not suitable for arable farming. Such straightforward suitability classifications are easy for planners to interpret and under-

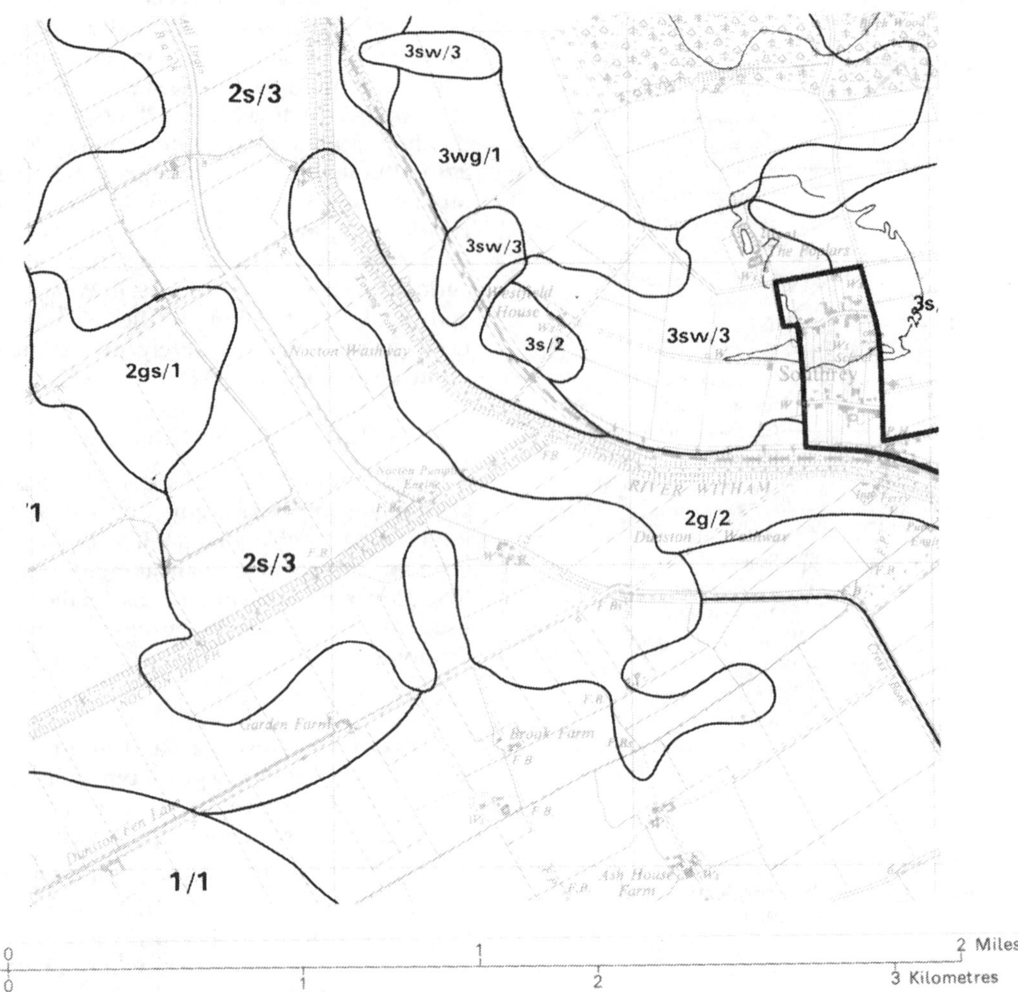

Fig. 12. Land capability map of part of Lincolnshire, England (from Robson *et al.*, 1974).

stand and are a useful guide to development in an area of fragile natural resources.

Suitability for specific crops

Land suitability classifications for crop production can be applied to types of farming as in the Rajasthan example or to individual tree species or field crops. For woodland many of the recent soil survey reports of the US Department of Agriculture give classifications of windthrow hazard, seedling mortality and figures for potential productivity. Ragg *et al.* (1984) describe a system which assesses the suitability of land in England and Wales for common field crops such as winter wheat, potatoes and sugar beet. Land quality parameters used include workability days in specified seasons, droughtiness, slope and accumulated temperature. The first of these is determined from soil factors, such as topsoil texture and duration of waterlogging at shallow depth, combined with the climatic parameter of field capacity days. Droughtiness is calculated from the available water reserves held by the soil and the rooting characteristics and potential moisture demand of the individual crop. Land is classified as well suited, moderately suited, marginally suited and unsuited, comparable to the FAO classes S1, S2, S3 and N. The system is continually being extended to less extensive or new crops as their requirements become better known, and offers much to the debate on changing land use policies in the countries of the European Community.

Suitability for specific techniques

The above examples relate suitability classifications for types of cropping. Suitabilities for agricultural techniques can also be devised. Cannell *et al.* (1975) describe a three class system for assessing the suitability of soils for the direct drilling of cereal crops. Ragg *et al.* (1984) classify soils on the 1:250 000 soil map of England and Wales according to their suitability for mole draining. However, one of the more common classifications for agricultural development is for irrigation suitability. A soil map and its interpretation are as vital as securing a supply of suitable water, assessing the cost-benefit of increased production and investigating social considerations. Griffiths (1975) describes a classification of land for irrigation in New Zealand, based on the system of United States Bureau of Reclamation (US Department of the Interior, 1953) but adapted for New Zealand soils and the requirement to irrigate for a dry season of only 3–5 months. To apply the irrigation suitability classification it is important that the soil map should be accompanied by data on soil drainage, permeability, infiltration rate and available water capacity or that these factors, important to water transmission and retention, should be capable of estimation from other soil properties (Reeve, 1986).

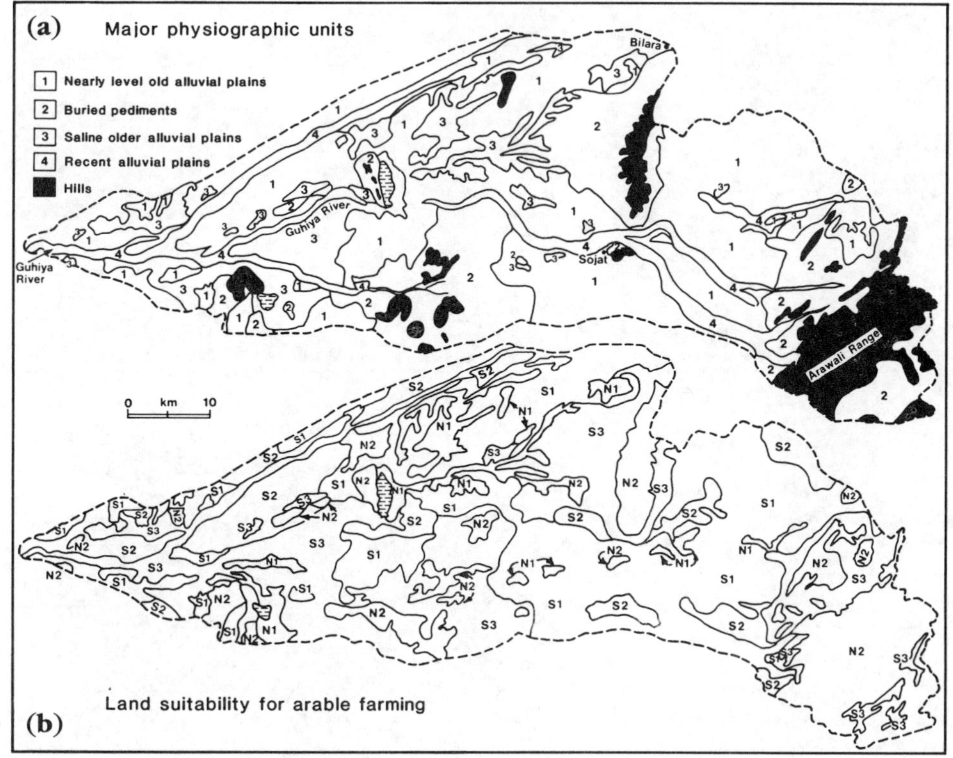

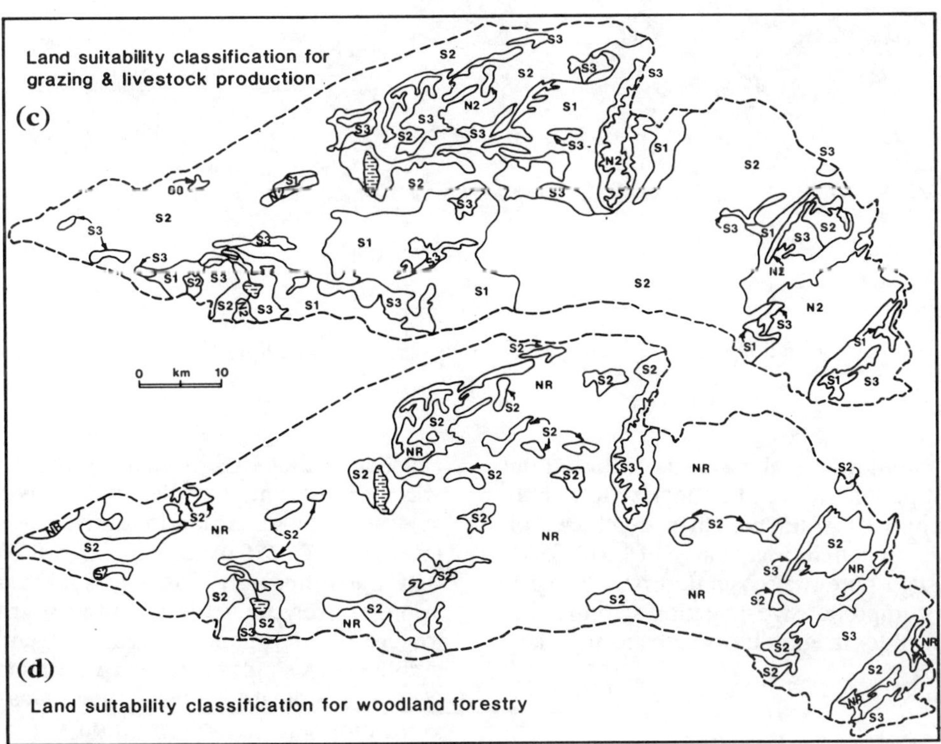

Fig. 13. Land suitability maps of Western Rajasthan (from Shankamarayan *et al.*, 1983).

Griffiths (1981) applied the New Zealand system to assess the suitability of soils for irrigation in the Ngatarawa area of the Heretaunga Plains near Hastings, North Island. This is a subhumid area, but, with summer moisture, deficits of 300 mm of water can occur which limit crop growth on most soils. Land use in the Ngatarawa area was sheep rearing, with some crops, but there was a potential for more intensive stocking or a change to higher value crops if irrigation was feasible. The 1:15 000 survey was of the general purpose type and delineated soils developed in loess-, greywacke- or volcanic ash-derived alluvium. Their properties relevant to irrigation were inferred from other soil characteristics or measured in the field (Cook and Watt, 1982) in order to arrive at a map of soil suitability for irrigation printed on an air-photo mosaic of the Plains (Fig. 14). Armed with this

Fig. 14. 1:15 000 irrigation suitability map of part of Ngatarawa Plains, New Zealand (from Griffiths, 1975).

map, the planner could then make a decision based on economic and soil parameters, and choose whether to make a limited supply of water benefit a small area of greatest water need, or benefit a large area of small need. In the event, the high potential costs deterred farmers from supporting a comprehensive irrigation scheme and only piecemeal sprinkler irrigation operates to this day.

Problem-solving

The previous sections describe the use of soil information to assess land during normal agricultural or forestry use or before agricultural or foresty development. Often, though, an agricultural development runs into problems or, in extreme situations, has to be abandoned leaving land in a derelict state. Usually problems can be attributed to insufficient care at the initial planning stage. Such is the case of the North Kent Marshes in England. Loveland *et al.* (1987) describe how much of this area of clayey marine alluvium, affected to some degree by salt, was drained with the help of government grants and converted to arable farming from the mid 1960s. By the late 1970s some fields were showing signs of drainage failure and the subsequent ponding of surface water was causing serious loss of yield. In the 1980s a soil survey was carried out, samples were taken for laboratory analysis and electrical conductivity was measured at the intersects of a 250 m rectilinear grid. The results showed that the inherently unstable sodium-rich subsoils become actively unstable as salt-rich groundwaters were lowered and organic matter content declined under arable use. A single-factor map of electrical conductivity (Fig. 15) indicated where there was a large salt concentration in the soil solution ($EC > 2\ mS\ cm^{-1}$) and thus where vulnerability to structural collapse was least. The information gained from the survey not only identified the areas with the greatest potential problems then and in the future but also enabled better-planned remedial treatments with gypsum on the worst-affected areas (Hazelden *et al.*, 1986). It must be concluded, however, that the original agricultural development was ill-advised.

Urban and industrial development

Many of the properties which influence plant growth are also properties that affect the engineering use of the upper layers of soil. A slowly permeable subsoil not only

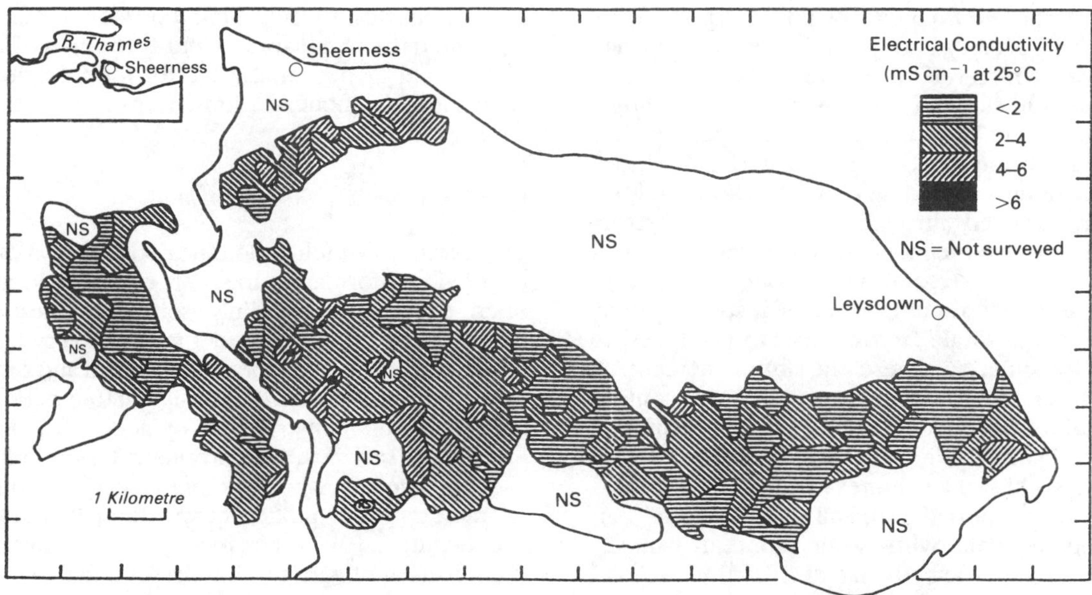

Fig. 15. Electrical conductivity of soils at 35–50 cm depth, Isle of Sheppey, England.

restricts root growth but also limits the use of the soil for septic tanks and makes a very poor subgrade for a highway (Bartelli, 1979). Bauer (1979) points out that the failure to recognize soil limitations in the planning and design stages of development projects not only results in additional site preparation and construction costs, but may also result in greatly increased maintenance and operating costs and the creation of slum and blighted areas. Thus, although the impetus for producing soil and interpretive maps in developed and undeveloped countries has largely been agricultural, there has been an increasing use of them for planning the expansion of urban communities, the siting of associated recreational facilities, the planning of routes for roads and pipelines, and the development of mineral reserves.

Urban expansion

With an increasing world population and, in many countries, a movement of the rural population into urban centres, there is a continual expansion of housing and industry onto land hitherto in agricultural use or under semi-natural vegetation. Ideally, urban expansion should be directed, whenever possible, onto land which:

(1) has the lowest intrinsic value for agriculture or conservation;
(2) enables structures to be safely founded and services to be efficiently provided; and
(3) is capable of providing a satisfactory living, working and recreational environment for the urban population.

Soils information in the form of pedological soil maps, single-factor soil maps and accompanying data to 1–2 m depth assists in all three of these evaluations. It is vital in assessing agricultural quality, it is useful in making preliminary assessments of site suitability and foundation conditions for one or two storey residences and services (though it does not eliminate the need for site-specific soil engineering investigations prior to construction) and it helps to ensure that amenity land is fit for the desired use. Barroso (1986) describes how fifteen map units on a pedological map were translated into eight geotechnical units for planning urban expansion west of Rio de Janeiro. In the USA, where soil resource surveys are widely used by town and country planners, Kellogg (1974) reported that the total benefits derived from soil surveys were: 50% from planning towns and suburban extensions; 25% from locating highways, airports, pipelines and other structures; and 25% from guiding use of soils in agriculture, forestry and recreation.

Protecting land from development

Many industrialized countries operate planning systems which ensure that non-agricultural development is directed away from the best quality agricultural land. In California, exclusive agricultural zoning was used to create districts in which farming is the only permitted use (Miller, 1979). Venn (1977) identified the need to identify prime agricultural land in South Africa which 'should be lost to food production only in exceptional circumstances'. The Town and Country Planning Amendment (1973, No. 95) in New Zealand states that 'the following matters are declared to be of national importance and shall be recognised and provided for in the preparation, implementation and administration of regional and district schemes ... the avoidance of encroachment of urban development on, and the protection of, land having a high actual or potential value for the production of food'. In England, Department of the Environment circular 16/87, Development Involving Agricultural Land (Department of the Environment, 1987), states that the general policy aim is to ensure that the best and most versatile land is not built on unless there is no other site suitable for the purpose.

Planners in England are required to consult the Ministry of Agriculture, Fisheries and Food before granting any planning application which would involve the loss of 20 hectares or more of Grades 1, 2 or 3a agricultural land. These grades are part of a five class agricultural land classification introduced in 1966, modified later and revised with more precise definitions in 1988 (MAFF, 1988). The grades are similar to the agricultural land capability classes of Bibby and Mackney (1977) described above, but with classes 5, 6 and 7 combined as one grade 5, and with grade 3 divided into a better subgrade 3a and a poorer subgrade 3b. Although the grades can be broadly interpreted from general-purpose pedological soil surveys, precision can only be achieved by a special-purpose survey which records particular factors such as stone content and slope in the way necessary to apply the classification. In planning the use of specific sites a determined developer with a large vested interest will want a low agricultural grading (e.g. 3b or 4) so that his venture can proceed through the planning system without objection. Sometimes a dispute can arise over the land grading and its resolution can depend on a single factor in the agricultural land classification system. Figure 16 shows a single factor soil map produced to resolve such a situation where the argument centred around whether the volumetric topsoil stone content was less than 15% (subgrade 3a) or greater than 15% (subgrade 3b) in a 20 ha site. Precise sampling on a 100 m grid and analysis of stone volumes in certain size grades enabled the production of a topsoil stone content map combining point data with stone content isolines. Because the survey showed that 80% of the site had a stone content greater than 15%, the agricultural objection to the development was withdrawn.

In broader planning terms, the use of soil information in planning urban expansion can be illustrated by reference to a study of the environs of Palmerston North, New Zealand (Cowie and Osborn, 1977). They felt that because the city is surrounded by soils of exceptional quality and was to expand at a rate well above the New Zealand average, early planning was essential to reduce haphazard urban sprawl and costly mistakes. Starting with a generalized soil map, Cowie and Osborn delineated land of four values for food production: high actual value, high potential value, moderate actual or potential value, and low actual or potential value. They then identified five categories of soil limitation for urban use and combined the two maps to identify the area of land not of high actual or potential value for food production and with slight to moderate limitations for urban use (Fig. 17).

Identifying constructional problems

The extent to which pedological soil surveys can be used to provide information to satisfy constructional needs is open to debate. Certainly soil investigations which barely extend to a depth of 1 m are of very limited use, but characterize the upper 1–2 m of soil and thus provide information relevant to the design of foundations for low buildings, the construction of access roads and the location of services. Two problems have hindered their use however. Firstly, pedologists classify soil physical characteristics in a different way to civil engineers (Lee and Griffiths, 1987). For example, they use a different classification of particle size classes and do not combine them with plasticity as in many of the engineering classifications (Fig. 18). Secondly, the sampling intensity for general purpose soil maps is often such that the soil map units might well contain small areas of contrasting soil where the particular development is sited. Because of this civil engineers prefer to rely on site specific analysis as an aid to design of structures. However, the pedological type of soil map can be an invaluable source of preliminary information in planning construction. For example, it can indicate areas of thin peat, high water-tables and acid, sulphate-rich or highly expansible subsoils not shown on any other type of earth science map. Lindsay *et al.* (1973) show how a soil map has been interpreted into four classes of potential corrosion

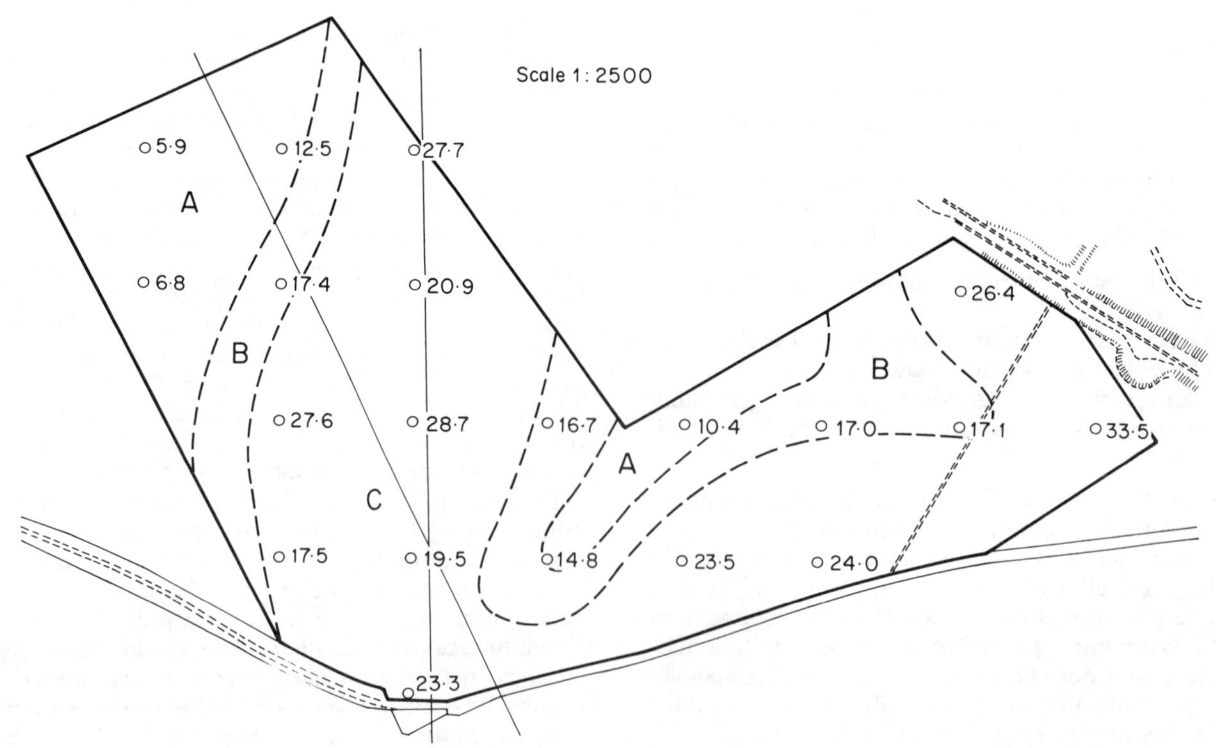

Fig. 16. Topsoil stone contents on the site of a proposed quarry extension in England.

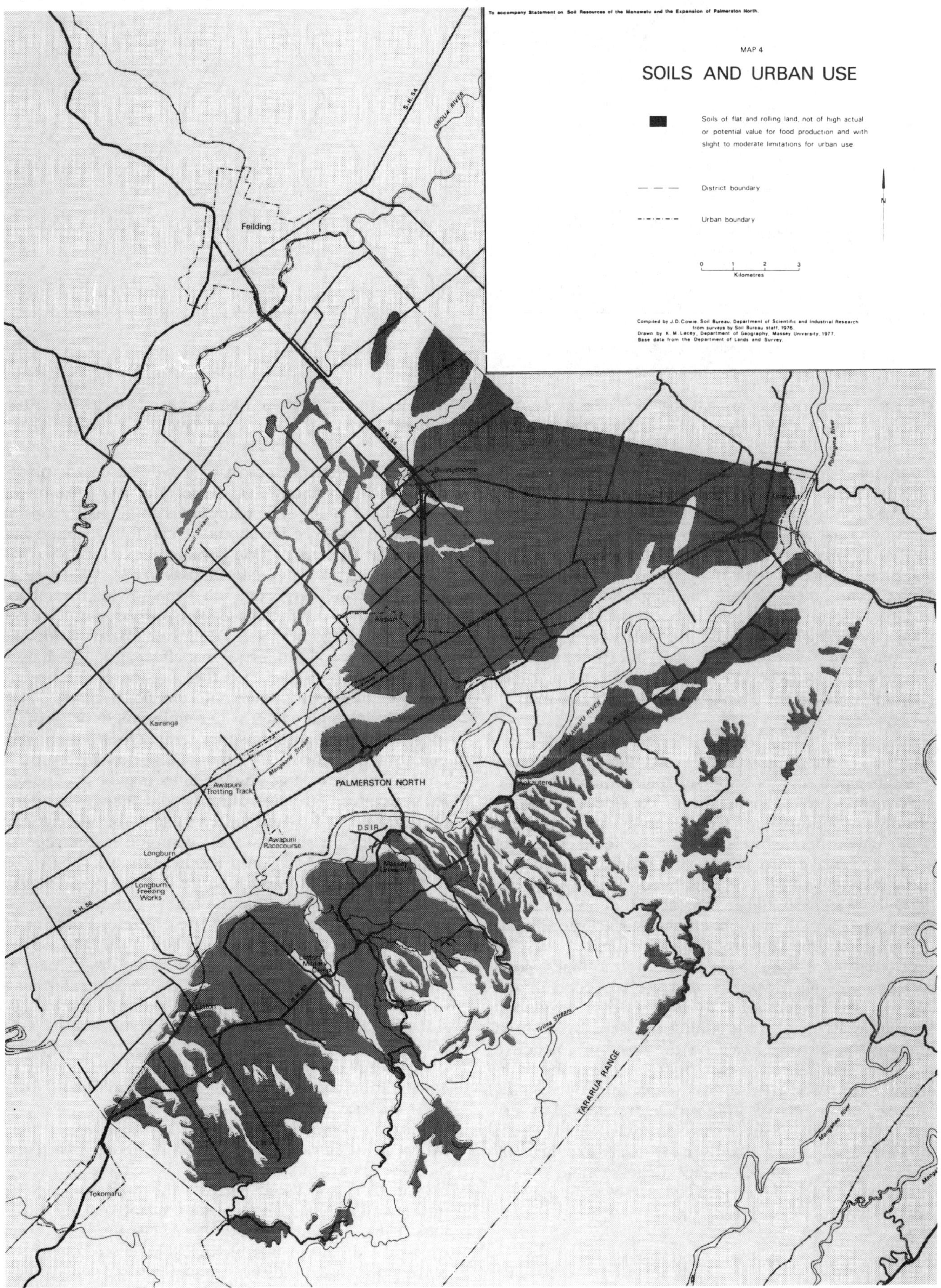

Fig. 17. The use of soils data in planning the expansion of Palmerston North, New Zealand (from Cowie and Osborn, 1977).

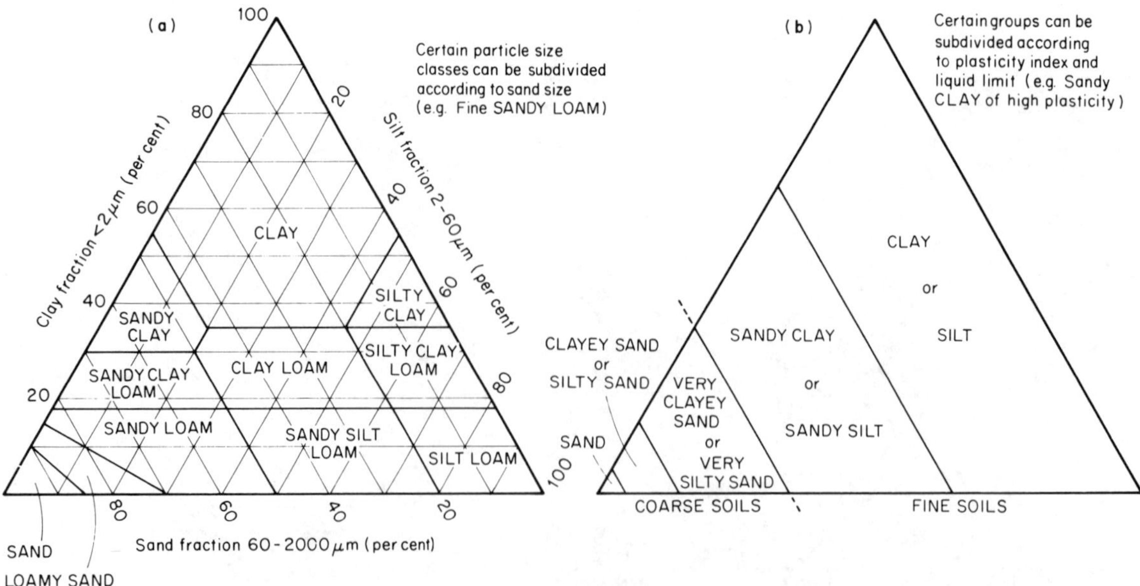

Fig. 18. Comparison of (a) pedological (Hodgson, 1976) and (b) engineering (Dumbleton, 1981) grading triangles for British mineral soils containing no stones.

hazard for concrete. A detailed soil survey of Montgomery County, Pennsylvania, was carried out with 'special emphasis on urban planning applications in addition to the traditional agricultural uses' (Witwer, 1966). Field sheets, at approximately 1:15 000, and accompanying mapping legends were then interpreted by the planners of the Montgomery County Planning Commission. This resulted in the production of several interpretative maps, including that in Fig. 19 which identifies five groupings of soils with different interpretations for construction. Witwer (1966) describes several other planning uses for the Montgomery County soil map.

On-site sewage disposal

Another common application of soil information in areas without piped sewers is the suitability for siting septic tanks. This can be an important consideration when planning developments remote from piped sewers. Septic tanks operate by distributing the liquid effluent of domestic sewage into an area of subsoil through a subsurface system of tiles or perforated pipes. A properly designed system should be capable of functioning at the designated capacity without effluent surfacing or polluting groundwater. The properties that limit the use of septic tanks are water-table depth, permeability, slope and flooding, all properties that are recorded in soil surveys. McCormack and Johnson (1982) developed a rating system for determining the severity of each limiting soil feature, based on the costs of corrective measures and the costs resulting from continuing limitations. They used this to group soils in Leon County, Florida, into five classes from very high potential to very low potential for on site sewage disposal. Rogoff (1982) encoded a soil map of Windsor township, Michigan, by a dot sampling technique and used a geographic information system to produce a grid cell map of limitations for home sewage disposal (Fig. 20).

Landscaping and recreation

Re-contouring, topsoiling urban gardens, tree-planting, provision of outdoor sports facilities and the creation of open spaces is as much a part of urban development as the built environment. The type of soil dictates to a large degree the range of trees that can be planted, the plants that can be established and the type and location of recreational facilities. In many areas good quality topsoil is a scarce resource and should be carefully stripped and stored during construction or mineral extraction so that it can be used at a later date. Lindsay *et al.* (1973) give an example of an interpretive soil map showing the suitability of surface soils for landscaping purposes in an area of Virginia. Montgomery and Edminster (1966) point out that the same soil properties that affect agricultural uses of soil are the ones that affect their use for recreation, but that the interpretations are different. For example, a clay soil suffering high winter water-tables might be satisfactory for growing autumn-sown cereal crops but entirely unsatisfactory for year-round nature trails. A topsoil containing a few large angular flints may be acceptable for agriculture but have painful consequences for participants of rigorous outdoor team games. Interpretations of soil maps for such uses are common in soil reports published by the Soil Conservation Service of the United States Department of Agriculture (e.g. Mettert, 1986).

The former West Riding County Council commissioned a detailed soil map of the Castleford district of Yorkshire, England (Hartnup and Jarvis, 1973) to aid the planning of an integrated reconstruction programme in an area partially blighted by the effects of coal-mining (opencast operations and deep-mine tipping and subsidence) and urban dereliction. Soil suitabilities as a source of topsoil and for various amenity uses (parks, gardens, golf courses and playing fields) were assessed in consultation with appropriate experts and ranked from most to least suitable (Table VI). Hartnup and Jarvis (1973) also commented about the development potential of certain localities based on their interpretive knowledge of soils. About one they stated 'The choice of uses is restricted by soil variability, and the predominance of poorly drained soils which make the land unsuitable for most uses ... The lakes and waterlogged ground, however, provide interesting biological habitats suitable for recreational uses; and the strips of freely drained Wick soils enhance this, since the main footpaths could be over dry land. The suggested use for this area is therefore a "passive" park or nature reserve. The freely drained soils could be used as (public) open spaces or alterna-

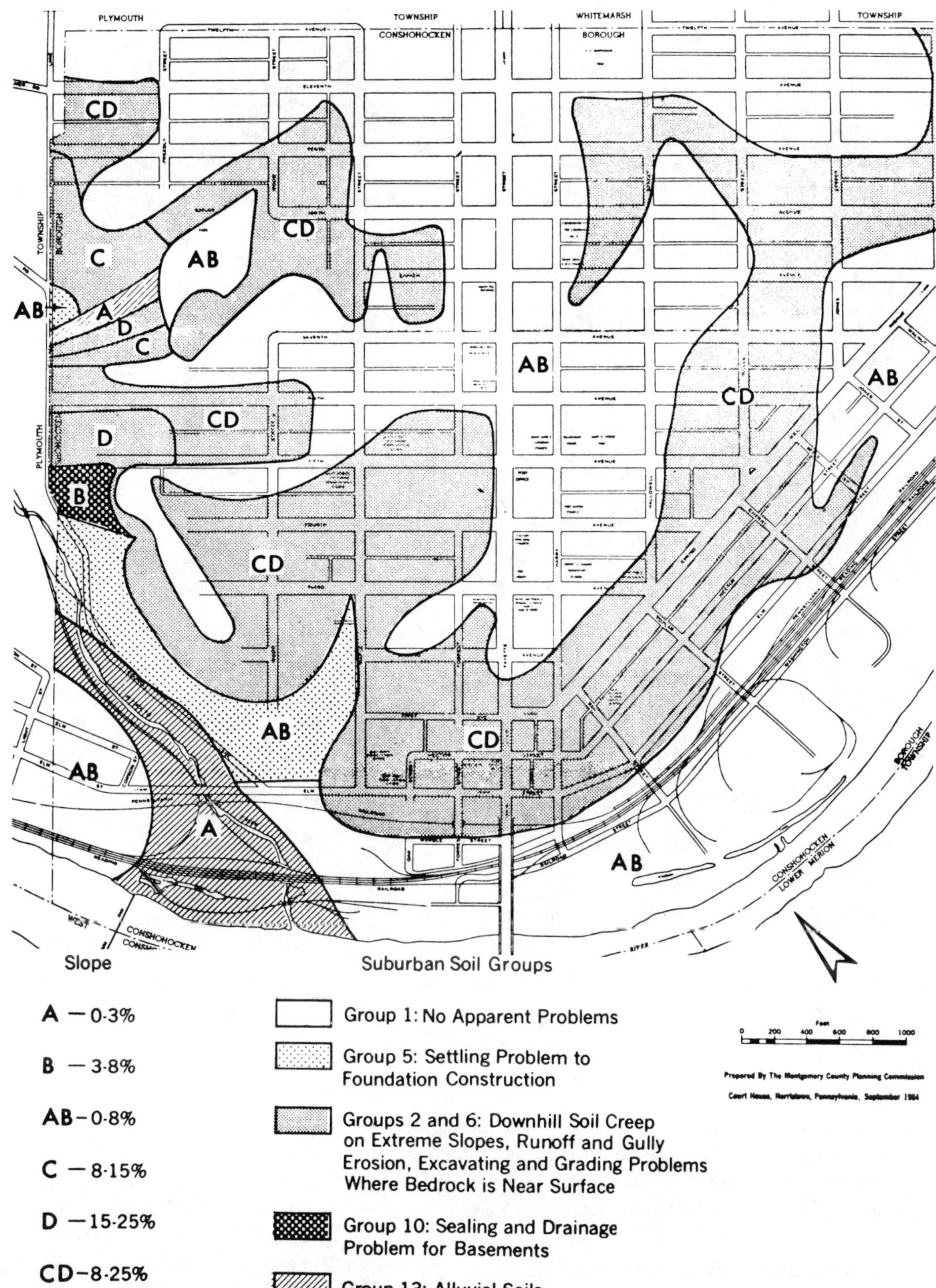

Fig. 19. Map of Conshøhocken Borough, showing some soil problems important to urban planning (from Witwer, 1966).

tively remain in agricultural use. Nature trails could be constructed, taking advantage of the wetland habitats; and if desired the lakes could be extended by removing material from waterlogged areas of the Quorndon series. The soil removed could serve as filling material for the various reconstruction programmes which are taking place in the area'. This interpretive use of soil survey data can be very useful in the formulation of local plans.

Planning roads and pipelines

Brink *et al.* (1982) state that 'Soil data are not required to as great a depth for planning roads, railways, canals and pipelines as for urban planning, and so the relatively shallow depths to which soils are examined in agricultural soil surveys is less of a disadvantage and data for these maps may help greatly in the selection of mapping units and the early assessment of the engineering properties of their soils'. In the USA extensive engineering testing has been carried out on the two to four major horizons of soil series (Fig. 21) and the results published as comprehensive reports (Thorburn *et al.*, 1966; Johnson, 1975). For road construction, Bartelli (1979) lists drainage, texture, type of clay and organic matter content as important soil characteristics for road subgrades and drainage, depth to water-table, texture and temperature as those important

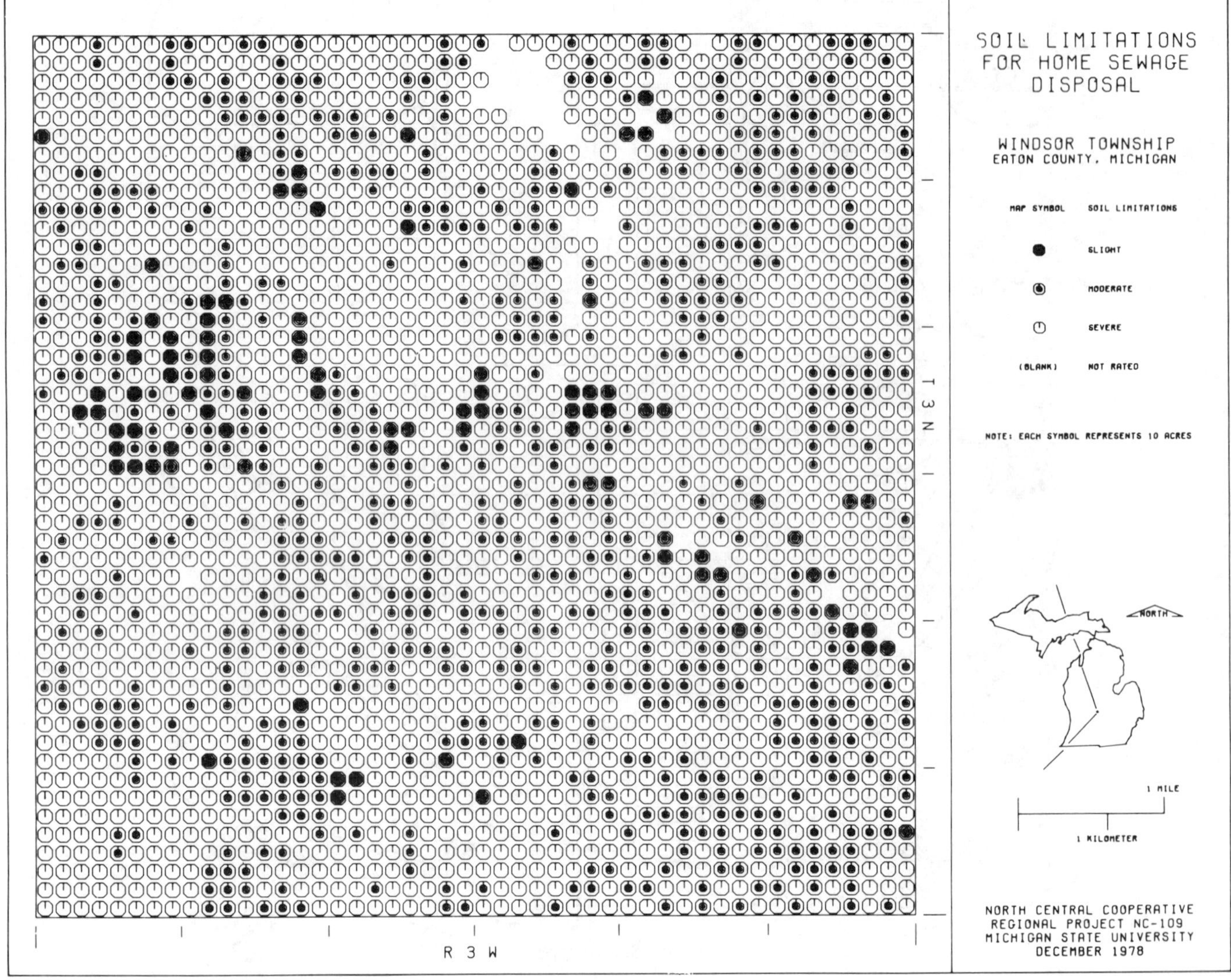

Fig. 20. Computer drawn interpretative map illustrating soil limitations for on-site waste disposal in Windsor Township, Michigan (from Rogoff, 1982).

for assessing susceptibility to frost action. McCormack and Fohs (1979) describe the extensive use made of pedological soil survey information by the Michigan Department of State Highways and state that soil surveys help in:

(1) Route selection—avoiding clays, organic materials and areas of high water-table reduces construction costs.
(2) Planning detailed soil sampling—avoiding ommission of detrimental soils or over-sampling of uniform tracts of soil.
(3) Planning drainage requirements—soil surveys are the only reliable source of information in most areas about seasonal high water-tables.
(4) Planning for erosion control—using special designs to minimize erosion in vulnerable soils.
(5) Recognizing excavation and grading problems—avoiding wet soils in winter, estimating for blasting requirements.
(6) Locating sources of sand and gravel for sub-base—soil surveys show sandy and gravelly horizons in the upper 2 m giving a clue to thicker resources.
(7) Selecting chemical stabilizers for fine-grained soils—soil-cement mixtures used in base courses depend on the soil series.
(8) Locating sources of topsoil—makes re-vegetating cuts and embankments easier.
(9) Locating rest areas—soils that are soggy and waterlogged are not desirable for picnic sites.
(10) Planning secondary roads, streets and car parks.

Allemeier (1973) describes how the specially commissioned strip maps of soil series are used in conjunction with a soils design chart to give recommendations concerning the engineering properties for highway design. The use of soil surveys for transportation planning has also been attempted in other countries; for example, in Scotland where McGown and Iley (1973) thought that soil maps would have been useful at the route planning stage and would have permitted site investigations to be more economically and more efficiently carried out. In England soil survey data is used in the Geotechnical Preliminary Sources Study (TRRL 403) stage of major road projects (in accordance with Dumbleton and West, 1976), but not to the extent that it is used in the United States.

Essential residential services, such as stormwater drains, freshwater mains, foul sewers and gas mains, and major regional pipelines such as for oil, natural gas, water or aviation fuel are nearly always placed underground fc economic and aesthetic reasons. They are usually installed

Table VI Suitability of soils for agriculture and amenity uses (from Hartnup and Jarvis 1973)

			Amenity							
	***Agriculture*[2]**		*Woodlands, parks and gardens*		*Golf courses*		*Playing fields*		*Borrow pits for*[3] *cladding tips etc.*	
			Main important attributes + ideal values							
	Texture: Loam, sandy loam or silt loam *Drainage:* Free to moderate *Depth:* 80 cm+ *Permeability:* Rapid but retaining sufficient moisture for plant growth in dry summers		*Texture:* As for agriculture *Drainage:* As for agriculture *Slope:* 4–8° *Variability:* High		*Drainage:* As for agriculture *Texture:* As for agriculture *Slope:* <4° *Variability:* High		*Drainage:* Free *Texture:* As for agriculture *Slope:* <1° *Permeability:* Rapid		*Texture:* Sandy loam or loam *Depth:* Deep sandy or loamy horizons	
Mapping units and symbols	*Limiting factors*	*Ranked position*	*Limiting factors*	*Ranked position*	*Limiting factors*	*Ranked position*	*Limiting factors*	*Ranked position*	*Limiting factors*	*Ranked position*
Aberford aF	*Depth:* Shallow in places	4	None	1	*Slope:* Often >4°	4	*Slope:* Often >1°	4	*Depth:* <80 cm	7
Wetherby Wu	*Depth:* <40 cm stony in places	7	*Depth:* <40 cm	3	*Slope:* Often >4°	7	*Slope:* Usually >3°	8	*Depth:* <40 cm	11
Watnall wA	*Texture:* Clayey *Drainage:* Poor	8	*Texture:* Clayey *Drainage:* Poor	10	*Drainage:* Poor *Texture:* Clayey	9	*Drainage:* Poor *Texture:* Clayey	9	*Texture:* Clayey	12
Kearby Ky	*Texture:* Clayey	6	*Texture:* Clayey *Slope:* <3°	9	*Texture:* Clayey	8	*Drainage:* Imperfect *Texture:* Clayey	6	*Texture:* Clayey	10
Dale Dl	*Texture:* Clayey *Permeability:* Slow	9	*Texture:* Clayey *Drainage:* Poor	11	*Drainage:* Poor *Texture:* Clayey	10	*Drainage:* Poor *Texture:* Clayey	10	*Texture:* Clayey	13
Swindon Bank sB	*Depth:* Shallow and stony in places	5	None	2	*Slope:* Often >4°	6	*Slope:* Often >2° *Stones:* Stony in places	5	*Depth:* <80 cm	6
Wigton Moor Ww	*Drainage:* Poor	11	*Drainage:* Poor	8	*Slope:* Often >4° *Drainage:* Poor	11	*Drainage:* Poor *Slope:* Often >2°	11	*Texture:* Loamy *Stones:* Stony	8
Wick wQ	*Permeability:* High, but with few fine pores for moisture retention in dry summers	2	*Slope:* <1°	5	None	1	None	2	None	1
Arrow aO	*Drainage:* Imperfect	3	*Slope:* <1° *Drainage:* Imperfect	6	*Drainage:* Imperfect	2	*Drainage:* Imperfect	3	None	2
Quorndon Qn	*Flooding risk:* High *Drainage:* Poor and very poor	14	Not recommended because of high flooding risk						Access difficult at flood times	3
Wharfe Wv	No flooding risk	1	*Slope:* <1°	4		3		1	*Texture:* Can have too much clay	4
	Flooding risk: High	12	Not recommended because of high flooding risk						Access difficult at flood times	5
Conway Ch	*Flooding risk:* High *Drainage:* Poor	13	Not recommended because of high flooding risk						*Texture:* Clayey and silty	14
Dale-Arrow Dl/aO	*Texture:* Often clayey *Variability:* High	10	*Texture:* Clay in places *Drainage:* Poor and imperfect	7	*Texture:* Clayey in places *Drainage:* Poor and imperfect	5	*Drainage:* Poor and imperfect *Texture:* Clayey in places	7	*Texture:* Clayey in places *Variability:* High	9

Notes: 1. The ranking positions of soils in this Table are based on contemporary experience. The suitability of a particular site for a specific purpose should be assessed by a more detailed site survey. See Table 9 for suitabilities for roads, buildings and services.
2. Agricultural land use capability is discussed in Chapter IV.
3. Soils with low rankings for other uses are ranked high for 'borrow pits'—*e.g.* Quorndon series.

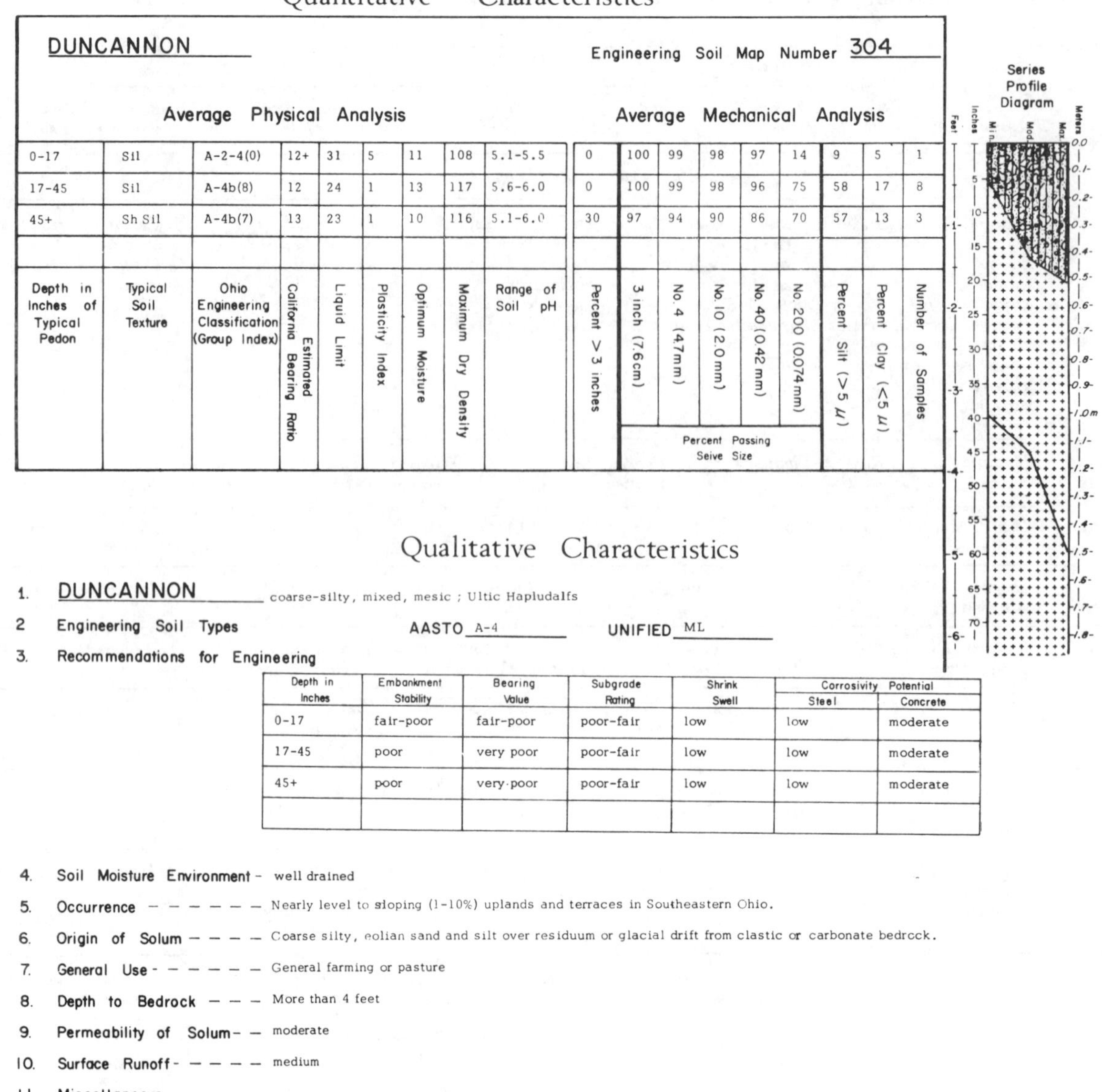

Quantitative Characteristics

DUNCANNON — Engineering Soil Map Number 304

Depth in Inches of Typical Pedon	Typical Soil Texture	Ohio Engineering Classification (Group Index)	Estimated California Bearing Ratio	Liquid Limit	Plasticity Index	Optimum Moisture	Maximum Dry Density	Range of Soil pH	Percent > 3 inches	3 inch (7.6 cm)	No. 4 (4.7 mm)	No. 10 (2.0 mm)	No. 40 (0.42 mm)	No. 200 (0.074 mm)	Percent Silt (>5 μ)	Percent Clay (<5 μ)	Number of Samples
		Average Physical Analysis							Average Mechanical Analysis (Percent Passing Seive Size)								
0-17	Sil	A-2-4(0)	12+	31	5	11	108	5.1-5.5	0	100	99	98	97	14	9	5	1
17-45	Sil	A-4b(8)	12	24	1	13	117	5.6-6.0	0	100	99	98	96	75	58	17	8
45+	Sh Sil	A-4b(7)	13	23	1	10	116	5.1-6.0	30	97	94	90	86	70	57	13	3

Qualitative Characteristics

1. DUNCANNON coarse-silty, mixed, mesic ; Ultic Hapludalfs
2. Engineering Soil Types AASTO A-4 UNIFIED ML
3. Recommendations for Engineering

Depth in Inches	Embankment Stability	Bearing Value	Subgrade Rating	Shrink Swell	Corrosivity Potential Steel	Corrosivity Potential Concrete
0-17	fair-poor	fair-poor	poor-fair	low	low	moderate
17-45	poor	very poor	poor-fair	low	low	moderate
45+	poor	very-poor	poor-fair	low	low	moderate

4. Soil Moisture Environment - well drained
5. Occurrence — — — — — — Nearly level to sloping (1-10%) uplands and terraces in Southeastern Ohio.
6. Origin of Solum — — — — Coarse silty, eolian sand and silt over residuum or glacial drift from clastic or carbonate bedrock.
7. General Use - — — — — — General farming or pasture
8. Depth to Bedrock — — — More than 4 feet
9. Permeability of Solum- — moderate
10. Surface Runoff - — — — — medium
11. Miscellaneous — — — — —

Fig. 21. Engineering interpretation for a soil series in Ohio (from Johnson, 1975).

at a depth of 1–3 m depending on the size of the pipe and the terrain they are crossing. The underground routing of pipes requires:

(1) soils that are easy to excavate;
(2) soils that are stable in the sides of trenches during excavation;
(3) soils that can support the pipes without appreciable movement; and
(4) soils that do not cause excessive corrosion.

Many of the soil properties that are collected during pedological soil surveys are factors that determine these four requirements. For example, the depth to (and the nature of) rock and the presence of cemented layers affects the ease of excavation.

High water-tables and soft materials exacerbate the likelihood of trench collapse while organic materials and swelling clays can result in fracture of pipes through differential movement. Soil stoniness is also an important factor as large stones can damage protective pipe coatings, necessitating the importation of sand to cover the pipes before backfilling commences.

Adverse soil properties not only slow down the rate of construction but increase the cost of materials needed to complete the installation successfully. Redding (1987) states that rock within 1 m of the surface can increase the costs of materials and slow pipeline construction by a quarter. Water entry below 0.5 m or the presence of organic soil can slow progress to a quarter of the original rate and increase the cost of materials by more than a third. When Redding (1987) described the merits of stereo air-photo interpretation, superficial geological mapping and field study to the planning of the route of a natural gas pipeline in the Irish Republic, he described the processes of pedological soil survey as carried out in many countries of the world, thus demonstrating its usefulness for planning pipeline routes.

Corrosion of pipelines is also soil-dependent, being affected by acidity (pH), drainage status, soil texture and the presence of soluble salts. Excessive corrosion greatly increases maintenance costs and can cause loss of life if, for example, a gas main fractures and the gas ignites. A knowledge of the corrosiveness of different soils not only influences the choice of original route, so that potentialy corrosive land can be avoided, but enables the utility authorities to better plan costly programmes of pipeline replacement and the location of anti-corrosion technology such as cathodic protection. Although a study by Corcoran *et al.* (1977) showed that an existing soil map was a good guide to the potential corrosiveness of land crossed by a natural gas pipeline in Oxfordshire, England, it is rare for published soil maps to be used for assessments of potential corrosivity in the United Kingdom. This contrasts with the United States where the risk of corrosion to concrete and uncoated steel is routinely assessed for all soils mapped by the Soil Conservation Service (e.g. Mettert, 1986) and regularly used for planning purposes. Thus, despite their limitation of only providing information to a depth of 1–2 m, it is clear that soil maps have a role to play in planning routes for transportation and utilities.

Environmental protection

As the twentieth century nears its close, man is becoming increasingly aware of the effects of his activities on the quality of the environment. In parts of Africa and South America, for example, catastrophic soil erosion and creeping desertification are often the consequences of ill-conceived clearance of natural vegetation and inappropriate development. In many countries, changing rural land use and vegetation is having a marked influence on patterns of rainfall, run-off from the land, river hydrographs and flood frequency. Intensive farming practices cause their own environmental and planning problems as animal effluents and artificial fertilizers find their way through the soil to water supplies. Industrialized countries suffer problems of disposal of toxic waste which, in certain areas, is already leading to a build-up in toxicity in soils. Airborne pollutants cause problems further from their source, increasing the acidity of rain and thus increasing the acidity of soils. Surface mining can lay waste large areas of land without the means of returning them to a beneficial after-use. Soil plays some part in all of these problems and man is still discovering many ways in which soil surveys can help solve them.

Soil erosion

Soil erosion is not a new problem. It has been continuing since man first started clearing the land for agriculture, and there are well documented instances as long as 5000 years ago (Davidson, 1982) from the countries bordering the Mediterranean Sea. Soil erosion is not, however, solely the concern of rural areas: severe cases can block roads, fill drainage ditches and cause the accelerated infilling of lakes and reservoirs. Attempts at revegetating badly eroded land can fail, preventing the re-establishment of a balanced ecosystem and resulting in many other environmental ills. Even slow rates of erosion, as in temperate climates, can result in significant loss of crop yield as the depth of soil available for rooting declines, and will eventually threaten agricultural use (Morgan, 1985).

Erosion increases as soil structure deteriorates and as the organic matter content of the soil declines. Land with convex slopes is particularly vulnerable to water erosion and seasonal cropping can exacerbate the problem through leaving the soil bare at times of the year when intense or prolonged rainfall is most likely. Other soils are particularly liable to erosion by wind, especially where shelter has been removed. Only by mapping the soils and classifying those vulnerable to erosion (Morgan, 1985) can critical areas be defined, so that developments can be planned to minimize the short-term and long-term detrimental effect on the environment.

Drainage and hydrology

Changing land use not only causes soil erosion but can have hydrological repercussions, sometimes causing flooding, unsightly colouration of drinking water supplies or affecting developmental potential downstream.

Soils are, in fact, an important component in watershed planning. Bauer (1973) shows how soils data are used in comprehensive watershed studies in Wisconsin—studies designed to produce a long-range plan for the development of water-related community facilities, including integrated proposals for pollution abatement, drainage and flood control, and land and water use.

Soil factors such as infiltration rate and storage capacity have long been used in the design of deterministic models for catchment hydrology (e.g. Chapman, 1968). In an attempt to extend this approach to flood studies on a regional scale, a scheme for interpreting soil maps for their winter rain acceptance potential was devised by Farquharson *et al.* (1978) and applied to the 1:1 M soil map of England and Wales. Later it was applied to 1:250 000 county soil maps (Fig. 22) and is currently being extended at this scale to the whole of the United Kingdom as the HOST (Hydrology of Soil Types) project.

As well as soils affecting regional hydrology, changes in regional and local hydrology also affect soils and their productivity. The construction of reservoirs and pumping of groundwater can have adverse effects on the water regime of soils in the vicinity, perhaps affecting crop yield. A large groundwater abstraction scheme in the Tern Valley of England caused fears among the farming community which have only been allayed by a detailed mapping of soils and allied studies of soil moisture regime.

Conservation and creation of wildlife habitats

The kind and abundance of wildlife depend largely on the amount and distribution of food, cover and water. Soils affect the type and amount of vegetation and the surface retention of water and are thus an important consideration in planning for wildlife. A knowledge of soils on a site is an important basis for:

(1) protecting and drawing up sympathetic management plans for existing sites; and
(2) assessing the suitabilities of new sites for the creation of habitats.

In Britain, soil surveys are often commissioned for those Sites of Special Scientific Interest where the interest is in a flora dependent on the medium in which it grows. A detailed knowledge of the soil and environmental properties which result in the development of the specific assemblage of plants helps management and conservation of the site. In the United States, all soils in recent surveys have been assessed for their potential for different habitats, such as wetland, and their potential for specific elements of a habitat, such as for supporting wild herbaceous plants (Table VII).

Restoration of mineral workings and mineral tips

Throughout the world, large areas of land are being drastically disturbed by excavation for near-surface mineral deposits or being covered by waste from deep mines. In many countries there are strong lobbies for

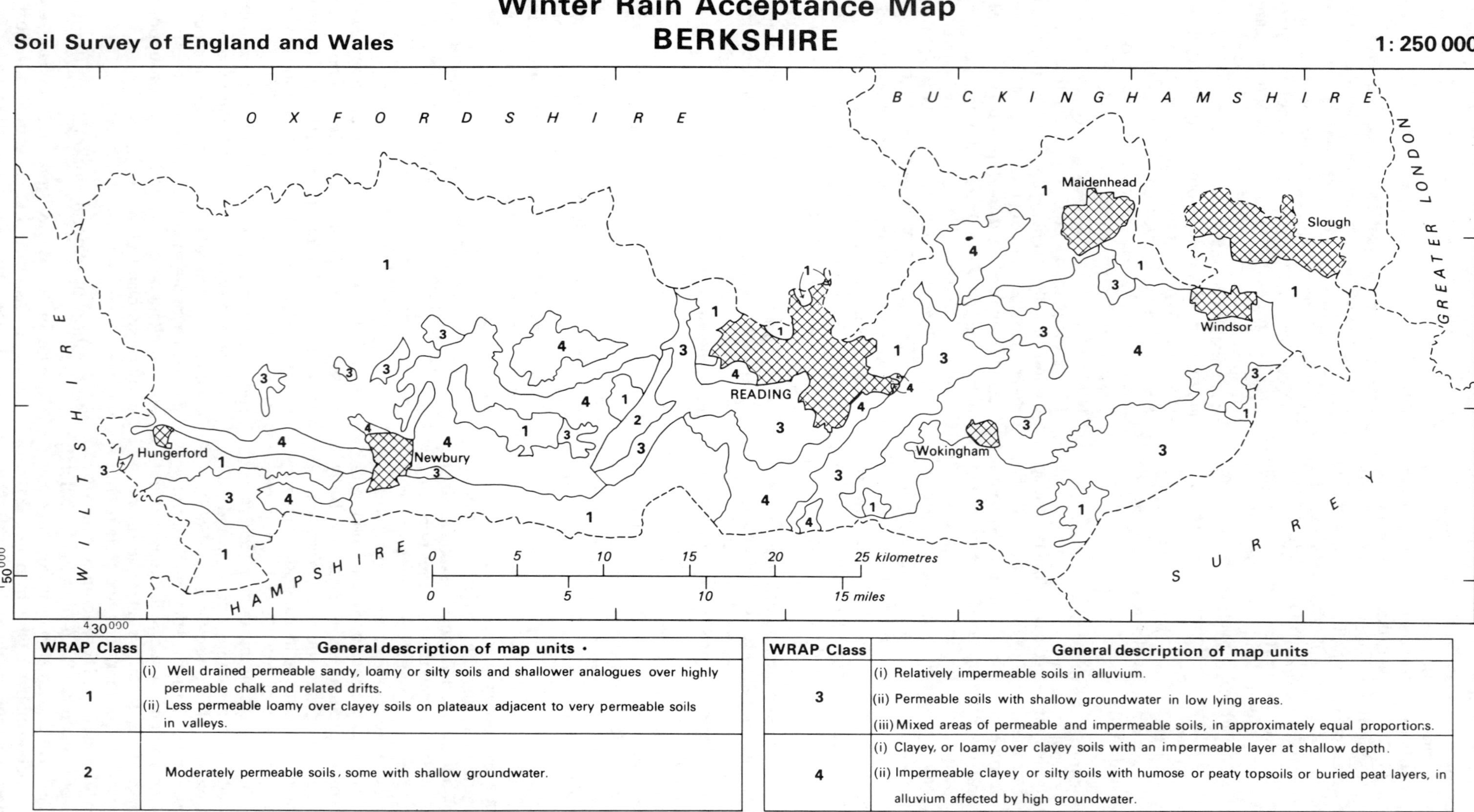

WRAP Class	General description of map units
1	(i) Well drained permeable sandy, loamy or silty soils and shallower analogues over highly permeable chalk and related drifts. (ii) Less permeable loamy over clayey soils on plateaux adjacent to very permeable soils in valleys.
2	Moderately permeable soils, some with shallow groundwater.
3	(i) Relatively impermeable soils in alluvium. (ii) Permeable soils with shallow groundwater in low lying areas. (iii) Mixed areas of permeable and impermeable soils, in approximately equal proportions.
4	(i) Clayey, or loamy over clayey soils with an impermeable layer at shallow depth. (ii) Impermeable clayey or silty soils with humose or peaty topsoils or buried peat layers, in alluvium affected by high groundwater.

Fig. 22. Winter rain acceptance potential (WRAP) map of Berkshire, England (scale 1:250 000) (from Jarvis *et al.*, 1979).

Table VII Habitat potential for five of the soils in Fig. 8 (after Mettert 1986)

	Potential for habitat elements							*Potential as habitat for*		
Soil name and map symbol	*Grain and seed crops*	*Grasses and legumes*	*Wild herbaceous plants*	*Hardwood trees*	*Coniferous plants*	*Wetland plants*	*Shallow water areas*	*Openland wildlife*	*Woodland wildlife*	*Wetland wildlife*
10B Metamora	Fair	Good	Good	Good	Good	Very poor	Poor	Good	Good	Very poor
12 Corunna	Poor	Fair	Fair	Fair	Fair	Good	Good	Fair	Fair	Good
25A Londo	Fair	Good	Good	Good	Good	Fair	Fair	Good	Good	Fair
36 Tappan	Good	Fair	Fair	Fair	Fair	Good	Good	Fair	Fair	Good
40B Chelsea	Poor	Fair	Fair	Poor	Poor	Very poor	Very poor	Fair	Poor	Very poor

minimizing the environmental impact of such mining activities, and for ensuring that the land can be eventually restored to a beneficial and visually acceptable after-use, rather than left in a derelict or neglected state.

In Great Britain, the Town and Country Planning Act 1971 (HMSO, 1971) and the Town and Country Planning (Minerals) Act 1981 (HMSO, 1981) provide the basis for control of mineral development. Development plans 'should include policies to ensure that the land from which minerals have been extracted or onto which mineral wastes have been deposited is landscaped or reclaimed to facilitate a beneficial after-use of the site' (DoE, 1988a). As in other forms of development (see above), one aim is to protect the best and most versatile agricultural land but, because minerals have to be worked where they occur, the feasibility of a high standard of restoration is also considered. To achieve this, planning permission to work land for minerals normally includes conditions which 'require the separate stripping, storage (where necessary) and respreading of topsoil, subsoil or any other soil making materials' (DoE, 1988b).Though general soil mapping can aid the preparation of development plans, a detailed soil survey of the upper 1–2 m, tailored to provide information to meet these aims, is usually required by the mineral developer.

The soil survey, if accompanied by the correct analyses, will ensure that:

(1) The agricultural land grades present can be determined.
(2) The properties of all soil layers on site are recorded before working commences.
(3) Individual soil layers can be assessed for their suitability for stripping, storage, replacement and the planned after-use.
(4) Volumes can be calculated for all soil materials to be stripped and retained.
(5) Adjustments (if any) to soil acidity or nutrient levels can be carried out prior to or during soil stripping.
(6) Soil handling can be designed to minimize deleterious effects on soil structure as a result of handling at inappropriate moisture contents.

These were the aims of a survey carried out for an area on which mineral wastes from a power station were to be deposited. The part of the report reproduced in Fig. 23 shows the three subdivisions made for soil stripping purposes.

In moist temperate climates such as in Great Britain, subsoils of high clay content and plasticity can rarely be stripped without suffering severe damage to their structure, and where such materials occur alternative soil-forming materials from below 2 m depth might have to be sought by trial pit or borehole or, at a later stage, during the working of the quarry. Restoration (see Chapter 5 on Mineral Resources) can be carefully planned by careful assessment of soil resources at an early stage of the development, but often the planning authorities are faced with older mineral workings for which little or no soils were stored or workings which have been restored to a state less than optimum for the desired after-use. Most problems of the latter type are caused by over-compaction of the soils during handling, resulting in excess surface wetness in winter and droughtiness, due to poor root ramification, in summer. These problems are best identified and corrected during the five years after-care period (HMSO, 1981) which is designed to bring the land to a required standard for the relevant after-use. Soil surveys during this period are designed to identify the specific defects (a pedological pattern in soil properties is not recognizable on restored land). They are likely to be carried out by sampling on a closely spaced grid or along random traverses. Such surveys can identify physical defects (Fig. 24), which can be remedied in various ways, or chemical defects which can be remedied by careful applications of fertilizer.

Chemical pollution

The interaction of soils and chemical pollution can take many forms. Disposal on the land of industrial waste or sludge from sewage works can result in a build-up of heavy metals or other deleterious compounds and restrict future development options. The processes of soil acidificaton, lake eutrofication and forest ill health in Europe is attributed partly to industrial emissions such as sulphur dioxide and nitrogen oxides ('acid rain') and, thus, has repercussions for broad-scale development planning. In intensive agriculture, soil must act as a filter for animal wastes and agrochemicals, both potential pollutants of

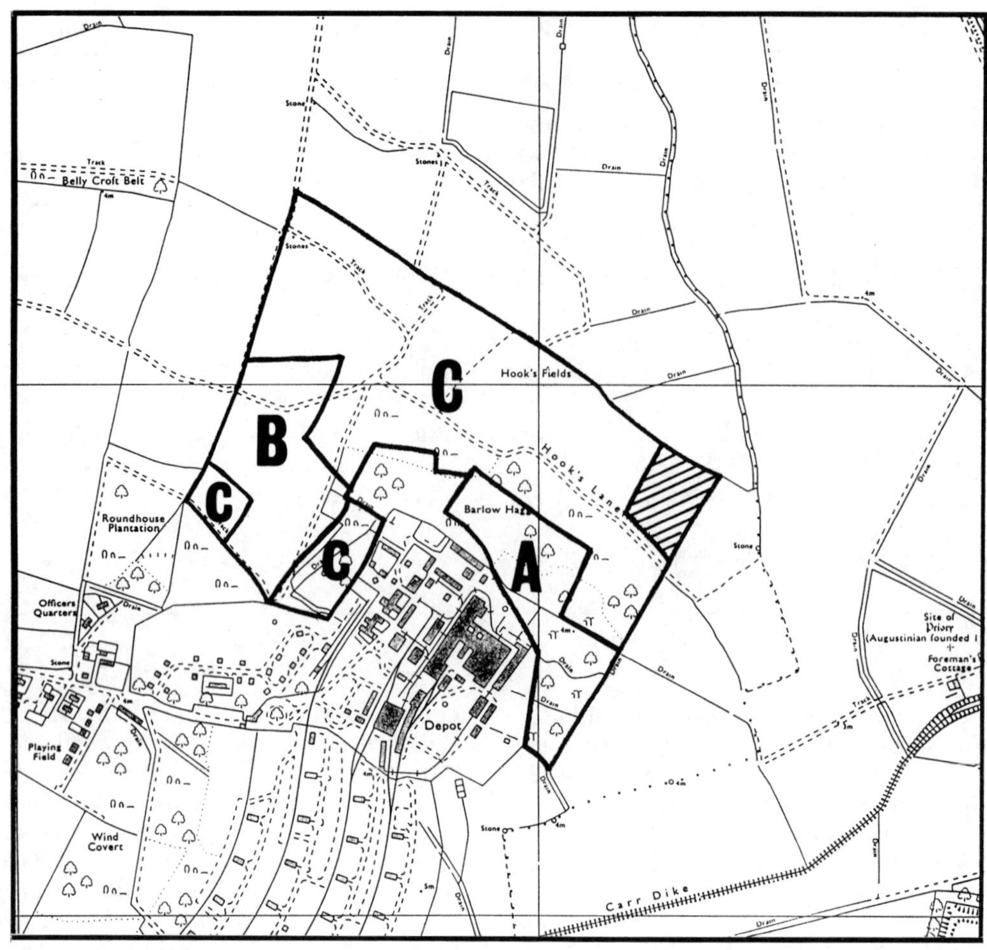

Resource block	Dominant soil(s)	Area (ha)	Topsoil resources Average thickness (m)	Topsoil resources Volume (m^3)	Subsoil resources Average thickness (m)	Subsoil resources Volume (m^3)
A	Everingham	7	0.23	16 100[1]	0.77	53 900
B	Portington	7.5	0.26	19 500	0.30	22 500
C	Foggathorpe Fenton	31	0.24[2]	74 400[1]	not suitable	
			TOTAL	110 000		76 400

[1] Tree roots and stumps might hinder topsoil recovery in scrubland areas.
[2] 0.17 m in the scrubland and 0.26 m in the farmland.

Fig. 23. Soil resource assessment of a proposed construction area in England.

aquifers and watercourses. Where it cannot effectively do so, agricultural development must be controlled.

Acceptance of agricultural wastes and chemicals

As populations grow larger, more land is taken out of housing and more food must come from a correspondingly small acreage of farmland. This leads to an intensification of agriculture and, for livestock farming in many areas of the world, farm effluent now presents serious disposal problems. Already this is leading to restrictions on the expansion of certain types of intensive livestock farming in the Netherlands. Traditionally, effluent has been applied to soil which is expected to filter and digest the material and release a filtrate free of pollutants while remaining relatively unaltered and still suitable for a beneficial use. There are two main aspects to assessing the suitability of soils for slurry disposal, their physical capacity to absorb the liquid component and their chemical capacity to absorb components of slurry which might cause pollution of soils and groundwater. Lea (1979) lists rockiness, stoniness, flood hazard, moisture deficit, winter rain acceptance potential, porosity of the upper 10 cm of soil and soil trafficability as properties limiting the physical capacity of soils to absorb slurry, and describes a slurry acceptance classification that can be applied to conventional soil maps to produce maps of soil suitability for slurry acceptance. However, his system ignores the difference between soils in their capacity to absorb the high levels of phosphate present in slurry. In intensively farmed areas where groundwater is close to the surface, pollution of water supplies by phosphate is a particular problem. Breeuwsma *et al.* (1986) interpreted the 1:50 000 soil maps of the Netherlands to show the

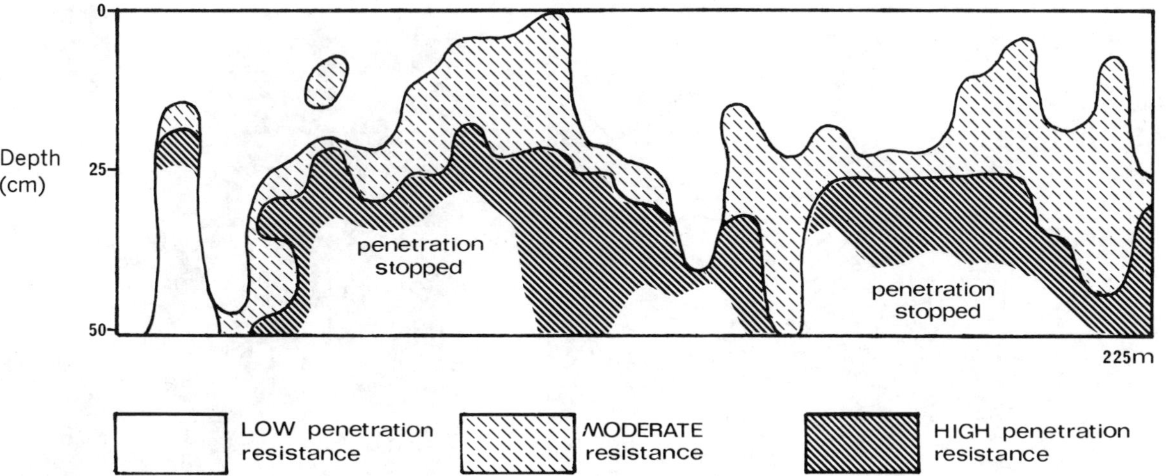

Fig. 24. Penetrometer transects used to identify over-compacted layers in soils restored after sand and gravel quarrying.

potential capacity of soil horizons to absorb phosphate to the depth of the average highest and lowest water-table (Fig. 25). Nitrate pollution of water supplies is caused not only by animal manures but also by artificial nitrogeneous fertilizers. Already deep groundwater-tables are being affected and pumping of some drinking water supplies is having to be suspended. There are also increasing worries about the migration of agricultural pesticides into water supplies. The mapping and display of relevant soil parameters (Fig. 25) is seen as a vital aspect of research while the establishment of groundwater protection zones is being discussed.

Disposal of sewage sludge

While much industrial waste is disposed of in deep excavations, urban waste from sewage works has been traditionally disposed of on the land surface. It provides nitrogen and phosphorus for plant growth and increases organic matter levels in the soil. However, industrial waste has, and in many parts of the world still does, find its way untreated into the normal urban sewer system bringing contamination to land on which sludge is spread and to water draining from the land. Heavy metals such as lead, nickel, copper, zinc, cadmium, chromium and mercury can accumulate in the soil and enter the food chain through insects and small mammals, through crops and through grazing cattle. Statutory limits being introduced (EC, 1986) for levels of individual metals in soils have repercussions for the long-term planning of sludge disposal and for the reclamation of some heavily sludge amended areas. With detailed soil sampling and analysis, grid maps can be prepared to show the distribution of metals and identify land where sludge disposal is still an option. Analysis of levels in current topsoils and in layers below can help formulate plans for reclamation by deep mixing. Furthermore, analysis of pH and delineation of pedological soil classes can identify areas where availability of and uptake of metals by crops is likely to be of greatest concern.

Soil acidification

Soil acidification, its effect on forest growth, crop growth and water quality is also of growing concern in Europe. Insomuch as man-made sources of acid-forming emissions are blamed for most of the acidification, an understanding of acid-soil interrelations is important in planning programmes which seek to reduce emissions by control of certain forms of development. Figure 26 shows the relationship of soil neutralizing capacity, as interpreted from the national soil map of England and Wales, to dry deposited acidity from SO_2 and can identify the areas of land most at risk to acidification. Holmberg *et al.* (1987) combined soil neutralization capacity with aquifer neutralizing capacity to produce an aquifer risk map for most of Europe.

Soil information can also be intricately involved with specific techniques for controlling emissions. Flue gas desulphurization of coal-fired power stations, for example, not only requires large quantities of lime from surface quarries, but produces large quantities of gypsum which could lower gypsum prices enough for it to be an economic option for reclamation of salt-affected soils, as in the coastal marshes of Eastern England (see above).

SOIL MAPPING IN INTEGRATED STUDIES

Most of the examples cited above relate to the mapping of soils to meet single planning and development objectives through a process of land evaluation concerned principally with the physical or ecological aspects of the land and its use. However, a major use of soil maps is in integrated studies of land use options under static or changing socio-economic factors.

Use of soil maps in integrated investigations is by no means new. Soil and vegetation maps formed an important component of the development survey carried out from Central African Rail Link (UK Colonial Office, 1952). Soils were just one facet of the study of the possible development of the areas to be served by the proposed link. However, many soil surveys and their interpretive applications have been carried out in a piecemeal fashion and, as Lee and Griffiths (1979) point out, their value to the planner would have been increased significantly if they could have been integrated into a more comprehensive approach. The Soil Surveys published by Soil Conservation Service of the United States Department of Agriculture go some way towards achieving this by presenting simple interpretive tables for a comprehensive range of uses including agriculture, woodland, environmental plantings, recreational development, wildlife habitat, building site development, sanitary facilities, water management and sources of

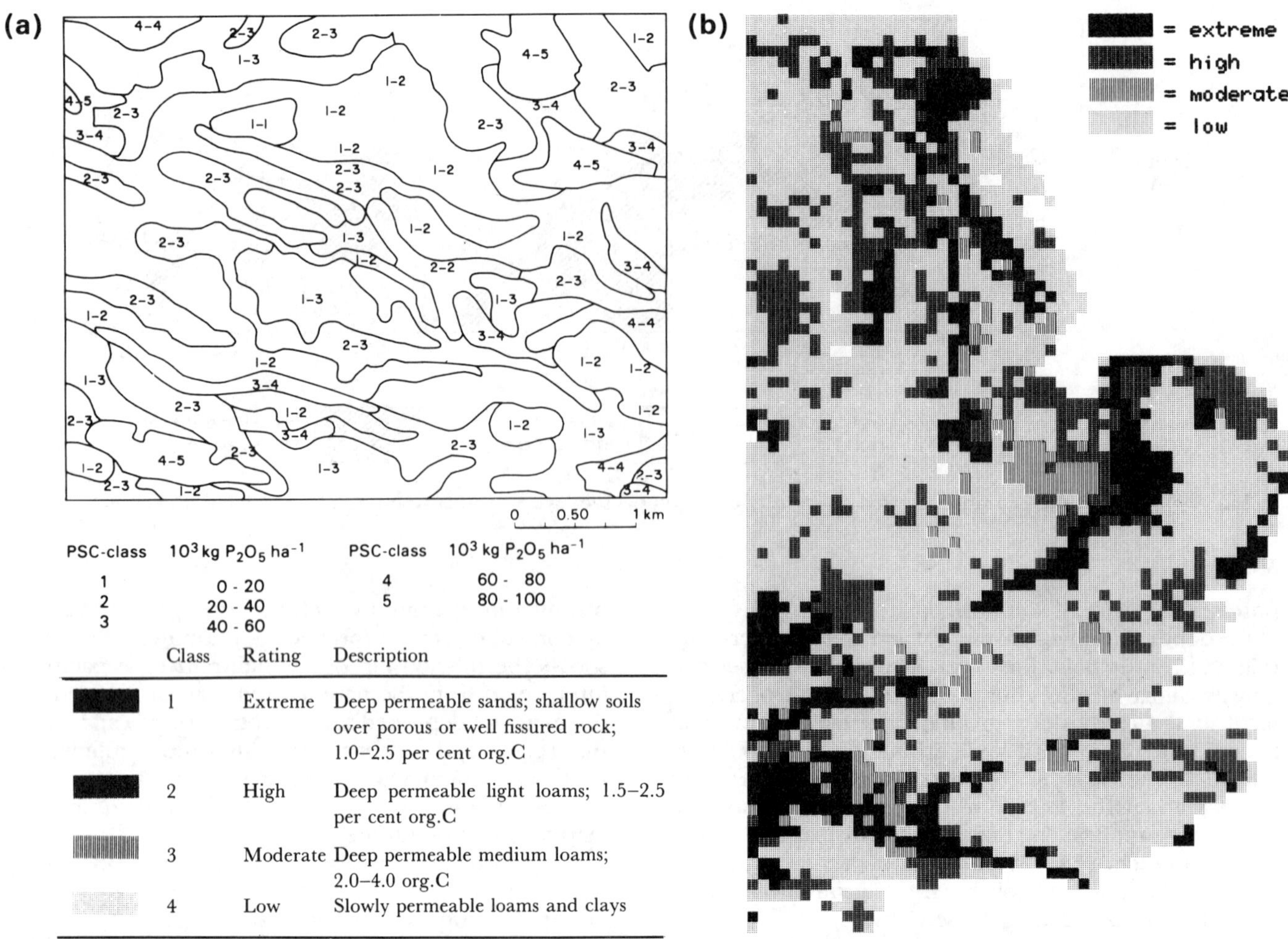

Fig. 25. Applications of soil mapping to groundwater protection. (**a**) Phosphate adsorption capacity classes in part of the Netherlands (from Breeuwsma *et al.*, 1986). (**b**) Nitrate leaching risk—soil factors in Eastern England.

construction material (e.g. Mettert, 1986). In the United Kingdom, the recent interest in environmental geological mapping (Geomorphological Services Ltd. 1988), including soil studies, is a useful start to adopting an integrated approach in areas of planning interest. Soil survey work in the Southampton area is generating spatial data which, when stored in vector or raster format in a geographic information system, can be interfaced with interpretive models and overlaid on other data to produce a series of maps, each targeted at a specific planning or land use issue.

Local planning documents often contain information on soil distribution or land quality in an integrated approach to planning land development. The Linn County Planning Department (1970) produced a comprehensive land use plan for Linn County, a rural area of Oregon. Using data largely derived from soil surveys, they assessed suitability for various opposing uses and produced an optimum land utilization map for the county. Bauer (1973) describes how soils data are fully integrated into the on-going regional planning programme for land use, transportation facilities and other public facilities and utilities in southeastern Wisconsin. Each kind of soil was rated for potential agricultural use, wildlife–soil relationships, non-farm plant–soil relationships, soil–water relationships and soil engineering properties to give interpretations for planning uses, including:

(1) Suitability ratings for potential intensive residential, extensive residential, commercial, industrial, transportational, natural and developed recreational, and agricultural land uses.
(2) Suitability ratings for septic tank disposal field, building foundation for low buildings, trafficability, surface stabilization, road and railway subgrade and earthwork uses.
(3) Suitability ratings for use as a source material for road base, backfill, sand or gravel, topsoil, and water reservoir embankments and linings.
(4) Rating with respect to flooding potential, watershed characteristics, susceptibility to erosion, and susceptibility to frost action.
(5) Suitability for wildlife habitat and habitat improvement, lawns, golf courses, playgrounds, and parks and related open areas requiring the maintenance of vegetation.

Soils information can be incorporated in other forms of planning. The Minerals Local Plan for Greater Manchester (Burns, 1986) essentially applies soils parameter (as agricultural land grade) to a sieving technique to determine cumulative constraints to development of sand and gravel resources.

The area of environmental protection is one which is encouraging an integrated approach. For example, in the Netherlands, (Breuuwsma and van Duijenbooden, 1987) and in Britain (Carter, Palmer and Monkhouse, 1987) a joint soils and geological approach has been taken for

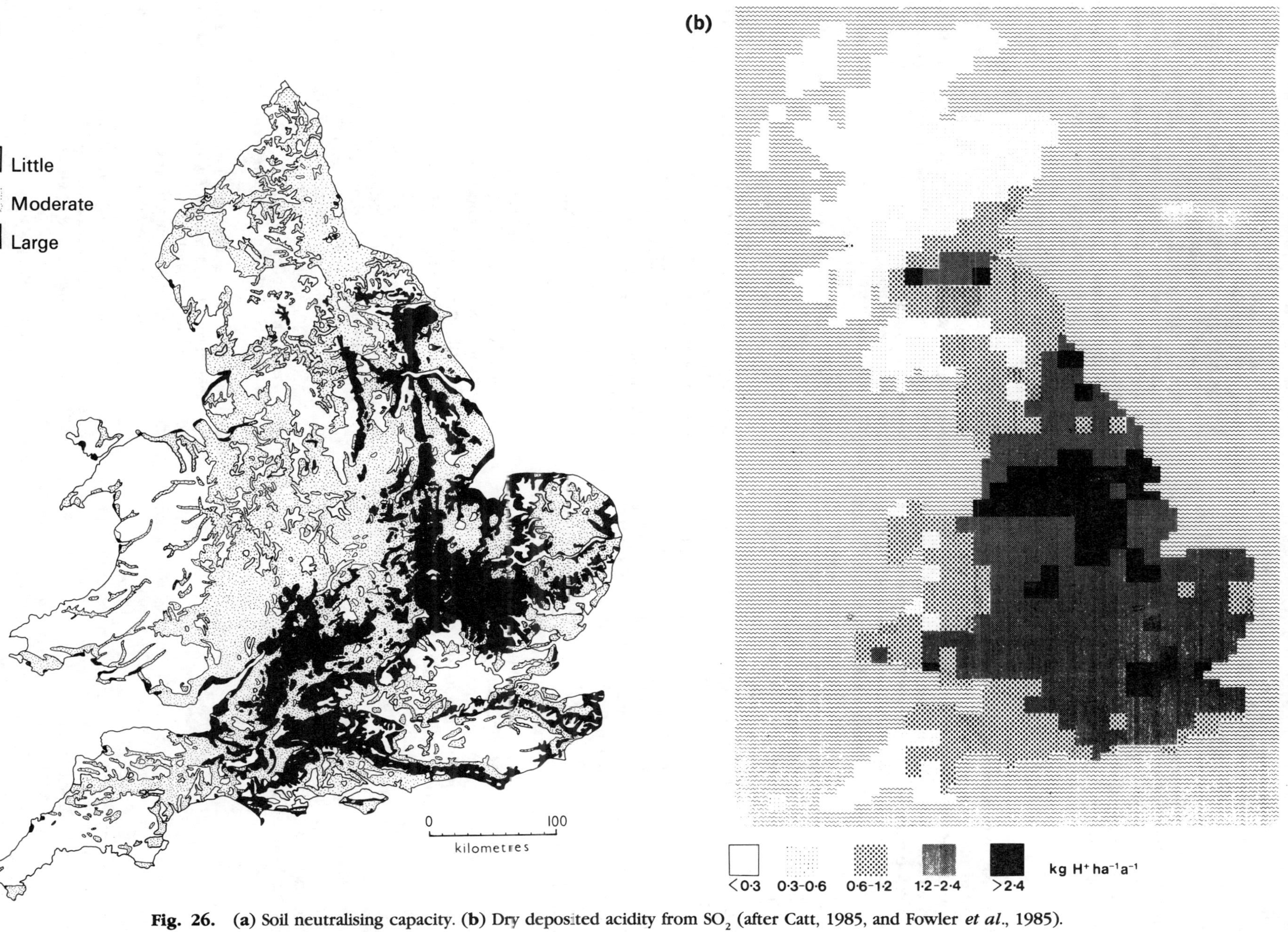

Fig. 26. **(a)** Soil neutralising capacity. **(b)** Dry deposited acidity from SO_2 (after Catt, 1985, and Fowler *et al.*, 1985).

mapping the vulnerability of groundwater to pollution from agricultural sources such as from nitrates and phosphates. Cramer and Vrba (1987) point out the usefulness of such maps to set priorities in policy development and implementation and to select sites for certain activities.

As development pressure increases on land and the population, it is certain that integrated studies will become more common, whether to meet broad planning objectives or, through environmental impact assessments, to settle the likely affect of precisely-defined development proposals.

CONCLUSIONS

In the last two decades, new mapping techniques, automated analytical methods, computerized data handling and the development of interpretive models have greatly increased the usefulness of soil maps in planning and development. Their value is undisputed in many countries. Klingebiel (1966) assessed the cost: benefit ratio of a soil survey in urban fringes of the United States as 1:120. Johannsen *et al.* (1979) described how a site selected for a school was shown by a soil map to have a severe soil limitation for building and, in comparison with the more favourable site eventually used, would have entailed US$80 000 of additional construction costs. In 1966 Thornburn *et al.* showed that use of a soil map could save US$810 000 in construction costs by indicating the more favourable of two potential alignments for a 10 mile interstate highway in Illinois. In the 1980s soil surveys have already shown their value in formulating plans for protecting the environment and as soil protection policies (e.g. Moen and Cramer, 1987) spread, the impact of soil mapping on development control will also spread. However, in many parts of the world, decision-makers are still unaware of the availability and usefulness of soil data. Even where it is known to planners, it is not always in an easily digestible form for non-technical users. However, where properly presented soils information can comprise important inputs into the formulation of goals and objectives, the preparation of planning standards, the analysis of existing land use, plan synthesis, test and evaluation and plan implementation (Bauer, 1973).

With growing environmental pressures on land development of all types, it is clear that a soil survey is one of the soundest investments in planning and formulation of policy for land use in the future.

REFERENCES

Allemeier, K.A. (1973). Application of pedological soil surveys to highway engineering in Michigan. *Geoderma*, **10**, 87–98.

Avery, B.W. (1980). *Soil Classification for England and Wales*. Soil Surv. Tech. Monogr. No. 14. Harpenden, England.

Avery, B.W. (1987). *Soil Survey Methods*. Soil Surv. Tech. Mongr. No. 18. Harpenden, England.

Barroso, J.A. (1986). *Pedological maps interpretation for soil planning in the west zone of Rio de Janeiro*. 5th International IAEG Congress, Buenos Aires.

Bartelli, L.J. (1979). Interpreting Soil Data. In *Planning the Uses and Management of Land*. (M.T. Beatty, G.W. Peterson and L.D. Swindale, eds), Agronomy Monogr. No. 21. American Society of Agronomy, Madison, USA.

Bauer, K.W. (1973). The use of soils data in regional planning. *Geoderma*, **10**, 1–26.

Bauer, K.W. (1979). Planning Metropolitan Land Uses in Relation to Natural Landscape Features. In *Planning the Uses and Management of Land* (M.T. Beatty, G.W. Peterson and L.D. Swindale, eds.), Agronomy Monogr. No. 21. American Society of Agronomy, Madison, Wisconsin.

Beckett, P.H.T. (1968). Method and scale of land resource surveys, in relation to precision and cost. In *Land Evaluation* (G.A. Stewart, ed.), Macmillan of Australia, 53–63.

Beckett, P.H.T. and Burrough, P.A. (1971). The relation between cost and utility in soil survey IV. Comparison of the utilities of soil maps produced by different survey procedures, and to different scales. *J. Soil Sci.* **22**, No. 4, 466–480.

Bibby, J.S. and Mackney, D. (1977). *Land Use Capability Classification.* Soil Surv. Tech. Monogr. No. 1. Harpenden, England.

Breeuwsma, A. and van Duijvenbooden, W. (1987). Mapping of groundwater vulnerability to pollutants in the Netherlands. In *Vulnerability of soil and groundwater to pollutants* (W. van Duijvenbooden and H.G. van Waegeningh eds.) CHO-TNO, The Hague, The Netherlands, 309–20.

Breeuwsma, A., Wosten, J.H.M., Vleeshouwer, J.J., Van Slobbe, A.M. and Bouma, J. (1986). Derivation of land qualities to assess environmental problems from soil surveys. *Soil Sci. Soc. Am. J.*, **50**, 186–190.

Bridges, E.M. and Burnham, C.P. (1980). Soils of the State of Bahrain. *J. Soil Sc.* **31**, 680–707.

Bridges, E.M. (1982). Techniques of Modern Soil Survey. In *Principles and Applications of Soil Geography.* (E.M. Bridges and D.A. Davidson, eds.), Longman, London, 28–57.

Brink, A.B.A., Mabbutt, J.A., Webster, R. and Beckett, P.H.T. (1966). *Report of the working group on land classification and data storage*. Report 940, Military Engineering Experimental Establishment, Christchurch, England.

Brink, A.B.A., Partridge, T.C. and Williams, A.A.B. (1982). *Soil survey for engineering.* Clarendon Press, Oxford.

Burns, D.J. (1986). *Minerals Local Plan.* Greater Manchester Council Planning Dept, Manchester, England.

Canada Land Inventory. (1965). *Report No. 2. Soil Capability Classification for Agriculture.* Department of Forestry and Rural Development, Ottawa, Canada.

Cannell, R.Q., Davies, D.B., Pidgeon, J.D. (1975). The suitability of soils for sequential direct drilling of combine-harvested crops in Britain: a provisional classification. In *Soil Survey Applications* (M.G. Jarvis and D. Mackney, eds.), Soil Surv. Tech. Monogr. No. 13. Harpenden, England.

Carter, A.D., Palmer, R.C. and Monkhouse, R.A. (1987). Mapping the vulnerability of groundwater to pollution from agricultural practice, particularly with respect to nitrate. In *Vulnerability of soil and groundwater to pollutants* (W. van Duijvenbooden and H.G. van Waegeningh, eds.), CHO-TNO, The Hague, The Netherlands, 333–42.

Catt, J.A., Weir, A.H., Norrish, R.E., Rayner, J.H., King, D.W., Hall, D.G.M. and Murphy, C.P. (1979). The Soils of Woburn Experimental Farm. III. Stackyard. *Rothamstead Report for* 1979, Part 2., Harpenden, England.

Catt, J.A. (1985). Natural soil acidity. *Soil Use and Management,* **1**, 1, 8–9.

Chapman, T.G. (1968). Catchment parameters for a deterministic rainfall-runoff model. *Land Evaluation* (G.A. Stewart, ed.), Macmillan of Australia.

Commission de Pédologie et de Cartographie des Sols. (1967). *Classification des sols.* Lab. Geol. Pedol., Ecole Nat. Super. Agron., Grignon, France.

Cook, F.J., and Watt, J.P.C. (1982). *Evaluation of a transient drainage method for measuring movement of water in soils in field situations.* N.Z. Soil Bureau Scientific Report 57, DSIR, N.Z.

Corcoran, P., Jarvis, M.G., Mackney, D. and Stevens, K.W. (1977). Soil corrosiveness in south Oxfordshire. *J. Soil Sci.* **28**, 473–84.

Cowie, J.D. and Osborn, W.L. (1977). *Soil resources of the Manawatu and the expansion of Palmerston North City.* Min. Ag. and Fish., Palmerston North, New Zealand.

Cramer, W. and Vrba, J. (1987). Conclusions on topic 3: Vulnerability mapping. In *Vulnerability of soil and ground-*

water to pollutants (W. van Duijvenbooden and H.G. van Waegeningh, eds.), CHO-TNO, The Hague, The Netherlands, 45–6.

Davidson, D.A. (1982). Soils and man in the past. *Principles and Applications of Soil Geography* (E.M. Bridges and D.A. Davidson, eds.), Longman, London.

de Bakker, H. and Schelling, J. (1966). *Systeem van bodemclassificatie voor Nederland.* (with English summary). Soil Survey Inst., Wageningen, The Netherlands.

Dent, D. and Young, A. (1981). *Soil Survey and Land Evaluation*. G. Allen and Unwin, London.

Department of the Environment (1987). *Development Involving Agricultural Land.* Circular 16/87. HMSO, London.

Department of the Environment (1988a). *Minerals Planning Guidance: General Considerations and the Development Plan System.* DoE/Welsh Office MPG 1, HMSO, London.

Department of the Environment (1988b). *Minerals Planning Guidance: Applications, Permissions and Conditions.* DoE/Welsh Office MPG 2, HMSO, London.

Dumbleton, M.J. and West, G. (1976). *Preliminary sources of information for site investigations in Britain.* TRRL Lab. Rep. 403 (Revised edition), Transport and Road Research Laboratory, Crowthorne, England.

Dumbleton, M.J. (1981). *The British Soil Classification System for Engineering Purposes: its development and relation to other comparable systems.* TRRL Lab. Rep. 1030. Transport and Road Research Laboratory, Crowthorne, England.

Eswaran, H. (1978). Evaluation of soil resource inventories. In *Soil Resource Inventories and Development Planning,* Proc. Workshop at Cornell Univ., Ithaca, New York, 37–48.

E.C. (1986). Council Directive of 12th June 1986 on the protection of the environment, and in particular of the soil, when sewage sludge is used in agriculture. *Official Journal of the European Communities,* L181/6, Brussels.

Farquharson, F.A.K., Mackney, D., Newson, M.D. and Thomasson, A.J. (1978). *Estimation of run-off potential of river catchments from soil surveys.* Soil Surv. Spec. Surv. No. 11. Harpenden, England.

FAO (1974). *Soil Map of the World, 1:5 000 000.* Vol. 1: Legend. Unesco, Paris.

FAO (1976). *A framework for land evaluation.* FAO Soils Bull. No. 32.

Fowler, D., Cape, J.N. and Leith, I.D. (1985). Acid inputs from the atmosphere in the United Kingdom. *Soil Use and Management,* **1**, 3–5.

Galloway, R.W., Gunn, R.H. and Story, R. (1970). *Lands of the Mitchell-Normanby Area, Queensland.* Land Research Series No. 26. CSIRO, Australia.

Ganssen, R. and Hadrich, F. (1965). *Atlas zur Bodenkunde.* 85 p. Bibliographisches Institut, Mannheim.

Geomorphological Services Ltd. (1988). *Applied earth science maps for planning and development, Torbay.* Geomorphological Services Ltd., Newport Pagnell, England.

Grant, K. (1976). *Terrain classification and evaluation for engineering purposes of the Canberra Area, Australian Capital Territory and New South Wales.* Div. Appl. Geomechs. Tech. Pap. No. 22, CSIRO, Australia.

Griffiths, E. (1975). *Classification of land for irrigation in New Zealand.* N.Z. Soil Bureau Scientific Report 22. DSIR, New Zealand.

Griffiths, E. (1981). *Suitability of land for irrigation in the Ngatarawa area, Heretaunga Plains, Hawke's Bay, New Zealand.* N.Z. Soil Survey Report 63. DSIR, New Zealand.

Hartnup, R. and Jarvis, M.G. (1973). Soils in civil engineering and planning. In *Soil Survey Applications.* (M.G. Jarvis and D. Mackney, eds.), Soil Surv. Tech. Mon. No. 13, Harpenden, England.

Hazelden, J., Loveland, P.J. and Sturdy, R.G. (1986). *Saline soils in north Kent.* Soil Surv. Spec. Surv. No. 14, Harpenden, England.

HMSO (1971). *Town and Country Planning Act 1971.* HMSO, London.

HMSO (1981). *Town and Country Planning (Minerals) Act 1981.* HMSO, London.

Hodgson, J.M. (ed). (1976). *Soil Survey Field Handbook.* Soil Surv. Tech. Monogr. No. 5. Harpenden, England.

Holmberg, M., Johnston, J. and Maxe, L. (1987). Assessing aquifer sensitivity to acid deposition. In *Vulnerability of soil and groundwater to pollutants* (W. van Duijvenbooden and H.G. van Waegeningh, eds.), CHO-TNO, The Hague, The Netherlands, 373–380.

Jarvis, M.G., Hazelden, J. and Mackney, D. (1979). *Soils of Berkshire.* Soil Surv. Bull. No. 8, Harpenden, England.

Johannsen, C.J., Barney, T.W. and Klingbiel, A.A. (1966). Residential, commercial and light industrial land uses. In *Planning the Uses and Management of Land* (M.T. Beatty, G.W. Petersen and L.D. Swindale, eds.), Agronomy Monogr. No. 21., American Society of Agronomy, Madison, Wisconsin.

Johnson, G.O. (1975). *Engineering Characteristics of Ohio Soil Series.* State of Ohio Department of Transportation, Columbus, Ohio.

Kellogg, C.E. (1974). Soil genesis, classification and cartography: 1924–1974. *Geoderma,* **12**, 347–362.

Klingebiel, A.A. and Montgomery, P.H. (1961). *Land Capability Classification.* US Dep. Agric., Soil Conserv. Serv. Agric. Handbk. No. 210, Washington, D.C.

Klingebiel, A.A. (1966). Cost and return of soil surveys. *Soil Conservation,* **32**, 3–6.

Lea, J.W. (1979). Slurry acceptance. In *Soil Survey Applications* (M.G. Jarvis and D. Mackney, eds.), Soil Surv. Tech. Mon. No. 13, Harpenden, England.

Leamy, M.L. (1974). *An improved method of assessing the soil factor in land valuation.* N.Z. Soil Bureau Scientific Report 16., DSIR, New Zealand.

Lee, E.M. and Griffiths, J.S. (1987). The importance of pedological soil survey in land use planning, resource assessment and site investigation. In *Planning and Engineering Geology* (M.G. Culshaw, F.G. Bell, J.C. Cripps and M. O'Hara, eds.), Geol. Soc. Eng. Geol. Sp. Pub. No. 4, London.

Lindsay, J.D., Scheelar, M.D. and Twardy, A.G. (1973). Soil survey for urban development. *Geoderma,* **10**, 35–45.

Linn County Planning Department. (1970). *Linn County physical characteristics and optimum land utilization.*

Loveland, P.J., Hazelden, J. and Sturdy, R.G. (1987). Chemical properties of salt-affected soils in north Kent and their relationship to soil instability. *J. Agric. Sci., Camb.,* **109**, 1–6.

Macvicar, C.N. and Loxton, R.F. (1967). *Soils of the Langkloof (Ongelegen to Heights).* Tech. Comm. No. 59, Dept. Agricultural Technical Services, Pretoria, South Africa.

Mansfield, J.E., Bennett, J.G., King, R.B., Land, D.M. and Lawton, R.M. (1975). *Land resources of the Northern and Luapula Provinces, Zambia–a reconnaissance assessment.* Land Resource Study 19, Min. Overseas Devel., Surbiton, England.

McCormack, D.E. and Fohs, D.G. (1979). Soil Considerations in Siting Highways, Airports and Utility Corridors. In *Planning the Uses and Management of Land* (M.T. Beatty, G.W. Petersen and L.D. Swindale, eds.), Agronomy Monogr. No. 21., American Society of Agronomy, Madison, USA, 531–554.

McCormack, D.E. and Johnson, R.W. (1982). Soil Potential for Onsite Sewage Disposal in Leon County, Florida, *Soil Survey and Land Evaluation,* **2**, 2–8.

McGown, A. and Iley, P. (1973). A comparison of data from agricultural soil surveys with engineering investigations for roadworks in Ayrshire. *J. Soil Sci.,* **24**, 145–156.

McRae, S.G. and Burnham, C.P. (1981). *Land Evaluation,* Clarendon Press, Oxford.

Mettert, W.K. (1986). *Soil Survey of Tuscola County, Michigan.* US Dept Agric., Soil Conserv. Serv., Washington, D.C.

Miller, F.P. (1979). Defining, delineating and designating uses for prime and unique agricultural lands. In *Planning the Uses and Management of Land* (M.T. Beatty, G.W. Petersen and L.D. Swindale, eds.), Agronomy Monogr. No. 21., American Society of Agronomy, Madison, Wisconsin, 291–316.

Milne, G. (1935). *Composite units for the mapping of complex soil associations.* Trans. 3 int. Congr. Soil Sci. **1**, 345–347.

Ministry of Agriculture, Fisheries and Food. (1988). *Agricultural Land Classification of England and Wales.* MAFF Publications, Alnwick, Northumberland, UK.

Moen, J.E.T. and Cramer, W. (1987). Implementation of a soil protection policy in the Netherlands. In *Vulnerability of soil and groundwater to pollutants* (W. van Duijvenbooden and H.G. van Waegeningh, eds.), CHO–TNO, The Hague, The Netherlands, 135–149.

Montgomery, P.H. and Edminister, F.C. (1966). Use of soil surveys in planning for recreation. In *Soil surveys and land use Planning* (L.J. Bartelli, A.A. Klingebiel, J.V. Baird, and M.R. Heddleson, eds.), Soil Sci. Soc. of America, Madison, Wisconsin, 104–112.

Morgan, R.P.C. (1985). Assessment of soil erosion risk in England and Wales. *Soil Use and Management*, **1**, 127–131.

Murthy, R.S., Jain, S.P. and Naga, B.S.R. (1968). Soil survey and land capability classification for sound watershed management in Kundah Project (Madras). *J. Indian Soc. Soil Sci.*, **16**, 223–227.

Obihara, C.H., Bawden, M.G. and Jungerius, P.D. (1964). *The Anambra-do rivers area.* Soil Surv. Mem. No. 1, Min. Agr., Nigeria.

Olding, A.B., Wicklund, R.E. and Richards, N.R. (1956). *Soil survey of Ontario County.* Ontario Soil Surv. Rep. No. 23, Canada Dept. Agr., Guelph.

Olson, G.W. (1982). The soil survey taxation system in New York State. *Soil Survey and Land Evaluation*, **2**, 14–17.

Oxford English Dictionary (1976). Oxford University Press, Oxford.

Pettry, D.E. and Coleman, C.S. (1973). Two decades of urban soil interpretations in Fairfax County, Virginia. *Geoderma*, **10**, 27–34.

Prickett, R.C. (1966). The land use capability system. *N.Z. Agric. Sci.*, **1**, 17–21.

Ragg, J.M., Beard, G.R., George, H., Heaven, F.W., Hollis, J.M., Jones, R.J.A., Palmer, R.C., Reeve, M.J., Robson, J.D. and Whitfield, W.A.D. (1984). *Soils and their use in Midland and Western England.* Bull. Soil Surv. Gt Br., Harpenden, England.

Redding, J.H. (1987). Route selection for natural gas pipelines in Ireland. In *Planning and Engineering Geology* (M.G. Culshaw, F.G. Bell, J.C. Cripps and M. O'Hara, eds.), Geol. Soc. Eng. Geol. Sp. Pub. No. 4, London, 579–584.

Reeve, M.J. (1986). Water retention, porosity and composition interrelationships of alluvial soils in mid Hawke's Bay and their relevance in irrigation planning. *N.Z. J. Agr. Res.*, **29**, 457–468.

Robson, J.D., George, H. and Heaven, F.W. (1974). *Soils in Lincolnshire I: Sheet TF16 (Woodhall Spa).* Soil Surv. Rec. No. 22, Harpenden, England.

Roeschmann, G. (1986). *Soil map of the Federal Republic of Germany 1 : 1 000 000.* Federal Institute for Geosciences and Natural Resources, Hannover.

Rogoff, M.J. (1982). Computer display of soil survey interpretations using a geographic information system. *Soil Survey and Land Evaluation*, **2**, 37–41.

Shankarnarayan, K.A., Kolarkar, A.S. and Singh, N. (1983). Application of the FAO framework for land evaluation to an arid area of Western Rajasthan, India. *Soil Survey and Land Evaluation*, **3**, 59–68.

Sibirtsiev, N.M. (1901). Soil science. In *Selected Works*, Vol. **1**, Israel Programme for Scientific Translations, Jerusalem.

Soil Survey Staff. (1975). *Soil Taxonomy*, US Dept. Agr. Soil Conserv. Service, Handbk No. 436.

Sombroek, W.G., Braun, H.M.H. and van der Pouw, B.J.Z. (1982). *Exploratory soil map and agro-climatic zone map of Kenya, 1980, Scale 1 : 1 000 000*, Exploratory Soil Surv. Rep. No. E1, Nairobi.

Stocking, M. (1983). Extremely rapid soil survey: an evaluation from Tanzania. *Soil Survey and Land Evaluation*, **3**, 59–68.

Stichting voor Bodemkartering. (1976). *1 : 50 000 Bodenkaart van Nederland, Blad 45 Oost's-Hertogenbosh en Blad 46 West-46 Oost Vierlingsbeek*, Wageningen, The Netherlands.

Thornburn, T.H., Morse, R.K. and Liu, T.K. (1966). *Engineering soil report Livingston County, Illinois.* Engineering Experiment Station Bulletin 482, Urbana, Illinois.

United Kingdom Government Colonial Office. (1952). *Report on Central African Rail Link Development Survey*, London.

United States Department of the Interior. (1953). *Bureau of Reclamation Manual. Vol. V–Irrigated land use, Part 2–Land classification.*

Valentine, K.W.G. and Schori, A. (1980). *Soils of the Lac La Hache-Clinton area, British Columbia.* Brit. Columbia Soil Surv. Rep. No. 25, Vancouver.

Venn, A.C. (1977). The position of agriculture in the competition for land. In *Proceedings 7th National Congress* (H.v.H. van der Watt, P.F. Fuls and C.N. MacVicar, eds.), Tech. Comm. No. 146, Dept. Agricultural Technical Services, Pretoria, 102–105.

Verheye, W.H. (1986). Principles of land appraisal and land use planning within the European Community. *Soil Use and Management*, **2**, 120–124.

Weiers, C.J. and Reid, I.G. (1974). *Soil classification, land valuation and taxation. The German experience.* Centre for European Agricultural Studies, Wye College, Ashford, England.

Witwer, D.B. (1966). Soils and their role in planning a suburban county. In *Soil Surveys and Land Use Planning* (L.J. Bartelli, A.A. Klingebiel, J.V. Baird and M.R. Heddleson, eds.), Soil Sci. Soc. America, Madison, Wisconsin, 15–30.

Chapter 7

Hazards

John C. Doornkamp

INTRODUCTION

A variety of contemporary earth processes cause sufficiently severe problems to be regarded as hazards to life and property. These range from relatively localized but sometimes costly ground subsidence problems to major disasters such as floods, earthquakes and volcanic eruptions. In the present context 'hazard' is taken to mean a hazard of a geological, hydrological or geomorphological nature which poses a threat to man or his activities. In particular, attention will be paid to the role of earth science mapping in the assessment of the hazards listed in Table I.

Table I The hazards discussed in this chapter

Landslip and avalanche
Natural ground subsidence
Hazards from quarrying and mining
Erosion and deposition
Flooding
Saline soils
Permafrost
Seismological hazards
Volcanic hazards

Most hazards occur when natural processes have crossed some kind of threshold and have changed into a rapid or catastrophic event, though, as many authors point out, there is no hazard without a concomitance in location between the event and mankind.

The concept of a 'natural' hazard is an ambiguous one, for many catastrophic events within the environment are man-induced, or at least made worse by the intervention of man. This reduces their 'naturalness' but not their impact, and gives added weight to the need for hazard mapping to include mapping which aims to identify those sites where there is potential for a hazardous event, rather than just identifying those sites where a hazardous event has already taken place.

Hazard maps, whatever form they take, provide a basic document for use in planning, engineering and development projects, and may pertain to a site, a region or a nation. Users include not only civil engineers, urban-, regional-, or national-planners, but also insurance and reinsurance companies, the development banks, and various environmental agencies (e.g. the FAO, the Food and Agricultural Organization of the United Nations). Flood hazard maps, for example, are used for land-use zoning whereby the land most liable to flood is reserved for recreational use and is not given over to urban construction programmes. Soil erosion assessments are used in agricultural development projects, including estimating the need for forestry development. Landslide hazard mapping is used (e.g. in California) as part of the building code which restricts high density housing development within areas prone to landsliding. One of the more formal schemes of hazard zonation is that used under the ZERMOS (zones exposées aux risques liés aux mouvement du sol et du sous-sol) scheme developed in France (Porcher and Guillope, 1979).

The aim of this chapter is to increase the user's awareness of the hazard maps that can be available. They may not exist, but they can be commissioned. However, such a commission first requires an awareness of what is possible. To that end this account also looks at the limitations and restrictions attached to hazard mapping,

Earth Science Mapping for Planning, Development and Conservation

since there is no value in raising expectations beyond those which are attainable. This chapter therefore provides both a review and an assessment.

Every hazard has three key properties, these are:

- magnitude;
- frequency;
- location.

It is the last of these which is most amenable to mapping (Fig. 1).

The mapping of a hazard may be undertaken for any of a number of reasons, the most important of which are:

- to define the nature and extent of an historical event;
- to define the nature and extent of a recent event;
- to define present conditions in order to assess the likelihood of a future event.

A full hazard study, however, cannot be achieved without giving consideration to the social and economic context in which it takes place. While this cannot be discussed in detail here, this very important theme is contained in the writings of Kates (1978), Hewitt (1983), Blaikie (1985) and Timberlake (1985).

Since whole textbooks have been written on each of the individual hazards to be touched upon in this chapter, justice can hardly be done here to such a full topic. This has meant that only the more important themes relating to the mapping of such hazards can be dealt with here. The choice has been, therefore, to select those points which the author feels are crucial, though sometimes neglected, in the mapping of hazards. Ample reference will be provided to published examples.

It is seldom explicitly expressed that in many situations it is not the *hazard* which is being mapped but rather the *results* of the hazard. In other words, most mapping of 'hazards' is actually the mapping of the resulting land forms (and possibly soil or rock materials). A hazard is a process, and a process is very difficult to map. Forms and materials on the other hand are easier to map. Mapping for predictive purposes is even more difficult than process mapping, for at that stage the process (hazard) has not even happened, so it cannot be mapped. This point will be discussed in more detail later in this chapter.

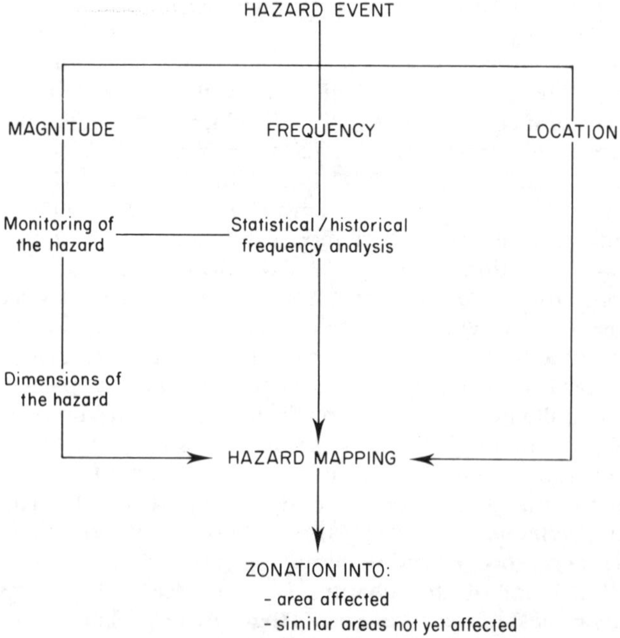

Fig. 1. The context of hazard mapping.

MAP SCALES

The question of map scale is crucial to the nature of the map produced for a hazard or hazard area. In general, it may be useful to think of map scale in terms of the classification given in Table II. The choice of map scale will depend on many factors including: the purpose of the map, the size of the project area, the nature and extent of the hazard and the amount of detail required. The scale at which a map is to be compiled also relates to the techniques by which the mapping is to be carried out. Table II includes a list of some of the data sources relevant to mapping at each of the scales identified.

AIDS TO HAZARD MAPPING

Many of the main data sources for hazard mapping are well known and well documented. The exceptions are perhaps satellite imagery, air-borne multi-spectral scanners, some forms of photogrammetry, and shallow geophysics.

Satellite imagery

Since the advent of satellite imagery on a systematic basis in 1972 a synoptic view of the earth has been available in a way that was not previously possible. In particular, data became generally available for a wide range of wavelengths of electromagnetic radiation. Since 1972 there has been a steady acquisition of space images for large parts of the earth's surface, including areas for which no adequate map cover exists. This has led to a proliferation of data, the usefulness of which, in terms of hazard mapping, has only been limited by the poor spatial resolution of the imagery. This shortcoming has tended to restrict the use of such imagery to those hazards which extend over very large areas (i.e. they supply data only useful at the synoptic map scale). However, over the years a succession of improvements in spatial resolution has taken place (Table III). As well as this improvement in spatial resolution there has been a series of changes in the wave-lengths of reflected radiation which is recorded by the sensors. Many data are recorded in the near infra-red, thus providing information outside the visible part of the spectrum.

Satellite imagery is available either as negatives or as paper prints (colour or monochrome in each case), or as tapes for computer processing. Such processing techniques are well described in American Society of Photogrammetry (1975), Curran (1985), Drury (1987), Harris (1987), Lillisand and Keifer (1979), Mather (1987), Sabins (1978), and Townshend (1981). These authors also describe the types of features best shown by each of the recorded wave-bands. Satellite imagery has proved to be especially useful (within the limits of its resolution) in providing (i) a synoptic view over a large area; (ii) an historical record of past states; (iii)

Table II A classification of map scales and data sources for use in hazard mapping

Type of map	*Scales*	*Level of detail*	*Usual data sources*
Synoptic maps	1:100 000 and smaller	boundaries of area containing hazard	satellite imagery photographically reduced air-photo mosaics
Medium scale	1:25 000–1:50 000	hazard sites	satellite imagery photographically reduced air-photo mosaics air-photo mosaics
Large scale	1:10 000–1: 5 000	detail of larger hazard areas	aerial photography field mapping
Detailed	1: 2 000–1: 5 000	detail within hazard areas	aerial photography air-borne multi-spectral scanner field mapping field surveying photogrammetry process monitoring
Site	1: 2 000 and larger	specific hazard form/material characteristics	aerial photography air-borne multi-spectral scanner field mapping field surveying photogrammetry shallow geophysics process monitoring sub-surface investigations

Table III Satellite sensors and their spatial resolution

LANDSAT	1972–present	Multi-spectral scanner (MSS) —effective resolution about 80 m
	1982–present	Thematic mapper (TM) —effective resolution about 30 m
SPOT	1986–present	Most sensors have effective resolution of about 20 m with 10 m available in one wave-band.

information on areas not readily accessible in the field; and (iv) data on surface reflectance characteristics in wave-lengths outside the visible part of the spectrum.

Air-borne multi-spectral scanner

In essence, air-borne multi-spectral scanner systems provide data similar to that supplied by satellite-borne scanners, except that the spatial resolution allows much more detail to be recorded. For example, depending on the flying height of the aircraft, resolutions of 1 metre are easily achieved. This allows data to be gathered at a scale commensurate with the needs of large-scale, detailed and even site maps (Table II).

Photogrammetry

Standard photogrammetric techniques are conventionally employed, for example, in the derivation of contour maps from aerial photography and limited field survey control points. This need not be further elaborated here. A new development, however, is the ability to monitor historical changes (e.g. of a hazard site) by the photogrammetric analysis of oblique photographs taken over a period of time. This can include a retrospective analysis of a hazard or a hazard site if suitable photography is available.

The technique is based on specialized equipment. It is dependent on a limited number of fixed points (three or more) being identifiable on each of the photographs. These fixed points are surveyed on the ground so that their positions and elevations are known. From this information and the geometrical characteristics of the photography, the elevation and position of every other point of interest can be mathematically calculated. By comparing two such photographs from different dates, changes (e.g. in the position of a landslide) can be calculated.

Shallow geophysics

Significant research is currently taking place in the development of shallow geophysical techniques which will improve the capability to predict sub-surface conditions. This has so far been essentially aimed at the detection of underground cavities, especially solution features in limestone (Geological Society Engineering Group Working Party, 1988). Information from such surveys are only available along the traverse lines adopted, although data from closely spaced adjacent lines can be automatically contoured to give an approximation to a spatial data set. Most applications for this technique lie in the range of large-scale, detailed, and site maps.

MAPPING AND CLASSIFICATION SYSTEMS

Every hazard map carries a classification of the hazard. This is necessary because it is the only basis on which a hazard map can be compiled. Seismic maps use the Modified Mercalli Scale; flood hazard maps use limits based on a statistical recurrence interval; soil erosion maps are based on selected soil properties or on a classification of rates of soil loss, and so on.

In a subtle way the methods of classification used can sometimes have a very strong influence on the nature of the final map produced. Users of hazard maps should spend as long thinking about the classification on which the map is based as on the distributions shown. An inappropriate classification system can seriously undermine the value of the whole map.

The heart of any classification is the theory on which it is based. For example, the landslide classification proposed by Varnes (1978) very much assumes that landslide morphology betrays the landslide process. In many cases this may indeed be so. It certainly is a seductive theory, because it allows a great deal to be said about landslides (whether right or wrong) without the need to wait for a subsurface investigation. More recently, however, Hutchinson (1988) has taken a different approach. While still relying on landslide morphology as far as possible, he nevertheless abandons morphology as a basis for landslide classification when available geotechnical data suggest that morphology alone is an inadequate basis for the classification process. The inference is that the morphological approach of Varnes, while attractive and easy to use, may (as Varnes himself recognizes) hide aspects that are very important in a fuller understanding of the hazard itself.

HAZARD MAPPING AND PERCEPTION

All hazard mapping tends to include an element of subjectivity. Once a classification system has been adopted, we must then depend on the mapper to recognise individual features and to place them 'correctly' within the classification system. This is a subjective process. Although subjectivity can be reduced by the use of tight and detailed rubrics, or by common training and experience (between mappers) on a world-wide scale, there is no doubt that much personal variation in perception of hazards exists. Hence, there are consequential differences in the resulting maps.

CARTOGRAPHICAL APPROACHES TO HAZARD MAPPING

It is useful to distinguish between maps displaying data (*primary* maps) and those showing the results of interpretations (*derived* maps). Primary maps present singular features or combinations of features that exist on the ground and relate to a particular hazard. Derived maps are those which involve analysis and interpretation (frequently of several sets of data) and typically include hazard zoning maps for planning or development purposes. Since hazard maps can display information either as dots, lines or areas, it is also important to accept the limitations inherent in each case, as discussed in Chapter 2. There is no accepted convention of hazard mapping symbols, though some geomorphological mapping legends include symbols that relate to natural hazards (e.g. Demek 1972; Demek and Embleton, 1978; see also Chapter 2).

In addition to these approaches, however, hazard mapping has also been carried out through a matrix approach (i.e. on a cell-by-cell basis of a grid placed across a map area) (see De Graff and Ronesburg, 1980). An area can be classified on this grid square basis according to a hazard risk classification. An example is shown in Fig. 2. This is sometimes derived through a sieve-mapping procedure in which a number of controlling parameters are combined on a pre-determined weighted basis, in order to define for each grid square the risk from a particular hazard.

Such a matrix approach is used not so much in relation to the mapping of a hazard as in the classification of an area in terms of vulnerability to that hazard. Such mapping takes into account those landscape controls which are thought to be influential in bringing about a hazard. The resulting map is therefore something of a predictive statement even if it includes specific information about sites where the hazard is, or has been, experienced. The same is true of some of the 'automatic' methods which have been proposed for the production of hazard zonation maps by computer (e.g. Huma and Radulescu, 1978).

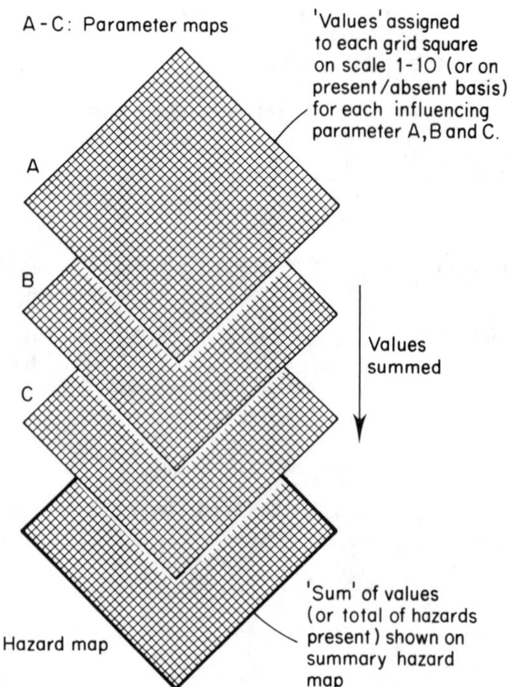

Fig. 2. A grid square approach to hazard mapping.

HAZARD MAPPING AND PLOTTING ACCURACY

The precision with which hazards can be mapped is a function of many constraints and opportunities. All maps should indicate the basis on which they were compiled and their sources of information. The user should evaluate these in order to see to what extent he can expect positions to be located within a tolerable range of accuracies.

The rest of this chapter considers each of the hazards listed in Table I. In each case there is a description of the basic concepts concerning the nature of the hazard, some examples of mapping, and an indication of the relevance to land use planning and development.

LANDSLIPS AND AVALANCHES

Landslips or landslides

Nature of movement

Landslips tend to occur as falls, slides or flows (Fig. 3). Falls tend to imply a lack of support from below. Slides

Mechanism ↓ / Material →		Rock	Debris	Soil
	Falls	Rock fall Rock topple	Debris fall Debris topple	Soil fall Soil topple
Slides:	Rotational	Single*, Multiple* or Successive*		
	Non-Rotational	Block slide	Slab slide	Spreading failure
	Planar	Rock slide	Debris slide	Mud slide
	Flows	Rock avalanche	Debris flow	Mud flow

* Any of these may occur in either rock, debris or soil.

Fig. 3. Landslide classification.

indicate the presence of a discontinuity along which movement can take place. Flows tend to indicate saturated conditions which generate a liquid, or at least fluid, state within the moving materials.

Main causes

Landslips occur because the forces that compel movement, within a slope, have overcome those forces that are resisting movement. Typical compelling forces are those of gravity and excess water pressures. Typical resisting forces are those of internal friction and cohesion within the slope materials.

Gravity always plays a crucial role because landslips have a downhill component. This is particularly evident when a slope is undercut (e.g. by erosion or by excavation); support is thereby removed and the upper slope fails. Conversely, a load exerted on the top of a slope (e.g. through the dumping of additional material, or the accumulation of a thick layer of snow) may also exert a gravitational force that will lead to instability in the slope below. However, gravity may not always be the primary cause of movement. In particular, movement can be a response to internal pressures (e.g. pore water pressures) within the slope materials (especially within clays).

The ability of the slope materials to resist movement is a function of their internal composition and physical characteristics (i.e. friction and cohesion). These in turn will be related to mineralogical composition, degree of weathering, internal physical structures, and the orientation and inclination of any bedding or joint structures. The extreme differences between clays and granites conjure up the image that clays are more likely to be prone to landslips than granites. There are also strong controls on slope failure exerted by the general environment in which the slope occurs. In fact, within a wet, hot, tropical environment, where granites are subject to very intense and deep weathering, these weathering products may experience very extensive landslips, especially if they are on steep slopes. Hong Kong is particularly prone to such failures because of the deeply weathered granite, steep slopes and intense rainfall that generates high pore water pressures within these slopes.

Another environment in which landslips are likely to be generated is where steep mountain slopes occur within an area subject to earthquakes. Vibrations generated by the earthquake may be capable of disturbing the slope so that landslides are triggered.

Mapping

Landslide maps exist at all of the scales listed in Table II. For example, Radbruch-Hall *et al.* (1976) has produced a synoptic map of the United States to show the boundaries of those areas prone to landsliding. Different levels of hazard exposure are identified. For planning purposes it may be enough to provide maps showing the zonation of ground with respect to its potential for slope failure, as has been done under the ZERMOS scheme (Landry, 1979) in France. In San Mateo County, California, landslide hazard zonation maps have been used to define limits to housing densities (Atwater, 1978). Whereas 1 house per 2 hectares is normally acceptable, in landslide prone areas it is 1 house per 16 hectares. In New Zealand (Stevenson and Sloane, 1980), where areas of marginal slope stability have been identified, either the public are warned and expected to act accordingly, or restrictions on development are imposed under the Local Government Act (1973). Varnes (1984) gives other examples of landslide hazard maps at a variety of scales, though most are either medium- or large-scale maps. Selected examples are also identified in Table IV. Further discussion is also supplied by Hanson (1984) and by Hearn and Fulton (1986).

Table IV Selected examples of landslide hazard (risk) maps

Location	*Scale*	*Notes*
Italy (Calabria)	1:10 000	Five risk classes superimposed on geomorphology and geology (Dumas *et al.* 1984)
Switzerland	1:10 000	Four risk classes combined with geomorphology (Kienholz, 1978)
France	1:25 000	Four risk classes and map of existing hazard sites (Meneroud and Calvino, 1976)

(Other examples are provided in Varnes, 1984.)

Avalanches

Nature of movement

Avalanche movement (Fig. 4) may be in the air, along the ground, or a mixture of the two. Velocities can be very high, and the run out distance is a function of slope height and steepness.

Main causes

Avalanches, which consist of the rapid downhill movement of detached masses of snow or ice under the influence of gravity, are only prevalent in those mountain regions with a source of snow and ice on or near the mountain summits. Ice avalanches usually occur by breaking away from a glacier snout, usually where it lies on a steep slope. Snow avalanches, though variable in detail, all tend to occur when, like some landslides, the internal forces resisting movement are exceeded by the gravitational forces that cause the snow mass to move down-hill. Meteorological conditions play a considerable part in providing conditions conducive to snow avalanching, but internal structural conditions and external topographical conditions such as slope steepness, length, orientation and roughness, also have a significant bearing on the process (see Embleton and King, 1975; and Embleton and Thornes, 1979).

Mapping

The approaches to avalanche mapping are similar to those described for landslides. Avalanche hazard zoning maps, which are produced on the basis of historical information and an assessment of the topographic con-

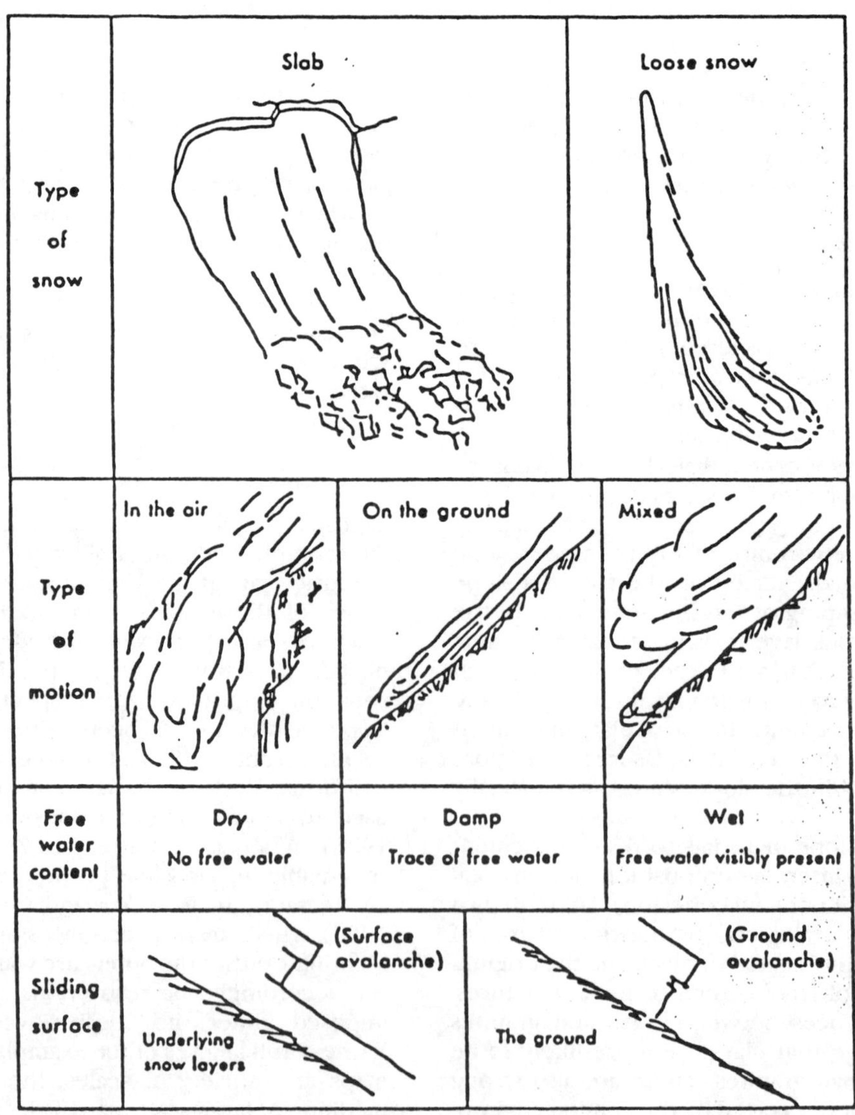

Fig. 4. Classification of avalanches (after Embleton and Thornes, 1979).

trols that influence avalanche location, are produced in Switzerland, France, Canada, the USA and Norway.

NATURAL GROUND SUBSIDENCE

Nature of natural ground subsidence

Natural ground subsidence leads to the formation of cavities or depressions. Cavities tend to be more localized and may relate, for example, to the movement of soil materials down into underlying pot-holes (as in karstic limestone) or into blow-holes such as those which occur above cave systems developed along coastal cliffs (especially in well-jointed rocks). Depressions, such as dolines developed in limestones, are formed more by solution weathering than by true subsidence.

Causes of natural ground subsidence

Subsidence through natural causes tends to be concentrated within limestones. Processes of underground solution can lead to the collapse of the surface above, simply because the underlying support has been removed (Culshaw and Waltham, 1987). A notable cause of subsidence in limestone terrain has been ground water withdrawal. In Alabama it is estimated that over 4000 surface depressions (some as much as 30 m in depth and 3.2 km in diameter) have been caused by this means (Newton, 1984).

Mapping of natural ground subsidence

In the mapping of limestone terrain, care has to be taken to make a correct diagnosis of the origin and cause of any depressions identified in order to identify those locations prone to subsidence through collapse. This is because depressions in a limestone surface can be caused by any one of a number of different factors. The following types of depressions need to be distinguished from each other:

- swallowholes: where surface streams disappear undergound
- dolines (sinkholes): or closed ground-surface depressions
- solution pipes: a pipe-like solution cavity typically infilled with surface wash deposits
- solution doline: formed by solution beneath a soil cover
- alluvial doline: where major discontinuities in the limestone are enlarged by solution beneath a superficial deposit
- collapse doline: collapse following solution below the top of the rock head

Clearly these different kinds of hollows provide different levels of hazard. Solution dolines pose little or no threat, although they can bring about additional costs to an engineering project developed within or across them. Collapse dolines, on the other hand, can be a major hazard, especially if collapse takes place after the site has been developed. The mapping of limestone-related subsidence hazard is made difficult by the fact that so much of the relevant evidence, being underground, is hidden from sight. While surface morphological characteristics and drainage conditions can be used as indicators (see Edmonds, 1983), it may also be useful to employ shallow geophysical techniques in an attempt to detect subsurface rock head behaviour (Geological Society Engineering Group Working Party, 1988).

QUARRYING AND MINING

Nature of hazards from quarrying and mining

The dominant geological hazards which result from quarrying and mining are rock fall within the quarry site, subsidence as a result of undermining, bad drainage, or even flooding (as a consequence of changed land slopes), earth tremors as underground mines collapse, and the hazards arising from the disposal of quarry waste.

Causes of hazards from quarrying and mining

A basic distinction has to be drawn between the following activities:

- quarrying of sand and gravel
- quarrying for rock
- quarrying for minerals
- mining for minerals

Within these main categories we must also distinguish between those activities that are essentially linear and those that extend across an area, whether it be above or below the ground. Even within these broad categories there are detailed variations that bring about differences in their consequences (Table V).

Table V Styles of mining and their surface expression

Mineral	*Style of mining*	*Surface expression*
coal	drift mine	
clays, ironstones, coal, chalk, flint	bell pits	crown hole*
coal, ironstone, fireclay, limestone	room and pillar	crown hole or general subsidence if deep
coal	longwall	general subsidence
metalliferous ores	vein stoping	crown holes
salt, potash	solution mining	general subsidence

*A crown hole represents the collapse of surface material down into the mined space below.

Mapping hazards from quarrying and mining

Some of these hazards, for example failure of faces of pits or quarries, tend to be localized. Although some factors which influence such movements can be mapped, such as joints and fissures in rock faces, work generally depends on detailed site surveys (Geoffrey Walton Practice, 1988). For this reason, these hazards will not be considered further here. Of much greater significance is the mapping and recording of major problems arising out of the extensive workings (whether at or below the surface) that affect large areas. The classic example is that of ground subsidence in association with oil extraction from the Wilmington Field causing subsidence in excess of 10 m in the Long Beach area of California (Mayuga and Allen, 1969). Similar examples occur in Japan where, as a result of groundwater extraction, significant amounts of subsidence have been recorded at more than 40 sites, and affecting a total of more than 7000 km^2 (Yamamoto, 1977).

Much more extensive, however, are the subsidence

problems that arise out of the removal of coal from underground. The accumulated experiences of the coal industries in a number of countries have allowed the consequences of coal mining to become not only well established but also amenable to detailed prediction (see National Coal Board (1975) *Subsidence Engineer's Handbook*). The task of hazard mapping in these cases is one of recording the amount, the rate and the extent of subsidence as it takes place. This relies on adequate means of recording changes in surface elevation. Thus detailed and repetitive field surveying has traditionally been carried out by levelling from precisely located bench-marks. With the advent of new techniques in photogrammetry it is now possible to obtain contours of changes in ground elevation from successive flights of oblique aerial photographs. By the very nature of the problem the mapping of subsidence tends to be one of depicting ground contours or iso-lines of changes in ground elevation between two moments in time.

EROSION AND DEPOSITION

Nature of hazards resulting from deposition and erosion

It is appropriate to see erosion and deposition as two ends of the same process. Erosion generates mobile sediments which then come to rest (albeit temporarily in some cases) as depositional forms. The hazards which occur in the erosional part of the system tend to fall into one of the following categories:

- soil loss leading to agricultural impoverishment
- gully erosion leading to structural damage
- slope undermining leading to landsliding.

The hazards within the depositional part of the system are associated with the accumulation of sediments where they cause (either directly or indirectly) economic loss or actual damage. This includes the siltation of reservoirs, harbours, navigable river channels and water intake sites. Frequently the deposition of sediment can also lead to the redirection of water flow which itself may have consequential damaging effects

Causes of hazards resulting from deposition and erosion

The main erosion–deposition systems occur in the following environments:

- glacial
- fluvial riverine
- fluvial marine
- fluvial estuarine
- aeolian

Within glacial systems the rate of transfer of sediment from source to end-point is comparatively slow and creates few hazards. In the other systems sediment transfer is much more rapid and tends to occur in areas of greater human population. The old distinction between geological rates of erosion and rates accelerated by the actions of man is crucial in the assessment of erosional and depositional systems. The key lies in understanding what causes accelerated erosion, for this consequentially leads to accelerated deposition.

Accelerated erosion usually stems from the clearance of natural vegetation. In some of the rapidly eroding areas of the Third World erosion follows the age-long need to gather wood for domestic burning. At the present time in the eastern Amazon region of Brazil, wood burning for industrial purposes (e.g. to produce charcoal) in connection with iron ore processing for the Carajas project is adding to the already severe deforestation of the rain forest. Sometimes the cause of erosion is indirect. For example, within those areas of the non-industrialized countries where labour is required in industry, the population tends to be drawn away from the rural areas. This severely decreases the man-power available for soil erosion prevention practices. For example, agricultural terraces fall into disrepair and soil erosion follows. In other cases the cause of soil erosion may be direct, such as when land is overstocked with grazing cattle and vegetation is thereby removed much more rapidly than it can be replenished.

Within the soil erosion system the two critical parameters are soil erodibility and rainfall erosivity. The first refers to the ease with which a particular soil can be eroded, and the latter refers to the ability of rain to achieve the erosion. The former can be mapped if selected criteria are used to define erodibility. The latter is variable and only a statistical property of past rainfall events can be mapped (such as maximum rainfall intensity expected within a 30-minute period).

Coastal erosion is a function of the direction, height and strength of waves impinging on the shoreline. It can occur as sediment removal from the beach, from the cliff behind the beach, or from both. Good accounts of these processes within coastal environments may be found in Hails and Carr (1975), Komar (1976), Bird (1985), Pethick (1984) and Carter (1988).

Aeolian erosion also depends on the availability of silts (and sometimes sands) to be transported. The actual transportation only occurs if the wind velocity exceeds a critical threshold value. The direction of transport is, of course, a function of wind direction. This is less easy to predict than the direction of sediment transport within a river, however, the consequences of mobile sand dunes can be serious when transportation routes, farmland or even settlements are over-run.

Mapping of hazards resulting from deposition and erosion

The mapping of erosional features is not the same as mapping the erosion itself. It may come close to it in the sense that the response to erosion is immediately apparent in the form of soil loss. Similarly, the identification of rates of cliff retreat as a result of coastal erosion indicates the direct result of the erosion process. The hazard becomes apparent when the location of human settlements and structures are added to the map.

Some maps attempt to show present rates of erosion. Russell (1981) attempted this for a drainage basin in Sri Lanka based on field measurements of past erosion amounts, and predictive (mathematical) models based on slope, cover and rainfall information. The result was expressed in the form of contours and the shading of areas apparently prone to excessive rates of erosion. On a predictive level, the map of soil erosion risk in Zimbabwe (Fig. 5) by Stocking and Elwell (1973) is not a map of erosion but of soil erodibility (as expressed through the vegetation cover and the steepness of slope) and rainfall erosivity. Mapping of wind-blown silts and sands, especially over very large areas, may benefit greatly from the regional perspective provided by satel-

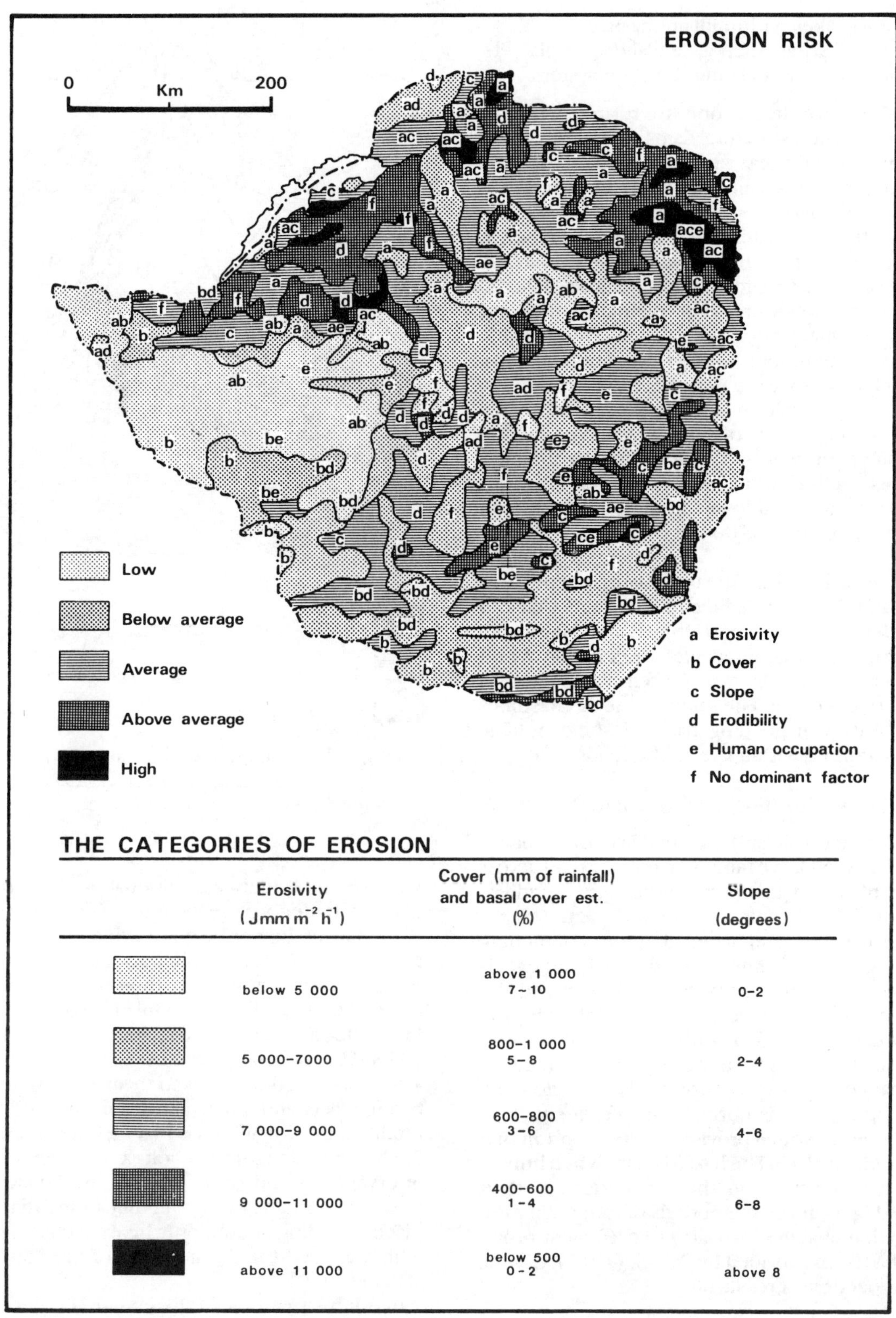

THE CATEGORIES OF EROSION

	Erosivity ($J mm m^{-2} h^{-1}$)	Cover (mm of rainfall) and basal cover est. (%)	Slope (degrees)
Low	below 5 000	above 1 000 7–10	0–2
Below average	5 000–7000	800–1 000 5–8	2–4
Average	7 000–9 000	600–800 3–6	4–6
Above average	9 000–11 000	400–600 1–4	6–8
High	above 11 000	below 500 0–2	above 8

Fig. 5. Erosion risk in Zimbabwe (after Stocking and Elwell, 1973).

lite imagery. Indeed, Jones *et al.* (1986) found that only by reference to such a regional perspective could the wind-erosion system be understood and defined.

FLOODING

Nature of the flood hazard

Flooding tends to occur in a number of different forms depending on the environmental location. In some instances they result from the failure of man-made structures such as dams, and in these circumstances they are unpredictable and catastrophic. The main types of 'natural' flooding (excluding reservoirs created by man, or areas of poor drainage resulting from land subsidence as a consequence of mining) are:

- on the flood plains of large alluvial rivers;
- on deltas;
- on alluvial fans;
- along the coast;

- as temporary lakes within inland basins;
- behind natural dams, such as landslides or glaciers, which block the pre-existing drainage system.

All of these sites, save the last one, are relatively easy to identify and map. The last, natural dams, tends to occur in generally uninhabited areas and thus seldom poses a hazard (though recent examples within the Himalayas have caused problems).

Since flood sites are so amenable to mapping it might at first sight appear odd that they should present a hazard. If these locations are prone to flooding, the hazard would be removed if the sites were to be avoided. This is not the case because many flood-prone sites often provide advantageous locations for a number of economic activities. This includes, for example, the food production areas of the Ganges Delta which are so critical to the survival of the population of Bangladesh. Yet, within the Indian sub-continent as a whole (which includes Bangladesh as well as other flood-prone areas) it is predicted that agricultural losses from flooding amount to more than 300 million US dollars per annum (Hewitt, 1983).

Set against these losses are the unmeasured benefits to soil fertility and the maintenance of wet lands that can only be achieved by flooding (Drijver and Marchand, 1985). Many of the large floodable rivers of the world have been 'engineered' in man's attempts to control flooding and river course. The same is true of coastlines where the control of flooding has had some notable success, for example within the Netherlands.

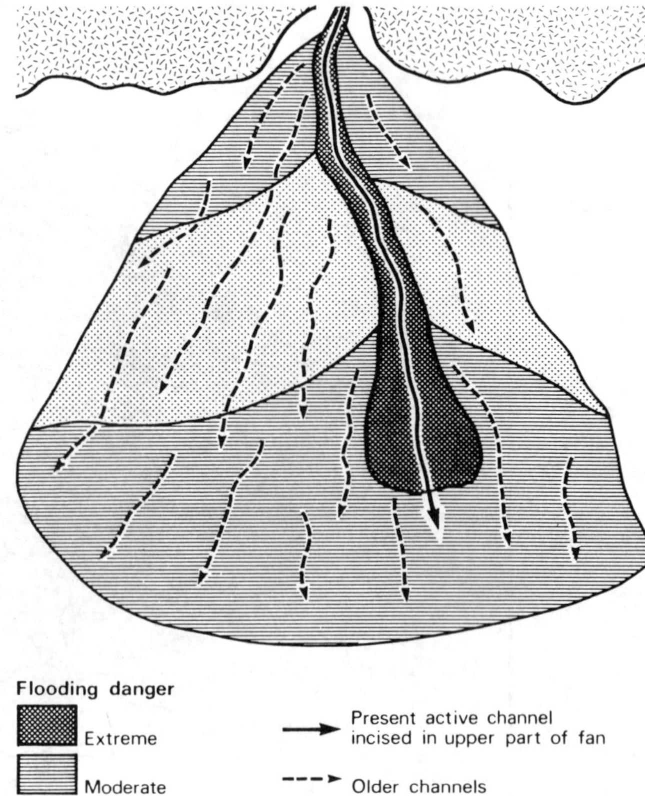

Fig. 6. Classification of alluvial fans in terms of danger from flooding (after Schick, 1971).

Causes of the flood hazard

Alluvial fans, especially in arid and semi-arid lands, occur where river channels leave mountains and extend across the adjacent plains. At this point the channel gradient flattens, channel flow becomes wider, sediment is deposited, and the main flow may dismember into a branching network of smaller channels. As more sediment is dropped these channels may become blocked and the water will change its course. Heavy rainfall in the mountains will cause flooding within the main channels and across those parts of the fan surface affected by changing channel flows. In a sense this style of flooding is not flooding at all. It is normal but infrequent flow. It is the infrequency which provides the perception that 'flooding' has taken place. The hazard arises when human use of the flow channels in the dry periods is then overwhelmed by water in the periods of mountain rain.

A classification of alluvial fan sites (Fig. 6) most prone to flooding has been provided by Schick (1971). Coastal flooding can occur as a result of:

- seasonal high tides;
- on-shore winds (especially at the time of high tides) especially in association with hurricanes and cyclones;
- tsunamis (high sea-waves generated by earthquakes);
- progressive rise in sea-level (e.g. in response to the 'green-house' effect).

Mapping the flood hazard

Because there has been a great deal of engineering in association with flood sites it has become more difficult to map the flood hazard solely on the basis of morphological evidence. The presence of a flood-plain, which is the evidence traditionally used within geomorphology to denote the location of floodable sites along a river, may no longer be good evidence that flooding will actually occur. Indeed, if a flood protection scheme has been properly engineered then there should no longer be any flooding of the 'floodplain', except as allowed for in the flood protection design.

Flood hazard mapping must therefore take account of any human modifications to the natural system that have taken place and must allow for the effects of these modifications. It will need to include an assessment of likely flood magnitudes on a frequency–recurrence interval basis and try to predict the behaviour of high flows in the light of any modifications that have taken place. Flooding should only be expected for flow levels which exceed the design limits of the flood protection scheme.

In other situations, such as where no flood protection work has been carried out, the mapping of flood hazard, as a basis for predicting where a flood will occur, is a geomorphological analysis of the landforms and the fluvial system, supported wherever possible by information concerning past floods. Wolman (1971) has provided a list of criteria by which areas subject to flooding may be mapped (Table VI). Past signs of flooding are used to indicate flood levels, flood volumes, and estimates of flood frequency (e.g. Brunsden *et al.* (1981) in Nepal). For planning and development purposes flood hazard zonation maps, such as that provided for urban development in the Suez region (Cooke *et al.* 1982), provide a useful summary of flood characteristics for use by non-specialists.

The approach to mapping coastal areas prone to

Table VI Techniques of mapping areas subject to flooding (based on Wolman, 1971)

Mapping method	*Mapping principles*	*Limitations*
Physiographic	Morphological and topographic features	Morphology may not coincide with actual flood characteristics
	Flood levels (observed)	
	River terraces (including height correlations)	Role in present-day flooding not always clear
Pedological	Soil characteristics	Distinction between soils of differing origins (including those formed by flooding) not always clear
Vegetational	Vegetation variations with flood heights/ characteristics	Assemblages of plants not directly or uniquely linked to flooding
Historical	Aerial photographs Satellite imagery	Availability Not available before 1972. Resolution poor
	Manuscript record of 'highest' floods	No spatial transposition
Analytical	Regional stage frequency (e.g., flood levels for 1 in *x* years floods)	Spatial transposition of stage data poor. Variability in hydraulic conditions not allowed for. Omission of backwater effects

flooding is not very different. Much reliance is placed on both elevation data and the geomorphological character of the areas. The latter includes the mapping of features likely to be associated with flooding (e.g. sites through which or over which sea water can gain entry to the land area, hollows within which the water can accumulate, sediments laid down by earlier flood waters, damage by earlier floods, and the location of salt-water tolerant vegetation).

The causes of coastal flooding are predominantly unmappable except to a very limited degree. While it is possible to map past tracks of hurricanes and cyclones, thus giving an element of location to the potential for further flooding from these causes, the level of predictability is very low. In the case of the progressive rise of sea-level currently being experienced as a result of the green-house effect, it is possible to record on a map the amounts of sea-level rise already experienced. It is also possible to predict, on the basis of a finely contoured map, how much additional land is vulnerable to flooding with every unit interval of sea-level rise.

Within arid and semi-arid areas it has become important to be able to map the flood hazard associated with alluvial fans because farmlands, travel routes and settlements are often found in these areas.

SALINE SOILS

Nature of saline soil hazards

High concentrations of salts in a soil can render the ground toxic to most plants, thus restricting agriculture. In some situations, such as where irrigation takes place in hot arid or semi-arid areas, salts have been mobilized within the soil system, concentrated at the surface and have completely destroyed the agricultural potential of the land. Millions of hectares of land have been affected in this way.

High concentrations of salts within the soil or the near-surface ground-water can also cause immense structural damage to concrete buildings and other structures. There is a growing list of buildings which are having to be demolished because processes of salt damage, especially at foundation level, have made them unsafe. Further damage may be caused if chlorides and sulphates enter the concrete, for chlorides in particular can react with the reinforcing bars, causing them to corrode, expand and produce strains within the concrete that cause it to crack and thereby to loose its fundamental strength.

Causes of saline soil hazards

The main causes of saline soil hazards are groundwater at or close to the surface, and high air temperatures. These high temperatures induce capillary rise (within loose sediments) to as much as 3 m above the water-table. This rise brings damaging salts towards the surface where they can evaporate to form damaging concentrations either within the agricultural soil or within concrete structures (Cooke *et al.* 1982).

Mapping saline soil hazards

The mapping of saline soil hazards tends to adopt one of two approaches. The first is to map the intensity and distribution of existing damage. This may be in relation to either crop damage or urban decay as building damage takes place. The second approach is to map the controlling parameters on saline attack; these are the depth of the groundwater and its saline concentrations. Since the latter is closely related to the value of the electrical conductivity, which is much easier to measure, this latter is used as a surrogate for saline concentrations (Cooke *et al.*, 1982).

It is often possible to detect the presence of saline areas by a quick examination of the ground surface and the surface soils. Here salt crystals will be evident, often as a halitic deposit across the surface. While this identifies where the problem occurs it does not quantify it. In some respects such an identification would obviate the need to quantify salinity. If salinity is so readily apparent, then the problem is undoubtedly at the hazard level and the ground should not be used for any construction work unless very effective methods are found to separate the structure from the saline soils. Such land is, of course, useless for agriculture.

An example of a salt hazard map is that produced for Port Said by Ibrahim and Doornkamp (1987), in which the combined information on depth to ground water and electrical conductivity are used to define hazard zones. The data used are all point data and the zones have been produced by hand contouring through these data. All of the observations made about the limitations of such methods (in Chapter 2) apply.

PERMAFROST

Nature of permafrost hazards

Permafrost is ground which is frozen for two or more consecutive years. The definition of freezing is based on temperature and not on the presence (or absence) of ice. The terminology of features associated with permafrost is given in Fig. 7 (Cooke and Doornkamp, in press). Permafrost can be divided into two types: *continuous permafrost* (mean annual temperature at 10–15 m of −5°C, and a permafrost table usually within 0.61 m of the surface), and *discontinuous permafrost* (mean annual temperature at 10–15 m of −5 to −1.5 °C,

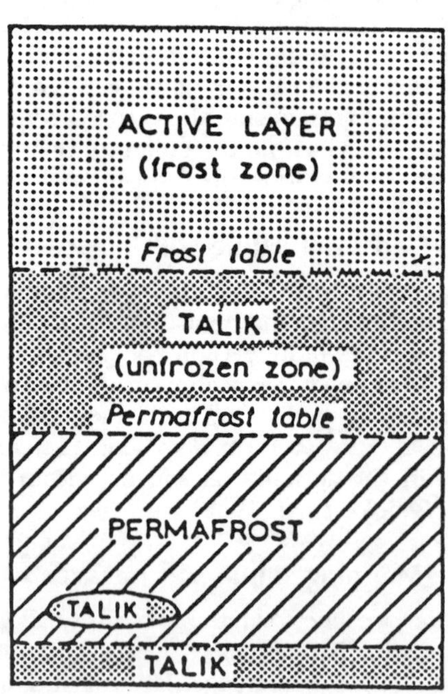

Fig.7. Terminology of some features associated with permafrost (from Cooke and Doornkamp, in press).

interrupted by thawed areas, and a deeper permafrost table). The active layer is the one in which intermittent freezing and thawing takes place, and it is this process which forms a hazard to certain of man's structures.

Over 22 million km^2 of the northern hemisphere is within the permafrost zone (Fig. 8). It occurs where there are cold long winters with little snow, short dry and relatively cool summers, and low precipitation during all seasons. Within this context the location of permafrost is also influenced by vegetation and ground cover, hydrological conditions, snow, terrain and surface material characteristics (Cooke and Doornkamp, in press).

The weathering processes of freezing and thawing can affect any structures placed within the active layer. Damage can occur to such structures (e.g. the footings of a house, the piles for a bridge). However, the effect of some structures (especially pipe-lines carrying hot oil) is to raise the temperature of the ground through which it passes, thereby destroying the permafrost. This changes the physical characteristics of the ground, and hence the stability of the pipe-line. For example, thawing can generate excess ground water which changes the medium through which the pipe passes to such an extent that the pipe may fracture. Conversely, chilled gas pipe-lines may produce local pockets of ice within the soil; frost heave may then take place and fracture the pipe. These and similar problems are discussed by Williams (1979), French (1980), and by Thomas and Ferrell (1983).

Causes of permafrost hazards

It is in this context that activities such as the clearance of vegetation, within permafrost areas, drainage of lakes, or other hydrological changes, the creation of roads (and hence an alteration in the surface materials) have to be seen. Any one of these can alter the permafrost environment and hence the nature of the depth and extent of the permafrost. The usual response is a degradation of the permafrost. The consequence is a change in ground stability conditions, foundation and load-bearing conditions. Herein lies the hazard.

Mapping permafrost hazards

The important parameters to map, in this case, include those of soil, vegetation, groundwater hydrology, surface hydrology and forms of the permafrost (such as pingos). The deeper elements of the permafrost system (such as the depths of the frost table and the permafrost table, Fig. 7) cannot be mapped directly. However, since it is now known that changes in the surface parameters which can be mapped will cause predictable changes in the depths of these tables, any mapped data on surface conditions will help to indicate the nature of the permafrost environment and how a planned action will cause it to respond.

Since much of the permafrost area is remote and inaccessible, initial mapping is most usefully carried out through the analysis of satellite imagery and aerial photography (e.g. Gaydos and Witmer, 1983). The study of *Landforms and Surface Materials of Canada* by Mollard (no date) includes examples of the data derived from satellite imagery within the permafrost areas.

The purpose of providing such maps is usually to:

- determine the nature of the terrain
- identify ground-ice conditions

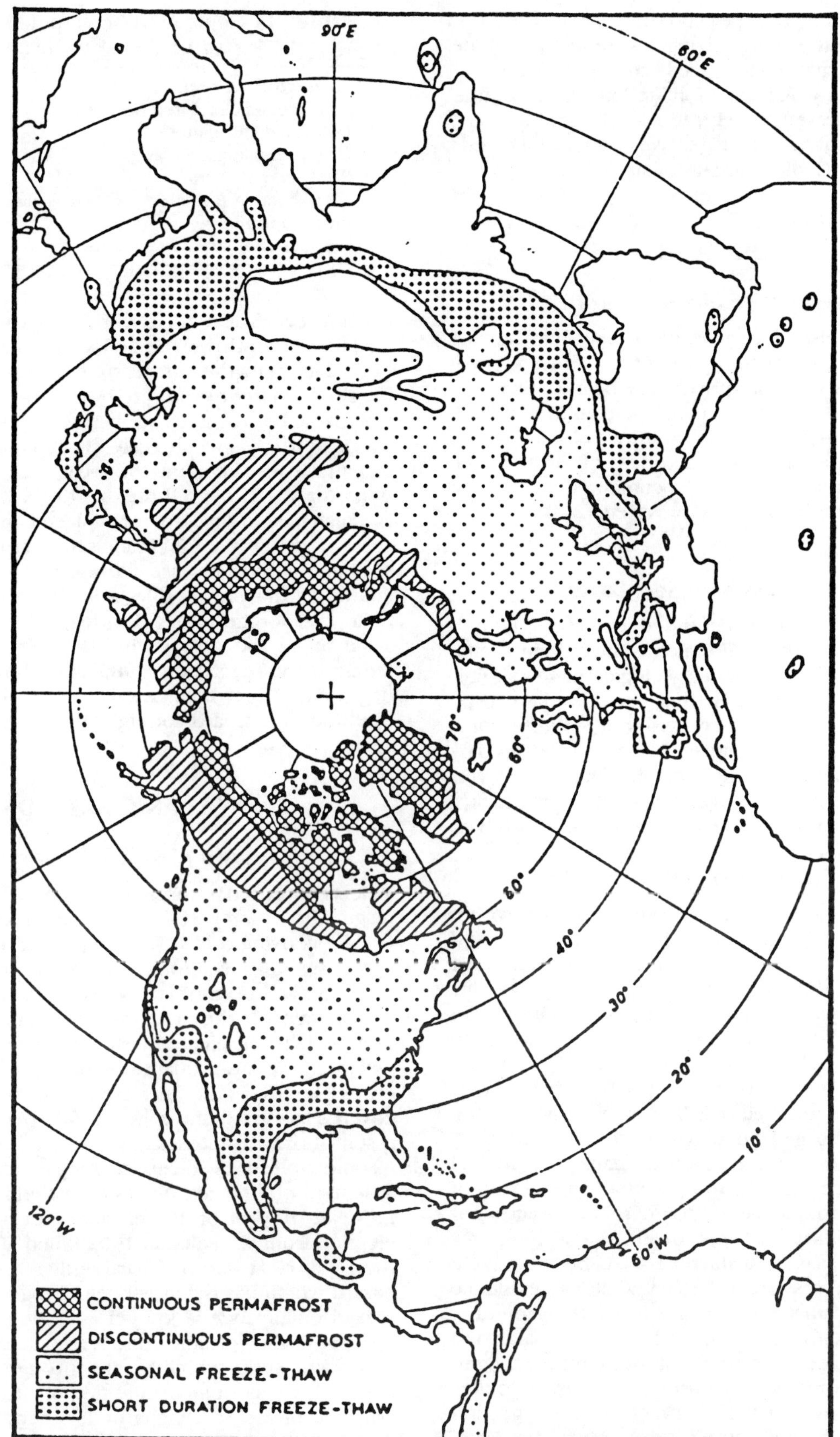

Fig. 8. Distribution of permafrost and related phenomena in the Northern Hemisphere (based on Corte, 1969, Mackay, 1972, Brown 1967, and others; from Cooke and Doornkamp, in press).

- define the nature of surface materials
- identify any sources for construction materials
- to plan access routes likely to do least damage to the environment
- to select sites for more detailed study on the ground.

More detailed mapping usually includes the same sets of surface parameters, though at a more site-specific scale. In the absence of cultural features and base maps, it is often necessary, and indeed advantageous, to map directly on to panchromatic aerial photography. The features most likely to appear on such maps, in addition to any pertinent ecological factors, tend to include: pingos, ice-wedges, high-centred and low-centred polygons, thermokarst features, solifluction lobes, solifluction terraces, gravel concentrations, silt, sand and clay deposits, areas of bad drainage, lakes, former lake beds, and peat bogs (Cooke and Doornkamp, in press).

SEISMOLOGICAL HAZARDS

Nature of seismological hazards

The dominant seismological hazard is that of earthquake activity. Associated with it are other consequential hazards such as volcanic activity (see below), tsunamis (coastal flooding arising from an earthquake) and landslides. These latter hazards may, in certain circumstances do more damage and cause more loss of life than the original earthquake. A well documented review of hazards from earthquakes is provided by Bolt *et al.* (1975).

Causes of seismological hazards

Many earthquakes are concentrated within the tectonically active belts of the world, such as the circum-Pacific rim, which coincide with active geological plate margins. At these plate margins it is usual for one plate to be passing under the other, thereby setting up a frictional stress between them. Whenever this stress is released by further movement (i.e. the frictional forces are overcome), there is an earthquake. However, earthquakes can and do occur within the 'stable' plates themselves.

Mapping seismological hazards

While it is true that a map can be made of the areas within which past earthquake activity has concentrated, and those areas within which future earthquakes are most likely to occur, the mapping of an earthquake is a silly concept. Earthquake motion is a process and all that can be mapped are the responses to that process. Such responses in the short term include mapping the fracture patterns at the surface which arise from dislocations within the soils and bedrock. It may also include the evidence of damaged buildings and other structures. However, much of the earthquake damage recorded within urban areas results from the swaying motion of buildings, a consequence of ground motion that is not expressed in linear fractures. Indeed, after the earthquake motion has passed there may be no visible topographic expression of the fact at all. The evidence usually lies in damaged structures or faulted ground.

More commonly, the mapping of earthquake and seismological events is carried out through the recognition of morphotectonic features. These are geomorphological characteristics that indicate ground movements in the geologically recent past. The list of morphotectonic features indicative of past earth movements is quite substantial (Cooke and Doornkamp, in press; Doornkamp, 1986). An abbreviated list is provided in Table VII, and all of these are capable of being mapped.

Evidence of this kind has been used in China not only to identify areas of past earthquake activity but also in attempts to identify the locations of future events (Doornkamp and Han, 1985). Much earthquake mapping, however, is concerned not with the identification and plotting of single morphotectonic features but with earthquake hazard zoning. This has been done on a global scale both by the Swiss Reinsurance Company (1978) in their *Atlas on Seismicity and Volcanism* and in a similar volume produced by the Munich Reinsurance Company. Both studies are based on historical data.

Table VII Some morphotectonic indicators of past earthquake activity

Emerged coral reefs
Displacement of dated beaches
Deformed shorelines
Distorted alluvial terraces
Fault scarps
Displacement of man-made structures
River capture
Responses of stream channels

At a more site-specific scale, earthquake hazard zoning maps take into account the local rock and soil conditions, for it is through these that the earthquake motion is transferred to any structures placed upon them. In the case of the 1906 San Francisco earthquake, ground shaking was most intense in the areas underlain by muds and other alluvium, and least in the areas underlain by bedrock (Borcherdt, 1975). Degg (1987) showed the same to be the case during the 1985 earthquake at Mexico City. Cross *et al.* (1985) show how various forms of ground and earthquake data can be combined to form the basis of a hazard zoning map (Fig. 9).

VOLCANIC HAZARDS

Nature of volcanic hazards

Volcanoes are fed by magma chambers out of which both magma (lava) and gases can escape through weaknesses in the overlying rocks. The magma can overwhelm and the gases may be toxic or asphixiating (or both). Explosive action within the volcano can cause fragments to be blown into the air and scatter over a very wide area. The weight of ejected ash can cause buildings to collapse, and large quantities of ash can destroy vital crops. In 1815 the ash from Tambora (Indonesia) covered farmlands and about 75 000 people starved as a result of the crop losses.

Some volcanoes eject glowing avalanches (nuées ardentes) of fluidized hot ash, dust and gas which can move at 100 km/hr for distances of 10 km. Such an ejection from Mt. Pelee in 1902 killed 30 000 people in the town of St. Pièrre (Martinique). A major volcanic hazard which has recently been recognized as of widespread occurrence is known as 'sector collapse'. This collapse of the volcanic cone is associated with a lateral pyroclastic surge which breaks through the side of the cone. This was dramatically demonstrated during the eruption of Mt. St. Helens in 1980 and appears to be common in andesitic volcanoes. Perhaps most damaging of all, however, are the volcanic mudflows (lahars) that are generated on the sides of volcanoes. Most dramatic are those which involve large masses of water from the melting of snow and ice (as temperatures rise) which becomes mixed with soil and debris on the volcano flank. A lahar was responsible for 20 000 deaths in Armero (Columbia) in 1985. In general, volcanic hazards occur at known sites, though, as the detailed studies of Mt. St. Helens showed, the precise prediction of the location of the consequential effects is very difficult.

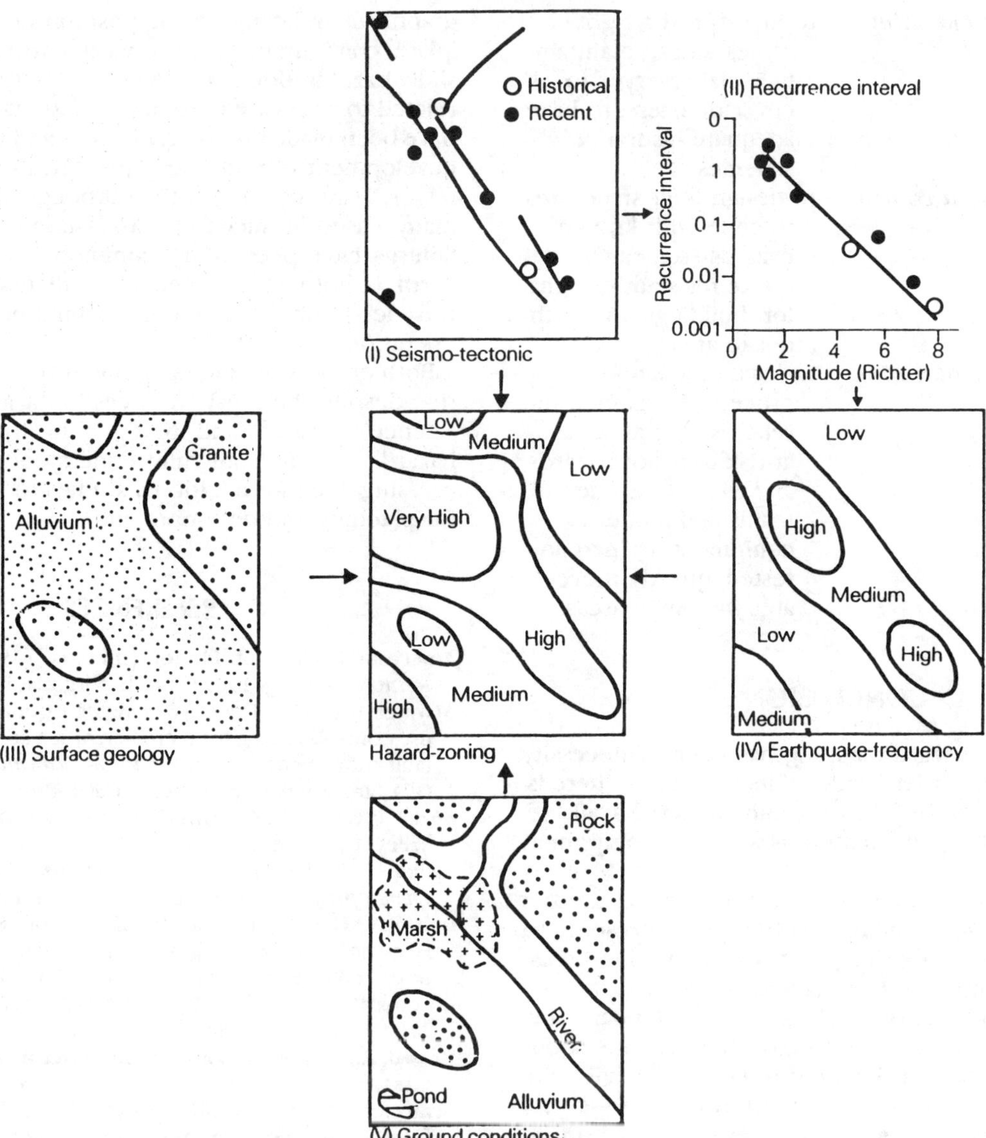

Fig. 9. Data sources for earthquake hazard zoning (after Cross *et al.*, 1985).

Causes of volcanic hazards

Volcanic events arise from geological conditions similar to those which determine earthquakes. Indeed, earthquakes and volcanic activity may occur together. Certainly there is a strong coincidence in their global concentrations.

Mapping volcanic hazards

Much of the mapping associated with volcanic hazards is the mapping of forms and materials associated with past eruptions. This does not provide a direct indication of where the hazards will concentrate during the next eruption. However, it does enable the nature of the events associated with that particular volcano to be defined, and allows the influence of the new forms on the direction of future lava flows to be assessed.

APPLICATION OF HAZARD MAPS

Despite the examples cited above, there is strong circumstantial evidence to indicate that the potential value of hazard mapping has not yet been realized. Fuller use would undoubtedly lead to better hazard avoidance and increased site mapping would lead to fewer cases of problems from hazards not previously identified. However, the mapper is frequently not the person who can bring about the implementation of his findings. These must be incorporated within an administrative and (frequently) political framework. One worthwhile planning (or engineering) response to hazard maps is to take them as a basis for hazard avoidance. This approach, however, may carry many complications, especially if few alternative sites for a development or engineering structure are available. In this case, a firm decision has to be made which includes the possibility that the project is abandoned. If development should proceed, the hazard map then becomes a relevant document in the engineering design process.

WIDER RESPONSES

There are wider responses, beyond mapping considerations that can be adopted to mitigate hazards. Among these may be cited:

(a) *Land use planning* —to place development in the least hazardous places.

(b) *Economic planning* —to invest in the right places and to maintain financial reserves to cover disasters; to have adequate insurance schemes.
(c) *Development control*—to design built structures to cope with known hazards (e.g. earthquake-resistant designs for buildings in earthquake areas).
(d) *Emergency planning* —to have evacuation procedures planned and rehearsed in all areas at-risk to major hazards, and to have rescue teams and modern equipment, trained and tested, quickly moveable into such areas.

CONCLUSION

The subject of hazard mapping has been of necessity condensed into a brief review of its essentials. There is, however, a wealth of further reading material available, covering both the general topic as well as its component aspects.

Two recent reviews outside Europe may be cited in conclusion. Johnson and Blong (1985) have presented a very concise summary of practices in the Southwest Pacific and Southeast Asia regions. In particular, they stress the importance of data files: '... more fundamental data are required in order to answer questions about where hazards will strike, how often impact will take place, and how big these events will be'. They cite several existing data recording systems covering earthquake epicentres, volcanoes and tsunamis, but conclude with the following observations:

(1) Satisfactory data files have not been assembled [in these regions] for landslides, floods, ground subsidence, coastal erosion, expansive soils or soil erosion.
(2) There is a general lack of compiled data such that the hazardousness of any particular place is difficult, if not impossible, to assess.
(3) Data files consisting only of information on individual hazards are less useful than those that allow assessment of the hazardousness of a place (for example there is no point in increasing the seismic resistance of structures, if they remain in flood, storm-surge or tsunami-prone sites, or in encouraging migration out of one hazard into another).
(4) Most data files cover far too short a period of record.
(5) A few very large events can distort the record and make quantification and prediction from it valueless.

In discussing vulnerability, these authors make the point that small countries are more vulnerable to hazard impacts than large, and small poor countries are the most vulnerable of all.

The second paper cited is by Wang Sijing (1988). He presents a remarkable account of Chinese practice. In China, the type of hazards experienced are geographically relatable to the position in the geotectonic 'plate' configuration: in the west, earthquake and landslide hazards dominate, whereas in the east, hazards related to rivers are predominant. Remarkable advances have been made in risk assessment and prediction, with development of numerical quantification systems. Professor Wang concludes that although there have been many cases of successful assessment and prediction, failures have been more common: 'the nature of the Earth is both elusive and unpredictable'. He stresses the need for international communication and cooperation.

Both of these accounts highlight the indisputable fact that despite the great advances in hazard-related earth science research and practice in the last decades, hazards are far from fully understood and there is a pressing need for additional studies to provide support for planning and development.

REFERENCES

American Society of Photogrammetry (1975). *Manual of Remote Sensing*. Falls Church, Virginia, USA.

Atwater, B. (1978). Central San Mateo County, California. Land use controls arising from erosion of sea cliffs, landsliding and fault movement. In *Nature to be Commanded* (Robinson, G.D., and Spieker, A.M., eds.), USGS Prof. Pap. 950, 11–20.

Bird, E.C.F. (1985). *Coastal Changes—a Global Review*. John Wiley, Chichester.

Blaikie, P. (1985). *The Political Economy of Soil Erosion in Developing Countries*. Longmans, London.

Bolt, B.A., Horn, W.L., Macdonald, G.A. and Scott, R.F. (1975). Hazards from earthquakes. In *Geological Hazards* (Bolt, B.A. *et al.*, eds.), Springer-Verlag, New York, 1–62.

Borcherdt, R.D. (ed.) (1975). *Studies for Seismic Zonation of the San Francisco Bay Region*. USGS Prof. Pap., No. 941a.

Brown, R.J.E. (1967). Comparison of permafrost conditions in Canada and the USSR. *Polar Record*, **13**, 741–751.

Brunsden, D., Jones, D.K.C., Martin, R.P. and Doornkamp, J.C. (1981). The geomorphological character of part of the low Himalaya of eastern Nepal. *Zeit. für Geomorphologie*, **37**, 25–42.

Carter, R.W.G. (1988). *Coastal Environments*. Academic Press, London.

Cooke, R.U. and Doornkamp, J.C. (in press). *Geomorphology in Environmental Management*. (2nd Ed.) Oxford University Press, Oxford.

Cooke, R.U., Brunsden, D., Doornkamp, J.C. and Jones, D.K.C. (1982). *Urban Geomorphology in Drylands*. Clarendon Press, Oxford.

Corte, A.E. (1969). Geocryology and engineering. *Reviews in Engineering Geology*, **2**.

Cross, M., Degg, M. and Johnson, J. (1985). The escalating earthquake risk in the Middle East. *Risk and Loss Management*, 16–19.

Culshaw, M.G. and Waltham, A.C. (1987). Natural and artificial cavities as ground engineering hazards. *Q.J. Eng. Geol.*, **20**, 139–150.

Curran, P.J. (1985). *Principles of Remote Sensing*. Longmans, Harlow, Essex.

Degg, M.R. (1987). The 1985 Mexican earthquake. *Modern Geology*, **11**, 109–131.

De Graff, J.V. and Ronesburg, H.C. (1980). Regional landslide—susceptibility assessment for wildland management: a matrix approach. In *Thresholds in Geomorphology* (Coates, D.R. and Vitek, J.D., eds.), Unwin, London, 401–404.

Demek, J., (ed.) (1972). *Manual of Detailed Geomorphological mapping*. Academia, Prague.

Demek, J. and Embleton, C. (eds.) (1978). *Guide to Medium-scale Geomorphological Mapping*. IGU, Stuttgart.

Doornkamp, J.C. (1986). Geomorphological approaches to the study of neotectonics. *J. Geol. Soc. London*, **143**, 335–342.

Doornkamp, J.C. and Han, M. (1985). Morphotectonic research in China and its application to earthquake prediction. *Progress in Physical Geography*, **9**, 353–381.

Drijver, C.A. and Marchand, M. (1985). *Taming the Floods: Environmental Aspects of Floodplain Development in Africa*. Centre for Environmental Studies, Leiden.

Drury, S.A. (1987). *Image Interpretation in Geology*. Allen and Unwin, London.

Dumas, B., Gueremy, P., Lhenaff, R. and Raffy, J. (1984). Mouvements de terrain et risques associés: presentation d'un essai cartographique. In *Mouvements de Terrain* (Flageollet, J.C., ed.), *Documents du BRGM*, **83**, 163–171.

Edmonds, C.N. (1983). Towards the prediction of subsidence risk upon the Chalk outcrop. *Quart. J. Eng. Geol.*, **16**, 261–266.

Embleton, C. and King, C.A.M. (1975). *Periglacial Geomorphology*. Arnold, London.

Embleton, C. and Thornes, J.B. (eds.) (1979). *Process in Geomorphology* Arnold, London.

French, H.M. (1980). Periglacial geology and permafrost. In *Progress in Physical Geography*, **4**, 254–261.

Gaydos, L. and Witmer, R.E. (1983). Mapping of arctic land cover utilising LANDSAT digital data. *Proc. 4th Permafrost Int. Conf.*, Nat. Academy Press, Washington, D.C. 343–346.

Geoffrey Walton Practice (1988). *Handbook on the hydrogeology and stability of excavated slopes in quarries*. Her Majesty's Stationery Office, London.

Geological Society of London Engineering Group Working Party (1988). Engineering geophysics. *Quart. J. Eng. Geol.*, **21**, 207–271.

Hails, J. and Carr, A. (eds.) (1975). *Nearshore Sediment Dynamics and Sedimentation*, Wiley, London.

Hanson, A. (1984). Landslide hazard analysis. In *Slope Instability* (Brunsden, D. and Prior, D.B., eds.), Wiley, Chichester, 523–602.

Harris, R. (1987). *Satellite Remote Sensing*. Routledge, Keegan Paul, London.

Hearn, G.J. and Fulton, A. (1986). Hazard assessment techniques for planning purposes. In *Planning and Engineering Geology* (Culshaw, M.G., ed.), Engineering Group, Geol. Soc. London, 22nd Annual Meeting, Plymouth, 8–12 September, 1986, Abstract Volume, 351–356.

Hewitt, K. (ed.) (1983). *Interpretations of Calamity*. Allen and Unwin, London.

Huma, I. and Radulescu, D. (1978). Automatic production of thematic maps of slope stability. *Bull. Int. Assoc. Eng. Geol.*, **17**, 95–99.

Hutchinson, J. (1988). Morphological and geotechnical parameters of landslides in relation to geology and hydrogeology. In *Proc. Int. Symp. Landslides*, Lausanne, in press.

Ibrahim, H.A.M. and Doornkamp, J.C. (1987). Constraints on urban development in Port Said. In *Third World Planning Review*, Liverpool University Press, Liverpool, UK, **9** (4), 325–343.

Johnson, R.W. and Blong, R.J. (1985). Vulnerability and the identification and assessment of geological hazards in the Southwest Pacific and Southeast Asia. In *Geosciences for Development—the Australian Role* (Katz, M.B. and Langevad, E.J., eds.), AGID Report Series, No. 11, 217–223.

Jones, D.K.C., Cooke, R.U. and Warren, A. (1986). Geomorphological investigation for engineering purposes, of blowing sand and dust hazard. *Q. J. Eng. Geol.*, **19**(3), 251–270.

Kates, R.W. (1978). *Risk assessment of environmental hazard*. Wiley, Chichester, (SCOPE 8).

Kienholz, H. (1978). Maps of geomorphology and natural hazards of Grindelwald, Switzerland, scale 1 : 10 000. *Arctic and Alpine Research*, **10**, 169–184.

Komar, P.D. (1976). *Beach Processes and Sedimentation*. Prentice-Hall, Englewood Cliff, New Jersey.

Landry, J. (1979). *Zones exposées à des Risques liés aux Mouvements du Sols et du Sous-sol*. Région de Lons le Saunier à Poligny (Jura). Bureau de Recherche Géologique et Minières, Orléans.

Lillesand, T. and Keifer, R. (1979). *Remote Sensing and Image Interpretation*, Wiley, New York.

Local Government Act (1973). HMSO.

Mackay, J.R. (1972). The world of underground ice. *Ann. Assoc. Am. Geogr.*, **62**, 1–22.

Mather, P.M. (1987). *Computer Processing of Remote Sensing Imagery: an Introduction*. Wiley, Chichester.

Mayuga, M.N. and Allen, D.R. (1969). Subsidence in the Wilmington Oil Field, Long Beach, California. *Publ. Inst. Sci Hydr.*, **88**, 66–79.

Meneroud, J-P. and Calvino, A. (1976). *Carte ZERMOS, Zones exposées à des risques liés aux mouvements du sol et du sous-sol à 1 : 25 000, Région de la Mayenne Verubie (Alpes Maritimes)*. Bureau de Recherches Géologiques et Minières, Orléans.

Mollard, J.D. (not dated). *Landforms and Surface Materials of Canada—a Stereoscopic Airphoto Atlas and Glossary*. J.D. Mollard, Regina, Saskatchewan.

Newton, J.G. (1984). Sinkholes resulting from groundwater withdrawals in carbonate terranes—an overview. *Geol. Soc. Am. Reviews in Engineering Geology*, **6**, 195–202.

Pethick, J. (1984). *An introduction to Coastal Geomorphology*. Arnold, London.

Porcher, M. and Guillope, P. (1979). Cartographie des risques ZERMOS appliquées à des plans d'occupation des sols en Normandie. *Bull. Mason Laboratoire des Ponts et Chaussées*, 99 Ref 2277, 44–54.

Radbruch-Hall, D.R., Colton, R.B., Davis, W.E., Skipp, B.A., Lucchitta, J. and Varnes, D.J. (1976). *Preliminary landslide overview map of the Conterminous United States*. US Geol. Surv. Misc. Field Studies, Map MF 771.

Russell, J.R. (1981). *Sedimentation in the proposed Kotmale Reservoir, Sri Lanka*. SE Asian Regional Symposium on Problems of Soil Erosion and Sedimentation, Bangkok, Thailand, 27–29 Jan. 1981, 405–418.

Sabins, F.F. (1978). *Remote Sensing*, Freeman, San Francisco.

Schick, A.P. (1971). A desert flood: physical characteristics: effects of man, geomorphic significances, human adaptation—a case study of the southern Arava watershed. *Jerusalem Studies in Geography*, **2**, 91–155.

Stevenson, P.C. and Sloane, D.J. (1980). *The evolution of a risk-zoning system for landslide areas in Tasmania, Australia*. Proc. 3rd Australian and New Zealand Geomechanics Conf., Wellington.

Stocking, M.A. and Elwell, H.A. (1973). Soil erosion hazard in Rhodesia. *Rhodesian Agric. J.*, **70**, 93–101.

Thomas, H.P. and Ferrell, J.E. (1983). Thermokarst features associated with buried sections of the Trans-Alaska pipeline. *Proc. Permafrost 4th Int. Conf.*, Nat. Academy Press, Washington, D.C., 1245–1250.

Timberlake, L. (1985). *Africa in Crisis*. Earthscan, London.

Townshend, J.R.G. (1981). *Terrain Analysis and Remote Sensing*. Allen and Unwin, London.

Varnes, D.J. (1978). Slope movement and types of processes in Landslide analysis and control. *Transportation Res. Board. Nat. Acad. Sci., Washington, Spec. Rep. 176*, 11–33.

Varnes, D.J. (1984). *Landslide Hazard Zonation: a Review of the Principles and Practice* UNESCO, Paris.

Wang Sijing (1988). Geological hazards in China. In *Geosciences in Development*. Proc. Conference at Nottingham University, UK, Sept. 26–30 1988. AGID, Geol. Soc. London, Inst. Geol. & Inst. Min. Metall., London, *in press*.

Williams P.J. (1979). *Pipelines and Permafrost: Physical Geography and Development in the Circumpolar North*. Longman, London.

Wolman, M.G. (1971). Evaluating alternative techniques of floodplain mapping. *Water Resources Research*, **7**, 1383–1392.

Yamamoto, S. (1977). Recent trend of land subsidence in Japan. *Science Reports of the Tokyo Kyioku Daigaku, C13, 1–7*.

Chapter 8

Construction

E.F.J. de Mulder and *J.G. Bakker*

INTRODUCTION

Earth resources are usually considered in terms of water, minerals and soils, but, in addition, good land for building and constructon is also a resource. Building and construction is essential to the fabric of modern society and, whilst led by the needs of society for houses, schools, factories, roads, pipelines, power stations and dams and many other installations, the costs are strongly influenced by the state of the ground on, or in which, development takes place.

In addition to construction on the land surface, works may take place underground (e.g., pipelines, sewers, tunnels, underground storage facilities) or in water (e.g., jetties, harbours, offshore oil platforms). Many projects involve the excavation of cuttings, trenches, basements and foundations as well as surface works.

In all of these settings the physical and chemical properties of the ground and its water content have a strong bearing on how construction should best proceed and on the design of structures and their foundations. In all cases those responsible for the works seek the best balance between the utility of the final structure, its stability and safety, and the cost-effectiveness of the operation.

Ground conditions vary greatly from place to place and depend on the physical and chemical properties of the ground materials, the topography, climate, ground and surface waters, and any previous human activity. The characteristics, for example, of an area with outcropping bedrock differ greatly from those of an area of thick unconsolidated superficial desposits. Man-made deposits such as tips and backfilled land may give rise to particularly difficult ground conditions. This is a matter of importance because such deposits are abundant in urban and urban-fringe settings where a great deal of building and construction activity may be concentrated.

This chapter describes those geological, geotechnical and geomorphological factors which may influence the design and stability of buildings and constructions, and the methods by which these factors can be presented to planners and engineers to aid the selection of optimum sites in the feasibility stage and to ensure safe, economical design of the construction. This is demonstrated by means of several case histories.

GROUND CONDITIONS— PROBLEMS AND RESPONSES

A wide variety of ground conditions may affect the methods, design and costs of constructions. These include physical factors, such as susceptibility to vertical or lateral movements of the ground, chemical effects including reactivity with construction materials and problems of contaminated land, and indirect, but sometimes very damaging, processes such as flooding or earthquakes. These factors range from local problems to major regional hazards of the types considered in the previous chapter. The engineer must be alert to all the possibilities since a single site may be subject to several potential difficulties. The main types of ground problems and some possible responses to them are summarized in Table I.

The stability of constructions is determined by the

Earth Science Mapping for Planning, Development and Conservation

Table I Summary of ground problems for construction and excavation

Potential problem	*Description*	*Remedial measures*	*Relevant maps*
Compressible ground	Loosely packed granular materials are liable to compress under load and may give rise to settling of structures. The settling may be even or, if ground conditions vary laterally, may cause tilting. Most materials compress to some extent but problems may be most significant when the load is put on sands, gravels, peat, clays and on man-made ground, especially putrescible landfills.	Compaction of the ground; digging out difficult layers if these are thin; covering by well-graded and compacted granular materials (e.g., sand, gravel, crushed rock); design foundations to accommodate movements; avoid early building on landfills.	Locations, extent and thicknesses of compressible materials. Load-bearing properties. Predicted settlement and land subsidence.
Shrinkage and swelling	Some deposits, notably clays and peat, shrink on losing water (e.g., because of pumping, drainage works, drought or the effects of vegetation). Many clays may swell on being moistened (e.g., by heavy rainfall, leaking water pipes, rising groundwater, or removal of vegetation). These volume changes may, respectively, cause subsidence or heaving of foundations. Clays of the type known as bentonites are particularly liable to volume changes.	Dig out suspect layers if these are thin; carefully design and maintain drainage works; avoid building too soon after clearing vegetation; design foundations to accommodate movements.	Locations, extent, types and thicknesses of suspect horizons. Depths below the surface; groundwater levels; drainage characteristics; vegetation changes.
Mass movement	Ground materials may be subject to lateral sliding or, if the water content is sufficient, to flowing. Such movements are normally down slopes but the critical angle for movement to commence depends on the type of material (lower angles for finer grained, more fluid deposits) and the pore-water pressures within it. Slides may be triggered by erosion or excavation of the lower part of the slope, deposition of materials or construction on the slope, removal of trees, earthquakes, or mine subsidence. Some sensitive clays are liable to liquify if disturbed by natural or artificial vibrations and to flow even if the ground is horizontal (glacial 'quick clays' of peri-polar regions). Saturated peats may also move if water pressure is great enough ('bog bursts').	Regrading and drainage works on slopes; avoidance of excavation or loading of suspect slopes; construction of retaining walls.	Extent, distribution, geological and climatic relationships, and frequency of past landslides; slope angles in relation to suspect horizons; groundwater levels and flows.
Erosion	Rain, rivers, ice, wind and the sea wear away the earth's surface and carry off the debris. Such removal of land may eventually threaten existing structures and is the cause of much expenditure on protective works. However, such works may change the pattern of erosion to affect other nearby areas.	Dykes and embankment strengthening. Construction of sea-walls and groynes. Beach replenishment.	Rates of erosion (e.g., from historical maps), locations of defensive works, resistance of materials to erosion, current velocities and directions.
Deposition	The products of erosion are carried away by water, ice or wind and will be deposited elsewhere when the means of transport decreases (slowing of air and water currents, melting of ice). This may lead to infilling of bays, harbours, reservoirs, lakes, rivers, canals and drainage channels and, consequently, costly maintenance dredging. Underground pipes may become clogged. Dune sands may be blown by the wind to bury roads or buildings.	Dredging of channels; cleaning and maintenance of pipes.	Existing patterns and rates of deposition, types and thicknesses of deposited sediments, trends over time (from historical maps).

Weathering	Physical and chemical alteration of the minerals which make up rocks lead to dissolving of soluble minerals, composition changes to others, and general softening and weakening of the rocks. Most rocks are affected to a few metres in depth, except where the weathered layer has been swept off by erosion. The depth of the weathered profile is strongly dependent on climate, biological activity and groundwater composition. The layer may be tens of metres thick in humid tropical settings. Cavities of variable sizes may be produced where minerals dissolve, ranging from small voids and widened joint planes, to large caverns and tunnels within some limestones. Such cavities, if interlinked, may give rise to underground drainage which leads to surface dry valleys. Such valleys may resume water flow in very wet seasons or if the ground is frozen. Weak rocks may give rise to foundation problems. Subsidence may occur in underground voids ('sink holes' or 'dolines' in limestones).	Take account of weakened rock when designing foundations; caution in areas with underground cavities, locate development away from cavities or infill and stabilize them.	Extent and depth of weathering, distribution and characteristics of solution cavities including depths and sizes, locations and frequency of subsidence events, patterns of underground drainage.
Permafrost	Permanently frozen ground occurs in circumpolar belts. The surface layer may be subjected to seasonal melting giving rise to marshy conditions with peat formation whilst the layer below behaves like hard rock even when consisting of superficial deposits. Constructions may cause heating of the ground sufficient to cause melting and subsidence.	Take precautions against seasonal melting; design structures to avoid heating the ground (e.g., raised insulated floors).	Extent and depth of permafrost, thickness and nature of seasonally melted layer, locations and sizes of major ice concentrations.
Mined ground and other man-made voids	Man-made underground voids include mines, tunnels, cellars shelters and pipes. Voids may be liable to collapse and give rise to surface subsidence especially when the cavities are at shallow depths. Access shafts and adits and wells may remain open or inadequately treated and can cause problems.	Dig out and backfill shallow voids; infill deeper voids; cap shafts and wells.	Types and locations of mined ground and other voids; types, locations and frequencies of subsidence events; extent and depths of minerals which may have been mined before mining records were kept.
Contaminated land	Toxic and other harmful substances may occur naturally in land (e.g., heavy metals, asbestos) but are more often associated with sites used for extraction of minerals and spoil disposal, industrial processes, sewage treatment and waste disposal. Some of these sites may contain harmful organic chemicals (e.g., phenols, polychlorinated biphenyls, tars) or, in the case of some old sites for disposal of wastes, biological problems (e.g., anthrax spores). Problems may be localized or can be more extensive when, for example, groundwater becomes contaminated by such sites or by herbicides, insecticides of fertilizers. Explosive gases such as methane may arise from natural sources (coal, shales) and be concentrated in mines, but are more often associated with putrescible landfills. Radon may be given off from some granites, uraniferous ores and black shales.	Dig out shallow contaminated layers and backfill; cap and isolate deeper sites. Chemical treatment and biodegradation of suitable materials; provide vents for gases, adequate under-floor ventilation and avoid sealed rooms; avoid building on landfills until gas emission has ceased.	Locations, depths, types and concentrations of contaminants; locations of suspect rock types, landfills, and industrial sites; groundwater behaviour.

Table I (*continued*)

Potential problem	*Description*	*Remedial measures*	*Relevant maps*
Underground combustion	Underground combustion may occur if carbon-rich materials, such as coal, are heated and a source of oxygen is available (e.g., in cracks, fissures, pores and other spaces) below the ground surface. Such fires can burn almost indefinitely until the carbonaceous material is used up and may give rise to subsidence of the surface, ground fissuring and death of vegetation. Such fires are almost common in abandoned coal workings and tipped or backfilled colliery spoil.	Difficult and dangerous to remedy– dig out the site (sometimes to considerable depths) and cool the materials.	Depths, extent and locations of potentially combustible materials and of actual instances of combustion.
Groundwater levels and flow	Water in cracks, fissures and pore spaces with the ground accumulates into a saturated zone. The upper surface of the zone (the groundwater table) fluctuates in level between wet and dry periods. Longer term changes may occur. The groundwater level may fall during water abstraction and this sometimes causes appreciable surface subsidence. Groundwater may rise if pumping ceases, if pipes leak or sea-level rises. This can cause flooding of tunnels and basements or even of the ground surface if the rise is great enough. Intersection of the groundwater table during excavation or tunneling may lead to collapses or to flushing in of loose sands ('running sands'). Such deposits may also clog boreholes or abstraction wells if the walls are not protected by a casing at suspect levels.	Design with falls in groundwater level in mind, especially in areas with underground cavities. Control rising groundwater by pumping and drainage. Protect subsurface structures with impervious barriers. Installation of recharge wells.	Level of the groundwater table, directions and rates of groundwater flow, topography, distribution of potential running sands, potential settlement and land subsidence.
Aggressive groundwater and chemicals in soils, rocks	Reactive chemicals which may affect construction materials, notably sulphates and chlorides, may be present in groundwater. Sulphates may come directly from rocks (gypsum and anhydrite) or from alteration of sulphides (e.g. pyrite). Alteration of sulphides to sulphates is accompanied by an increase in volume. Such changes in bulk can cause heaving of foundations, notably when these are constructed on pyritic clays or shales. Sulphates and chlorides can cause particularly severe problems in arid areas such as the plains bordering the Gulf area of the Middle East.	Careful selection of construction materials. Allow for volume changes in the design, or dig out suspect materials if these are thin.	Groundwater composition and level of groundwater table, locations and distribution of suspect minerals.
Flooding	Flooding is most common in coastal lowlands, lower courses of rivers (deltas and floodplains), upstream areas where several tributaries meet, and enclosed basins floored by impermeable rocks. Destructive flooding may also result from the failure of dams or reservoirs. Widespread flooding may follow melting of snow or ice both in the spring and as a result of volcanic heating. Urbanization may increase flood potential since engineered surfaces commonly prevent water from soaking away into the ground.	Avoid placing vulnerable development in zones of high flood risk. Construct defensive works such as barriers, dykes, and flood-water reception tanks.	Surface and groundwater hydrogeology, topography, distribution of permeable and impermeable strata, recorded flood levels and frequencies, climatic data.

Geological structures	Most rock types contain one or more sets of structural surfaces (e.g., bedding planes between successive sediment layers; cleavages which are parallel partings produced by crustal compression; joints produced by compressional or tensional forces). In addition, displacements on faults often bring materials with contrasting physical properties into juxtaposition. Physical discontinuities in rock influence the ease of excavation of the material as well as the potential for rock-fall from cliffs, cuttings and quarry faces, roofs and walls of underground cavities.	Avoid placing development to straddle major fault lines, take design precautions across minor faults. Avoid placing development beneath unstable rock faces. Stabilize faces by means of retaining walls, rock bolting, netting, etc.	Positions and characteristics of faults, bedding and cleavage dips; and characteristics, trends and frequencies of principal joint sets.
Seismicity	Sudden release of stress in the earth's crust gives rise to tremors which may damage constructions directly, or cause fissuring of the ground, subsidence, landslides, liquefaction of sensitive sediments or tsunamis (destructive water waves). Natural seismicity arises from movements on faults, usually at depth. Tremors can also result from mining activity, injection of fluids into the ground (e.g., disposal of liquid wastes in boreholes), or large works which significantly load the crust (e.g., major lakes behind dams).	Design structures to survive the effects of significant tremors.	Intensities, frequency and locations of past seismic events, types and locations of potentially vulnerable structures and ground features (e.g., landslides, rock falls, and mined ground); extent of thick sequences of unconsolidated sediments which respond unfavourably to tremors.

bearing capacity of the subsoil. This bearing capacity depends on the composition of the subsoil but is also influenced by earth pressures which, in turn, relate to the topography of the area surrounding the foundations. Geomorphology therefore plays an important role, but bearing properties may also be influenced by underground voids, whether natural caves or mines. Water pressures are affected by fluctuations in groundwater levels in the vicinity of the foundations.

In some cases the stability of the soil mass, which has to take the foundation loads, may be affected by settling, or by shrinkage or swelling of clays. Other effects which might be relevant are landslides or seismic events, and in some instances these two are related. The resistance against slides depends on the composition of the subsoil and the weight and friction properties of each of its layers. Groundwater conditions and fluctuations play a major part in this respect (see Chapter 4).

In coastal areas or alluvial plains which contain unconsolidated sands, cyclic loading due to seismic events may liquify loosely packed, waterlogged sand layers.

Climatic factors which may influence site selection are:

- precipitation (e.g. snow-bearing capacity)
- sunshine
- temperature

Frost heaving has to be avoided by choosing a frost-free foundation level or by careful use of construction materials. Special problems may arise in permanently frozen ground ('permafrost') of polar regions where water-saturated melted layers may develop at the surface during summer melting, whilst deeper layers remain firmly frozen.

Man-induced factors, such as surface subsidence due to underground mining for minerals or extraction of fluids such as groundwater, oil or brine, can affect site location and design of development. Similarly backfilled areas, such as old pits and quarries, may have difficult ground conditions.

Harmful chemicals may be associated with landfills and waste tips as well as with a range of industrial sites. These may need to be dealt with before sites can be safely developed. In addition some compounds, notably sulphates and carbonates, may react chemically with concrete and other construction materials.

Sulphates occur naturally, notably in coastal areas, reclaimed coastal land, and arid environments where similar problems may occur. Sulphides, which occur in many types of clay but also in decaying organic materials such as peat and in some volcanic soils, may alter chemically to sulphates with changes of volume which may disrupt foundations.

A wide variety of factors which are relevant to site selection, design and construction are summarized in Table I, together with some of the measures which can be taken to overcome problems. Sometimes, for example, ground conditions have to be improved by exchanging weak soils for materials of better quality such as sand, gravel or crushed rock. In other circumstances drainage works may be required to reduce the possibility of land instability due to high pressures of water in pore spaces. Whilst embankments, roads, dams, dykes, light buildings and some other installations may be directly founded on engineering soils, most constructions require foundations designed to take the load exerted by the structure.

Soils that are used as construction materials have to meet specific requirements; thus, sand used as fill or in embankments must be compactible in order to gain sufficient strength. Materials for the sub-base of roads must have only limited capillary rise so that disruption by frost heaving does not occur in cold or temperate climates. Sand used for drainage must have satisfactory permeability.

Lightweight volcanic soils are sometimes used as fill to resist soil deformation. Clays used as revetment material for dykes need to be sufficiently impervious and erosion resistant. Gravel, crushed rock and clay used as construction fill must be easily compactible as well as having sufficient compressive strength and high permeability. Aggregates used in concrete must also be free of aggressive chemical constituents such as sulphates.

For large projects an inventory of construction materials in the vicinity of the site is needed and the availability of suitable materials may influence the choice of site (see Chapter 5 on Mineral Resources).

SITE INVESTIGATION

If problems are to be avoided and foundations are to be properly designed, careful investigation of the ground is needed before development. This involves:

- topographical, geological and geomorphological survey of sites;
- investigation of ground properties by means of boreholes, trial pits, trenches and, less commonly, shafts and adits (with tests carried out on the surface or down boreholes, etc.);
- taking samples for physical and chemical testing;
- undertaking indirect investigations using geophysical techniques.

Some methods of ground investigation are summarized in Table II. The majority are concerned with the resistance of the ground to penetration or to circular movements, and with the levels and rates of inflow of groundwater. Samples of ground materials may be taken in a disturbed or an undisturbed state. Only a very rough description of an engineering soil can be gained from disturbed samples, since these are suitable for only very simple laboratory tests (grain size analysis, plasticity, etc.). Undisturbed samples are required for more sophisticated tests to determine the parameters for foundation, settlement or stability analyses. Undisturbed samples are obtained by cutting a borehole core or by ramming steel sample tubes into the bottoms of boreholes. Cores allow a very accurate description of soils.

Geochemical tests are commonly, but not exclusively, carried out on water, rather than rock or soil, samples. These tests are designed to detect toxic and chemically reactive substances. In some cases chemical composition may also influence susceptibility to landsliding. Some of the relevant tests are summarized in Table III.

Geophysical techniques are used mainly for reconnaissance purposes in the preliminary stages of site investigation and to aid interpolation between direct observations in boreholes and trial pits. These properties include:

- propagation velocities of vibrations (seismic surveys);
- electrical resistivity;
- electrical conductivity.

Results are presented as cross-sections of the ground,

Table II Direct methods of ground investigation

Test	*Method*	*Purpose(s)*
Standard penetration test	SPTs are made by ramming a split tube sampler into the bottom of a borehole. The number of blows needed for a 1 foot (30.5 cm) penetration is an indication of the stiffness of the soil.	From the number of blows the following items can be calculated: —bearing capacity of shallow foundation and driven piles; —relative density; —undrained shear strength of cohesive soils; —angle of internal friction of cohesionless soils; —compressibility.
Cone penetration test	CPTs are made by thrusting a cone vertically into the ground at a constant velocity. The necessary force to perform this is measured as an oil pressure or electrically, and converted into a cone resistance. Also the friction between an adhesion sleeve, installed at a short distance above the cone, and the soil is measured and converted into the local friction. The quotient of the local friction and the cone resistance at the same depth, the so-called friction number, is characteristic of the soil type occurring at that depth.	—from frictional variations the numbers and sequence of the various layers, and their lithology, can be determined; —cone resistances in cohesionless soils like sand are dependant on the friction number (Φ); therefore bearing capacities of shallow and piled foundations can be calculated from the cone resistances; —compression coefficients can be estimated from the cone resistances on the basis of empirical formulae; —determination of undrained shear strength of cohesive soils.
Vane test	A cruciform four-bladed vane is carefully forced in the undisturbed bottom of the borehole and then rotated about its vertical axis. By this the cylindrical soil mass within the vane is caused to shear off. The torque needed for this is measured.	The object of the vane test is the determination of the (apparent) cohesion of cohesive soils.
Plate test	Measurement of the vertical penetration of a plate placed either on the surface or within a borehole under a given load.	Determination of the compressibility of the ground.
Piezometer	Device placed in borehole to determine the head of water pressure.	Determination of the level to which water would rise if unconfined by the ground.
Slug and packer tests	Measurement of rates of water flow into or out of a borehole by injecting water (packer test) or withdrawing water suddenly (slug test).	Determination of ground permeability.

Table III Examples of laboratory tests*

Type of test	*Used on*	*Bearing capacity*	*Settlement under load*	*Ease of excavation*	*Earth pressures, stability*	*Frost heave susceptibility*	*Chemical reactivity*	*Toxicity*
Moisture content (weight before and after drying)†	clay, rock	×	×	×	×	×		
Liquid, plastic and shrinkage limits (from observations on changes of state)‡	clays	×	×	×	×			
Particle size distribution (by sieving)†	sand, gravel	×			×			
Bulk density (weight per unit volume)	clays, weak rocks	×	×	×	×			
Maximum and minimum density	sand, gravel	×						
Specific gravity	weak rocks	×	×					
Porosity (proportion of voids within rock)	rocks				×	×		
Point load test (penetration of hammer blows)	rock	×						
Disc test (penetration of disc under load)	rocks	×		×				
Triaxial compression test (sample compressed by piston; surrounding water pressure is measured)	clays	×	×	×	×			
Uniaxial compression test (cylinder of rock compressed to failure)	rocks	×	×	×				
Shear box test (split box filled with sample; force applied for displacing the two parts is measured)	sand, gravel, weak rocks[§]	×			×			
Mineralogy (examination under microscope, X-ray diffraction, etc.)	rocks	×	×			×		
Permeability (rate of water movement through sample)	all		×	×	×	×		
Organic content (by oxidation or combustion)	clays, silts sands	×	×		×			
Acidity	water samples						×	

Table III (*Continued*)

Type of test	*Used on*	*Bearing capacity*	*Settlement under load*	*Ease of excavation*	*Earth pressures, stability*	*Frost heave susceptibility*	*Chemical reactivity*	*Toxicity*
Content of significant chemicals such as carbonates, sulphates, chlorides, heavy metals, etc. (chemical analysis e.g. spectrometry)	water samples, leaching of soil or rock samples				×		×	×

* Adapted and generalized from tables in Weltman and Head (1983) and British Standards Institution (1981).
† These tests can be carried out on disturbed samples. With the exception of tests carried out on water samples, the remainder of the tests are carried out on undisturbed samples.
‡ Liquid limit—water content when behaviour changes from liquid to semi-solid.
Plastic limit—water content when behaviour changes from plastic to semi-solid.
Shrinkage limit—water content when behaviour changes from semi-solid to solid.
§ Reflects the existence of discontinuities in weak rock.

with layers differentiated by their indicated physical properties, and contour maps. A range of techniques is summarized in Table IV.

The surveying of sites aids the design of foundations, constructions and the optimum site layout. Surveys may extend beyond the specific site to set it in a broader context. This is particularly important where site works might trigger slope instability.

Ground investigations are preceeded by studies of existing geological and geotechnical literature and documents including investigation reports on nearby sites if these can be traced (see Chapter 3). This gives forewarning of likely problems and allows the direct ground investigations to be efficiently designed. Use of existing site reports may remove the need for more costly new ground investigations.

SITE PREPARATION

The Site investigation may demonstrate a range of ground problems of varying severity. Modern engineering techniques can overcome almost all of these problems. However, large-scale ground preparation and precautionary measures may add substantially to the cost of development and, if the real extent of problems is underestimated, to considerable delays. Necessary measures may include:

- infilling or grouting of voids;
- digging out and replacing weak surface layers (e.g., peat, soft clays, some types of landfill) and heavily contaminated deposits;
- capping off more deeply contaminated sites;
- strengthening ground by carefully emplacing layers of good quality fill;
- installing drainage to aid stability or to reduce flooding;
- reducing slope angles and making retaining walls to decrease landslide potential;
- netting and bolting rock outcrops to prevent rock falls.

DESIGN OF STRUCTURES

Foundations are required to transmit and distribute the load exerted by a construction evenly onto a surface within the ground which will support it. In some cases sufficiently strong layers are close to the surface (e.g. outcropping bedrock), whilst in other settings adequate strata may be very deep in the ground (e.g., thick estuary infills, river deltas). Some examples of the bearing strengths of ground materials are given in Table V.

In addition, foundations have to be designed to accommodate the size and distribution of the load exerted by the structure and to take account of any lateral variations in ground properties. In practice, there is a wide choice of foundation types ranging from those only a few tens of centimetres below the surface to deep piles 50 or more metres in length—for heavy constructions on thick, weak deposits. Some examples of foundations are shown in Fig. 1.

SPECIAL CONSIDERATIONS RELATING TO EXCAVATIONS AND TUNNELS

Varied temporary excavations may be required for site preparation and development, and laying-in services and pipelines. For some development, however, permanent excavations such as road and railway cuttings may be needed. For all of these, the ease of excavation is a major factor influencing costs. In the UK, for example, claims for recosting of works are frequently made by contractors when ground conditions prove to be more difficult than originally expected. In addition, excavations form slopes which need to be carefully designed and treated in order to avoid problems of landsliding or rock-fall (see Chapter 7).

The construction of tunnels and other underground spaces in rock is closely analogous to some of the processes of mining reviewed already in the mineral resources chapter (Chapter 5). Measures may be needed to ensure the stability of the structure or to exclude the inflow of water through joints, fissures or fault planes.

Table IV Geophysical methods used in ground investigation (slightly adapted from British Standards Institution, 1981)

Problem		*Examples*	*Method and remarks*
Geological	Stratigraphical	Sediments over bedrock	*Land*
		(i) Sands and gravels over bedrock, water-table low in sands and gravels	Seismic refraction
		(ii) Sands and gravels overlying clay, water-table high in sands and gravels	Resistivity
		(iii) Clay over bedrock	Resistivity or seismic refraction
		Sediments over bedrock generally	*Marine* Continuous seismic reflection profiling
	Erosional	Buried channel	Seismic refraction Resistivity for feature wider than depth of cover.
		Buried surface of limestone	Resistivity
	Structural	Buried fault, dykes	Resistivity Seismic reflection or refraction Magnetic gravimetric (large faults)
Engineering parameters	Elasticity, density and porosity	Dynamic deformation modulus Check on effects of ground treatment	Seismic velocity at surface, or with single or multiple boreholes. (Cross hole transmission.) Borehole geophysics
	Rock rippability	Choice of excavation method	Seismic (velocities at surface)
	Corrosivity of soils	Pipeline surveys	Surface resistivity. Redox potential
Buried artifacts	Cables	Trenches on land	Magnetometer Electromagnetic field detectors Ground radar
	Pipes	Submarine trenches	Echo sounding, side scan sonar
		Submarine pipelines	Side scan sonar, magnetic, continuous seismic profiling (especially if thought to be partially buried) with high frequency pinger
	Shafts, adits and caverns	Shaft, sink holes, mine workings	Resistivity. Magnetometer contouring, infra-red air photography on clear areas Cross hole transmission. Detailed gravity for large systems Ground radar

This may be achieved by lining the structure. Tunnels in poorly consolidated deposits will normally need to be lined immediately after excavation. Particular problems may be encountered where the excavation passes between strongly contrasting rock types; for example, for unconsolidated sediments into bedrock or between very different types of rock brought into contact by faulting. It is important to know, therefore, the types and thicknesses of drift deposits, the depth of solid rock, and the types and locations of geological structures such as faults. Any features controlling water flow are also very important.

PRESENTATION OF INFORMATION

Information about a specific data point can be obtained from a borehole or trial pit, together with any physical or chemical tests carried out either in the field or on samples in a laboratory. Site investigations will usually consist of large numbers of investigated points often distributed in a grid pattern through the site. The results can be plotted as profiles across the site or as maps showing specific features as areas or systems of contours. The form of the results depends to some extent on the type of development being considered—thus, for example, strip maps are prepared for lines of roads, canals, dykes, cables and pipelines or these are shown on cross-sections of the ground usually with additional short profiles perpendicular to the main line, where special works are contemplated or ground conditions are particularly difficult. Usually a number of lines are examined so that a choice of route, taking account of environmental impact, ground conditions, and general cost-effectiveness, is possible.

The results of earlier site investigations can be compiled as engineering geological thematic maps and profiles before new site investigations are designed. If the earlier results are well distributed and well recorded they may aid the engineer a great deal. Sometimes information will be good enough to reduce the number of observation points in the later investigations. However it is important that earlier information is not used to the exclusion of new investigations, since sites often

Table V Examples of presumed bearing values under static loading (slightly adapted from Weltman and Head, 1983)

Group	*Types of rock and soils*	*Presumed bearing value (kN/m^2)**	*Remarks*
Rock	Strong unweathered igneous and gneissic rocks	10 000	These values are based on the assumption that the foundations are carried down to unweathered rock
	Unweathered limestones and sandstones	4 000	
	Schists and slates	3 000	
	Weak shales, mudstones and sandstones	2 000	
	Weak weathered shales and mudstones	600–1 000	
	Weathered chalk and limestone	600	
	Thinly bedded limestones, sandstones, shales	To be assessed after inspection	
	Heavily shattered rocks		
Non-cohesive soils	Dense gravel, or dense sand and gravel	>600	Width of foundation, *B*, not less than 1 m (3 ft). Groundwater level assumed to be a depth not less than *B* below the base of the foundation
	Medium dense gravel, or medium dense sand and gravel	200–600	
	Loose gravel, or loose sand and gravel	>200	
	Dense sand	>300	
	Medium dense sand	100–300	
	Loose sand	<100	
Cohesive soils	Very stiff boulder clays and hard clays	300–600	Susceptible to long-term consolidation settlement
	Stiff clays	150–300	
	Firm clays	75–150	
	Soft clays and silts	<75	
	Very soft clays and silts	Not applicable	
Peat and organic soils		Must be individually assessed	
Made ground or fill			

* These values are for preliminary design purposes only, and may need amendment. No addition has been made for the depth of embedment of the foundation. Reference should be made to CP2004[63] when using this table.

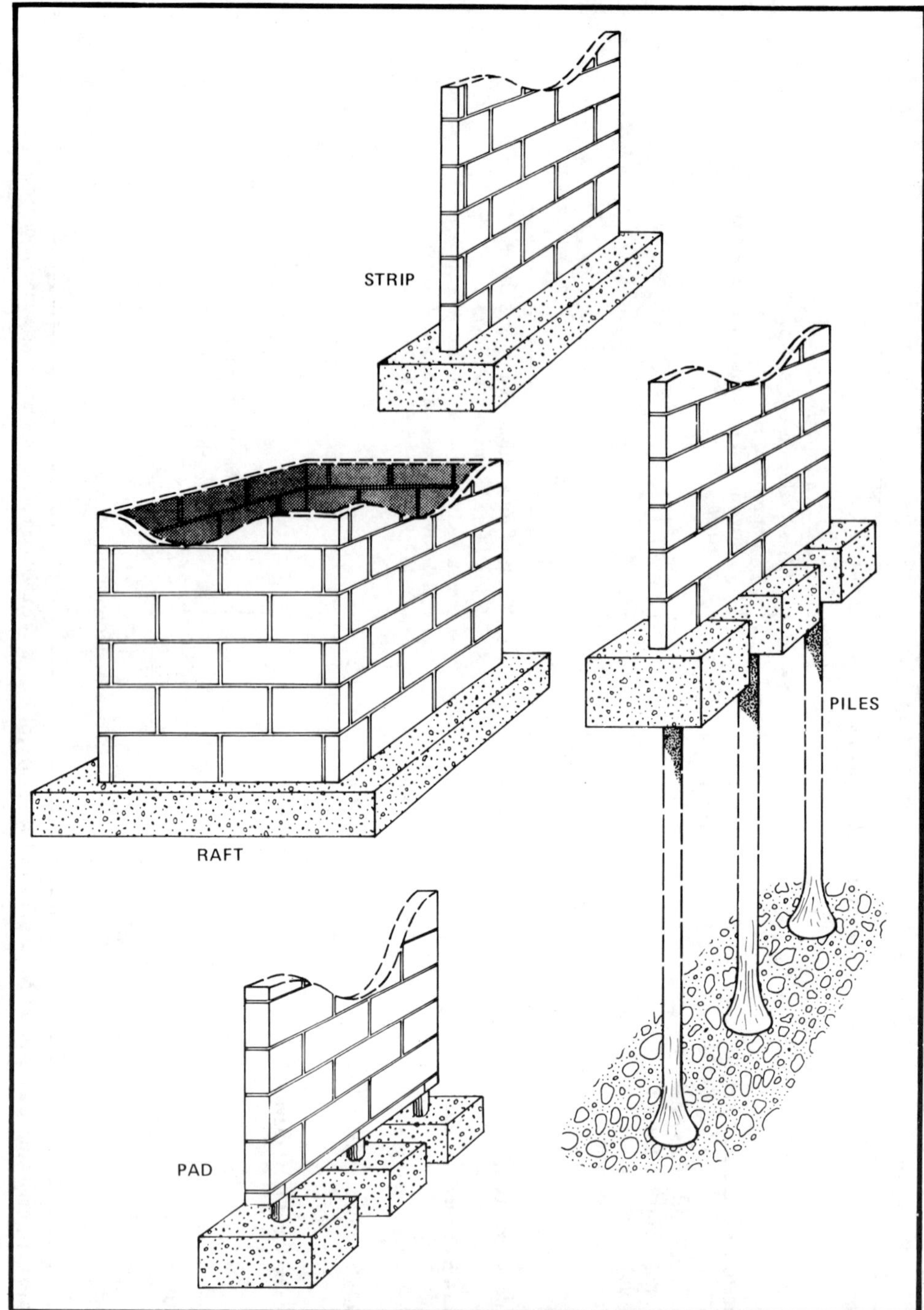

Fig. 1. Some types of foundations.

show great variations even over short distances. Even so, good collection and collation of earlier information may lead to substantial financial savings to the construction project.

Several examples of data presentation are given below. They have been selected to cover a range of geological and physiographical settings, and to range from site specific, local, regional and provincial to national scale. However, the main emphasis is placed on the coastal lowlands and alluvial plains of the Netherlands which is the authors' main field of expertise.

CASE HISTORIES

Storm surge barrier in the mouth of the Eastern Scheldt

The storm surge barrier of the Eastern Scheldt is a part of the so-called Deltaplan in which dykes and sluices are built to protect the southwest part of the Netherlands against storm floods. The mouth of the Eastern Scheldt is divided in three tidal channels: the Hammen, Schaar and Roompot.

As a result of the enormous velocities with which the water will flow through channels and between the piers of the barrier, when it is open, the foundations of these piers will have to meet very heavy requirements. Therefore the possible foundation layers in the subsoil of the channels had to be studied in a very thorough programme of site investigation. The investigations included tens of borings and many hundreds of cone penetration tests. These were executed along the axis of the location of the barrier and in broad strips at both sides of the location. To reach also the deeper layers, a large number of the cone penetration tests were made starting from the bottom of shell and auger borings or by using a diving-bell. By these methods depths of 40–50 m could be reached.

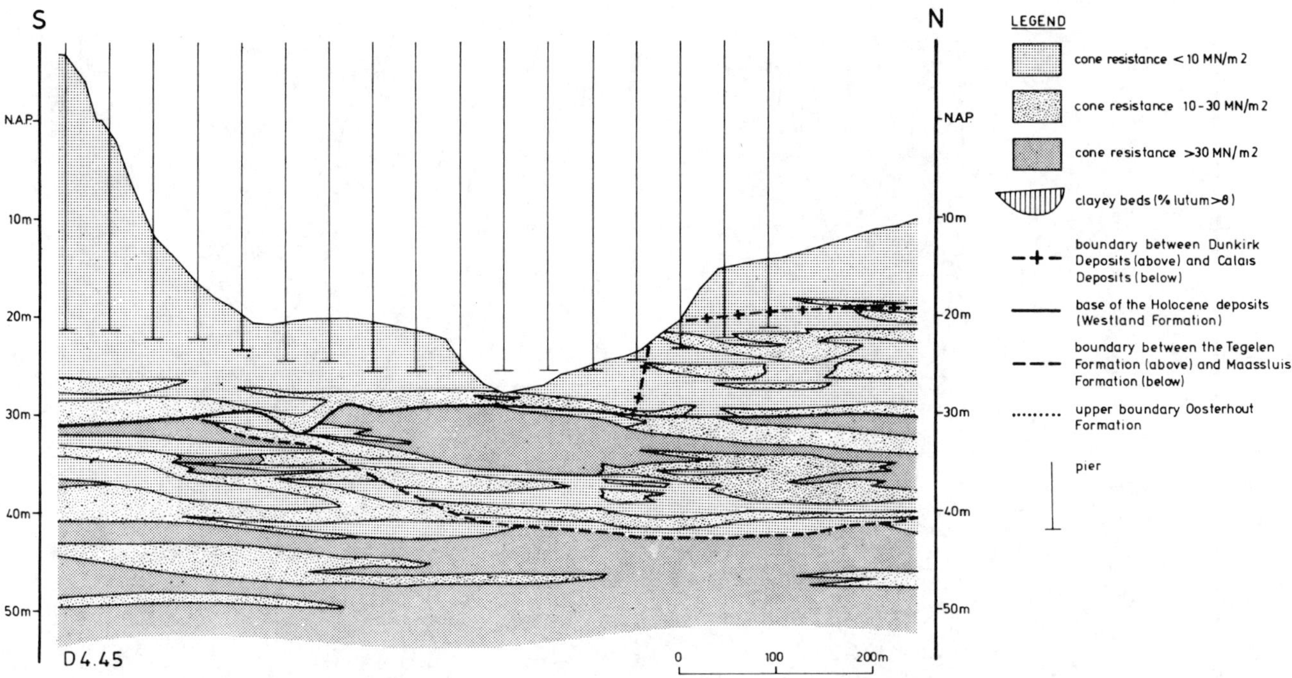

Fig. 2. Section showing the distribution pattern of cone resistance zones in the subsurface of one of the three tidal inlets in the Eastern Scheldt mouth.

On the basis of the results of these field tests several sections across the location of the barrier were constructed. As an example, the section over one of the tidal channels is given in Fig. 2. In this section the distribution pattern of cone resistances zones beneath the floor of the Schaar channel is shown.

With the results of the cone penetration tests, the borings were correlated in terms of lithology and lithostratigraphy. For this purpose the cone resistances were schematized into three categories: namely, 0–10 MN/m^2, 10–30 MN/m^2 and larger than 30 MN/m^2. Roughly speaking, the categories represent clay beds or loosely packed sand, sands of medium density and dense sand, respectively. By correlation of these categories of cone resistances it was possible to establish the cone resistance zones shown in the section. The spatial configuration of the zones is entirely determined by the geological conditions.

Based on such sections and other relevant information, foundation depths for each of the piers of the barrier were chosen. Whenever the piers had to be founded on Holocene sand layers vibrocompaction was applied to obtain the required bearing capacity. Intercalated soft, Holocene clay beds under planned piers were removed and replaced by sand-fill.

Geo-statistical risk analyses of subsea canyon instability at Gabon

In an area of about 28 km^2 west of Cap Lopez (Gabon) the risk of local instability had to be mapped in connection with the future location of an offshore structure for a petroleum company. For this purpose a special three-dimensional probability model based on loss of equilibrium of soil masses was developed by Delft Geotechnics. With this model a large number of locations were analysed.

The risk of instability of the sea bottom could be calculated on the basis of:

- the local slope of the sea bottom determined from a bathymetric survey; and
- an estimation of the expected mean value and standard deviation of the parameters, that were determined from the geotechnical investigation of the area by means of boreholes and testing of undisturbed samples, and in which the geological model was also considered.

The expected mean value and variability of the available shear strength in each location was estimated. From the driving moment of the soil masses involved, and the moment of resisting forces resulting from the shear strength, a safety factor against local loss of stability (the so-called 'stability factor') can be calculated.

For individual locations the parameters for the shear strength can be expressed in terms of an expected mean value and standard deviation, indicating a degree of certainty. This means that the stability factor also has to be considered as a statistical value (with its own expected mean value and standard deviation). With statistical methods a probability of failure can be determined for each location.

By drawing lines of equal probability of failure on a map, as shown in Fig. 3, a risk map can be constructed, in which safe and unsafe areas are indicated. If there is a large risk of local instability, two aspects can be of importance:

(1) an intrinsic unsafety due to a steep slope of the sea bottom, and a low shear strength; and
(2) a large uncertainty about the actual values of the parameters of the shear strength; for instance, due to large distances to the boreholes where the parameters were measured.

As a result of this study it was concluded that the selected site for the jacket construction (S2 on the map) was sufficiently safe with respect to the risk of submarine canyon instability.

Settlement map of the city of Leiden

An example of data presentation on a local scale deals with the preparation of a thematic applied geological

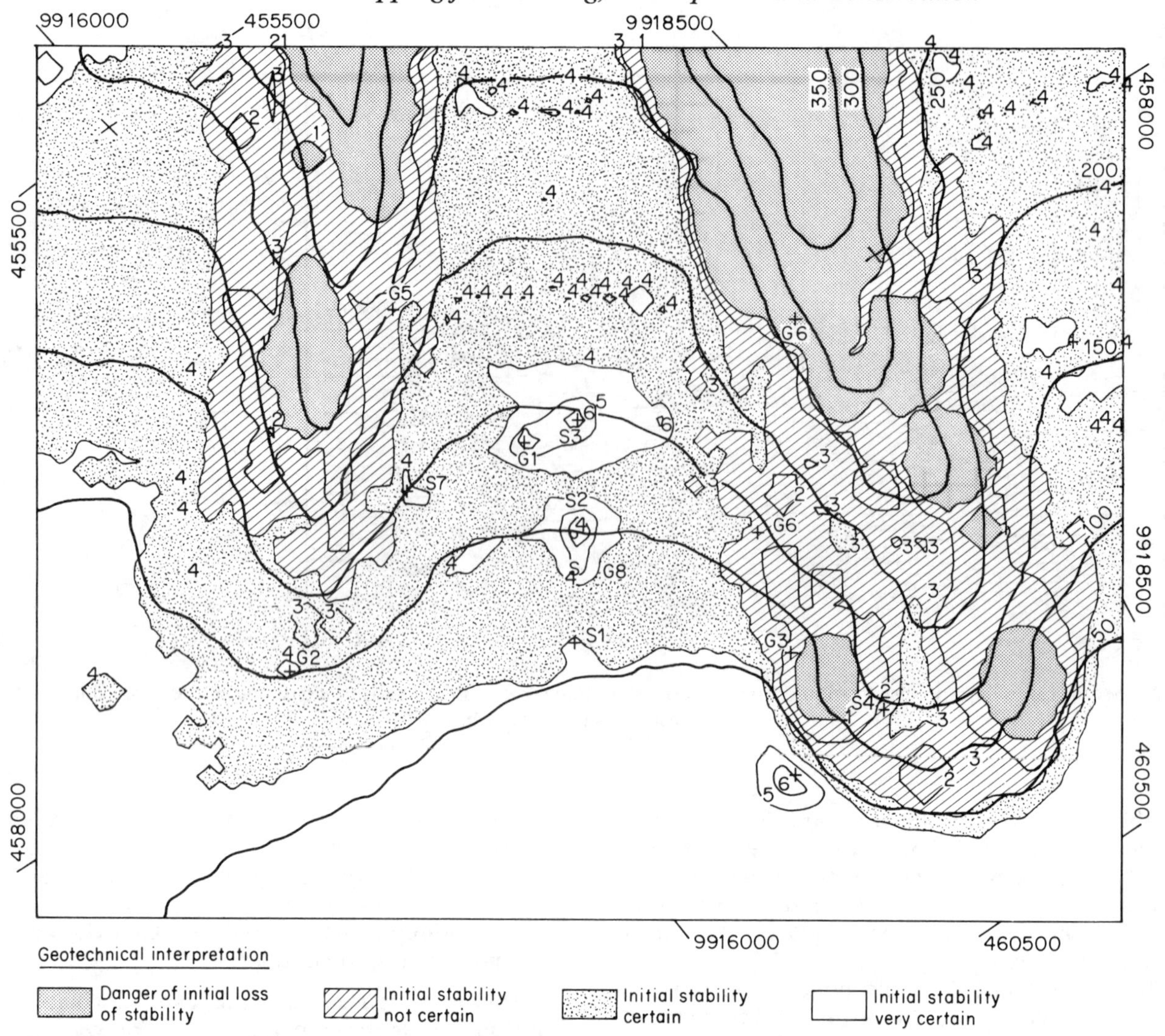

Fig. 3. Instability risk map showing probability of failure for a planned offshore jacket structure West of Cap Lopez (Gabon).

map of the centre of the city of Leiden (The Netherlands).

A map indicating the potential settlement of the surface in this part of the city was constructed for the municipal authority. Maintenance costs of infrastructures in urban areas are strongly related to the amount of settlement that may take place in the compressible beds under the surface. A special method for the preparation of settlement maps, based solely on existing subsoil data, was jointly developed by the Geological Survey of The Netherlands and Delft Geotechnics, as part of a set of engineering geological thematic maps. The procedure for preparing thematic applied geological maps, such as a settlement map, may consist of the following steps (see Fig. 4):

(1) Inventory of all available geological, geophysical, geotechnical, hydrogeological, and geochemical data, presented on the location maps.
(2) Preparation of several cross-sections, illustrating the local geology and giving insight into the soil layers of geotechnical or hydrogeological importance.
(3) Uniform interpretation of all data and preparation of a set of single-value maps indicating the depth and thickness of individual soil layers.
(4) Preparation of comprehensive multi-purpose maps indicating the sequence of soil layers as well as the thickness and depth of important individual layers.
(5) Characterization of soil layers on the basis of geotechnical, hydrogeological and geochemical properties.
(6) Preparation of special-purpose maps, such as settlement maps, foundation-depths maps, industrial-aggregate maps, hydrological maps, and environmental geology maps.

Some basic elements of the set of engineering geological maps of the city of Leiden are shown in Fig. 5 as an example. Leiden is situated along a former course of the River Rhine, behind a Holocene coastal barrier system. For the preparation of the engineering geological maps of 9 km^2 of the city centre, use was made of about 2730 borehole logs and cone penetration tests.

The cross-section shows both the marine and the fluviatile sediments. The cone resistance is illustrated by some characteristic graphs of cone penetration tests. The settlement-bar shown above the section makes a direct correlation possible between the vertical sequence of soil layers and the susceptibility to settlement.

The multi-purpose comprehensive map in the same figure shows the distribution of the older Holocene soil layers; the depth of the Pleistocene sand is indicated by contour lines, the thickness of the Holocene Calais clayey deposits can be seen in the accompanying

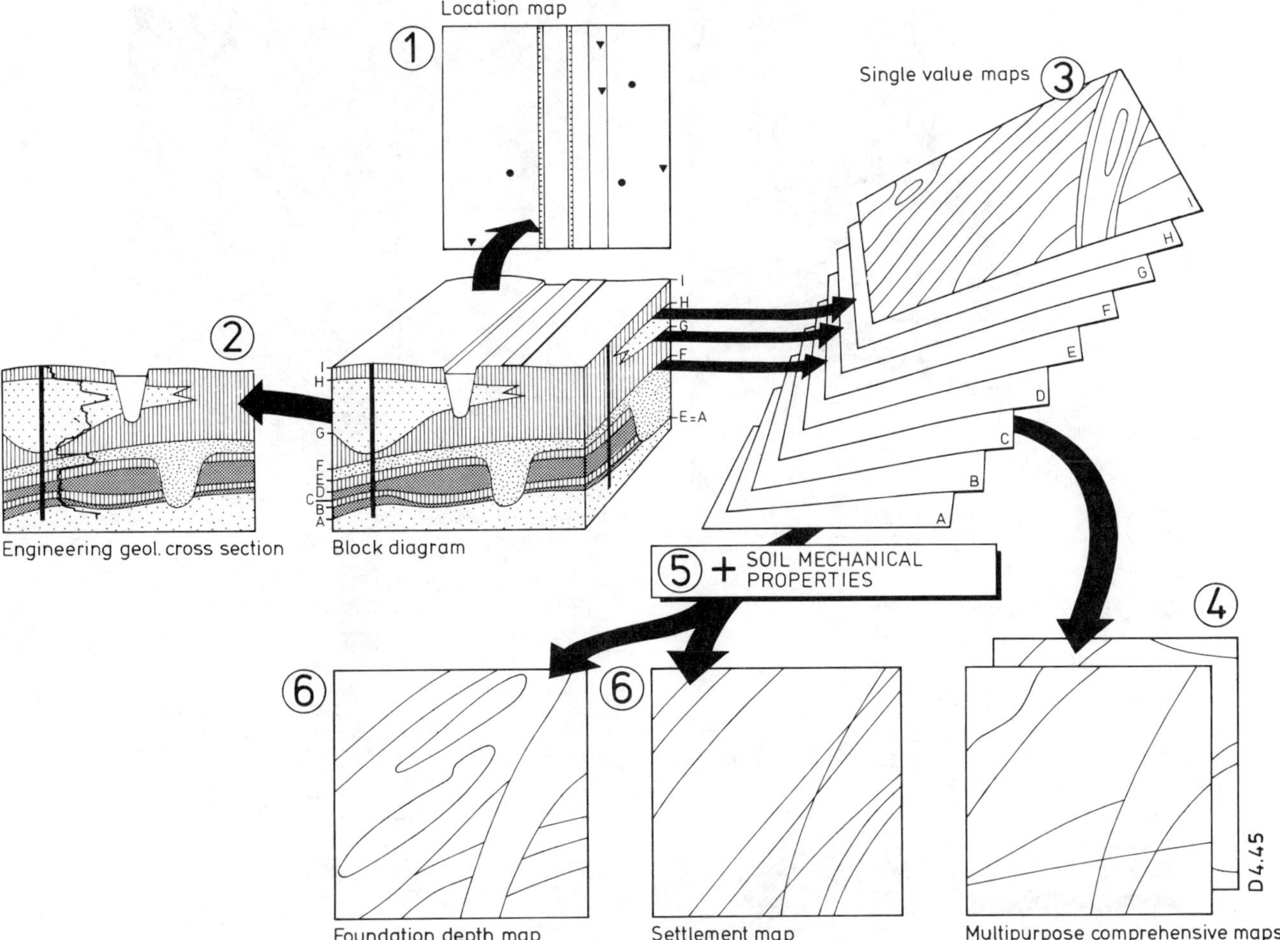

Fig. 4. Six-step procedure for the preparation of thematic applied geological maps or engineering geological maps.

isopach map. The original set of engineering geological maps of Leiden includes another multi-purpose comprehensive map, dealing with the distribution and depth of the younger Holocene deposits.

The settlement map shows the calculated total settlement of the Holocene soil layers under an evenly distributed imaginary surface load. The differences in calculated settlement can be easily related to the subsoil conditions; the highest settlement has been calculated for the northern part of the city where the peat and clay layers are the thickest. These maps are made on a large scale (1:5000 or 1:10 000).

Cost benefit analysis was used to justify the preparation of similar maps for Amsterdam.

Foundation depth map of the city of The Hague

Another example on a local scale deals with the prediction of foundation conditions in three possible sites for the construction of about 9000 houses to benefit citizens in The Hague, the residence of Dutch government. This study has been made on request of the National Planning Service, part of the Ministry of 'VROM' (Public Housing, Planning and the Environment). In broad terms, the procedure followed for this project was similar to that described in the previous example.

In The Hague, foundation conditions may differ widely because of geological variations. The geological setting of the area is determined by the distribution and thickness of highly compressible peat and clay beds, lying between elongated sand bars corresponding with fossil beach barriers (Fig. 6). No pile foundation is needed for houses built on top of the sandy beach barriers, but piles up to 23 m long would be needed for the same buildings in the southern parts of the area beyond the beach barriers (Fig. 7).

The difference in average pile length between sites A and C is about 5 m, which corresponds with a difference in foundation costs of about 3000 Dutch guilders per house in favour of site A. Use of site A would reduce the total foundation costs of the project by 27 million guilders (US$ 13 million in 1987). These figures are even more interesting because the costs of foundation and soil improvement for housing for low-income families (i.e. the majority) are financed by the State, which at present is attempting to make stringent budget reductions.

As the quantity of existing subsoil data in the potential building sites was much lower than in the already urbanized city centre in the previous example, the scale of these maps had to be considerably smaller (1:50 000).

Map indicating measures to be taken for the preparation of building sites

Although foundation costs may be very high and amount to about 5% of the total costs for house construction in large areas in the western part of The Netherlands, another geoscientific factor may contribute considerably to the total soil costs as well. This factor relates to both the bearing capacity of the top soil transport of building materials within the building site, and to the drainage capacity of the area to keep the air-space under the houses

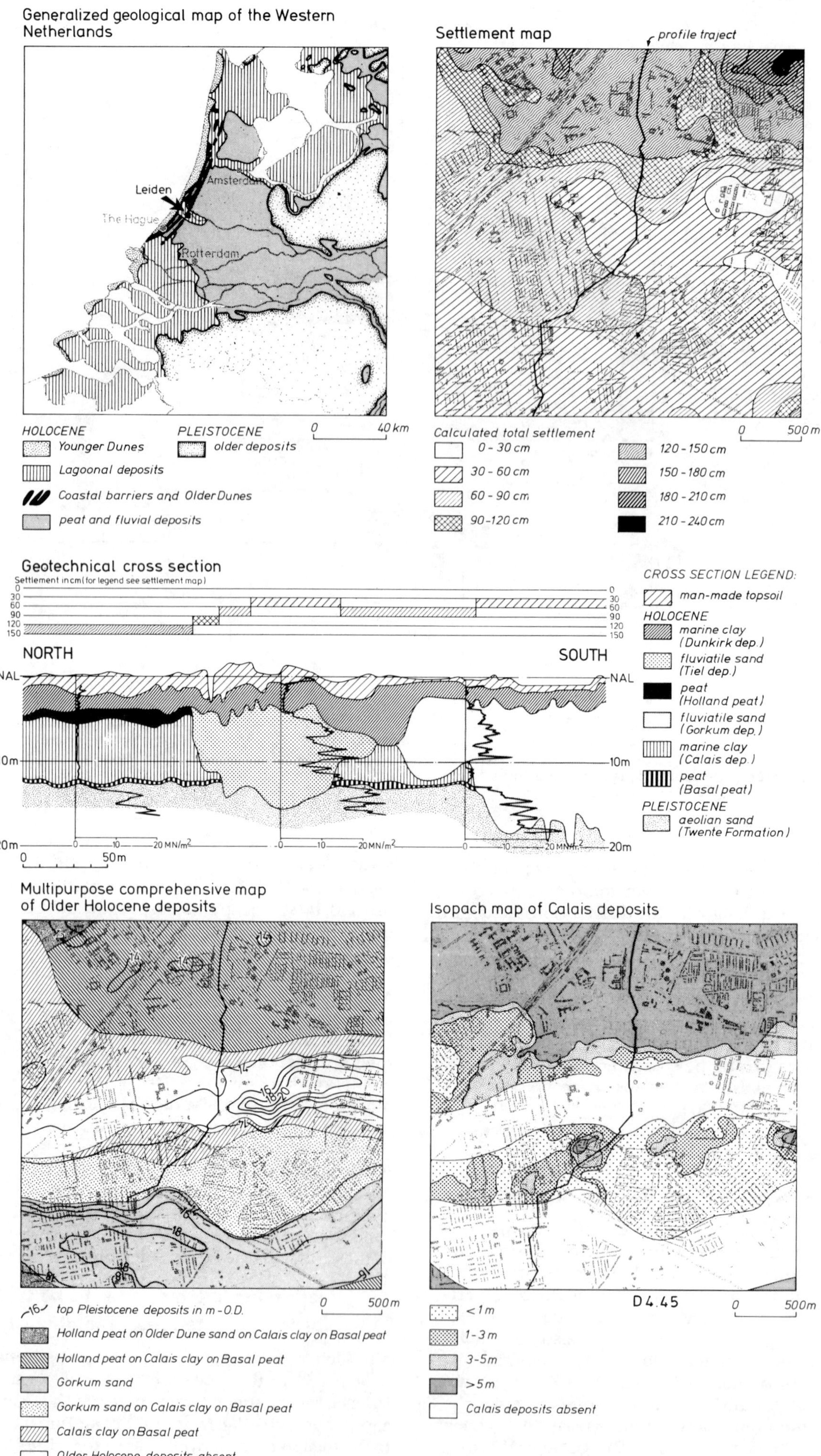

Fig. 5. Results of settlement map production concerning the city centre of Leiden (The Netherlands).

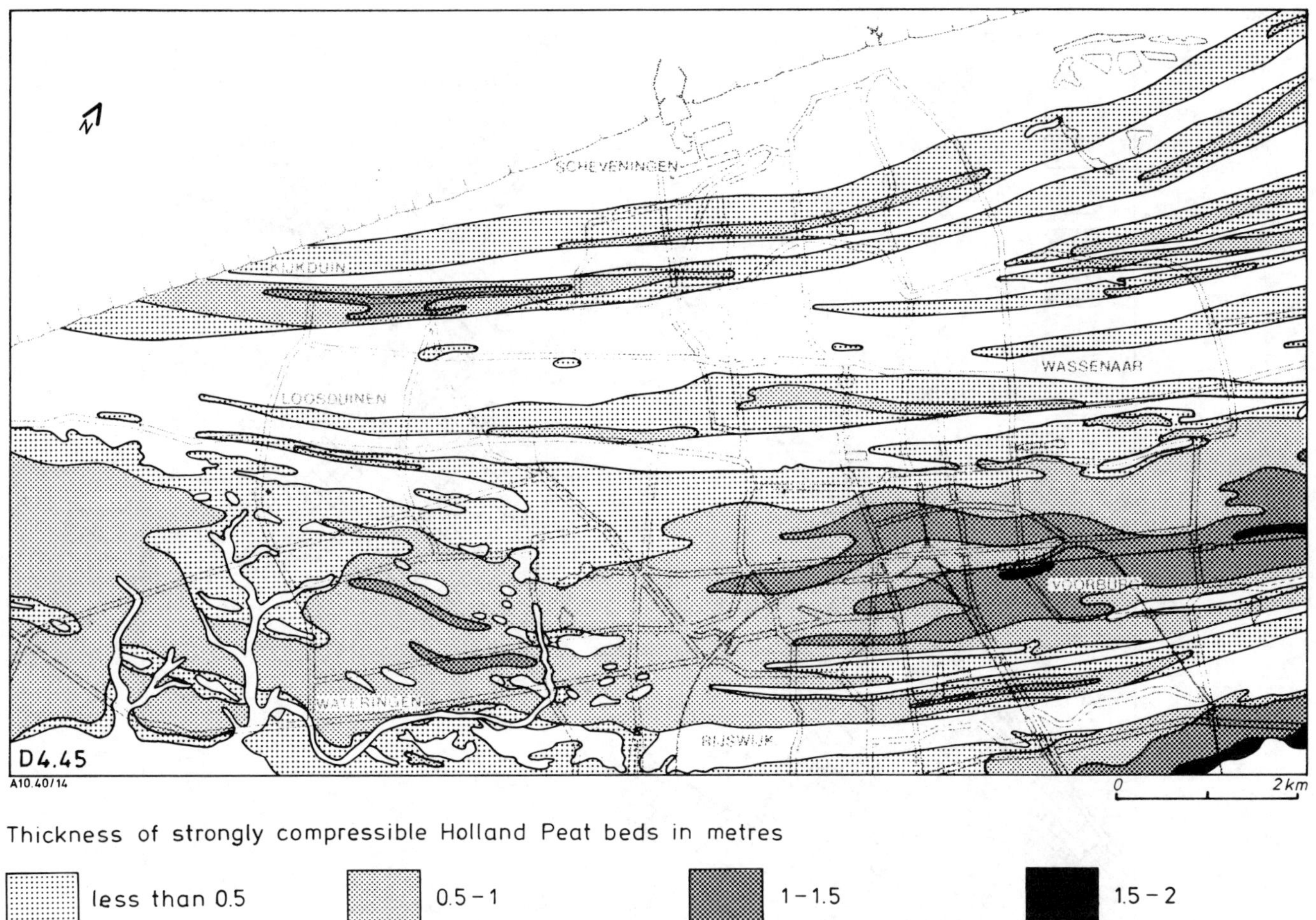

Fig. 6. Distribution of strongly compressible clay and peat layers in depressions in between coastal barriers in the city of The Hague.

dry, and to prevent frost-damage of the streets. The bearing capacity of the ground is usually improved by the application of a sand-fill. Adequate drainage can also be obtained by this method or by lowering the level of the surface water by pumping, installation of drainage pipes or a combination of these methods.

The purpose-directed, thematic map shown in Fig. 8 is based on existing data, compiled in the basic geological map, soil map, the hydrogeological map and the settlement map (for method of preparation, see Leiden example). The method of making the map on the preparation building sites is more extensively described by Schokking (in press). This map enables the planners and decision-makers to make reliable cost calculations during the feasibility phase and to learn where the construction of expensive drainage systems can be reduced by the application of the cheaper sand-fills. In contrast to general practice, the thickness of the sand-fills can be varied from place to place using the information on the map. This leads to a reduction in the quantities and costs of the sand used for filling on the one hand, and a reduction in the number of failures because of insufficient sand-fill on the other. The next step might be to apply a cost/benefit analysis to the application and thickness of sand-fills against a reduction in infrastructure (e.g. sewerage systems) maintenance costs. This is strongly correlated with the settlement potentials in the uppermost beds.

Map indicating soil costs for housing in the Province of Utrecht

For the prediction of costs for foundation and preparation of building sites, a special soil cost classification system has been developed in The Netherlands by Grontmy Soil Engineering Consultants in Zeist. A soil costs map of the entire Province of Utrecht (in the centre of The Netherlands) has been made at the request of the Provincial Planning authorities.

The soil costs map is based on two thematic maps: the first described foundation depth map and the second map indicated the measures to be taken for preparing building sites. Both maps are made according to the procedure described in the example of the city of Leiden and are based solely on data already existing. Maps indicating the depth of the top of firm sand, the thickness of both compressible clay and peat layers covering the firm sands, and the groundwater level were prepared. The bearing capacity of the sands resulting in the pile lengths was verified by comparison with earlier cone penetration tests. The compressibility of the clays and peats was calculated by means of pre-existing results of compressibility tests.

The 1:50 000 scale soil costs map (Fig. 9) shows three areas with different soil costs (including higher foundation costs and costs for preparation of the building sites) expressed in Dutch guilders per house. Soil costs

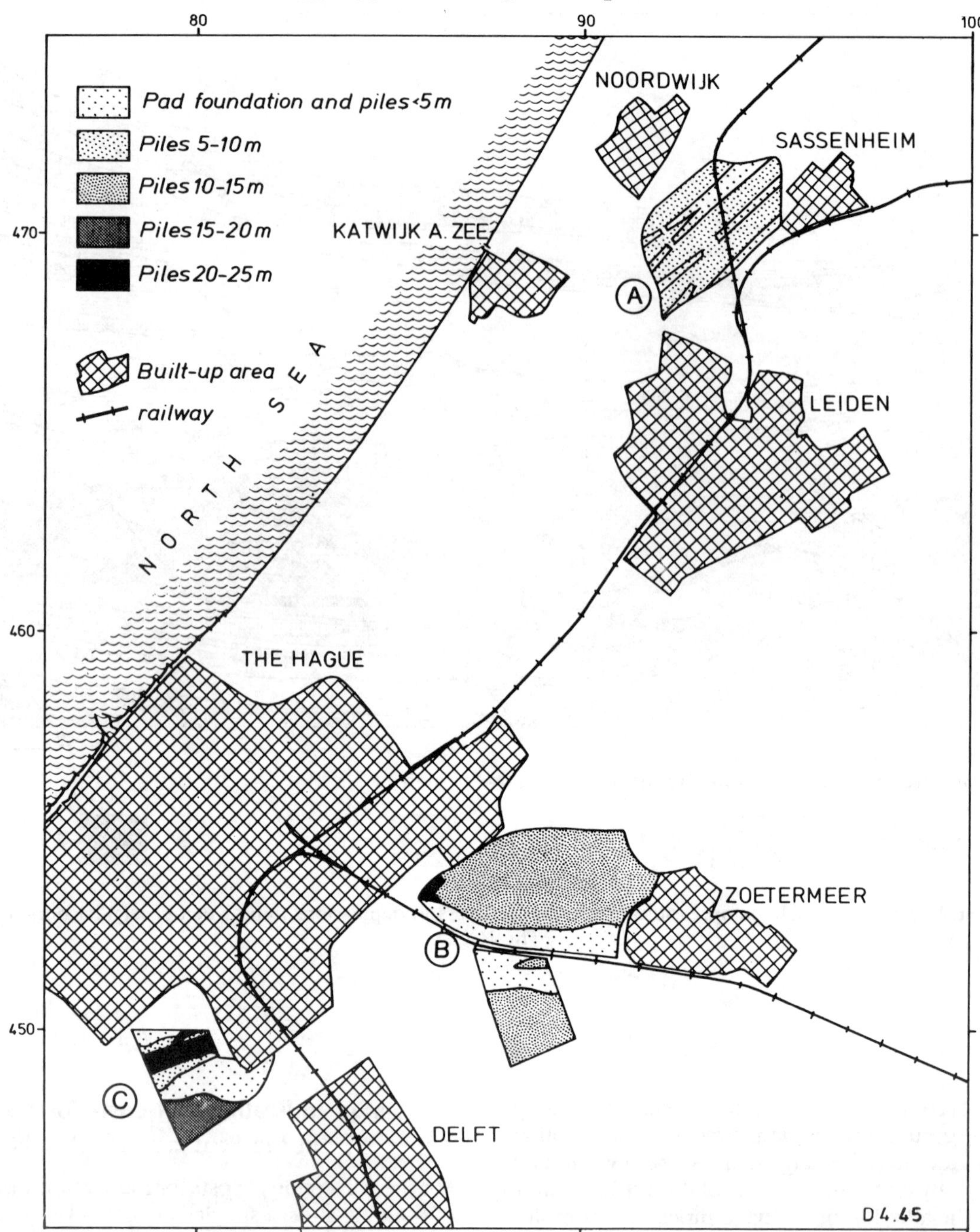

Fig. 7. Foundation-depth map showing pile lengths for low houses in potential building sites around the city of The Hague (The Netherlands). According to the calculations, the average foundation costs per house would be US$ 1500 higher for the southwestern site (C) than for the northern location (A).

maps are among the most appropriate tools for planners and decision-makers dealing with site selection and cost calculation in the feasibility phase of urban expansion and urban renewal planning.

Relative ease of excavation, Salina Quadrangle, Utah

A map showing the relative ease of excavation in part of Utah was prepared as part of a multi-theme study (Williams, 1972). The succession was classified into categories which could generally be excavated by:

(a) hand tools and light machinery (very easy and easy categories);
(b) blasting and heavy machinery for soft rocks and hand tools and light machinery for soft rocks (intermediate to difficult categories);
(c) blasting and heavy machinery (difficult to very difficult categories).

Superficial deposits, principally those of the river valleys, fall within the very easy category. Elsewhere the classification is based on bedrock conditions, even where thin superficial deposits overlie the bedrock. This study was presented at 1:250 000 for a broad area but the principal can be applied to smaller areas if the strata are sufficiently variable.

Geotechnical properties of bedrock, Bath, England

One of a suite of maps of the Bath area in southwest England summarized geotechnical properties of the

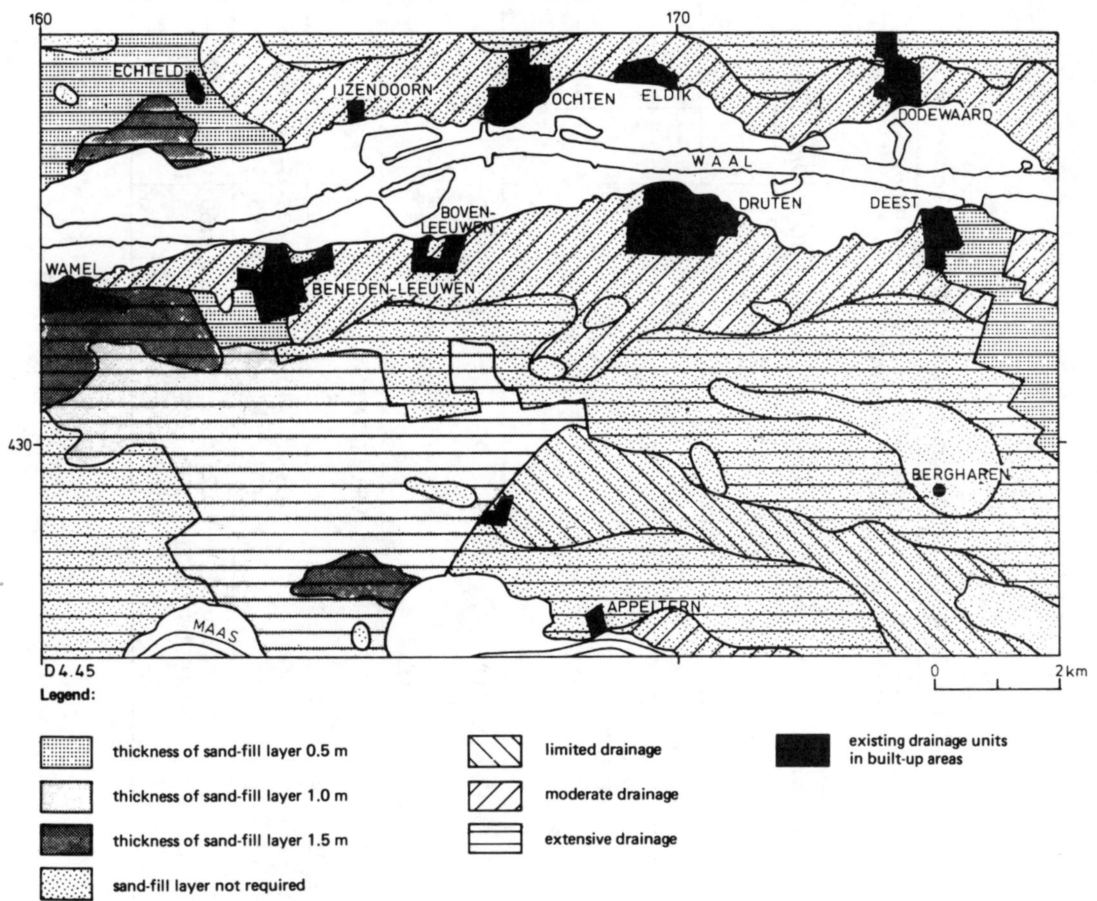

Fig. 8. Map showing the measures to be taken for the preparation of building sites.

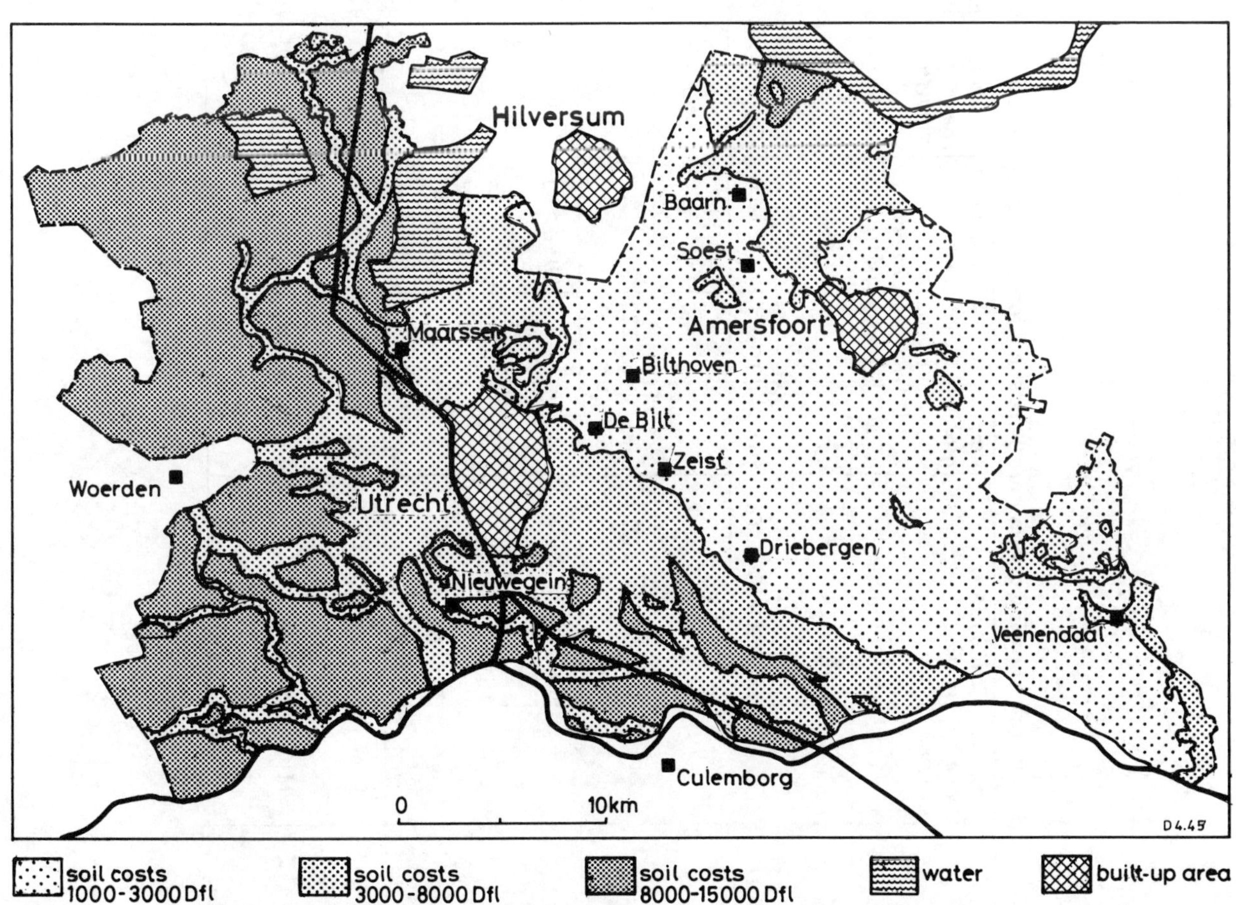

Fig. 9. Soil costs map of the Province of Utrecht (The Netherlands) showing three areas with different soil-costs (foundation costs plus costs for preparation of the building site) expressed in Dutch guilders (1 Dfl = US$0.5 in 1987) per house. Note that the indicated soil costs are very rough approximations of the real costs.

Table VI Section of table showing geotechnical classification of bedrock for parts of West Wiltshire and South East Avon, UK (from Forster *et al.*, 1985). For key see page 196

Geotechnical unit	*Geotechnical sub-unit*	*Stratigraphy name*	*Rock quality*		*Undrained total shear strength*	*Residual shear strength*	*Elastic modulus*	*S.P.T.*	*Liquid limit*	*Vertical compressibility (soils)*	*Chemical analysis*			
		(not in order)	*R.Q.D. [C.R.%]*	*F.I. (mm)*	*Su[Øu]*	*C′r & O′r*	*Ei (MPa)*	*N (blows/30 cm)*	*(B.S.5930)*	*Mv(m²/MN)*	*PH*	*SO_3%*	*ORG%*	*CARB%*
L	L	GREAT OOLITE			7.6–19.0 comp $_{9}$		12 000–24 000 $_{2}$	30–300 mm RPT			7.7w $_{2}$			
		INFERIOR OOLITE	50w $_{1}$	9.9w				58–65 w			7.8 $_{1}$	0.040 $_{1}$		
		CORNBRASH			0.7–15.9 comp $_{7}$			20–50 RPT 40–60 RPT				trace*		
		FOREST MARBLE LIMESTONE	0–85 $_{14}$	60–220				8–12 w						
S		MIDFORD SANDS	[43] $_{1}$		Øu = [38°] $_{1}$	C′ = 0 Ø′ = 32 $_{1}$		18–32 $_{6}$	LOW $_{2}$		7.6 $_{1}$	0.010 $_{1}$		
	Ss/Sw	HINTON SANDS							LOW*					
	Sw	DOWNEND FORMATION			100–200* comp					VERY LOW*				
C,M	C/M	OXFORD CLAY			20–150 $_{4}$	C′ = 0 Ø′ = 13–17*	40–140*	34 $_{1}$	HIGH $_{11}$	0.08–0.5 (LOW to HIGH) $_{8}$	7.0–8.0 $_{9}$	0.075–0.195 $_{6}$	0–10*	0–50*
	M/C	MERCIA MUDSTONE			72–478 $_{19}$	Ø′ = 18–30*	100–1 200* 10–100*w		LOW to INTERMEDIATE $_{8}$	0.025–0.30 (VERY LOW to MEDIUM)w $_{3}$	6.8–8.2 $_{2}$	0.160 $_{1}$	0.01–1.0*	(MOD–HIGH) 5 to 20*

Geo-technical unit	*Geo-technical sub-unit*	*Stratigraphy name (not in order)*	*Effect of water on engineering behaviour*		*Weathering*	*Excavation*	*Foundations*	*Slope conditions*
L	L	GREAT OOLITE	SOLUTION	Generally free draining (joints/gulls)	Very susceptible to cavities: dissolution in fissures and gulls	Hard Oolitic LIMESTONE softens on exposure. Thin surface 'brash' (rubble in soft matrix)	Very good. Bridging of gulls/fissure may be required. Thin surface 'brash' (rubble in soft matrix)	Possibility of cambering near outcrop edge. Block slides/rock fall
		INFERIOR OOLITE		Small artesian pressure possible	May be strongly weathered to rubbly LIMESTONE in soft clayey matrix	Very variable due to slope conditions	Very variable	Stronger possibility of cambering and infilled gull: Block/slab slides
		CORNBRASH		k1 = Very low k2 = High*	Weathering results in flaggy zone with soft joint infill	High overbreaks Flaggy	Good Good fill Poor aggregate	Forms scree on steep slopes
		FOREST MARBLE LIMESTONE		Very variable due to clayey pockets	Very variable Weathers to 'brash'	May be very hard Very variable	Very variable due to clayey pockets	No evidence of movement
S		MIDFORD SANDS		Particle size varies locally from coarse SILT to fine SAND k = Very low to low (10^{-5} to 10^{-3})*	Weather from grey to yellow Spring line at base	Possibly running sand in excavations below water-table. Boiling/piping under artesian pressures. Well sorted (poorly graded)	Fair to good (very good if well cemented or with major SANDSTONE bands)	Little movement when undisturbed and above water-table but liable to camber, flow slides and spring sapping when below water table
	Ss/Sw	HINTON SANDS		ditto	ditto	ditto	ditto	No evidence of movement
	Sw	DOWNEND FORMATION		k1 = Very low* k2 = High*	Resistant to weathering except along joints*	May be very hard*	Very good*	No evidence of any movement
C,M	C/M	OXFORD CLAY		Low to moderate swelling/shrinkage potential k1 = Low k2 = Low*	Considerable reduction in shear strength due to weathering particularly for mudstones and shales. Increase in strength anisotropy due to weathering	Possibly stress-relief fissuring resulting in long-term instability Overconsolidated	Good (Approximate depth of weathering 5m) Anisotropic compressibility	No evidence of significant movement in the study area
	M/C	MERCIA MUDSTONE		Possible loss of strength due to dissolution of carbonates in marly bands. Clay forms agglomerations which may break down	Considerable reduction in shear strength and deformation modulus with weathering. Increase in plasticity; loss of brittleness effects of over consolidation	Heavily overconsolidated Softens and heaves on stress relief	Good. Elastic settlement when fresh May be sensitive to disturbance Good fill if stabilized with cement	No evidence of movement in the study area

Table VI (cont.) Key

L Limestone

S Sand/Sandstone { S_S Sand; S_H Sandstone }

C Clay

M Mudstone Siltstone Shales

Rock quality definitions

R.Q.D. (%) = $100 \times \frac{\text{lengths of rock recovered in sound lengths} > 100\text{ m}}{\text{length of core run}}$

C.R. (%) = $100 \times \frac{\text{lengths of core recovered}}{\text{length of core run}}$

F.I. (mm) = $\frac{\text{unit length}}{\text{number of fractures}}$

0–25	very poor
25–50	poor
50–75	fair
75–90	good
90–100	excellent

Compressive strength of rock

Compressive strength (MN/m²)	Description
<1.25	very weak
1.25–5	weak
5–12.5	moderately weak
12.5–50	moderately strong
50–100	strong
100–200	very strong
>200	extremely strong

kPa = kN/m²
Mpa = 1 000 kN/m²

SO_3 total sulphates (CP2004) in soil

Class 1	<0.2%
2	0.2–0.5%
3	0.5–1%
4	1.0–2.0%
5	>2.0%

Compressibility (soils)

Description	Mv(m²/MN)
very low	<0.05
low	0.05–0.1
medium	0.1–0.3
high	0.3–1.5

	Relative density	N(blows/ foot)
Sand and Silt	very loose	<4
	loose	4–10
	medium	10–30
	dense	30–50
	very dense	>50

	Consistency	
Clay	very soft	<2
	soft	2–4
	medium	4–8
	stiff	8–15
	very stiff	15–30
	hard	>30

F.I.	Fracture spacing Index (mm)
R.Q.D.	Rock Quality Designation (%)
C.R.	Core Recovery (%)
S.P.T.	Standard Penetration Test
N	Number of blows for 30 cm penetration
R.P.T.	Penetration in mm for 50 blows
Su	Undrained shear strength (kPa)
Øu	Undrained angle of internal friction (degrees)
C′r	Effective residual cohesion (kPa)
Ø′r	Effective residual angle of internal friction (degrees)
Ei	Youngs Modulus of Elasticity (initial) (MPa)
Mv	Coefficient of volume compressibility (m_2/MN)
k	Permeability (m/s)
k1	primary permeability
k2	secondary permeability
SO_3	Total sulphate content (%) (B.S. 1377, 1975 Test 9)
ORG.	Total organic content (%) (B.S. 1377, 1975 Test 8)
CARB.	Total carbonate content (%) (J. Sedimentary Petrology, 1974)
pH	pH of soil suspension (B.S. 1377, 1975 Test 1)
M	Mudstone component
L	Limestone component
‡	Su for Dyrham Silt member
w	Weathered
*	Extrapolation from data outside the study area
comp	Compressive strength (MPa) (see table)
4	Number of observation or test results
††	Cotham Bed (shear plane Clay)

bedrock (Forster *et al.*, 1985). The local succession consists of varied lower and middle Jurassic rocks and poorly lithified clays, sands and mudstones and relatively hard limestones in an interbedded sequence. The horizons were classified into five groups based on properties and these were then described in terms of existing geotechnical data. Comments were then made on the effects of water on engineering behaviour, weathering characteristics, ease of excavation, foundation conditions and slope conditions (Table VI). This represents a compromise between giving detailed technical data and a more general interpretation which can be used by a wider audience.

CONCLUSIONS

A wide variety of general problems may be experienced in planning and carrying out construction, although almost all of these can be overcome, at a price, by modern engineering methods. The variability of ground conditions, even on small sites, means that careful site investigations are needed before every development.

It makes common sense, however, to compile the results of site investigations systematically to aid the planning of future investigations and to alert planners and developers to potential problems experienced in the vicinity. Appropriate maps will, therefore, speed up desk studies and will reduce the likelihood of unexpected problems which might cause delays to development or substantially increase costs.

Whilst the maps give general guidance, geologists and engineers will need to refer to the original site records. Preparation of carefully indexed data bases help them to gain quick access to the information.

It is strongly recommended that special, purpose-designed data bases, which should be continuously up-dated, should be prepared. This gives the possibility of producing computer-made maps directly from the data base (see Chapter 3). These have only temporary value, but this is even more the case with paper maps.

It is often very difficult to quantify the benefits of a geoscientific study, but it is worthwhile calculating the reduction in additional site investigation which can be achieved when the results of the proposed study are available. It has also proved to be possible to calculate the savings in foundation costs and the amounts of sand for filling when the optimum site is selected using such maps. Even more difficult is the calculation of savings in maintenance costs of infrastructural works or the costs of remedial works, which may be saved by the prediction of all kinds of natural or man-induced hazards (for example, land subsidence). Such cost/benefit analyses should be conservatively designed and may result in establishing the earn-back time of the costs of a study project. A procedure of making such cost benefit analyses has been described recently (de Mulder, 1988).

When a report contains a proposal for further investigations, a cost/benefit analysis together with a time schedule should be presented.

Because so many natural and man-induced factors with respect to site selection and design of constructions are involved, a multi-disciplinary approach for the investigation of the ground is highly recommended. In order to demonstrate the present possibilities in applied geological data presentation to planners and decision-makers, the Geological Survey of The Netherlands, together with the Soil Survey Institute and the Institute of Applied Geosciences, issued a publication entitled 'Subsoil Uncovered' (Geological Survey of the Netherlands, 1986).

REFERENCES

British Standards Institution (1981). *Code of practice for site investigations, BS 5930.* London.

De Mulder, E.F.J. (1979). Engineering-geological investigations in the mouth of the Eastern Scheldt (SW Netherlands). *Geol. Mijnbouw*, **58**, 471–476.

De Mulder, E.F.J. (1988). Thematic applied Quaternary maps: a profitable investment or expensive wallpaper? In *Applied Quaternary* (De Mulder, E.F.J. and Hageman, B.P., eds.), Balkema, Rotterdam.

De Mulder, E.F.J. and Hillen, R. Preparation and application of thematic engineering and environmental geological maps in the Netherlands. *Geol. Mijnbouw, in press.*

Forster, A. *et al.* (1985). Environmental geology maps of Bath and the surrounding area. In *Planning and Engineering Geology* (Culshaw, M. and Bell, F.G., eds.), Geol. Soc. London, Eng. Geol. Spec. Publ. No. 4.

Geological Survey of the Netherlands, Institute for Applied Geosciences DGV-TNO and Soil Survey Institute (1986). *Subsoil Uncovered.* Geol. Surv. the Netherlands, Haarlem.

Lüttig, G. (1978). Geoscientific maps for land-use planning: a certain approach how to communicate by new types of maps. *Intern. Yearb. Cartogr.*, **18**, Bonn, Bad Godesberg, 95–101.

Lüttig, G. (1984). Geologie und Rohstoffsicherung. In: *Kartographie der Gegenwart in der Bundesrepublik Deutschland.*

Schokking, F. Engineering Geological Thematic Maps of urbanizable areas in the Netherlands. Proc. Symp. Dev. Quat. Geol. East and Southeast Asia during the last Decade, Bangkok, *in press.*

Weltman, A.J. and Head, J.M. (1983). Site investigation manual: PSA civil engineering technical guide No 35. *CIRIA Special Publication*, **25**, London.

Williams, P.L. (1972). Map showing relative ease of excavation in the Salina Quadrangle, Utah. *Map I-591-J.* United States Geological Survey. Washington D.C.

Part IV

Current Status and Future Prospects

Chapter 9

Environmental Geology Mapping

Brian Marker and *Joseph McCall*

INTRODUCTION

The previous chapters have reviewed many types of earth science information which can be illustrated graphically and which are relevant to planning of land use and to development. It has been demonstrated in Chapter 1 that applied earth science mapping is a vigorously growing and world-wide field of activity. Subsequent chapters show the wide variety of approaches which have been taken to many themes.

The earth's physical resources, such as water, minerals and soils, provide for many of the material needs of society, and construction gives the fabric of modern society. Set against such facts are the wide range of hazards and problems which limit human activities or make them more costly. Each of these subjects has been considered separately, but the reader will have found many cross-references to information which has a bearing on a number of these different aspects. That is not altogether surprising, since many of the earth's dynamic processes interact and the composition of the crust of the earth and distribution of materials within it reflect the cumulative effects of geological history.

The characteristics of superficial geological deposits, to take one example, may define mineral resources, such as sand and gravel or peat. The properties and thicknesses of these same materials influence ground conditions for building and construction. The spatial relationships of the constituent units, coupled with their various porosities and chemical characteristics, control the occurrence and chemistry of groundwater. Soil profiles developed on the upper surfaces, and proximity of water to the surface will influence natural ecosystems and agricultural potential. Slopes developed on the surface may link to land instability and to erosion of soils.

An earth scientist may, therefore, get many general clues to the characteristics of an area by examining the geological map of such deposits and comparing it with a topographical base. Even so, the amount of interpretation and its relevance can be greatly enhanced if individual theme maps isolating and expanding upon relevant factors are prepared. Using the present example, these might include maps showing the minerals, groundwater, surface water, slope angles, thicknesses of geological units, soils and so on. Such maps, however, give only part of the story. Soil erosion and land instability, for example, are influenced by factors such as vegetation cover and rainfall, and information on these factors may be required. These effects may be enhanced by removal of vegetation cover for agriculture or construction and by installation of drainage works. Farming practices may make areas more vulnerable to natural processes, for instance by leaving fields ploughed at times of heavy rainfall or by removal of hedges. Geological information alone is often not adequate if a full picture is to be developed, and maps of existing land use and climate might also be needed.

The precise combination of requirements will depend on the purpose of the investigation but will commonly be multi-disciplinary. However, many of the themes will result in information which can only be fully interpreted by specialists in the various disciplines: other specialists may misinterpret some information and non-specialists may find it difficult to gain a useful impression from the results. The problem of emphasizing the relevant information for various audiences is at the heart of environmental geology mapping.

Earth Science Mapping for Planning, Development and Conservation

DEFINITIONS

The term 'environmental geology mapping' has been used in many senses. It has been used both for studies covering only a few themes and for research incorporating sets of a dozen or more maps addressing a broad spectrum of earth science (and sometimes other) interests. The term 'engineering geology mapping' has been treated sometimes as a synonym of environmental geology mapping, although engineering issues are only one section of the total span of interests which have been covered in such studies. 'Geoscience mapping' has likewise been treated as a synonym.

The preceding chapters are concentrated on studies with topics aimed at individual themes. This chapter uses environmental geology mapping in its widest possible sense and considers therefore:

(a) studies incorporating diverse (and often numerous) themes in individual maps;
(b) studies which include wide ranging summary maps designed specifically for the use of non-specialists.

REASONS FOR UNDERTAKING ENVIRONMENTAL GEOLOGY MAPPING

Environmental geology exercises have been undertaken for a wide variety of reasons. In general terms, studies so far have been carried out:

(a) as geologist-led exercises intended to demonstrate to planners and developers the usefulness of the technique;
(b) because of perceived major problems, often hazards, in populated areas which give rise to public and official concern, sometimes after disasters;
(c) in areas designated for widespread development or re-development, such as new towns and cities, or, less commonly, for environmental conservation;
(d) as strategic exercises, for example, systematically prepared to give well-organized information relevant to planning and development decisions for major towns and cities.

The first of these motives is promotional, the next two are reactive to specific problems, and cover most circumstances, whilst the final approach is forward looking, even though all are intended in some way to aid land use planning and development.

It is generally the case that earth scientists consider the physical characteristics of an area, infer what resources and problems are likely to be present, and then select which are likely to be significant for planning and development. Planners and developers, however, will normally consider the social, economic and environmental requirements of an area and will then ask what information is available or needed to aid the planning and development objectives. Obviously much of this information will concern matters like population, supply and demand, and these aspects will often outweigh the earth science information in decision-making. This is often rightly the case since, for example, ground conditions will usually affect the costs of development rather than preclude it. However, the significance of earth science information may be systematically underplayed to the detriment of planning and development aspirations.

The reverse approaches to thinking about areas taken by earth scientists, on the one hand, and by planners and developers, on the other, may lead to misunderstandings. Results produced by the scientists may be poorly designed for answering the questions which planners and developers regard as critical. On the other hand, studies specified by non-earth scientists may not highlight the aspects which are most important to achieving the objectives. This can be readily avoided in several ways. There is an immediate advantage if earth scientists involved in such work have training in planning and development or experience in dealing with these fields. Conversely, it is helpful if planners and developers have sufficient knowledge of earth science matters to make full use of results. These are matters of long-term education and experience. In the shorter term there are clear advantages if teams including both earth scientists and planners work on specific studies so that problems can be solved as investigations progress.

In general, if planning and development objectives are to be met in the most economical ways these must be the starting point for defining the necessary supporting studies, whether these are earth science investigations or research on wider issues.

PLANNING AND DEVELOPMENT OBJECTIVES

Planning objectives will vary, but in general will be aimed at ensuring adequate supplies of materials and development to meet the needs of society at the best possible balance of social, economic and environmental costs. The nature of the balance will vary from country to country and is largely a matter for political decision. In some circumstances the balance will be shifted more firmly towards development and supply of materials, whilst in others a greater emphasis may be given to environmental protection and conservation. Earlier chapters have already demonstrated how earth science information may be important for both of these aspects.

Planning strategies and policies will normally be stated in terms of factors such as those listed in Table I. Although the table is illustrative (no attempt has been made to make it comprehensive because of the large number of inter-linkages and amount of repetition involved), it is immediately obvious that even if only one topic of planning interest is addressed, several earth science aspects may be relevant. Secondly, the form in which the information is needed for direct use often involves a considerable amount of interpretation. To take one example, the distribution of mineral resources may be of interest but the production of a mineral resource map may require:

(a) mapping of the extent of possible minerals;
(b) assessment of the quality of the mineral;
(c) determining the depth and thickness of the mineral;
(d) establishing the ease of excavation of the material (e.g. groundwater, excavation characteristics).

These factors may be mapped separately but are most conveniently presented as a single synthesis, as described in Chapter 5.

Similarly many technical factors enter into the preparation of a suite of maps describing landsliding, but the most convenient presentation for use in planning and development is a single hazard, risk or potential map. Such summaries, if carefully presented, can be easily appreciated by non-earth scientists and are of direct application in the planning process (Chapter 7).

Table I Illustrative list of some relationships between planning topics and earth science information

Planning interests	*Earth science information*
Water supply and protection of water resources from derogation	Nature and extent of groundwater and surface water resources; quantities and flow patterns; vulnerability to pollution
Minerals supply and protection of mineral resources from sterilization by virtue of other forms of development	Types, extent, quality & uses of mineral resources; ease of extraction (e.g., depth, groundwater, excavation, problems)
Agriculture and forestry	Physical and chemical properties of soils; availability of water; soil erosion potential; chemical contamination
Fisheries	Surface water characteristics, composition and susceptibility to pollution
Housing, industry and infrastructure	Ground conditions including problems due to land instability and chemical reactivity with construction materials; rising or falling groundwater; chemical contamination; ease of excavation
Waste disposal	Suitability of ground for landfill and risks to ground and surface water
Reclamation, restoration and urban renewal	Ground conditions (physical and chemical); availability of materials suitable for infilling and landscaping; groundwater behaviour
Conservation	Potential for ground and water pollution; potential effects of new developments; land instability (e.g., damage to historic buildings; rising or falling groundwater)
Recreation and tourism	Potential effects of activities (e.g., erosion to footpaths); ground instability (e.g., open mine shafts, avalanches near ski resorts); coastal erosion and beach accretion

The processes of development and of development control (i.e. the mechanism for making decisions on whatever specific developments should go ahead or, when such developments are allowed, what conditions or limitations should be set) are pitched at two levels:

(1) the search for sites in a broad area which might be suitable for a specific type of development so that the most promising of these can be shortlisted for more detailed consideration;
(2) detailed investigation of the selected site as a precursor to design and construction.

The broader level of consideration is often led by economic, transport and infrastructure factors rather than earth science information. This is partly because of the assumption that problems can be overcome, for a price, by appropriate design. Hawkins and Tomlinson (1987), for example, describe a proposed shopping facility on mined ground where a precise location dictated by the best marketing position completely overrode the cost penalties incurred by making the ground safe. This is not always the case: mineral extraction and waste disposal proposals are much more tightly tied to geological factors, although good access to the market is still important.

At this broader or preliminary level of assessment, the types of maps and detail included in environmental geology maps may be of assistance. However, the coverage of maps in most countries is still so restricted or confined to a limited range of themes that they are comparatively rarely available for this purpose. Except in the case of very major, often centrally funded, development to be carried out on a relatively long time-scale (e.g. major highways), it is seldom possible to carry out such studies in advance. The implication is that where money is available for environmental geology mapping it needs to be targetted at areas where substantial development is expected in the future. Such areas should be identified in the land use planning process. A corollary is that studies should be defined within planning boundaries rather than within some arbitrary framework such as a National Grid.

At the site specific level, the amounts of detailed information required are much greater and can only be obtained by appropriate site investigations, including trenching, drilling and the taking and testing of samples. Site investigations normally take place in two steps:

(a) preliminary desk study based on collection, collation and interpretation of earlier information (maps, publications and earlier nearby site investigation reports);
(b) physical investigation of the site and testing of materials on a pattern partly designed on the basis of the preliminary study.

Adequate site investigations should always be undertaken before development. Generalized information can never be a substitute because of the complicated local variations in physical and chemical properties of ground materials. Not surprisingly, there is a general tendency for larger site investigation budgets to be allocated for more expensive developments, but the corollary of this is that greater risks may be taken in small-scale developments, such as a single house. Ideally the budget of the investigations should reflect the complexity of the ground conditions and not just the value of the development. In some cases that might make small-scale exercises impractical and lead to inhibition of development in a particularly difficult terrain.

As a result of the importance of site investigation to site development, it has been said that environmental geology maps are of no particular value to that level of activity (see the discussion in de Mulder, 1988). The present authors do not agree with that view for several reasons. These are that environmental geology exercises, if properly carried out, achieve the following:

(1) Accumulate and organize in one place, and so make more accessible, the pre-existing published and manuscript material on an area: if thoroughly sought out and properly organized such data bases can make preliminary studies more complete and cost-

effective, since otherwise much time is wasted in seeking relevant records.

(2) Preserve valuable data, which might otherwise be lost—many of the present archiving organizations are not necessarily permanent, and records would be lost with their closing down. The changes in local government boundaries in Britain have, in fact, caused the irretrievable loss of valuable archives. There is a strong case for a statutory lodgement of all site investigation and related data in regional archives (see Chapter 3) and environmental geology mapping would contribute to such a process.
(3) Act as indexes to the back-up data bases.
(4) Draw attention to the full range of problems which have been, or might be, experienced in a particular area, thus encouraging the site investigator to look out for warning signs.
(5) Emphasize parts of areas where it is very imprudent to carry out limited, rather than detailed, site investigations on sites associated with minor development.

Site investigations are, of course, matters for professional earth scientists with appropriate experience. These experts will require a full technical treatment of the characteristics of an area and will place much more emphasis on specialist maps and original records. The general treatment needed in broader scale planning, however, requires communication with a wider audience. The immediate consequence of this is that results need to be prepared for different purposes and in different styles in order to be suitable for all purposes.

LEVELS OF DETAIL IN PRESENTATION OF RESULTS

Results are normally presented as maps accompanied by reports describing the characteristics of the study area. If a map is to be intelligible it is important not to overburden it with detail, so the report often performs the important function of clearly stating the limitations on the meaning of the material shown. These limitations are very varied and will depend partly on the amounts of data available and the sources from which it has been obtained, and also on the amount of generalization needed for the scale at which it is shown. The level of detail of planning provisions will also be relevant. In general, the basic maps which show the more or less factual data on various mapped themes (often called 'element maps') will contain more detail than the maps which show a higher degree of inference based on combinations of several themes ('derived maps' and 'potential maps').

The steps towards generalization can be taken one stage further if summary maps are used to draw together the critical information from other types of maps onto a single sheet. Maps of this type, which are often the most useful for a rapid and general planning overview, include those separately showing:

(a) resources for development;
(b) constraints on, and opportunities for, development;
(c) areas within which site investigations prior to development should be more detailed than usual;
(d) development potential.

These and various others will be described later in this chapter. Such maps should normally be used only to alert users to opportunities and problems, and reference to both the more detailed maps and to accompanying descriptions is important.

The scale of treatment used depends very much on the purpose of the study, the extent of the terrain to be covered and the amount of detailed information available. Some studies are regional in character and may be presented at small scales whilst others are much more local and appear at fairly large scales. Examples of scales are indicated in Table II. Relatively small and well mapped countries such as the UK, Netherlands, Denmark and Belgium have an obvious advantage for gaining a good coverage.

Table II Some examples of scales of presentation of environmental geology maps

Level of treatment	*Example scales*
National	1:1 000 000–1:625 000
Regional	1:625 000 –1:100 000
Area	1:100 000 –1:10 000
Site*	1:2 500 –1:1 250

*Outside the field of environmental geology.

A particularly interesting approach has been taken in the San Francisco Bay area of the USA, where a series of environmental geology mapping and specialized studies on different scales have been undertaken in an integrated programme (Association of Bay Area Governments, 1976; 1978). Within the broad programme, for example, nine 'flatlands' of the Bay area have been studied comprehensively (Helley and Lajoie, 1979).

The choice of scale may relate to the size of area to be surveyed. In large, relatively remote countries mapping at less than regional scale may be impractical. Conversely, in a small country the 'regional scale' may be adequate to cover a large proportion of the territory.

An ideal approach is to build up a block of local maps which can then be further generalized into regional or national coverage. In practice this is seldom possible. Either generalized coverage is produced first and, then, more localized studies are carried out or, alternatively, scattered studies of targetted areas are completed and results are interpolated across the intervening ground. If funds are limited it is, perhaps, best to progressively update smaller-scale coverage.

A further factor which may influence the scale of presentation is the cost of results. It is often possible in environmental geology studies to produce very large suites of maps and sizeable reports even for comparatively small areas. The costs of reproduction of such sets, especially since use of colour may be essential for clarity, may be very high indeed. Except where costs are defrayed in some way, there is a real danger that the results will be priced beyond the means of those who might or should use them. An alternative is to sell only summary maps with a summary report, but if this is done the user will have to be cautioned more strongly about the limitations on the use of the information.

COMMISSIONING AGENCIES AND CONTRACTORS

The accessibility and terms of scale of results depend very much on the nature and role of agencies which commission studies. From the material reviewed here it seems that the majority of environmental geology studies have been commissioned by central or regional government, a lesser proportion by local government or municipalities, and very few by private industry. This is far from surprising given the general, rather than site specific, nature of the work.

Studies have mainly been undertaken by national or state geological surveys or university departments, although some of the latter act as government surveys where no central organization has been set up. A smaller proportion have been carried out by private companies and these have been principally devoted to themes involving engineering geology, geomorphology and soil science. This pattern partly reflects the way in which the state sector is used in many countries but also the spread of expertise which is available in some state surveys. The United States Geological Survey, for example, covers a much broader field of activities than the geological surveys in many other countries.

Leaving aside political differences, reasons for the dominance of the government sector in commissioning and updating work seem to relate, on the one hand, to the very wide range of beneficiaries of the work, and, therefore, difficulties in identifying key funding interests and, on the other hand, to the improbability that sale of results could ever cover the cost of the work without pricing the results beyond the means of purchasers. The major use of government agencies in carrying out the work reflects the role and status of such bodies and the traditional methods for planning government research expenditure. A number of countries, including the UK, have been experimenting in recent years with using a much greater variety of contractors on work of this type, such as engineering consultants.

Whatever system is used, however, it is clear that a wide range of expertise has to be brought to bear on environmental geology studies, sometimes including, not only land-use planners but ecologists, agriculturists and others alongside earth scientists. The organization and management of such teams may have its own problems. Equally the compilation of results into a meaningful synthesis requires a high degree of collaboration and a very broad understanding by the manager or editor; the best specialists do not necessarily make the best communicators of the results of these studies.

DATA AND UPDATING

A perpetual problem with maps is that they become rapidly out of date after printing. Changes to the topographical base occur, additional information becomes available, and advances are made in understanding and interpreting the basic information. The collected data base can, if organized for rapid storage and retrieval of information, have a crucial role in updating.

The factual data are collected from examination of existing documents, from fieldwork and sample testing. The types of data collected will depend on the objectives of the study; however, there is an economy of handling and investigation in retrieving data on several themes at once rather than only one or two. It is prudent to consider whether additional information should be captured, even if not for immediate use, to avoid duplicate searches, drilling and mapping. There may be substantial savings in the long run.

The plotting of additional themes can give rise to unexpected benefits. In northeast England, for example, the closure of a number of coal mines and the cessation of pumping has led to a rise of groundwater within the mines. This has flushed methane upwards through the mine voids. The gas may become trapped beneath caps on treated mineshafts and under glacial tills where these rest above the coal-bearing strata. It may also pass however, through fractured, jointed, porous sandstones and can emerge at the surface where the sandstones outcrop, or where porous sands and gravels rest on the sandstones. Since dwellings may exist above such zones there is sometimes a risk of the methane entering foundations. The juxtaposition, therefore, of information on mined ground, shafts, bedrock and superficial deposits gave indications of ground locations to which the gas might be channelled (I. Jackson, personal communication).

Data may be held for a wide variety of organizations, such as government departments, local authorities, geological and soil surveys, libraries, museums, consultants and other private companies. Considerable skill and tact may be required to obtain co-operation in providing data. Sensitivity to the legal risks of releasing information may be extreme, and some may be commercially valuable, or even subject to government security classifications. Such information can be generalized before use in environmental geology mapping, thus protecting the confidential details. Particularly intractable problems may arise, however, when data is already contained in a commercial data base. Nevertheless, the completeness of a study and, thus, value to the community at large, depends in large measure on goodwill and co-operation. For all of these reasons (as noted in Chapter 3) data collection is best supervised and, often, best carried out by reasonably senior experienced personnel who are well-versed in the skills of negotiation. Some of these problems may, of course, be less severe in countries where the land and resources are owned by the state and mapping exercises are also conducted by the state survey.

For reasons of speed and convenience there may be temptations to rely on a few large national collections of data, as was the case with some of the earlier UK studies. Great improvements can be made, however, by searching more widely in records of local authorities, academic institutions, scientific and historical societies, commercial companies, and even private collections. Even so, a point is reached where extra effort produces little additional information. In general, therefore, data collection exercises will usually fall short of the ideal and are cut off by a time or cost limit rather than by completion of the task.

Despite the importance of maps showing locations of original sampling and testing points, these are omitted from many environmental geology exercises. Often they only include unannotated points to show, for example, the general 'density' of borehole data. In some studies they are annotated and differentiated to indicate the types of source document (Fig. 1), but only rarely are they fully annotated to lead the user back to the information source (Fig. 2). Annotation of such maps is essential if they are supported by a manual data base. It is perhaps less necessary if the full details are mounted on a computer which his been given a capability, for example, for retrieval by grid or other map co-ordinates. Even so, there may still be some virtue in labelling sites of observations to give some idea of the type of

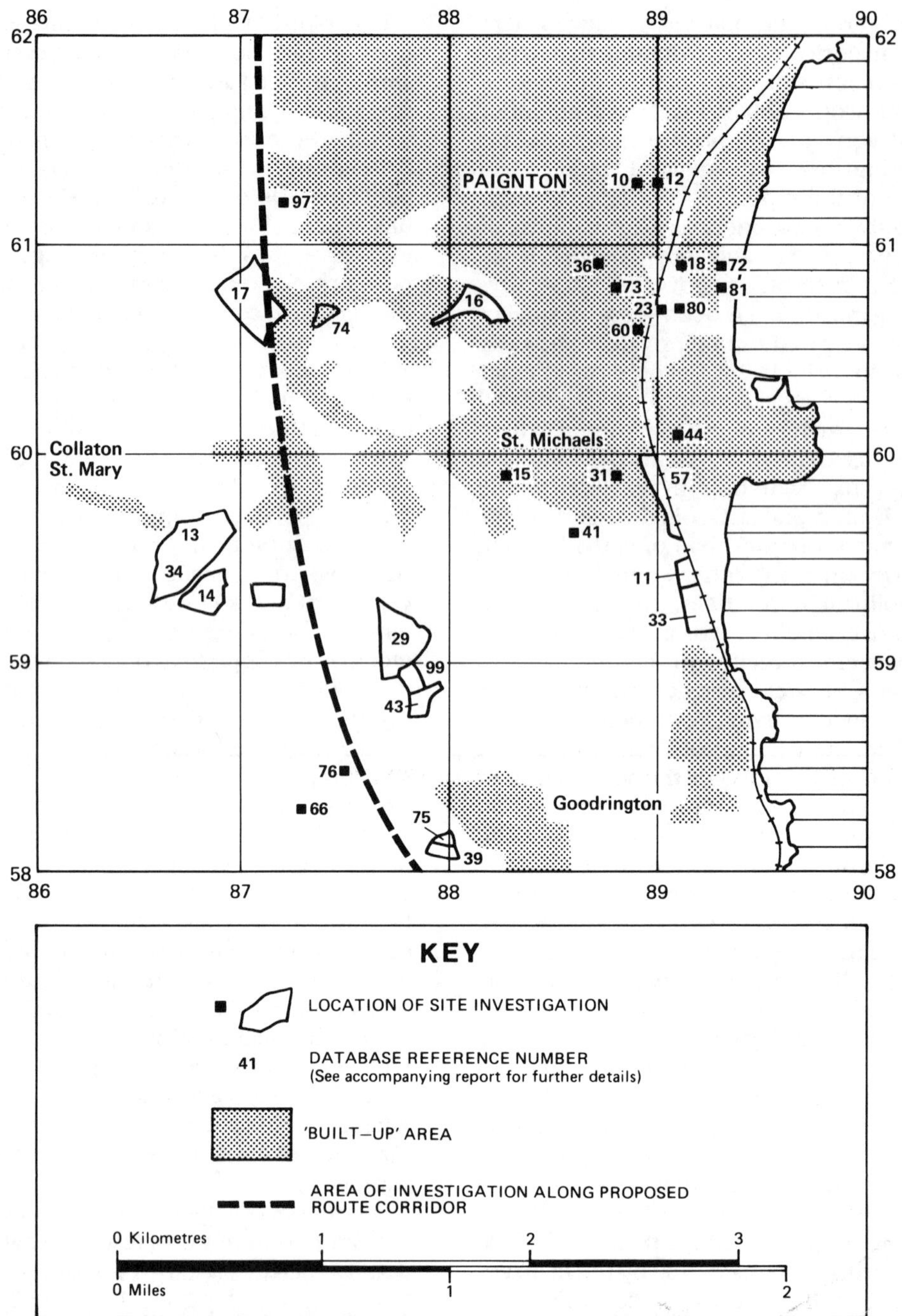

Fig. 1. Extract from a map showing the locations of site investigation and other basic data on which an environmental geology study of Torbay, Devon, England, was founded (from Geomorphological Services Ltd, 1988).

investigation and, for example, whether sampling and testing was undertaken.

Collected data can be organized into manual or computerized data bases, or merely used and discarded. It is argued here that careful construction of a data base is an investment of lasting value. Computer data bases are justified if large data sets are to be stored or manipulated—that will usually be true of environmental geology studies (Freeman, Fox Ltd, 1988).

Recent developments in digital mapping are moving rapidly towards the point where preparation and editing of trial theme maps can be carried out on a screen. Similarly, updating the date base leads automatically to updating the digital maps. The point may conceivably be reached where the redundancy inherent in conventional maps is avoided and up-to-date maps can be produced when required. Once the initial data collection has been carried out, therefore, the maintenance of the data base becomes a crucial issue.

Corollaries of this are that capacity is needed to continue capturing data or for those carrying out the site investigations, for example, to lodge copies of results as a matter of course with designated reception points. Suitable points could be, for example, government geological surveys, designated university departments, commercial data bases, or local government depending on what system of responsibilities exists. To be effective, the places to be consulted must be well-known and access to the system must be straightforward. One line of argument suggests that local government is the best repository since this is where, in many systems, planning policies are made, discussions are held on planning applications, and decisions are made on planning proposals. However, some authorities may

Legend

Opencast prospecting boreholes

- Diamond, drill rock cores taken
- Water flush—chip samples
- Hand auger in soil

Engineering site investigation boreholes

- Shell and auger
- Shell and auger with rotary core in rock
- Shell and auger with rotary in rock
- Rotary—rock roller

Research boreholes

- Power auger—disturbed samples
- W Well, backfilled or inaccessible
- Mine shaft, abandoned
- Trial pit
- Sewer trench
- Geophysics—constant separation resistivity traverse
- 12 Area of site investigation, with reference number
- 6 oc Opencast prospecting area, with reference number

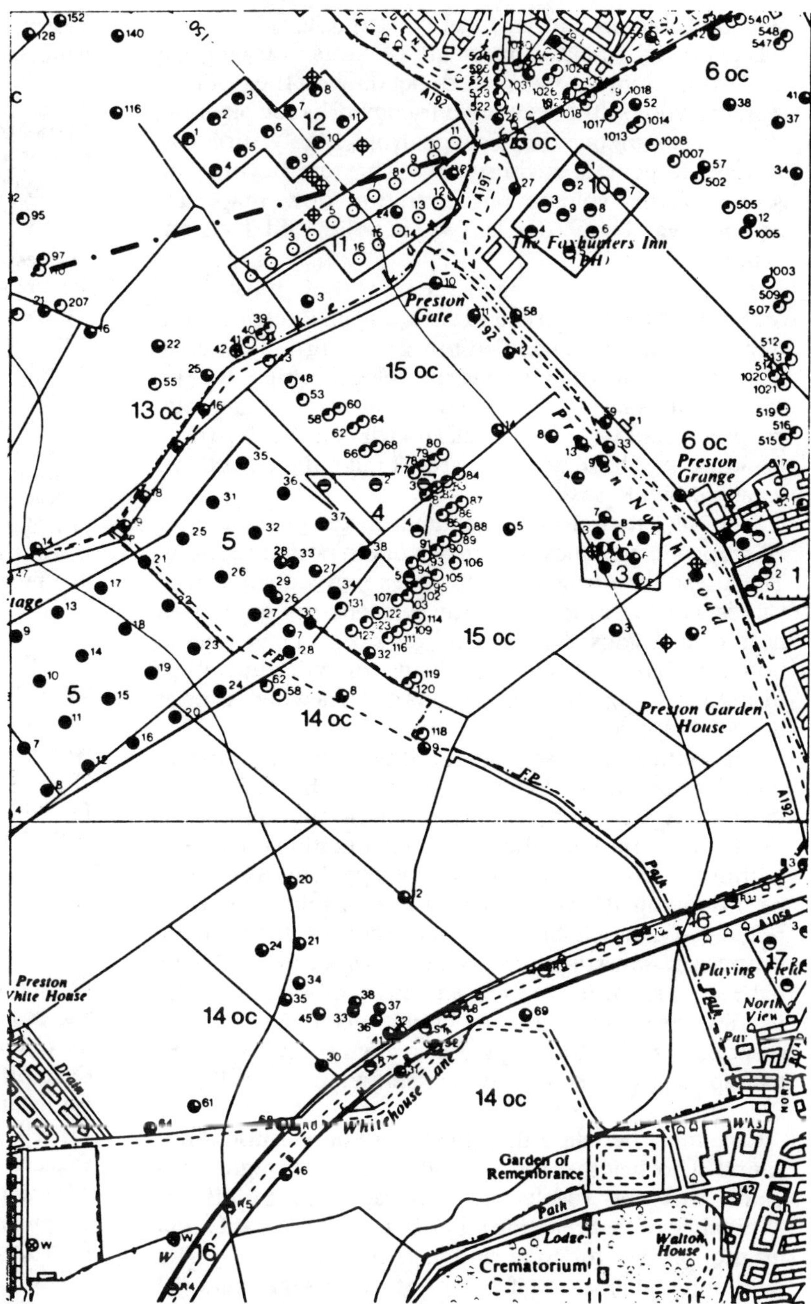

Fig. 2. An extract from a site documentation map with fully referenced data points (from Dearman *et al.*, 1973).

lack the resources and expertise needed to maintain such facilities, and there is the problem (already mentioned) of changing local government boundaries. Centralized data stores may be less accessible, but as on-line computer facilities expand this may be less of a problem (see Chapter 3). Even if data bases are dispersed, there is much to said for a central indexing system—a 'data bank of data banks'—to direct users to the right place (Clayton *et al.*, 1987).

Further development of computer data facilities to handle larger quantities of data will progressively increase the possibilities of simultaneous handling of multiple themes of information to produce derived maps more widely based.

A word of caution is needed since, at present, digital mapping may not produce logical results. It is difficult to define the complex steps of inference involved in producing some map types in terms which can easily be coped with by existing computers. The results of digital mapping should, therefore, be carefully examined for unlikely representations. An interactive cartography facility is essential if sound results are to be obtained.

In the meantime, paper maps will continue to be the norm, although increasingly underpinned by digital rather than manual data bases, since facilities suitable for studies of modest size can now be obtained very cheaply (Freeman, Fox Ltd, 1987, 1988).

LIMITATIONS AND LIABILITIES

Apart from the technical limitations in the available data coverage and in observation or sample points, there may be limitations in the presentation of data. The most obvious of these is cartographical (i.e. the constraints of scale). However, other limitations relate to matters such as legal liabilities and financial consequences.

Legal and financial aspects will of course vary a great deal from country to country depending on the political systems, and on the legislative frameworks for planning and development. These range from highly directive systems to general advisory ones. They are beyond the scope of this volume but condition how maps are presented and how they are used in policy and decision-making.

If maps are to be implemented they must have technical credibility, a sound basis, and an interpretation which is not over-extended but can be firmly justified. Many systems contain a concept of legal liability for the results presented. For instance, a map negligently prepared could lead to financial loss or damage to people or property. Those who prepared it might be placed in a legally difficult position.

Secondly, technical judgements made on land may affect property values. This may work in both directions. The presence of mineral resources may enhance land values. On the other hand, the existence of geological hazards may cause loss of value, decrease in investment, non-availability of financial loans or withdrawal of insurance cover. These are serious matters which cannot be side-stepped.

The solution of these problems is twofold. First, studies must be carried out competently. It must be made quite clear what the results contain and how they should be used. They must be the best interpretations available within the current state of knowledge. The user must be discouraged from misusing the results either by reading too much into them or, perhaps, by avoiding proper site investigations because no specific problems have been recorded in the area. This can be achieved by making clear statements of limitations and the user's responsibility on the maps and in any reports. Selected examples of such statements are shown in Tables III and IV.

Care must be taken that these statements should not sound so restrictive that any non-specialist might conclude that the results are too vague to be useful. Avoidance of that reaction is partly a matter of education in the proper function and use of the results. The possibility of misuse of these, or any other, material cannot be completely avoided but the responsibility can be firmly placed on the user. With technical material, that can be a heavy burden; therefore, it is sensible to stress the circumstances where the user, if a non-specialist, may need to seek advice from a competent person and what sort of expertise is required to obtain sound answers.

Table III Statement of limitations from an applied geological study (from Freeman, Fox Ltd, 1987).

This study has been produced by collation and interpretation of, and interpolation from, geological, and related data from a wide variety of sources.

The Reports, Maps, Appendices and Annexes provide only a general description of the nature and extent of factors relevant to the planning of land use and development.

The data on which this study is based is not comprehensive and its quality is variable, and the results reflect the limitations of that data. Localized or anomalous features and conditions may not be represented, and any boundaries shown on maps are only approximate. No information made available after December 1986 has been taken into account. For these reasons:

This study provides only general indications of ground conditions and must not be relied upon as a source of detailed information about specific areas, or as a substitute for site investigations or ground surveys. Users must satisfy themselves, by seeking appropriate professional advice and carrying out ground surveys and site investigations if necessary, that ground conditions are suitable for any particular land use or development.

Table IV Statement of the responsibilities of those using the results of an applied geological mapping study (from Forster *et al.*, 1986).

THIS MAP IS TO BE USED FOR PRELIMINARY STUDIES ONLY AND IS NOT INTENDED AS A SUBSTITUTE FOR ON-SITE INVESTIGATION.

USER'S RESPONSIBILITY

The responsibility for assuring that geological and mining data for any given site are as indicated in these maps and in the figures and text of the report must remain solely that of the user. Care has been taken in selecting, plotting and interpreting data. Emphasis has been placed on data obtained by mining surveyors, geotechnical engineers and geologists, and on observations made in the field.

Nevertheless, the possibilities of undetected anomalous conditions and uncharted shallow workings and mineshafts are always present. There is no substitute for the knowledge provided by detailed surveys and the drilling of trial boreholes that take into account the extent, nature and location of a proposed development.

IMPLEMENTATION

No matter how technically sound or how well presented the results of a study may be, it will not have achieved its objectives unless it is taken into the planning and development process. This requires an acceptance of the value and appropriate uses of the results by government—both officials and elected representatives. Whilst this can be done through *ad hoc* use of results, it is most effectively achieved if the process is linked to legislation or government advice and guidance as appropriate (see, for example, Department of the Environment, 1989).

The precise method will depend on the existing framework of legislation and the planning policy framework—matters which will vary greatly from country to country. It is beyond the scope of this book to review the great variety of measures around the world. In general terms, however, it needs to be established that:

(a) applied earth science factors need to be taken into account, alongside all other relevant information, when formulating planning policies or assessing specific development proposals;
(b) applied earth science maps can be compared with proposed planning allocations and can be used to modify these to reduce environmental impact and avoid locating development towards the most difficult or dangerous areas, or land containing resources which might be studied;
(c) applied geology studies provide background which should be used in design of site investigations and in interpretation of the results of such investigations by setting them into the regional context.

Where committees are used to advise on the specification and conduct of studies they may be uniquely placed to advise policy-makers on the implementation of findings. The researchers who carried out the studies should

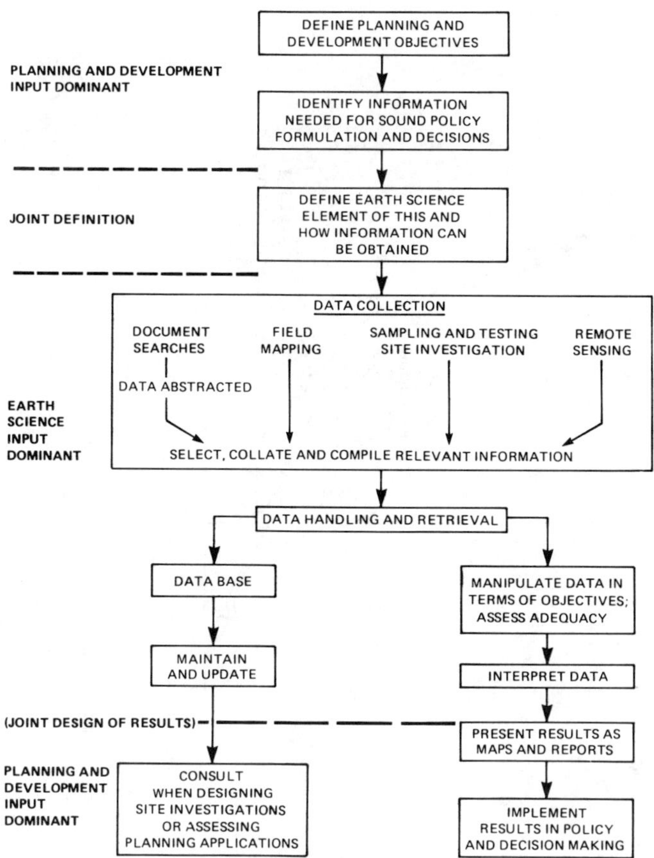

Fig. 3. Environmental geology mapping study with optimum planning input.

also be in an excellent position to give technical advice to planners and developers in the study areas.

The whole process of carrying out an environmental geology mapping study with the optimum planning input is illustrated in Fig. 3. In some cases the implementation of the results in the planning process may lead to identification of the need for further information or for revised planning and development objectives. Where the collected data base is maintained and updated after the initial studies have been completed, further problems or earth resources may be identified later.

The maps prepared are often used for straightforward comparison with planning and development proposals as illustrated in Fig. 4.

EXAMPLES OF ENVIRONMENTAL GEOLOGY EXERCISES

Within the definition of environmental geology maps stated earlier, a large number of studies have been completed in many countries. A limited number are reviewed here to give an impression of the range of approaches taken. By and large these reflect the contrasting environments in the study areas and the varying balances of interest in, for example, planning, engineering and conservation.

Europe

Pioneer work on engineering geology mapping in Europe stemmed from Czechoslovakia in the early 1960s (Pasek and Rybar, 1961; Anon. 1965; Matula, 1965; 1969), and at the same time these techniques were also being developed in Spain (Echevarria, 1967). A review of European studies (Doornkamp *et al.*, 1985) revealed the dominance of engineering geology approaches and most European studies are entitled 'Engineering Geology Mapping', whether narrow, or broad and multi-thematic. However, lately in Germany and Spain, the title 'Geoscience Mapping' has been adopted for the broader environmental geology mapping studies, and in the UK the term 'Applied Geological Mapping' has been used in some recent studies.

Examples of European studies are summarized in Table V. From this it can be seen that the multi-thematic approach is best exemplified in Spain, but elsewhere, in Belgium, Czechoslovakia, France, Hungary, Italy, UK, West Germany and Poland slightly narrower multi-thematic studies have been carried out, directed primarily at the land use planner. Some countries such as Sweden (Viberg and Adestam, 1980) appear to have only quite recently considered the multi-thematic approach. The study of Valencia province (Cendrero *et al.*, 1986) can be cited as a good model for a multi-thematic mapping programme designed for environmental and land use purposes. Another Spanish study, of part of the Basque country, is unusual in combining climatic and geological hazards in a single set and in being designed specifically for a minority audience—an important step towards getting the results implemented in a strongly ethnocentric area (Cendrero, personal communication).

Elsewhere, for example in Norway, a strong emphasis has been put on natural hazards. This has been taken further in Scandinavia, where a good deal of work has been done on the relationships between geochemistry and the occurrences of diseases (Glattre *et al.*, 1985; Finne *et al.*, 1987). Spain and West Germany have initiated systematic programmes of multi-thematic mapping, province by province or area by area, the aim being to effect a systematic and more or less uniform country-wide coverage. Other countries, such as Belgium, also aim at a uniform, systematic country-wide coverage but have adopted a rather narrower engineering geology related approach. Some countries, such as France, favour a decentralized approach, the responsibility for these studies resting with regional authorities; they do not form part of an integrated network of mapping studies, but rather are isolated. Work in the UK has been aimed mainly at target areas, with strong planning and development prospects, and known to have mineral resources or geological problems.

The use of summary maps has been relatively limited, especially in the UK, although some have been commissioned in recent studies. Amongst the many examples prepared in Europe a number can be used as examples of presenting advice on potential for development of land.

A broad view has been taken in West Germany where planning atlases have been prepared to cover some of the administrative areas known as Länder at scales varying from 1:200 000–1:800 000. Maps within these sets cover topics as diverse as statistical data, land use, economic and population trends, planning and conservation areas, flood zones, and geological factors. Treatment of the earth science aspects includes basic geology, minerals, hydrogeology, climatic data, topography, and soils. When combined with vegetation maps these lead to summary classifications of the natural environment. Comparisons of land use conflicts revealed in these map sets give the basis for formal planning maps (Wille, 1987).

Amongst these studies is one of Lower Saxony, at

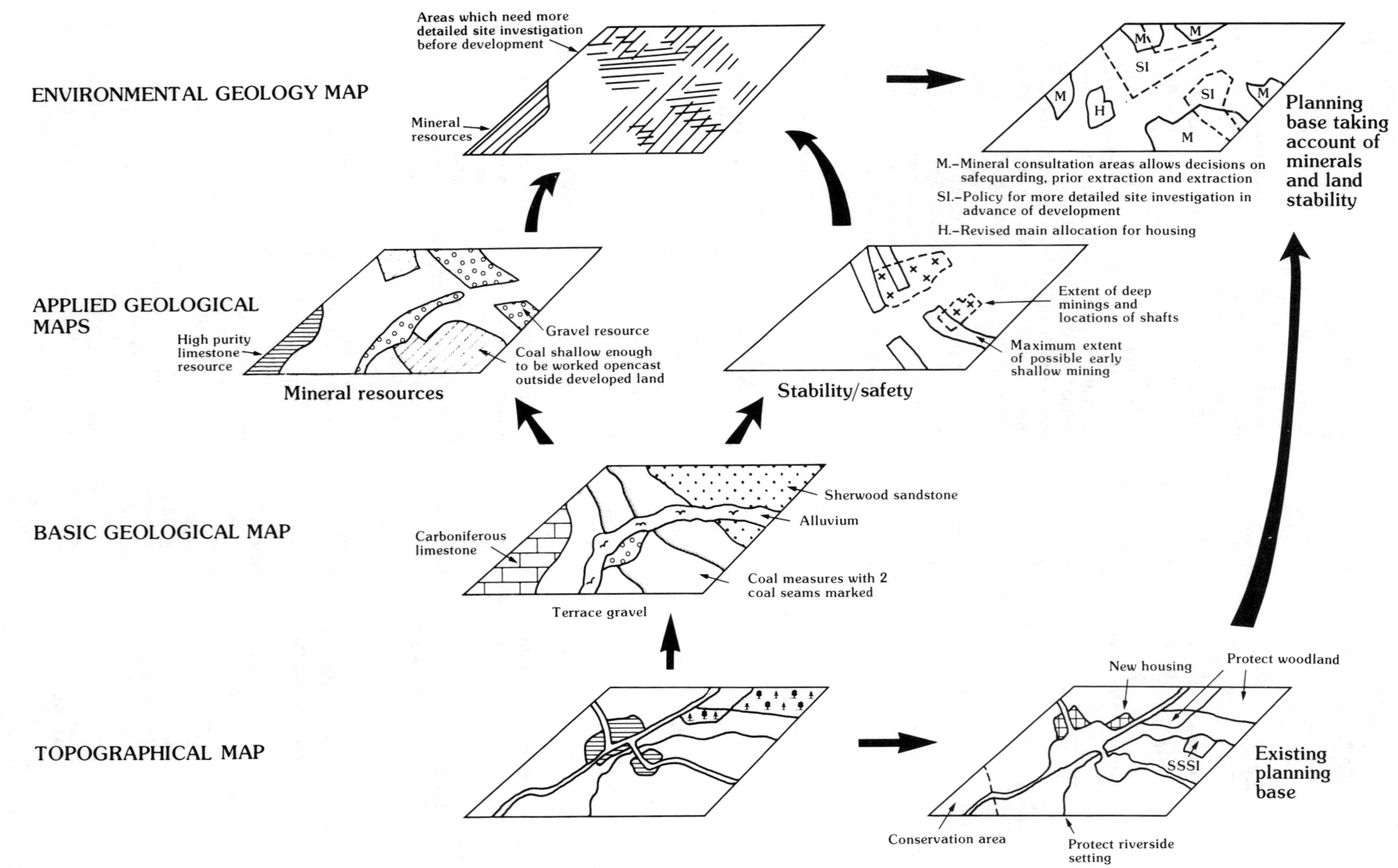

Fig. 4. Applied geology mapping in land use planning using mineral resources and land instability as examples.

Table V Summaries of selected European Environmental Geology Mapping Studies

Country	*Area*	*Coverage*	*Scale*	*Map Topics*	*Description*
SPAIN	Tenerife, Canary Islands (González de Vallejo, 1977) (Doornkamp *et al.*, 1985)	Entire Island	1:25 000	Geology Geomorphology Engineering geology Urban suitability (hazards, constraints and hydrogeology included)	Described as 'Engineering Geology Mapping', this study has only a limited range of element topics but there is a strong interpretative (potential) component.
	Madrid (Gilsanz *et al.*, 1985)	Metropolitan Area	1:200 000	Physiographic divisions (colour) Superficial deposits (stipple) * Bedrock geology * Bioclimatic divisions * Altitude * Land use	Termed a 'Physiographic Map', this is aimed at planning, and is supported by text with well designed block diagrams. Data is presented on a single map sheet with topics marked * being covered by inset maps. There is no interpretative (potential) component but this constitutes an invaluable base for planners.
	Valencia (Cendrero *et al.*, 1987) (Nieto Salvatierra and Obarti Segrera, 1987)	Province	1:200 000	Preliminary geological mapping Geoenvironmental units Vulnerability of water to pollution and landslide risk flooding Subsidence Collapse Seismic risk * Risk of potential erosion * Conservation capacity * Orientation and limitations on land use	This is termed a 'Geoscience Map' and it fits into a national province by province systematic programme of coverage involving all levels of government. The lack of hydrogeology among the topics is surprising, but this data may have been supplied by another organization. The last three maps (*) are potential maps. There was also a site-specific pilot study of solid waste disposal in the same area. It included seven themes: mines, quarries, hazards, settlement, tracks, nature reserves and archaeological sites.
CZECHOSLOVAKIA	Bratislava (Matula and Letko, 1980) (Doornkamp *et al.*, 1985)	Metropolitan Area	1:25 000	Trends in urban development The main engineering problems Geoenvironmental factors: (a) seismicity (b) geodynamic processes (c) hydrogeology (d) foundation conditions (e) geomorphology Protection of resources for development Engineering zones Zones of suitability for development	This comprises a combination of Element, Derived and Potential Maps, and is aimed at: (a) ensuring the optimum use of land; (b) avoiding hazards; and (c) optimizing the exploitation of mineral resources at the same time.

(*continued over*)

Table V (*continued*)

Country	*Area*	*Coverage*	*Scale*	*Map Topics*	*Description*
HUNGARY	Budapest	Metropolitan area	1:25 000	Topography Relief Hydrography Geology Climate Hydrology Bedrock Soils Foundation conditions Slope stability Seismic risk	The author presented what he believed should comprise engineering geology maps for general use, as opposed to specific site investigations. The different demands of flat and mountainous terrain and of urban and rural areas were covered, the rural area being flat and the urban area part hilly. Both element and derived maps are presented. The coverage is broad and the topics are treated factually rather than in an interpretative manner.
	Part of the Great Plains (Ronai, 1979) (Doornkamp *et al.*, 1985)	Province	1:100 000		
POLAND	(Lozinska-Stepien, 1979) (Doornkamp *et al.*, 1985)	Regional	1:25 000 and 1:50 000	Topography Geomorphology Lithogenesis of deposits Hydrogeology Geodynamic processes: (a) aeolian (b) fluvial (c) ravine erosion (d) landslides Geotechnical properties of soils and rocks	These maps are designed as engineering geology maps, for regional planning in urban areas. This account describes what are essentially element maps, which are used to produce two potential maps showing favourable and unfavourable areas for structure foundation.
ITALY	(Merla, Merlo and Oliveri, 1976) (Doornkamp *et al.*, 1985)	Regional	1:25 000 to 1:50 000	Subzones of geological conditions Hydrogeological conditions Geomorphology (soil, land use forestry and pasture maps) Engineering geology zoning	This was a series of studies of selected areas of hazard potential. The element and derived maps, based on data on soils, bedrock, geomorphology, hydrology and protective engineering works, provide a factual basis of constraints to be interpreted by land use planners. The engineering geology zoning map, based on lithology, morphology, overburden character and adverse geodynamic conditions is particularly useful, though preliminary and in no way a substitute for site investigation.
FRANCE	Paris (Arnould *et al.*, 1979) (Doornkamp *et al.*, 1985)	Metropolitan area	1:5 000 to 1:10 000	Landforms Thickness of deposits Depths to formation tops Erosion Foundation conditions Limestone workings Karst Geotechnical characteristics (zoned)	This comprises an atlas of maps defining the sedimentological and structural character of the city area and the main geotechnical problems. These are factual element and derived maps. Data sources include existing maps, borehole records, excavation records, available geological and geotechnical information.

	Carcassonne (Mazeas and du Mouza, 1979) (Doornkamp *et al.*, 1985)	Urban area	1:25 000 (1:50 000)	Lithology and facies Alluvium, Colluvium Glacis Screes Residual Soils Made Ground Hydrogeology Stability Erosion Thickness of deposits	This is a scheme for incorporating a large number of variables onto a single map (cf. Madrid above). In this case, in lieu of inset maps, there are additional map sheets, each covering a topic on the second list. Though termed 'Geotechnical', the approach is essentially geomorphological, with much greater emphasis on superficial deposits than in traditional geological mapping.
FEDERAL REPUBLIC OF GERMANY	Lower Saxony, Bremen (Tönnies *et al.*, 1987) (Becker-Platen *et al.*, 1987) (Doornkamp *et al.*, 1987)	Province (Länder)	1:200 000	Recommendations for land use Groundwater utilization Groundwater and mineral spring protection areas Recharge areas Large reservoirs for drinking water supply Mineral resource exploitation (areas of near-surface, first priority mineral deposits) Soils (productivity, general yield) Ground unsuitable for construction on account of: (a) organic soils liable to subsidence above salt diapirs, sink holes, cavern fields, solution mining, underground repositories for radioactive and toxic wastes (b) sites with geoscience protection value	This is one of a series of 'Geoscientific Maps' in a systematic coverage. The approach is unusual in that it is highly subjective and interpretative.
BELGIUM	Liege, Brussels, Mons, Ghent, Charleroi (De Beer *et al.*, 1980) (Doornkamp *et al.*, 1985)	Cities and large towns	1:5 000 to 1:10 000	Topography Site locations Thickness of geological formations Depths to tops of specific formations Hydrogeology Engineering geology zones Geological profiles (supported by geotechnical properties of the mapped formations)	Atlas of maps. Part of a systematic programme covering 160 such urban areas. Termed 'Engineering Geology Maps', they are somewhat limited in topic scope compared with many 'EGM' treatments, and appear to be entirely factual with little interpretative (potential) component.
NETHERLANDS	Druten (Rijks Geologische Dienst *et al.*, undated)	Area of flat level former flood basin terrain bordering the River Waal around the town	1:50 000	*Element maps include*: Industrial minerals Aggregate minerals *Derived maps include*: Foundation conditions Preparation of building sites Settlement	This is a series of thematic maps with an emphasis on foundation conditions and settlement of engineering soils, and mineral resources related to construction and local industrial needs.

(*continued over*)

Table V (*continued*)

Country	*Area*	*Coverage*	*Scale*	*Map Topics*	*Description*
UNITED KINGDOM	Glenrothes Fife, Scotland (Nickless *et al.*, 1982) (Doornkamp *et al.*, 1985)	A single Ordnance Survey Sheet, No. 20	1:55 556 (bound in) 1:25 000 (potential maps)	*Element maps* (*18*): Borehole sites Unconsolidated deposits (a) distribution (b) lithology (c) engineering properties (d) thickness (e) depth to water-table Sand and Gravel thickness Bedrock Geology Rockhead contours Shallow undermining Natural landslip potential Opencast workings Resources (a) Hard rock for aggregate (b) Brick and tile clay (c) Mudstone for brickmaking (d) Limestone Hydrogeology	This was a pilot EGM study within the UK to investigate the validity of the 'EGM' technique within the UK setting and to assess implications for manpower and finance. The area had been very well investigated previously and this study (entirely a desk study) utilised: (a) Geological maps (1856–1867; 1926–1946) (b) National Coal Board mine plans (c) Recent survey of drift deposits (d) Recent hydrogeological survey (e) Site investigation reports for Glenrothes New Town (f) Well records (g) Historic topographic maps (h) Geochemical data The aim was to compile element maps in a simple cartographic form from these data.
				Derived maps (*4*): Underground Storage Potential within 100 m of the surface Sand and gravel potential Foundation conditions Groundwater resources	This study was initiated by geologists to demonstrate the usefulness of EGM's for planners and developers.
				Potential maps (*5*): Development potential Priority areas for on-site investigation Mineral resources (a) near-surface (b) buried (opencast) (c) buried (pumping, mining)	
	South Humberside, England (Williams, I., 1983) (Doornkamp *et al.*, 1985)	A quite small area of less than county size, much built up.	1:25 000	*Element maps* (*7*): Topography and drainage Drift and outcrop geology Drift thickness Rockhead contours Opencast workings and mined land Area underlain by the buried Kirmington channel *Potential maps* (*1*): Development potential	The aim was to produce a series of maps and accompanying report for planning authorities, providing guidance on the geological and geotechnical conditions that might affect siting of large-scale future developments opened up by the Humber Bridge. There was a strong interpretative (potential) component in this study.

Edinburgh (Floyd, ***et al.***, 1982) (Doornkamp ***et al.***, 1985)	Area southeast of the city	1:25 000	*Element maps*: Topography and drainage Drift and outcrop geology (stratigraphy, lithology) Bedrock geology and structure Rockhead contours Borehole locations Drift thickness Opencast workings/landfill sites Underground mining (coal) Areas of shallow workings (subsidence prone) Hydrogeology Areas for which the British Geological Survey holds site investigation reports *Derived maps*: Subsidence potential Predicted foundation conditions Mineral resource potential *Potential maps*: Summary of geological constraints	This is a study of an area with strong urban development pressures. The area is one in which coal mining has been extensively carried out, and the emphasis is on the effect of underground workings on foundations, while at the same time evaluating the residual mineral resource potential.
Bridgend, South Glamorgan (Wilson and Smith, 1985)	Small urban area	1:25 000	*Element maps*: Topography and drainage Drift and outcrop geology (stratigraphy, lithology) Bedrock geology and structure Rockhead contours/borehole locations *Derived maps*: Foundation conditions Mineral resource potential (a) Limestone (b) Sand and gravel Potential for coast erosion	This was a study of a small non-mining area in South Wales with strong pressures for urban development.
Chacewater, Cornwall, England (Freeman, Fox Ltd. 1987) (Doornkamp ***et al.***, 1985)	A small area in which there had been intensive copper mining	1:10 000	*Element maps*: Topography and drainage Drift and outcrop geology Bedrock geology Underground mining Hydrogeology Surface hydrology Mine property boundaries Mineral mining consultation area Derelict land *Derived map*: Underground storage potential Existing and proposed development sites Nature conservation sites Critical and non-critical areas of mine-related hazards	The aim of this study was to present the various constraints on redevelopment for other land due to the residual hazards of underground mining in this part of Cornwall. There was a strong emphasis on subsidence potential. The study was carried out in discussion with officers of government and local authorities who would make use of the results.

(*continued over*)

Table V *(continued)*

Country	*Area*	*Coverage*	*Scale*	*Map Topics*	*Description*
UNITED KINGDOM (*cont.*)	West Wiltshire and South East Avon, England (Forster *et al.*, 1986, 1987) (Doornkamp *et al.*, 1985)	A small area around the city of Bath	1:10 560 base maps (reduced to 1:25 000)	*Element* and *Derived maps*: Topography and drainage Solid lithostratigraphy Drift deposits (extent, lithogy, thickness) Made ground/infilled land Distribution of Great Oolite freestone Groundwater Ground conditions related to groundwater Landslipped and cambered strata Slope angle distribution Geotechnical properties (bedrock, superficials) Mining (extent, location of shafts) Location of geotechnical data sources	This was a desk study with a limited field survey of foundered strata. The study was aimed at potential surface and subsurface development. The emphasis was on freestone and fuller's earth resources and man-made/natural land instability. The study concentrated on a very factual approach, and little interpretative (potential) material was included.
	Plymouth, Devon, England (McCall *et al.*, 1985) (Chandler and McCall, 1986)	Metropolitan area and its fringes	1:10 000 and 1:25 000	*Element and derived maps*: Topography (and contour pluck) Solid geology Superficial geology Scale sections of superficial deposits down to rockhead Hydrology (natural drainage and man-made conduits) Hydrogeology (borehole and well distribution; the Plymouth Limestone; marine incursion) Location of quarries, mines and caverns Location of mineral workings Depth to rockhead Thickness of weathered zone below rockhead Depth to engineering defined bedrock Engineering properties of soils and bedrock Made ground, reclaimed land, landscaped and terraced areas	This was a study of a city of a quarter of a million people, where the main problems are: (a) conflict between limestone quarrying and urban development; (b) stability of limestone cliffs and quarry faces; (c) made ground, reclaimed land including bomb damage fill; (d) foundation problems on steep slopes, marine silts and made ground. This study was unusual in the very large number of site reports included in the data base which had a microcomputerized search system. This allowed very full treatment of the engineering properties of superficials and bedrock from existing data. Treatment of the limestone resource was restricted because of restrictions on release of technical data held by commercial operators. The waste disposal aspect was similarly restricted by the sensitive nature of information held the Local Authority on potential sites for disposal.

Upper Forth Estuary Scotland (Gostelow & Browne, 1986)	An area of 35 × 20 km	1:50 000	*Element maps*: Drift thickness/depth to rockhead Contours on the upper surface of glacial deposits Drift geology Distribution of mine workings *Derived maps*: Engineering geology of solid rocks Engineering geology classification of soft sediments Geotechnical cross-section *Potential map*: Geotechnical Planning for heavy structures	Though termed an engineering geology study, this was very broadly designed and may reasonably be regarded as an EGM study, with a strong constructional emphasis (heavy foundations). It is perhaps surprising that there is no detailed treatment of hydrogeology but it is stated that it is less important where heavy constructions founded on superficial deposits are concerned (and the study is primarily aimed at these). There is an interpretative (potential map) component. This study well illustrated the 'join' between 'Engineering' and 'Environmental' Geology Mapping (multi-thematic) and the fact that the two are semantic terminological choices and in practice are overlapping and not distinctly separable.
Torbay, Devon, England (Geomorphological Services Ltd., 1988)	The borough of Torbay: a large coastal residential and tourist area	1:50 000	*Element maps*: Bedrock geology Superficial geology Geomorphology Slope steepness Soils Geotechnical conditions Site geotechnical investigations Sites of mineral working Land use planning provisions *Planning map*: Ground characteristics for planning and development	This was the first study in the UK which included soils and geomorphology in detail alongside the more traditional geological themes. Attention was paid to limitations on development set by slope instability, mainly near the coast; variable foundation conditions (backfilled quarries and solution pipes within Devonian limestones). The geological base was compared with current land use planning provisions and a summary map for planners was prepared in dialogue with representatives of Government and Local Authorities.

1 : 300 000 scale, which takes agriculture, forestry, mineral resources, recreation, wildlife, landscape and water resources as major themes. Summary maps show priority areas for resource exploration, industry and waste disposal (Tönnies, 1987). Another of Hanover emphasizes, as is appropriate in that area, additional aspects such as air pollution (Lüttig, 1987). An extract from the key of the Geoscientific Map of Natural Environmental Potential of Lower Saxony and Bremen is given as an example in Fig. 5 (Becker-Platen *et al.*, 1987).

A further example of a systematic approach is shown by the 12–15 year programme to produce 1 : 50 000 scale maps of natural environmental potential in Czechoslovakia based mainly on extensive existing data.

MINERAL DEPOSITS AND OCCURENCES

Area with 1st. priority mineral deposits (prime economic importance) Involvement of the relevent authorities (Geological Survey of Lower Saxony or the local Mining Authority) is essential from the initial stage of all planning involving land use.

Area with 2nd. priority mineral deposits. All plans involving land use in this area should be coordinated with the relevent authorities (Geological Survey of Lower Saxony or the local Mining Authority).

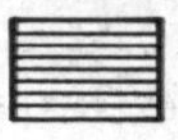

Area with valuable mineral resources, not yet fully investigated, exact extent not known. Relevent authorities (Geological Survey of Lower Saxony or the local Mining Authority) should be informed at an early stage of all planning involving land use in this area.

Sizeable open-pit mine

B Raw materials for buiding and civil engineering construction

E Energy raw material

I Raw materials for the iron and steel and chemical industries etc.

K Raw materials for the ceramic industry

T Peat

MISCELLANEOUS

Cavern field

A Underground repository for radioactive wastes

Solution-mining field

R Underground repository for toxic wastes

Blank areas denote that, from the geoscientific point of view, no recommendation is made, in some cases due to lack of information

GROUNDWATER UTILIZATION

Groundwater or medicinal spring protection area, authorized or application submitted; recharge area of waterworks in operation, under construction or planned.

Large groundwater reservoir exploitable for drinking water supply

GROUND UNSUITABLE FOR CONSTRUCTION (RESTRICTED USE)

Organic soils, very liable to settlement

Area above salt diapir, where its top lies higher than 200m. below m.s.l.

* Area of sinkholes and subsidence

• Sinkhole

MINERAL RESOURCES EXPLOITATION

Area of near surface, 1st. priority mineral deposit

Soils predominantly with high to very high productivity and high sustained yield: a decrease in yield is possible in particularly wet years.

Soils with medium to high productivity and fairly sustained yield: a decrease in yield is possible in dry years on dry soils and in wet years on moist soils and on those containing perched water.

SITES WORTH PROTECTING FOR THEIR GEOSCIENTIFIC VALUE

● Single site, protected or recommended for protection

△ Several sites, protected or recommended for protection

Areal feature, protected or recommended for protection

Fig. 5. Extract from the key of the Geoscientific map of Natural Environmental Potential of Lower Saxony and Bremen (adapted from Becker-Platen *et al.*, 1987).

Nine maps are prepared for each sheet area covering:

- solid and drift geology with a simplified key;
- geophysical map emphasizing fault zones, seismology and radioactivity;
- minerals;
- hydrogeology;
- engineering geology including the significance of rocks and soils for construction and erosion by wind and water;
- pedology, including information on forests; geochemistry of rocks and water including contamination and reactivity, based on analyses for between 30 and 40 elements;
- important landscape phenomena;
- overlapping or competing land uses.

Different colours are used for features of the geology, soils, water, atmosphere, biological aspects, and human activity, and red is reserved for hazards. Positive features for land use are shown by surfaces and contours whilst negative features are ornamented by various hatchings (Moldan *et al.*, 1987).

Some particularly elaborate work has been carried out on the Santander area of northern Spain (Cendrero, 1976; Environmental Analysis Group, 1980). The approach taken was to carry out mapping at 1 : 20 000 and to generalize the results at 1 : 50 000. The exercise was fully multi-disciplinary covering earth science, biological and cultural aspects. Rock types were ranked and classified (high, medium and low significance, or not applicable) in terms of permeability, corrosion potential, slope instability, ease of excavation, shrink/swell potential and bearing capacity. Morphological settings included estuarine and coastal, slopes and hills, karst features and man-made landforms such as tips. Processes and hazards such as landslides and rockfalls, compaction and subsidence, flooding, erosion and sedimentation, sand transportation by wind and water recharge were indicated. The mass of detail was then classified into eight environmental geology units of basically morphological origin. The significance of each unit was allocated, on a scale of 1–3, in terms of heavy construction, light construction, laying of cables and pipelines, road making, and solid and liquid waste disposal. This resulted in scores of between 5 and 15 for the various units and these were used to put each unit into a class for building and construction. A basically similar system was used for each of agriculture, mining and quarrying, conservation and recreation. Comparable processes were used to classify the ground in terms of biological factors (vegetation, faunas, ecology) and cultural features (artistic, archaeological, historic and architectural). Results were then integrated using weightings to reflect uniqueness of some features at the world, national and regional levels of significance and of fragility or sensitivity of other features to changes and development. The final products were maps of capability for development, limitations for building and construction, and of quality and fragility for conservation.

More restricted interpretations of development potential, reflecting the characteristics of study areas, are taken, for example, in the studies of Glenrothes (Scotland) and South Humberside (England). In the former study, emphasis was placed on suitability of ground for various types of structures based on extent of past mining and quarrying, foundation conditions and landslip potential. The resulting map of development potential indicated areas where ground conditions are likely to be acceptable for lightweight structures and those where ground conditions require detailed site investigation and, possibly, ground treatment for such structures. However, slope steepness and topographical features, such as reservoirs, were not taken into account (Nickless *et al.*, 1982). In the south Humberside study emphasis was also placed on the level of site investigation required (Table VI).

Table VI Types of area defined in a development potential map of part of South Humberside, England (based on Williams, 1983)

(a) Areas where ground conditions are likely to be acceptable for the construction of lightweight structures. More detailed site investigations may be warranted for heavier structures and where preliminary investigations reveal unexpected and largely unpredictable geological hazards.

(b) Areas requiring detailed site investigation and possibly ground treatment because of potential geological hazards.

(c) Areas where ground conditions are likely to be acceptable for the construction of most structures but where there is a substantial thickness of buried deposits of variable strength which may necessitate detailed site investigation.

(d) Areas where ground conditions are such that detailed site investigations would be required and which are underlain by thick deposits of variable engineering properties. These areas appear on geological grounds to be the least suited to development but the hazards are not insurmountable and indeed much of the present industry within the study area lies within them.

(e) Areas of Made-Ground or Fill of varying composition and thickness. Detailed site investigations may be required depending upon which of the four categories above, these areas rest.

Boreholes in which peat had been recorded were also plotted. Each of the areas (a–e) were plotted with distinctive colours and ornaments. This interpretative map was based on element maps showing engineering properties of unconsolidated deposits, rockhead contours, and the distribution of glacial deposits.

A still more limited focus was included in the engineering geology study of the Upper Forth Estuary where the target was the classification of ground stability for heavy structures (Table VII) in terms of foundation types, bearing values and site investigation requirements.

Turning to urban development, a study of Tenerife has set out urban suitability of ground in terms of engineering geology factors, classifying units into categories termed 'favourable', 'restricted' and 'unfavourable' (Fig. 6).

A study of part of the Druten area in the Netherlands focuses on the preparation of building sites in terms of the amounts of fill and extent of drainage works required (see Chapter 8, Fig. 8) (Rijks Geologische Dienst, undated).

A somewhat different approach was taken in the study of Torbay, England, where the final map of a set shows constraints on development which can be directly compared with an additional sheet showing currently designated planning areas, such as mineral resource consultation zones, areas of landscape value and sites of special scientific interest (Fig. 7).

A different approach to landscape considerations is taken in the physiographic map of Madrid, which is an unusual amalgam of landscape zones based on topography, vegetation, and climate directly printed onto a geological base (Table VIII).

Overall, therefore, experience in Europe was initially closely tied to engineering geology mapping, but over the last ten years or so the diversity of studies and associated innovation of presentation has developed greatly. It will be interesting to see if a greater emphasis on environmental assessment of proper development within the European Community has an effect on the content and presentation of environmental geology studies. A consolidated list of themes included in the selected European and USA studies is presented in Table IX.

USSR

An important published summary by Melnikov (1979) outlines the approach to multi-thematic mapping in the USSR. The USSR has a national programme of 'engineering geology mapping' carried out by the Ministry of Geology. Its aims are to:

(1) evaluate geological conditions which determine the site of 'large terrestrial complexes of construction units of shallow location' (industrial, civil, road, piping, reclamation);
(2) evaluate conditions of development areas involved in economic activity;
(3) geologically substantiate environmental protection measures.

Mapping is usually on 1 : 200 000 scale initially, later reduced to 1 : 500 000, in simple areas, but a scale of 1 : 100 000 is adopted in more complex areas. The scale of coverage may be compared therefore with an English or American county. The themes considered are relief, geological structure, permafrost, hydrology, hydrogeology, recent geological and engineering geology processes, and areas for prospecting for building materials.

The approach is to erect a number of contrasting model or 'key' areas. The procedure is:

(1) archival and literature compilation;
(2) field work, in which the nature and engineering properties of the materials are first evaluated, and testing of samples recovered is carried out;
(3) landscape evaluation and geophysical programmes;
(4) drilling and testing of critical areas;
(5) office compilation.

In the USSR the practice is exceptional in the amount of new field investigations incorporated in multi-thematic programmes, as against reliance on existing information. The amount of money spent on these programmes probably matches or exceeds expenditure in the USA, and must greatly outweigh the funding in any west European country.

USA

Environmental geology mapping (EGM) multi-theme studies were published in the United States from the end of the 1960s onwards, as a response to the growing awareness of the need for environmentally informed planning decisions and to the legislation supporting this (for example, the National Environmental Policy Act, 1969). The USGS and the State Surveys were the main agencies involved. Early studies tended to emphasize the traditional interests of the Geological Surveys (topographical and geological mapping, minerals and water resources) and to subordinate the geomorphological and soils aspects. More recent studies have adopted a more

Table VII Ground classification for planning of locations of heavy structures on the Upper Forth Estuary, Scotland (from Gostelow and Browne, 1986)

Ground classification for planning heavy foundation sites	*Normal foundation type*	*Material group and classes*	*Site investigation/ testing requirements*	*Presumed bearing value*
A – Very good	Shallow	Group I, class 1–8, within 3 m of ground surface	Rotary cored boring, estimated RQD from logs. *In situ* plate loading, tests for very heavy structures. Point load testing on rock core.	600–10 000 kN/m^2
B – Good	Shallow and ground improvement for low presumed bearing values	Group II, classes 9–10, 12, 13. Group III class 15 and 16 within 3 m of ground surface (with a reliability of I, not underlain by soft or loose sediments)	Shell and auger boring, undisturbed drive samples or SPTs. Laboratory tests to include undrained triaxial and oedometer. Dutch cone soundings for granular deposits.	200–600 kN/m^2
C – Fair	Deep/Intermediate	Bearing stratum of A and B groups for driven piles between 3 and 20 m of the ground surface (with a reliability of I, not underlain by soft or loose sediments)	Shell and auger boring through overburden with drive or piston samples. Dutch cone soundings where appropriate. Laboratory tests to include undrained triaxial, oedometer, index properties. Rotary coring to prove condition of bearing stratum.	200–10 000 kN/m^2
D – Poor	Deep/Intermediate	Bearing stratum of A and B groups greater than 20 m. Structures often requiring friction piles, or buoyant foundations	Shell and auger boring, with drive or piston samples, Dutch cone soundings. Laboratory tests to include undrained triaxial, oedometer, index properties.	75–200 kN/m^2
E – Unpredictable (very poor to very good)	Shallow + ground improvement	Assessment zone material A and B within 3 m of ground surface, with shallow mine workings (longwall or stoop and room) within 20 m of foundation level	Geophysical techniques, rotary cored boring, careful examination of mine plans and relevant geological maps for cavities, RQD evaluation, point load testing on rock core. Plate bearing tests for very heavy structures. Consideration of ground improvement schemes.	Less than 200 kN/m^2 but assessed after inspection, check data reliability possibly better ground than B in parts.
F – Unpredictable (very poor to fair)	Deep/Intermediate + ground improvement	Assessment zone material A and B greater than 3 m from ground surface, underlain by mine workings (longwall or stoop and room within 20 m of foundation level). Landslipped areas	Geophysics techniques, where appropriate. Shell and auger boring, rotary boring. Dutch cone soundings. Examination of mine plans and relevant geological maps. Laboratory tests to include undrained trixial, oedometer. Consideration of ground improvements schemes.	Less than 200 kN/m^2 but assessed after inspection, check data reliability, possibly better than D and E in parts.

Zones were also given a reliability classification of: I = reliable, O = assessment based on limited geotechnical data. The depths below the ground surface of various classes of materials were also indicated.

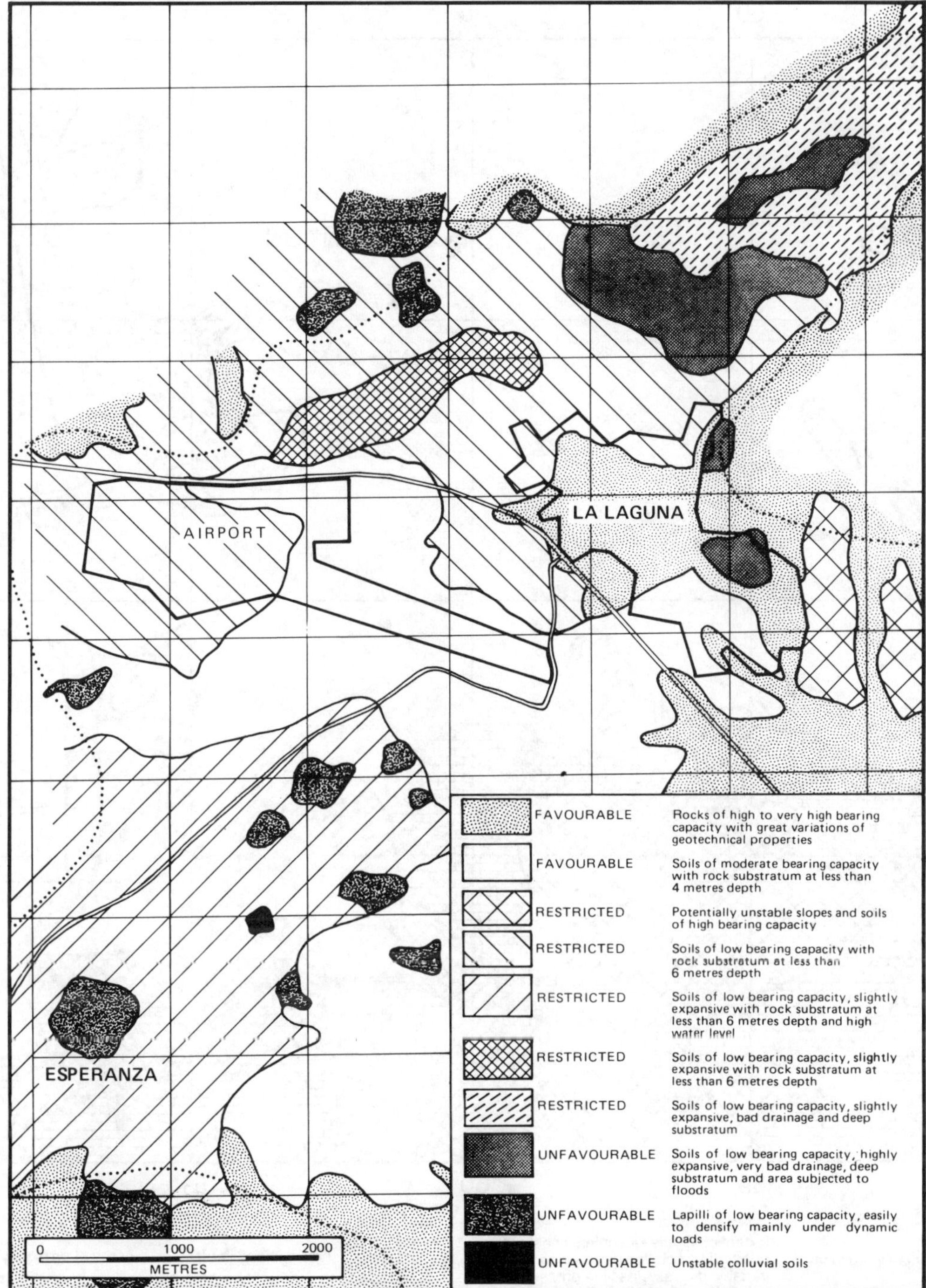

Fig. 6. Urban suitability map based on engineering geology conditions for part of Tenerife, Canary Islands (adapted from Gonsález de Vallejo, 1977).

comprehensive format, in some cases quite novel and original, and have also extensively utilized computerized cartography and data base methodology. A number of studies have reflected the importance of environmental management assessment for major projects.

Doornkamp *et al.* (1985) remarked that environmental geology is perhaps better developed in California, and Los Angeles County in particular, than anywhere else in the world. Indeed, the area is also exceptional for the numbers of earth scientists directly involved in land use planning. The same might well be said of the San Francisco Bay area in California, where a very broad, comprehensive approach has been adopted, in a series of related studies which go far beyond the hazard aspects. The amount of money available for these studies would appear to be quite exceptional. The United States in general is far advanced in environmental geology mapping, and in Table X a comparison is made between a number of multi-theme studies from a number of different States and types of terrain.

In the USA, as in Europe, studies may be entitled 'Environmental Geology Mapping' or 'Engineering Geology Mapping', but the latter title may embrace very narrow specialized investigations or very broad multi-theme studies which are more correctly termed 'Environmental Geology Mapping'. For example, a study

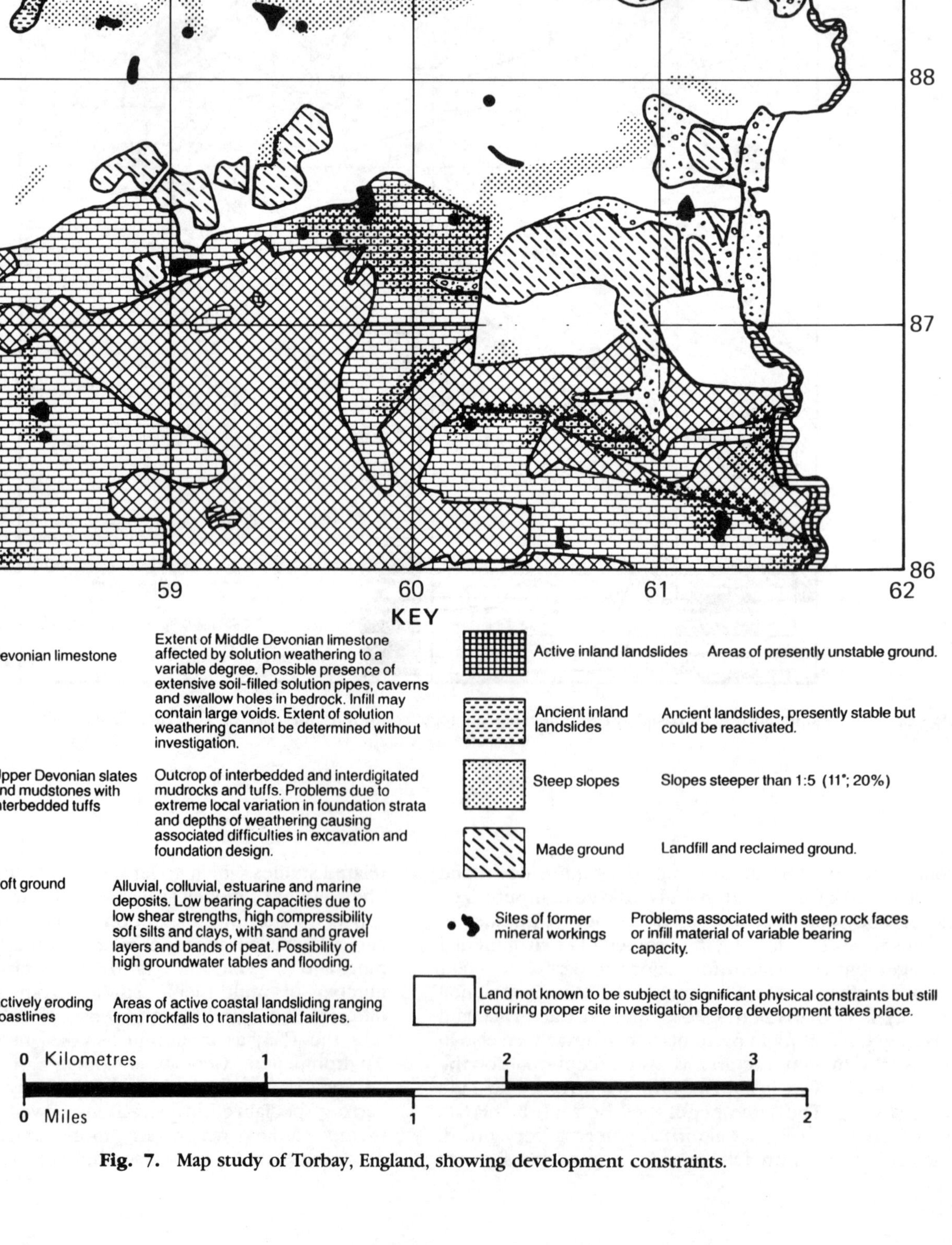

Fig. 7. Map study of Torbay, England, showing development constraints.

Table VIII Classification of areas used on a 1 : 200 000 scale 'physiographical' map of Greater Madrid (from Gilsanz *et al.*, 1985)

Physiographical units	*Lithological basis*
(1) Peaks	(1) Superficial formations
(2) High moorland	—alluvial deposits (sands, gravels, clays and muds)
(3) Mountain slopes	—mixed sediments (alluvial and colluvial clays and muds)
(4) Piedmont	—conglomerates
—inclined	(2) Bedrock
—depression or trough	—limestones
(5) Open countryside	—detrital deposits (arkoses)
—Tajo River valley	—chemical and transitional deposits (marls, clays, gypsum)
—valley depressions	—calcareous rocks, detrital sediments, marls and gypsum
—'substitution surfaces' of the plain	—granitic rocks
(6) Plain	—low- to medium-grade metamorphic rocks (slates, quartzites)
(7) River plains	—high-grade metamorphic rocks (gneiss, mica schists)
—valley of the Tajo River	
—depression in the crystalline massif	

The physiographical units were defined by means of coloured areas on the map, whilst the geological information was printed over these using varied ornamentation printed in red or blue.

of the San Francisco Bay flatlands (Helley and Lajoie, 1979) is essentially a multi-theme environmental geology mapping study.

The range of themes covered is more extensive than those in the European studies in two respects. First, there is, not surprisingly, a strong emphasis on seismicity and on slope instability in the western USA—a tectonically active continental margin with relatively frequent large-scale problems of this type. Second, there is a much broader range of interpretive and summary maps.

A number of other interesting features emerge from some studies, including the emphasis on inland wetlands (an important conservation topic) in the Connecticut River valley study. The maps of arid Maricopa County, Arizona, include the distribution of 'caliche' (hard masses of calcium carbonate which form within the soil profile) which has significance for excavation of sites and for design of foundations.

The Allegheny County study classes together two main groups of themes. The first of these concerns the legacy of dereliction, subsidence and underground combustion left by the mining of coal in this, the earliest, coal mining area in North America. The second covers the relationship between water impoundment in mountainous country, land instability and heavy storms. The full significance of this topic is apparent if one remembers that not very far away one hundred years ago the collapse of an earth-fill dam at Johnstown, Pennsylvania, took over 2000 lives (McCullough, 1968).

Work on Ann Arundel County, Maryland, includes a strong emphasis on coastal processes such as shoreline erosion and protection—a topic which is of widespread interest in many areas and may become of still greater significance if recent predictions of rising sea-levels are fulfilled.

The study of the Alaskan north slope is unique amongst those reviewed in plotting mineral resource information in direct juxtaposition with ecological data, including ranges of vulnerable mammals such as wolves, bears and foxes. Other associated studies covered ground engineering characteristics including problems of construction and maintenance of a pipeline on permanently frozen ground, and the potential effects on migrating herds of caribou. This was probably the most extensive environmental impact assessment yet mounted (Brew, 1974; Brew and Gryc, 1974; EIS Taskforce, 1976; Environmental Protection Board of Canada, 1974; Kreig and Reger, 1976; Moening, 1974).

The general emphasis on interpretative maps is widely coupled with attempts to quantify potential problems. These are often expressed in terms of a 'stop: caution: go' type, using colour coding for 'high, medium and low' development potential; in some cases even more complicated coding systems have been used.

Particular attention is drawn here to three compilation maps. The first is the natural land use intensity map included in the Connecticut Valley study. In this, information unconsolidated materials, depth to bed rock, slope angles and ground water potential are combined into an advisory map which distinguishes wetlands, high yield aquifers and potential sewers. A flood plain is emphasized and the designations 'high', 'moderate' and 'limited' are used to distinguish zones in which varying degrees of site problems are likely to be experienced (Fig. 8).

The second is a map of Franconia, Virginia, which shows inferred capability for housing development. In this, information on ground conditions interpreted from maps of clay distribution, slopes and drainage characteristics were used to assign 'scores' to grid squares throughout the area. The scores reflect coincidence of problems, thus higher values indicate more site difficulties. The resulting map is at first sight unfamiliar since it does not depict lines and coloured or shaded areas. However, its form is one ideally suited to digital cartography (Fig. 9).

The third is the waste disposal map included in the study of the McDowell Mountains, Maricopa County, Arizona. In this, the suitability for sanitary landfills is classified on a seven point scale (I—least favourable to VII—most favourable). The scale is described in Table XI. This study contains a good deal of technical information, but it also emphasizes the relative suitability of areas for septic tanks, waste stabilization ponds, sanitary landfills, and in terms of available materials for capping landfills. The resulting table is useful, but is likely to be a little daunting to non-specialists.

A final aspect of one study worth noting in terms of communication with users of results is part of the Franconia study. Earth science results are firmly based on three-dimensional systems. Maps, however, are two-dimensional surfaces. Those familiar with using earth

Table IX Themes incorporated in EGM studies (USA and Europe)

Construction

Depth to bedrock/rockhead contours
Depth of weathering
Groundwater chemistry/piezometer readings
Unconsolidated materials
Hillside materials/engineering properties
Engineering properties of superficials
Engineering properties of bedrock
Geotechnical cross-sections
Liquefaction susceptibility of superficial deposits*
Excavation conditions
Foundation conditions
Recommendations for preparation of building sites
Engineering zones
Soil creep*
Settlement*
Landfill (earth and rock)
Reclaimed land
Made ground/artificially landscaped and terraced areas
Rock ledges
Shrink and swell
Shoreline protection structures
Ground unsuitable for constructions (above caverns, salt diapirs, sinkholes, underground repositories for radioactive waste or toxic waste, on old gasworks or other polluted industrial sites—old mine workings)*
Location of underground water, gas, oil pipelines, tunnels, and electricity conduits

Geological

Solid geology
Superficial geology
Lithostratigraphy
Thickness of deposits
Depths to tops of stratigraphic units
Structure
Active faults*
Caliche
Alluvium, colluvium
Geological cross-sections

Hazards

Flood prone areas/flooding risk
Landslipped and cambered strata
Landslide incidence/susceptibility
Earthquake incidence/risk
Earthquake epicentres
Stability during quakes
Predicted seismic intensities
Accumulated damage potential from seismic shakings
Potential damage to buildings
Seismic response
Seismic amplification
Tsunamis
Stability of cliffs and quarry faces
Water impoundments
Heavy storm precipitation
Vulnerability of water to pollution
Risk of natural or other subsidence
Risk of collapse/shallow undermining
Critical and non-critical areas of mine-related hazards
Risk of potential erosion, including coast erosion

Minerals and mining/quarrying

Mineral resources
Mineral conservation
Mine waste control
Mined land
Mine sites, worked out and operative pits and quarries
Sand and gravel resources
Distribution of building stone
Brickmaking clays etc.
Hard rock resources, potential uses
Limestone resources
Thickness and nature of overburden
Mine tips and slag dumps*
Mine shafts*
Mine subsidence*
Mine property boundaries
Mineral mining consultation areas etc.

Hydrogeology

Groundwater availability
Depth to water-table
Piezometric surfaces
Groundwater potential
Areas with seasonally high groundwater tables*
Groundwater chemistry
Groundwater conservation
Groundwater and mineral spring protection area
Recharge areas
Reservoirs
Surface water conservation
Water pollution
Sewerage
Waste disposal
Saltwater intrusion
Location of water pipelines, surface conduits, boreholes and wells

Planning

Existing land use
Derelict land
Natural land use intensities/suitability for development
Geological factors affecting land modification
Urban development
Redevelopment
Capability for planned housing development
Capability for industrial development
Capability for heavy constructions
Potential for agricultural/forestry use
Potential for recreational use
Mineral resource potential
Water supply potential
Potential waste disposal sites
Conservation capacity
Sites with geoscience protection value
Underground storage potential
Summary of geological constraints

Geomorphological

Landforms/topography/relief
Hillside development
Slope/slope angle distribution
Altitude/contours
Geodynamic processes—aeolian, fluvial, ravine, landslip
Erosion
Weathering
Residual soils
Flood plains
Wetlands
Water courses
Topography of bedrock surface
Hydrology
Drainage
Historic coast erosion rates
Beach distribution
Historic shorelines
Channel change
Karst
Glacial features
Screes
Permafrost

Soil science

Soils and their classification
Suitability of soils for agriculture
Suitability of soils for construction
Soil productivity and general yield

Data distribution

Data points/borehole sites/test sites/site report localities
Location of geotechnical data sources

Others

Ecological areas
Habitats for wild life in need of protection
Climate
Bioclimatic divisions

*also have a hazard factor

Table X Summaries of selected North American Environmental Geology Mapping Studies

Location	*Coverage*	*Scale*	*Themes*	*Description*
Greater Anchorage, Alaska (Dubrovolny and Schmoll, 1968)	Borough and City area within it (1 730 square miles)	1:125 000 and 1:24 000 (special selected topics)	Topography Geology Slopes Construction materials Foundation and excavation conditions Stability of natural slopes Recreation areas Gravel and surface water (areas of potential development) Groundwater availability Depth of unconfined aquifers Water-table contours Piezometric surface Saturated thickness Principal recharge areas Chemical quality of water	This was among the first systematic multi-theme programmes of mapping to be mounted in the United States. It was a response to the catastrophic earthquake of 1964. The city was quite remote from the epicentre but, as had been previously warned, the Bootlegger Cove Clay Unit was highly susceptible to shock-induced slump and flow. There is a discussion in the text cited of the reason for ignoring this warning by planners and the lack of communication between them and earth scientists prior to the catastrophe. The integrated multi-theme study initiated by planners and earth scientists in collaboration after the event was used for: (a) Development of land for agricultural, residential, industrial and recreational use; (b) Site selection in planning major engineering works (dams, bridges, highways, etc.); (c) Assessment and development of the hydrologic system for the effective exploitation of water resources.
Tucson, Arizona (Davidson *et al.*, 1978)	County	1:48 000 and 1:250 000	Hillside development Resource conservation (minerals, water) Control of waste from mining Hazards from old mine workings	This was a study aimed primarily at the conflict between mining for copper and potential copper resources, waste disposal from mining, hazards from mining; and pressures for urban development. It is also cited in Chapter 5 and thus illustrates the overlap between mineral resource mapping for planning and development purposes and multi-theme EGM studies.
Front Range Urban (Crosby *et al.*, 1978)	County	1:100 000	Hillside development and slope stability Resources and their development (minerals; sand and gravel; Water: shrink and swell in soils and rocks) Subsidence due to coal mining	This study covered two areas in Colorado with extreme pressures for urban development. The themes are restricted, but the maps provide interesting models.
Connecticut River Valley, Connecticut, (East Granby) (Langer and Johnson, 1978)	Rural Township	1:12 000	Unconsolidated materials Depth to bedrock Slope steepness Areas with seasonal high water-tables Groundwater potential Floodplains Inland wetlands and water courses Natural land use intensities (suitability for development)	This is an excellent example of a very small area study. It was aimed at the potential for land use and the constraints on land use in a single township. The natural land use intensities are quantified on a trimodal coding (high, medium, low).

(*continued over*)

Table X (*continued*)

Location	*Coverage*	*Scale*	*Themes*	*Description*
San Mateo County, California Phase 1 1970 (Atwater, B., 1978) (Doornkamp *et al.*, 1978)	County	Mostly 1:62 500	Basic geology Flood prone areas Physical character (flatland, hillside) Landslide (incidence, susceptibility) Slope Active faults Earthquake epicentres Tsunamis Sewerage Water supply Pollution Waste disposal Groundwater chemistry	More than 160 maps were produced in Phase 1. The pronounced hazard and pollution emphasis reflects the nature of the subject terrain. The study was used to: (a) reduce density of development; (b) design flooding set-backs; (c) ensure seismic safety; (d) ensure general public safety; (e) plan open space conservation; (f) assess environmental risks; (g) identify general problem areas.
San Mateo County, California Phase 2 (Atwater, B., 1978) (Doornkamp *et al.*, 1985)	County	Mostly 1:62 500	Maps cover: Basic geology Bedding dip/slope steepness Hillside materials/ Engineering character Stability during earthquakes Liquefaction susceptibility Predicted seismic shaking intensities Accumulated damage potential/ground shaking Potential earthquake damage to buildings	This illustrates the manner in which more highly specialized follow-up programmes can be mounted after completion of a broad multi-thematic study. In this case a selection of a number of hazard-related themes was made. Digital computer cartography was used. Quantification was attempted. The emphasis was upon earthquake and landslide risks, and foundation and building design related to these.
McDowell Mountains Area, Maricopa Country, Arizona (Christenson, Pewe *et al.*,1979)	Part of a county	1:24 000	Geology, landforms, Caliche Conditions of excavation Waste disposal Materials resource Development Hazardous areas (unstable slopes, flooding) Groundwater (depths, quality)	This is a study of a desert area and is related to the encroachment of residential and commercial development onto the foothill slopes of the range. The report has a strong emphasis on hazard problems. It includes element, derived and potential maps. Although the quantification specification is not strong, the simple methodology is effective.

St Clair County, Illinois (Jacobs, 1971)	County	1:125 000	Geological regions Thickness of superficial deposits Topography of bedrock Surface Resources and minerals (a) limestone (b) sand and gravel (c) clay and shale (d) coal, petroleum water Geological conditions affecting construction: (a) topography (b) surface materials (c) water-table (d) flooding (e) bedrock (f) soil creep Geological conditions affecting solid waste disposal	This is a study of a rural area adjacent to the St Louis metropolitan area, and now being transformed into urban and suburban development. The most important maps are potential maps of geological conditions affecting construction and waste disposal. The first map listed colour codes 10 major regions within the study area. The waste disposal map relates to water movement and risks of pollution through superficials into limestone aquifers.
Allegheny County, Pennsylvania (Briggs, 1977)	County	1:24 000 (Base maps) 1:50 000 and 1:125 000 (Final maps)	Earth disturbance inventory: (a) Land modified by man: Mining Mining related subsidence Mine fires Refuse banks Quarries, pits Slag dumps Urban development Pre-development modification Land modified for transportation Earth and rock fill Near-vertical cuts and cliffs Water impoundments (b) Areas affected by Natural Hazards: Landsliding Flooding Heavy storm precipitation and related mass movement Natural subsidence	This study had an emphasis on hazards. It is highly specialized and deals with the overriding hazards of the region and topics related to them. Mineral resources, urban development engineering and water works are all covered as well as natural hazards. This study again illustrates the way a broad study embraces a number of themes, covering as it does water, mineral resources, construction and hazards.

(*continued over*)

Table X (*continued*)

Location	*Coverage*	*Scale*	*Themes*	*Description*
Ann Arundel County, Maryland (Glaser, 1976) (Doornkamp *et al.*, 1985)	County	1:62 500	Geology Mineral resources/inventory of mined land Geological factors affecting land Modification (combination of environmental hazard variables) (a) Steep slopes (b) Rock ledges (c) Sand and gravel distribution (d) Clay distribution (e) Fine clay/sand/silt distribution (f) Flood plains Shoreline protection Structures/beach Distribution/historical erosion rates	This series of maps (atlas) is designed for land use planning purposes in an area with rather limited constraints. There is an emphasis on coastal processes and superficial geology. The mineral resources to sand and gravel, brick and other clays, diatomite. This includes potential maps. The shoreline aspects emphasize geomorphology and coastal engineering and are quantitatively assessed. The study illustrates how superficial geology and geomorphology assume a much greater proportional significance than in most traditional geological mapping programmes.
Towson Quadrangle, Maryland (Cleaves, Crowley and Kuff, 1974)	Quadrangle	1:24 000	Geology Landforms: (a) Upland (b) Lowland (c) Floodplain (d) Steepness of slope (e) Relief Estimated thickness of overburden Geological factors affecting land modification Mineral resources/mined land inventory	This is a site-specific (small area) equivalent study to that presented in the Atlas of 4 maps referred to above. It is more detailed. The estimated thickness of overburden is a critical parameter both for foundation design and mineral resource evaluation. It is surprising that neither of the two related Maryland studies includes hydrology or hydrogeology (this may be due to the fact that responsibility for water rests with another organization). The omission illustrates well the need for the overriding of both interdisciplinary and departmental boundaries to obtain the maximum potential from EGM studies.
Franconia Area, Fairfax County, Virginia (Froelich *et al.*, 1978)	Urban Community	1:24 000	Topography Landforms Geology Hydrology (including drainage, flood prone areas) Surface materials Subcrop (gravel, sand, clay, bedrock) Slope stability Capability for planned housing development	This is another study of a small subject area. The aims were very limited: to establish the suitability of land for housing development in an area close to the District of Columbia, where there are obvious strong pressures for urban development. The element and derived maps build up into a final computer composite potential map evaluating all the relevant constraint factors and effects on the environment.

Location	*Coverage*	*Scale*	*Themes*	*Description*
San Francisco Bay Area, California, Flatland Areas (Helley and Lajoie, 1979)	Region	1:125 000	Basic geology (Pleistocene and Recent—mainly unconsolidated, superficial deposits) Historic Shorelines of the Bay Water-related problems (a) Flooding (b) Channel change (c) Salt water intrusion (d) Subsidence (e) Settlement (f) Shrink and swell Earthquake problems (a) Response (b) Amplification (c) Liquefaction potential Resource potential (a) Water (b) Soil (c) Minerals: sand and gravel, clay, salines, shale, peat	The studies cited above in this table cover county-sized areas or parts of them. This study covers a much larger subject area including nine counties (San Mateo County cited above is included). These flatlands around the bay have maximum pressures for urban development. This study is aimed at superseding reliance on soil maps (inadequate because of shallow depths treated therein). It will ensure the avoidance of many hazards. It includes potential maps quantifying the suitability of areas for development ('Stop–Caution–Go' maps with a three-fold coding for high moderate, and low development, potential). There are several other companion studies of the Bay Area, each dealing with specific aspects (e.g., seismic safety—Blair *et al.*, 1979; slope stability—Nilsen *et al.*, 1979; and flood prone areas—Waananen *et al.*, 1977)
North Slope Alaska (Summaries by: Radbruch-Hall, 1979; and Brew, 1974) (Additional references: Brew and Gryc, 1974; EIS Taskforce, 1976; Environmental Protection Board of Canada, 1974; Kreig and Reger, 1976; and Moening, 1974)	Immense Area, (the National Petroleum Reserve) Companion Study of a large area of the McKenzie District of Canada	1:100 000	Drift geology Bedrock Mineral resources (coal, metals, construction minerals, vein minerals, industrial minerals) Ecological areas Areas important for arctic foxes, wolf packs, etc.	This was by far the largest area ever covered in an EGM study. It embraced many other disciplines as well as the earth sciences. It was aimed primarily at conservation of wildlife and the natural environment during major mineral resource development. It combines ecological information directly with mineral resources to highlight problem areas. It was prepared as part of the extensive environmental impact programme carried out during the construction of the Trans-Alaska Oil Pipeline.

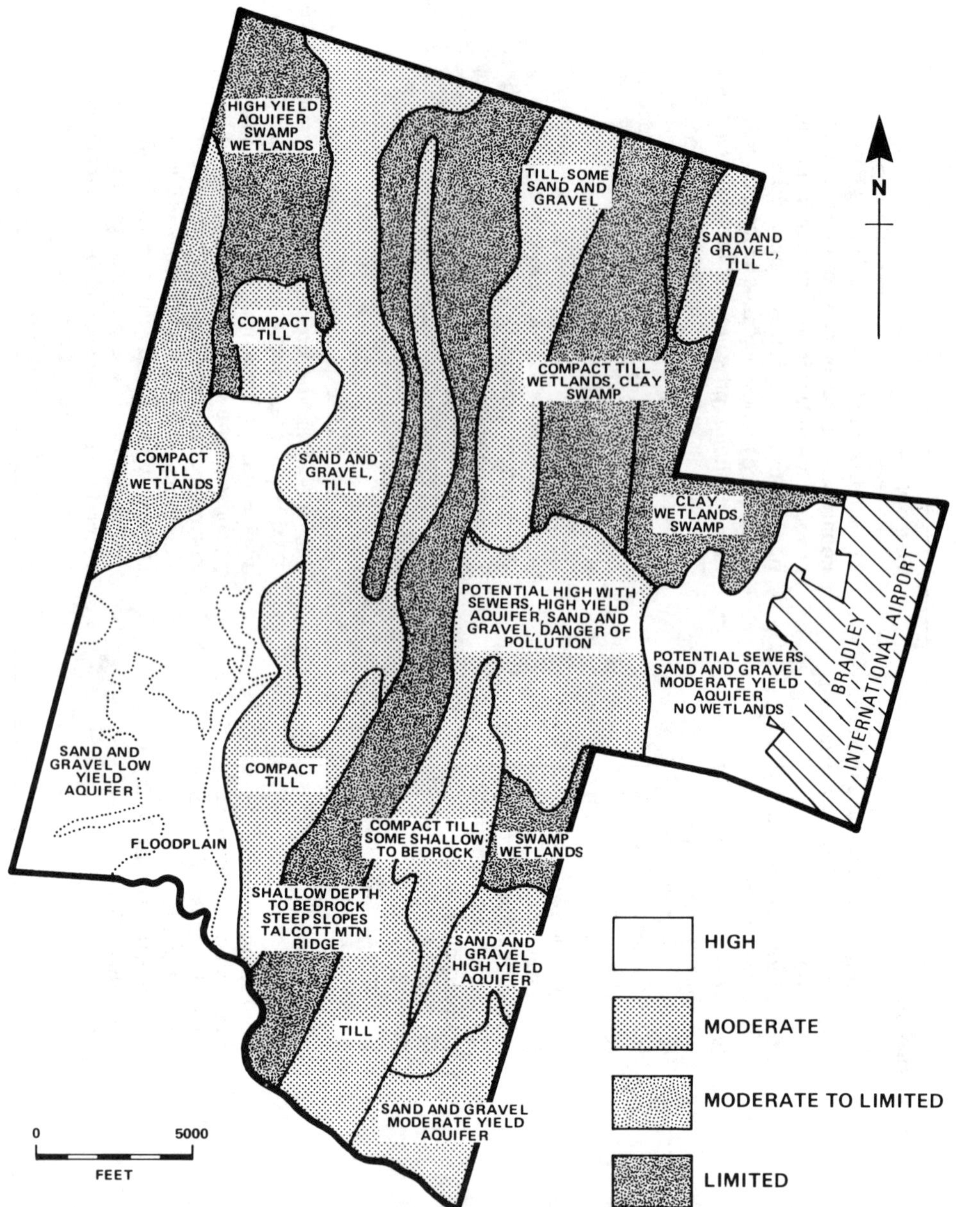

Fig. 8. Map of natural land use intensities for part of the Connecticut Valley, USA (from Crosby *et al.*, 1978). Areas designed as: high intensity have few site problems; moderate intensity are likely to have problems which require careful planning; low or limited intensity have severe problems which require extensive, costly site preparation.

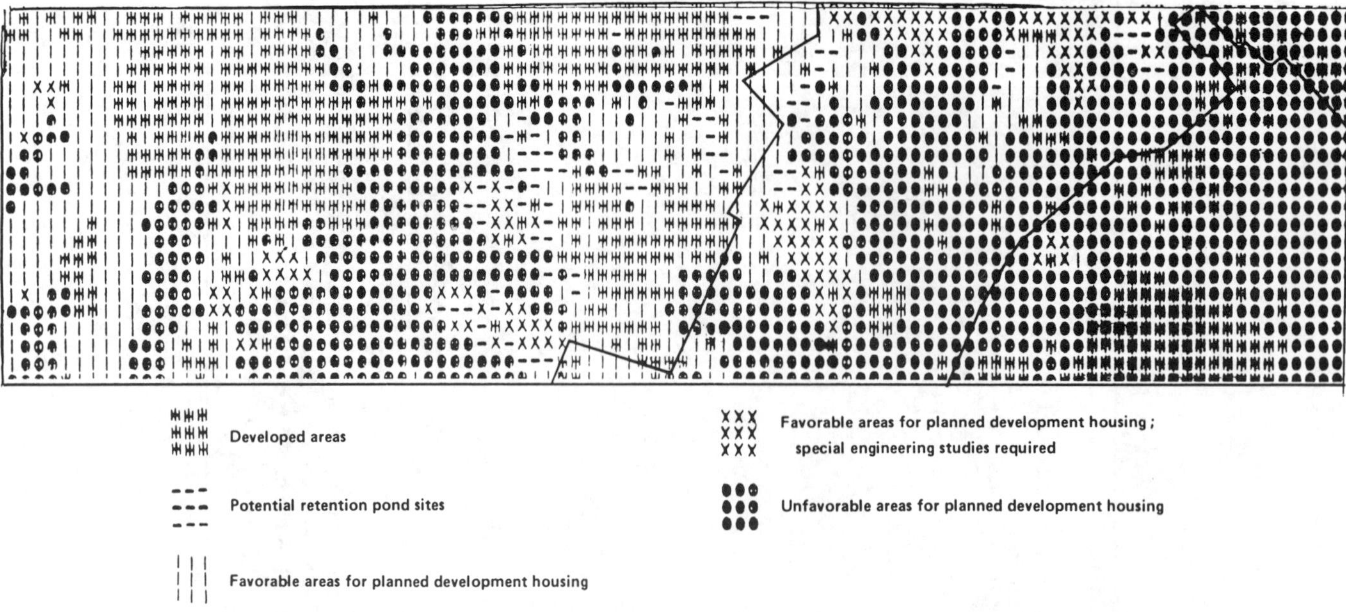

Fig. 9. Composite capability map for planned housing development, Franconia, Virginia (from Froelich *et al.*, 1978).

Table XI Extract from the legend of a map reviewing suitability of land for disposal of wastes, Maricopa County, Arizona (from Christenson *et al.*, 1979)

Unit	*Suitability for septic tanks*	*Suitability for waste stabilization ponds*	*Suitability for sanitary landfills*	*Suitability for use as landfill cover material*
(1)	Good. Permeability sufficient; depth to bedrock great; easily excavated; no caliche problem; flood hazard great; particularly near washes.	Poor. Slopes and relief are sufficiently low, but permeability and flood hazard great; easily excavated but lining of ponds necessary.	Fair to good. Local sources of cover; depth to bedrock great; excavation easy; flood hazard locally severe near streams; permeability sufficiently low.	Good to poor. Good to poor compaction. Impermeable when compacted; difficult to work when wet.
(2)	Good to fair. Permeability sufficient; depth to bedrock great; moderately easy 1.4–2.4 m; difficult below this due to strong caliche; flood hazard moderate.	Poor. Slopes and relief moderate; permeability too great and flood hazard moderate; excavation easy to depths of 1.5–2.4 m.	Fair to poor. Permeability sufficiently low below shallow depths; excavation difficult below shallow depths due to caliche and large boulder size; flood hazard moderate; source of of cover within 3.2 km.	Poor. Impermeable when compacted except compaction fair where gravelly.
(3)	Fair to poor. Permeability locally sufficient but generally impaired by near-surface caliche; bedrock locally near surface; excavation difficult; flood hazard slight.	Fair. Slopes and relief locally prohibitive and excavation difficult, but permeability low due to caliche and flood hazard slight.	Poor. Permeability low but excavation difficult due caliche and large boulder size, flood hazard slight; local near-surface bedrock; source of cover within 8 km.	Poor. Impermeable when compacted but gravelly and too 'calichified', excavation difficult, due to boulder content and caliche binder.
(4)	Poor. Impermeable except in open fractures with piping and plugging possible; no filtering or cleansing of effluent; very difficult to excavate; flood hazard slight.	Poor. Slopes prohibitive, excavation very difficult; impervious except for fractures; no flood hazard.	Extremely poor. Impermeable but piping of leachate possible; very difficult to excavate, slopes prohibitive.	Extremely poor. Very difficult to excavate; crushing necessary.

The study includes a wider range of units and describes these in terms of landforms, geological unit and technical data such as percolation rates, depths to caliche horizons, average slopes and soil types before setting out interpretative assessments of the type shown above.

science maps are trained to read the third dimension into them. Whilst the majority of users will be used to this process in handling, for example, topographical notations on geographical maps, the perception of depth on earth science maps may be elusive. The Franconia study presents block diagrams showing relationships between deposits, but in which various units are emphasized in turn by separating the block at different geological surfaces. (Fig. 10). The use of this device might be helpful in many other studies.

A consolidated list of themes included in the European and USA studies reviewed is given in Table IX.

Africa

Planning related mapping has been undertaken of the federal capital site in Nigeria (Ege *et al.*, 1985). This study embraced bedrock geology, superficial geology, engineering geology, soil properties, bedrock properties, study of structural lineaments, construction, materials, hazards, (landslide, flood, earthquakes) and a final evaluation of the suitability of the area for proposed urbanization.

In Morocco multi-disciplinary studies have led to the preparation of 'habitability maps' of areas such as Fez and Tangier (Hafdi, 1987). Pressure on resources has been increased by rapid population growth; therefore, mapping contributions both to planning for the greater use of land and for economic development of agriculture and mineral deposits. However, the suite of resource maps extends, when appropriate, to tourism and fisheries. Morocco has considerable areas of arid land; thus, considerable emphasis was placed on water resources. Other maps covered ground conditions, notably slope stability and seismic risk. In parallel with this work, further maps were made of coastal installations, transportation, energy supply, population distribution, locations of hospitals and schools, industries and administrative organizations. Groups of these varied maps formed the basis for summary sheets entitled:

(a) aptitude of the area for land planning;
(b) utilization of natural resources;
(c) socio-economic potential.

The physical aspects of areas were classified using a generalized ranking system (Fig. 11).

Middle East

The Ministry of Development and Engineering Services of the Government of Bahrain commissioned a surface

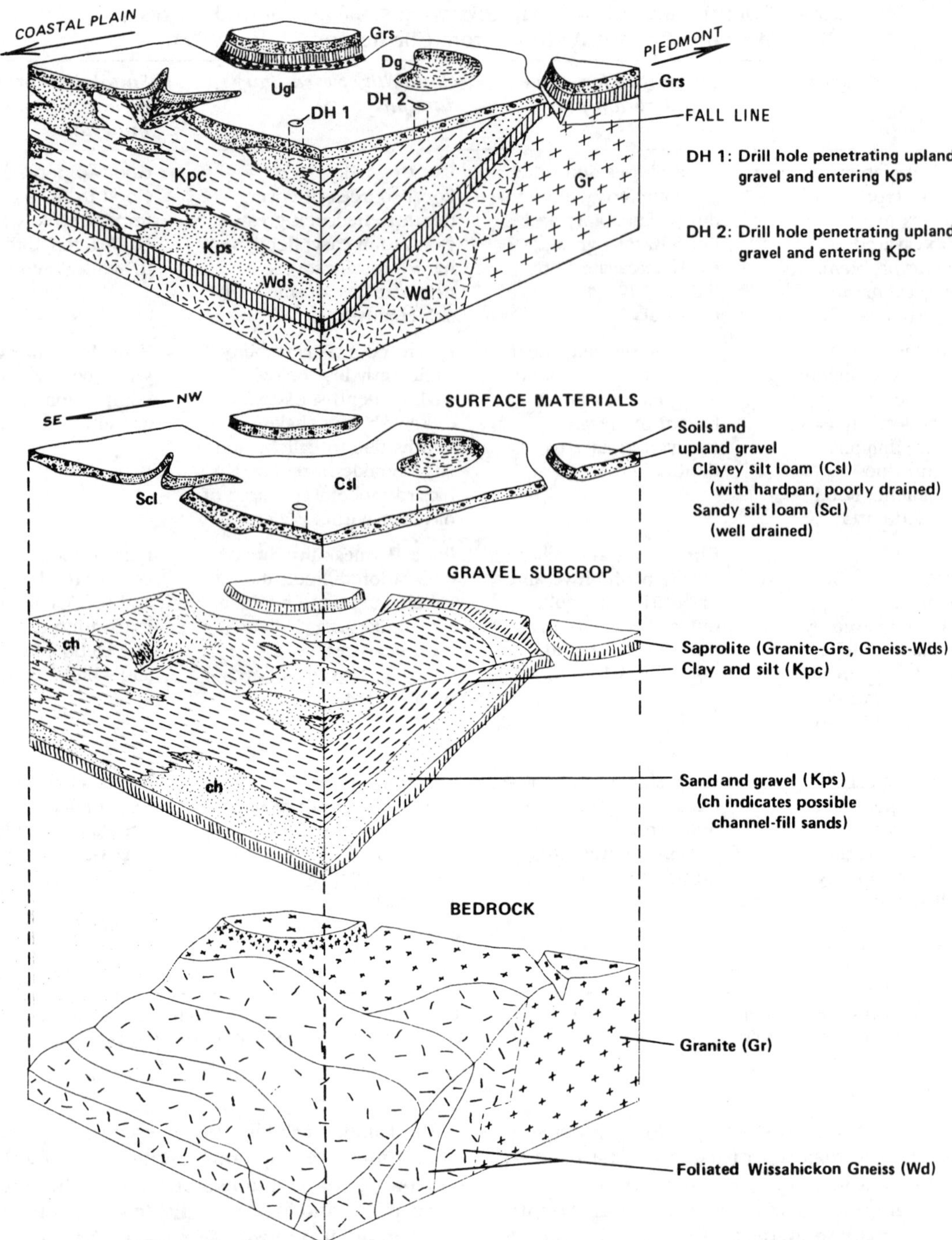

Fig. 10. Block diagram to demonstrate spatial relationship between geological strata in the Franconia area, Virginia (from Froelich *et al.*, 1978). Note: symbols are identified in the labelling except for: Dg—disturbed ground; Kpc—clay and silt; Ugl—upland gravel.

materials resources survey between 1974 and 1976. The study covered the major islands of the group and included studies of the solid geology and geological structure, geomorphology and surface materials, soils, local capability for agriculture and hazards to construction (Brunsden *et al.*, 1979). The set of maps included scales of 1 : 10 000, 1 : 30 000 and 1 : 50 000 depending on the levels of development interest. Amongst the resulting summary maps were sheets defining possible conservation areas such as landforms of regional significance, areas of scientific value and of archeological value. These are set alongside land of good and moderate agricultural potential and potential areas for urban expansion and aggregate mineral extraction (Fig. 12).

Another multi-theme study was that carried out on the Greater Amman Master Plan (McCrae and Al-Ammouri, 1987). The subject area was about 5000 km^2 in extent and within it dwell half Jordan's population and most of the economic activity of the country is located there. Doubling of the population is expected within 20 years. The geological and geotechnical aspects of the master plan for land use, transportation and infrastructure covered:

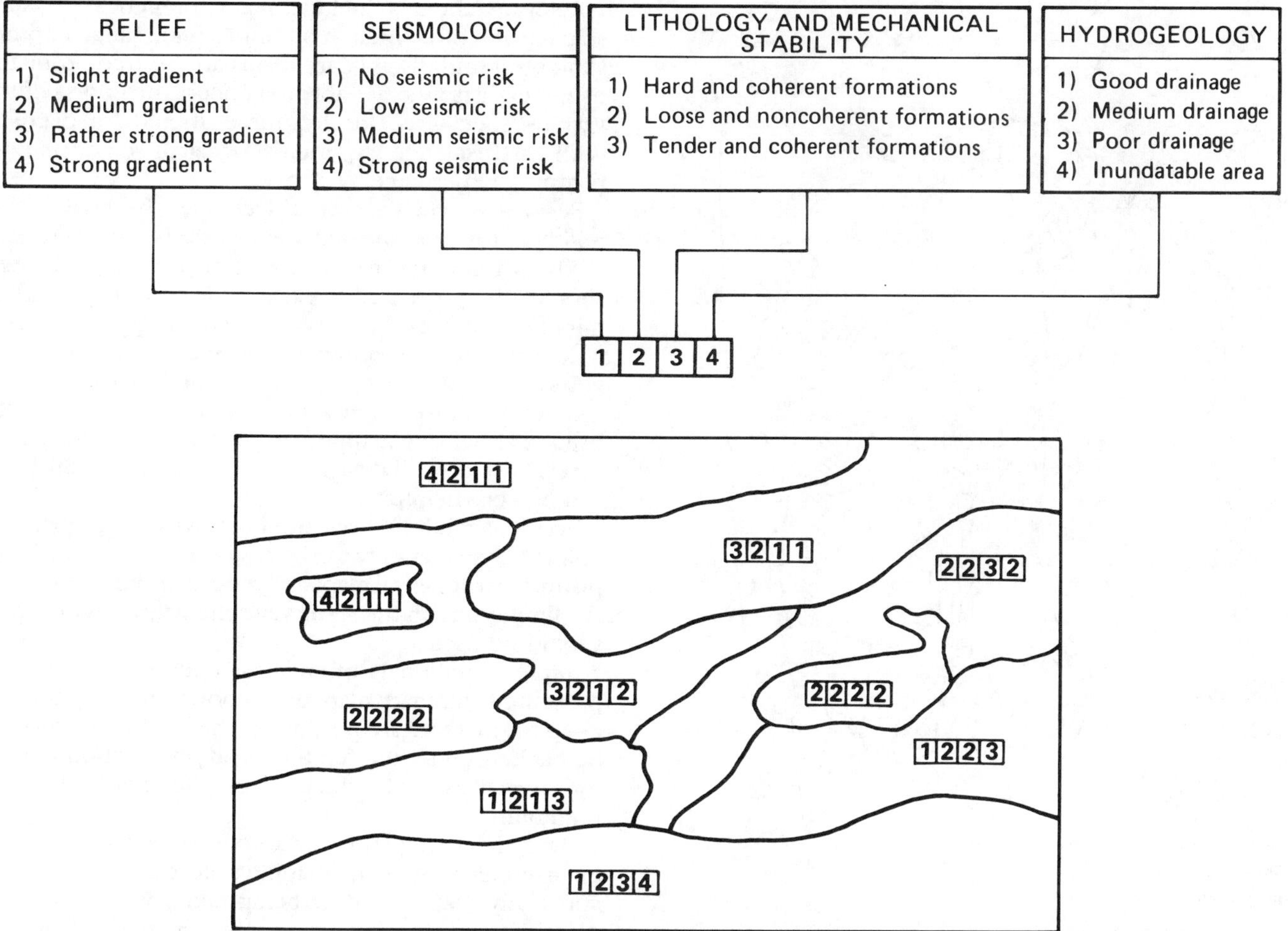

Fig. 11. Method of ranking used in preparing a map of zones based on relief, seismology, lithology and stability, and hydrogeology in various areas of Morocco (from Hafdi, 1987)

- relief;
- geology;
- tectonics;
- geotechnics and planning:
  (a) hydrogeology;
  (b) landslides;
  (c) foundation problems.

A geotechnical data bank was compiled. Major problems are protection of the scarce groundwater resource from damage, sterilization and pollution, and tackling the problem of dangerous man-made cliffs in worked-out quarries. Swelling of clays was another major problem.

Asia

The Geological Survey of India carried out a number of environmental geology mapping exercises aimed primarily at mineral resources and agricultural needs (Raju, 1987). Work on the drought-prone Anantapur area included maps showing geology, geomorphology, hydrogeology, engineering geology and soils. Other sheets depicted an unusual range of mineral resources including gold, diamonds, corundum and steatite, as well as barytes and limestones. Zones of active surface erosion were also defined. Final summary maps gave a land capability classification emphasizing, in particular, areas suitable for irrigated agriculture, dry land farming, pasture, forestry and quarrying. A second study concerned an urban area in Hyderabad which had a comparable scope but also included a geophysical map interpreted to indicate the depth of bedrock below the surface.

CONCLUSIONS

It has been demonstrated that a wide variety of environmental geology mapping exercises have been carried out for a wide range of purposes in many different terrains. The resulting information ranges from highly technical theme maps to very generalized advisory maps. There is a role for both with specialist and generalist audiences respectively.

If studies are to be aimed at land use planning, however, the objectives need to be defined in planning terms, for areas outlined on a planning basis, so that they can be readily implemented in the planning process. Thus, a planner should always be involved in the study from the outset, including the design of the programme, asking the question 'what is and what is not of use to the users in this particular subject area?'. Ideally, planners and earth scientists should work together on studies if the best compromise is to be reached and the significance of the results fully appreciated. However, the balance of work is likely to vary at different stages of the study.

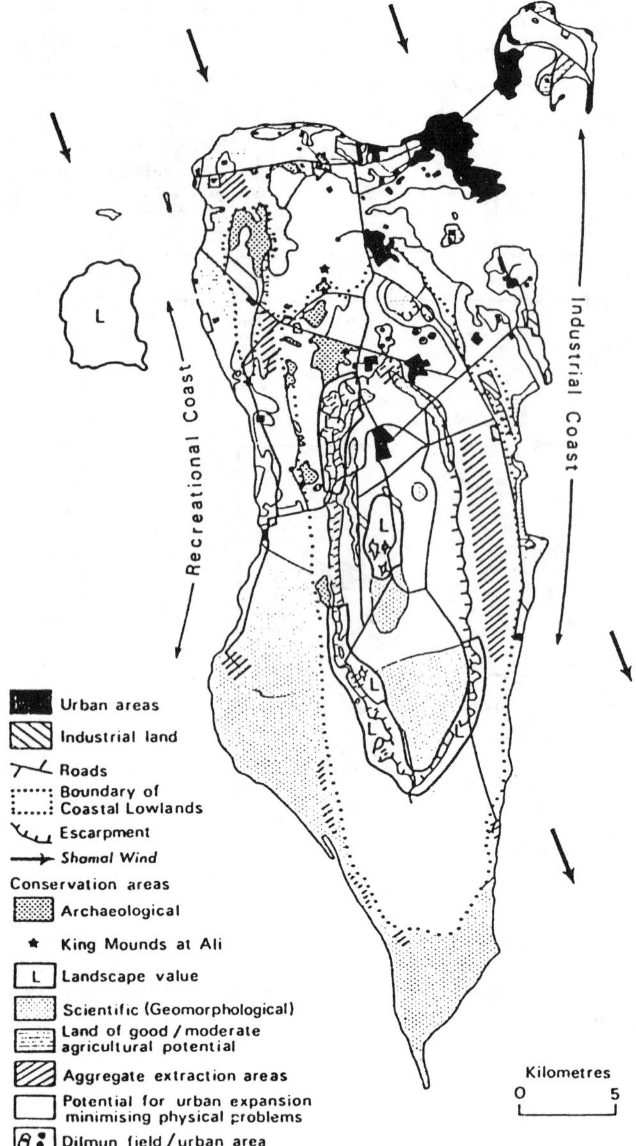

Fig. 12. Regional planning implications for Bahrain Island (from Brunsden *et al.*, 1987).

Assiduous collection of existing data is often an essential aspect of studies and this needs to be carried out vigorously and systematically. Ideally, carefully indexed archives or data bases of information should be built up and maintained for long-term benefit.

Field mapping, ground investigations, sampling and testing of samples may be required for the whole study or to fill in gaps in the pre-existing data. There are obvious economic advantages if field and site work can be carried out after the main collection of existing data. New data should be integrated into the same data base but all information should be carefully labelled to indicate the reliability for use in later stages of the work.

There are obvious cost advantages in carrying out work on several earth science themes simultaneously rather than repeating expensive searches or field work.

Data retrieval and handling is essential to the interpretative phase of the work. In the case of large data sets the use of a computer is cost effective. Computerized data bases are likely to be linked increasingly to digital mapping facilities. These, however, cannot replace interpretation by experts. Interactive mapping, using the machine as an aid, is the best compromise.

Relevant themes include many branches of earth science and often branch out into a wide range of other subjects. Multi-disciplinary teams are best co-ordinated by generalists who are able to concert the efforts of the various specialists. Ideally, these should be planners with earth science experience or earth scientists with planning experience.

Besides succinctness and accuracy, the quality that militates towards the success or otherwise of an EGM study is relevance to the needs of the users. Results should be presented clearly and concisely, with a minimum of jargon, and with essential technical terminology and background concepts well defined (e.g. in a glossary) if the widest possible audience is to be reached.

Given that interested general audiences might not be used to visualizing mapped features in three dimensions, careful use of well designed block and ribbon-diagrams can be very helpful.

Results should be presented in terms of the planning objectives so that they can be readily implemented in the planning process. All maps and reports should make clear the limitations on the results and the ways in which they should be used.

Implementation is often most effective if only well presented summary maps and reports concentrating on essential material are produced. These should, however, be backed up by the full technical information needed by specialists and which gives the study scientific credibility.

Over the last quarter of a century the approach to environmental geology mapping has developed greatly and many maps are now being taken into account in planning and development. However, the response to them has been patchy and all too often their value is not perceived by potential users, or the prepared results fail to meet the users' requirements as well as they might.

Whilst it is fairly straightforward to explain the relevance of studies is subjective terms, relatively little information is available on the real savings which can be made if adequate base information is available. More work and discussion is needed. It is hoped that this book might be a useful contribution to that debate.

ACKNOWLEDGEMENT

Much of the thinking in this chapter comes from the experience of one of the authors (BRM) in specifying and managing applied earth science mapping studies and the other (GJHMcC) in carrying out a study of the Plymouth area, England. However, we are indebted to discussions with other earth scientists too numerous to be mentioned individually and, particularly, to a study carried out by Geomorphological Services Ltd (Doornkamp *et al.*, 1985).

REFERENCES

Anon, (1965). *Engineering Geology maps of Prague, Scale 1 : 5000*.

Arndt, P. and Lüttig, G. (eds.) (1987). *Mineral Resources Extraction, Environmental Protection and Land Use Planning in the Industrial and Developing Countries*. E. Schweitzerbart, Stuttgart.

Arnould, M., Broquet, J.F., Deveughele, M. and Usseglio Polaterra, J.M. (1979). Cartographie géotechnique de la ville de Paris—premières réalisations (13ème, 19ème and 20ème Arrondissements). *Bull. Int. Assoc. Eng. Geol.*, **19**, 109–115.

Association of Bay Area Governments (1976). *Areas of Critical Environmental Concern*. Report to the Regional Planning Committee by the Regional Open Space Task Force. Berkeley, California.

Association of Bay Area Governments (1978). *San Francisco Bay Area, Environmental Management Plan*. Berkeley, California.

Atwater, B. (1978). Central San Mateo County, California. Land use controls arising from erosion sea cliffs, landsliding and fault movement. In *Nature to be Commanded* (Robinson, G.D. and Spieker, A.M., eds.), USGS Prof. Paper. No. 950, 11–20.

Becker-Platen, J.D., Dorn, M. and Look, E-R. (1987). An introduction to the geoscientific map of the Natural Environment's Potential (GMNEP) of Lower Saxony and Bremen. In *Mineral Resources Extraction, Environmental Protection and Land Use Planning* (Arndt, P. and Lüttig, G., eds.), Stuttgart, 119–126.

Blair, M.L. and Spangle, W.E. (1979). *Seismic Safety and Land-use Planning: Selected Examples from California (Basis for Reduction of Earthquake Hazards on the San Francisco Bay Region)*. USGS Prof. Pap. 941-B.

Brew, D.A. (1974). *Environmental Impact Analysis, the Example of the Proposed Trans-Alaska Pipeline*. USGS Circular No. 695, 16 pp.

Brew, D.A. and Gryc, G. (1974). The analysis of the impact of oil and gas pipeline systems on the Alaskan environment. Internat. Congress Foundation Français d'Etudes Nordiques: *Arctic Oil and Gas—Problems and Possibilities*. 5ème, Le Havre, May 2–5, 1973 (Proceedings).

Briggs, R.P. (1977). *Environmental Geology, Allegheny County and Vicinity, Pennsylvania: Description of a Program and its Results*. USGS Circular, No. 747.

Brunsden, D. *et al.* (1979). The Bahrain surface materials resources survey and its application to regional planning. *Geographical Journal*, **145**(1), 1–35.

Cendrero, A. (1975). Environmental geology of the Santander Area, Northern Spain. *Environmental Geology*, **1**, 97–114.

Cendrero, A., Díaz de Teran, J.R. and Saiz de Omenaca, J. (1976). A technique for the definition of environmental geologic units and for evaluating their environmental value. *Landscape Planning*, **3**, 35–66.

Cendrero, A., Nieto, M., Robles, F. and Sanchez, J. (1986). Geoscientific map of the province of Valencia. Diputación Provincial de Valencia, Servicio de Recursos Geológicos.

Chandler, P. and McCall, G.J.H. (1986). Stratigraphy and structure of the Plymouth–Plymstock area, a preliminary revision. *Proc. Ussher Society*, **6**, 253–257.

Clayton, A.R., Cannell, B., Adlam, K.A.M., Loudon, T.V. and Robson, P.G. (1987). *Geological Databank Pilot Study*. British Geol. Surv. Research Report. No. ICSO 87/1.

Cleaves, E.T., Crowley, W.P. and Kuff, K.R. (1974). *Geologic and Environmental Atlas—Towson Quadrangle*. Maryland Geological Survey.

Christenson, G.E., Pewe, T.L. *et al.* (1979). *Environmental Geology of the McDowell Mountains Area, Maricopa County, Arizona*. State of Arizona Bureau of Geology and Mineral Technology, Geologic Investigation Series, Maps G1 series.

Cratchley, C.R. and Denness, B. (1972). Engineering Geology in urban planning with an example from the New Town of Milton Keynes. *Proc. 24th Int. Geol. Congress, Montreal*, **13**, 13–22.

Crosby, E.J., Hansen, W.R. and Pendleton, J.A. (1978). Guiding development of gravel deposits and unstable ground. In *Nature to be Commanded* (Robinson, G.D. and Spieker, A.M., eds.), USGS Prof. Pap. No. 950, 30–41.

Davidson, E.S., Feth, J.H. and Spieker, A.M. (1978). Tucson, Arizona: controlling hillside development; protecting and conserving mineral and water resources. In *Nature to be Commanded* (Robinson, G.D. and Spieker, A.M., eds.), USGS Prof. Pap. No. 950, 30–41.

Dearman, W.R. *et al.* (1973). Techniques of engineering geological mapping with examples from Tyneside. In *Engineering Geology of Reclamation and Development*, Engineering Group of the Geol. Soc. London, Regional Meeting (Durham), (Proceedings), 31–34.

De Beer, E., Fagnoul, A., Lousberg, E., Nuyens, J. and Maertens, J. (1980). A review of engineering geology mapping in Belgium. *Bull. Int. Assoc. Eng. Geol.*, **21**, 91–98.

De Mulder, E.F.J. (1988). Thematic applied Quaternary maps—a profitable investment or expensive wallpaper? In *Applied Quaternary* (De Mulder, E.F.J. and Hageman, B.P., eds.), Balkema, Rotterdam.

Department of the Environment (1988). *Review of the Geological and Mineral Planning Research Programme, 1987*. Department of the Environment (London, England).

Department of the Environment (1989). *Development on Unstable Land*. PPG6, HMSO, London.

Doornkamp, J.D. *et al.* (1985). *The GSL Report on Environmental Geology Mapping*. Geomorphological Services Ltd. (2 vols), Department of the Environment, London, Open File Report.

Dubrovolny, E and Schmoll, H.R. (1968). Geology as applied to urban planning: an example from the Greater Anchorage Area Borough, Alaska. *Proc. 23 International Geological Congress*, **12**, 39–56.

Echevarria, C.M.R. (1967). Geotécnica y mapas geotécnicos. *Not. Y Com. del Instituto Geológico y Minero de España*, 97–98, Madrid.

Ege, J.R., Griffitts, W.R. and Overstreet, W.C. (1985). Engineering geology site appraisal for the federal capital of Nigeria. *Bull. Int. Assoc. Eng. Geol.*, **31**, 71–79.

EIS Taskforce (1976). *Alaska Natural Gas Transportation System, Final Environmental Impact Statement*. US Dept of the Interior.

Environmental Analysis Group, University of Santander (1980). Environmental survey along the Santander–Unquera coastal strip, northern Spain—its capacity for development. *Landscape Planning*, **7**, 23–56.

Environmental Protection Board of Canada (1974). Environmental Impact Assessment of the portion of the Mckenzie Gas Pipeline from Alaska to Alberta, *Environmental Atlas, Volume 3*.

Finne, T.E., Glattre, E., Langmark, F. and Oleson, O. (1987). A Norwegian Cancer Atlas. In *Geology for Environmental Planning*. (Wolff, F.C., ed.), Norges Geologiske Undersøkelse, Trondheim, 119.

Floyd, J.D. *et al.* (1982). *Environmental geology mapping, South East Edinburgh*. British Geological Survey, Edinburgh, Open File Report.

Forster, A., Hobbs, P.R.N., Wyatt, R.J. and Entwisle, D.C. (1986). *An environmental geology study of parts of West Wiltshire and Southeast Avon*. British Geological Survey, Keyworth, Open File Report.

Forster, A., Hobbs, P.R.N., Wyatt, R.J. and Entwisle, D.C. (1987). Environmental geology maps of Bath and the surrounding area for engineers and planners. In *Planning and Engineering Geology* (Culshaw, M.G. *et al.*, eds.), Geol. Soc. London, Eng. Geol. Spec. Publ. No. 4, 221–235.

Freeman, Fox Ltd. (1987). *Investigation of ground characteristics in the area around Chacewater and St Day in the county of Cornwall* (5 vols). Dept. of the Environment, London, Open File Report.

Freeman, Fox Ltd. (1988). *Methods of compilation, storage and retrieval of data on disused mine openings*. HMSO, London.

Froelich, A.J., Garnaas, A.D. and Van Driel, J.N. (1978). Franconia Area, Fairfax County, Virginia, planning a new community in an urban setting—Lehigh. In *Nature to be Commanded* (Robinson, G.D. and Spieker, A.M., eds.), USGS Prof. Pap. No. 950, 69–89.

Geomorphological Services Ltd. (1988). *Earth Science Mapping for Planning and Development, Torbay*. Department of the Environment Open File Report.

Gilsanz, J. de P., Carvillo, J. de D.C., Ruíz, L.I.O., Alonso, S.G., Deltell, E.A. Bombín, R.E. and Martínez, C.P. (1985). *Mapa fisiográfico de Madrid. Escale 1 : 200 000*. Communidad de Madrid. Consejería de Agricultura y Granadería.

Glaser, J.D. (1976). *Ann Arundel County: Geology, Mineral*

Resources, Land Modifications and Structural Conditions. Maryland Geological Survey, County Atlas No. 1.

Glattre, E., Finne, T.E., Olesen, O. and Langmark, F. (1985). *Atlas over Kreftinsidens i Norge 1970–1979 (Atlas of Cancer Incidence in Norway, 1970–1979).* Landsforeningen mot Kreft, Oslo.

Gonsález de Vallejo (1977). Engineering Geology for Urban Planning and Development with an example from Tenerife, Canary Islands. *Bull. Int. Assoc. Eng. Geol.* **15**, 37–43.

Gostelow, T.P. and Browne, M.A.E. (1986). Engineering Geology of the Upper Forth Estuary. *British Geological Survey Report,* **16**, No. 8, HMSO, London.

Hafdi, A. (1987). Approach of a methodology for drawing up a habitability map. In *Mineral Resources Extraction, Environmental Protection and Land-use Planning in the Industrial and Developing Countries* (Arndt, P. and Lüttig, G., eds.), E. Schweitzerbart, Stuttgart.

Hawkins, A.B. and Tomlinson, D. (1987). Planning and construction of an industrial site in the Bristol Coalfield. In *Planning and Engineering Geology* (Culshaw, M.G. *et al.*, eds.), Geol. Soc. London, Eng. Geol. Spec. Publ. No. 4, 439–446.

Helley, E.S., Lajoie, K.R. *et al.* (1979). *Flatland Deposits: their Geology and Engineering Properties and their Importance to Comprehensive Planning.* USGS Prof. Pap. 943.

Jacobs, A.M. (1971). Geology and planning in St. Clair County, Illinois. *Illinois State Geological Survey Circular No. 165.*

Kreig, R.A. and Reger, R.D. (1976). Preconstruction terrain evaluation for the Trans-Alaska pipeline project. In *Geomorphology and Engineering* (Coates, D.R., ed.), Dowden, Hutchinson and Ross, Stroudsburg, Pennsylvania, 55–76.

Langer, W.H. and Johnson, L.H. (1978). East Granby: a plan for development of a rural community. In *Nature to be Commanded* (Robinson, G.D. and Spieker, A.M., eds.), USGS Prof. Pap. No. 950, 47–51.

Lozinska-Stepien, H. (1979). Engineering geology maps at a scale of 1 : 25 000 for regional planning purposes. *Bull. Int. Assoc. Eng. Geol.*, **19**, 69–72.

Lüttig, G. (1987). The subcommission of maps on environmental geology (SEMEG) of the IUGS Commission for the International Geological Map of the World and its work in the industrialised and developing countries. In *Mineral Resources Extraction, Environmental Protection and Land Use Planning in the Industrial and Developing Countries* (Arndt, P. and Lüttig, G., eds.), E. Schweitzerbart, Stuttgart, 5–6.

McCall, G.J.H., Chandler, P., Crimes, T.P., Coote, V.J. and Roxburgh, I.S. (1985). *Environmental Geology Study of the Area around Plymouth and Plymstock.* Department of the Environment, London, Open File Report, 2 vols and maps.

McCrae, R.W. and Al-Ammouri, A.F. (1987). Engineering Geology input to the Greater Amman master plan. In *Planning and Engineering Geology* (Culshaw, M.G. *et al.* eds.), Geol. Soc. London, Eng. Geol. Spec. Publ. No. 4, 179–184.

McCullough, D.G. (1968). *The Johnstown Flood.* Hutchinson, London.

Matula, M. (1965). Regional engineering geological characteristics of the Zvolen Basin. *Acta Geol. et Geograph., UC, Geologica,* **10**, Bratislava.

Matula, M. (1969). Regional engineering geology of the Czechoslovak Carpathians. *Slovenskej Akadamievied,* Bratislava, 225.

Matula, M. and Letko, V. (1980). Engineering geology in planning the Metropolitan Region of Bratislava. *Bull. Int. Assoc. Eng. Geol.*, **22**, 139–145.

Mazeas, H. and du Mouza, J. (1979). Cartographie géotechnique des formations superficièlles et zones non-urbanisées. *Bull. Int. Assoc. Eng. Geol.*, **19**, 47–52.

Melnikov, E.S. (1979). The main principles of procedures for the natural engineering geological of the USSR. *Bull. Int. Assoc. Eng. Geol.*, **19**, 93–95.

Melvin, R.L. (1978). Use of drainage area maps in evaluating waste disposal conditions, assessing the impact of highway salting, and designing of bridges and culverts. In *Nature to be Commanded* (Robinson, G.D. and Spieker, A.M., eds.), USGS Prof. Pap. No. 950, 52–54.

Merla, A., Merlo, G. and Olivieri, F. (1976). Detailed engineering geology mapping in selected Italian mountain areas: methodology and examples. *Bull. Int. Assoc. Eng. Geol.*, **14**, 129–135.

Moening, H.J. (1974). An overview of the Trans-Alaska Pipeline: engineering and environmental geology problems. In *Geology, Seismicity and Environmental Impact* (Moran D.E. *et al.*, eds.), Assoc. Eng. Geol. Spec. Publ., University Publishers, Los Angeles, 405–409.

Moldan, B., Cicha, I., Bednar, J., Cadek, J., Jetel, J., Kovanda, J., Opletal, M., Pokorny, J., Tomasedk, M., Vesely, J., Volsan, V., Zeman, M. and Zikmund, J. (1987). 1 : 50 000 maps of the natural environmental potential of the CSSR. In *Mineral Resources Extraction, Environmental Protection and Land Use Planning in the Industrial and Developing Countries* (Arndt, P. and Lüttig, G., eds.), E. Schweitzerbart, Stuttgart, 183–196.

Nickless, E.F.P. *et al.* (1982). *Environmental Geology of the Glenrothes District, Fife Region.* Description of 1 : 25 000 sheet NO 20, Report of the Institute of Geological Sciences. HMSO, London.

Nieto Salvatierra, M. and Obarti Segrera, J. (1988). *Environmental Cartography and Management of Natural Resources.* Service of Geological Resources, Provincial Diputation of Valencia.

Nilsen, T.H., Wright, R.H., Vlasic, T.C. and Spangle, W.E. (1979). *Relative Slope Stability and Land Use Planning in the San Francisco Bay Region, California.* USGS Prof. Pap. No 944.

Pasek, J. and Rybar, J. (1961). An illustration of engineering-geological conditions in the map 1 : 25 000. *Geol. Exploration,* **3**, 45–47, Prague.

Radbruch-Hall, D.H. (1979). Environmental aspects of engineering geological mapping in the United States. *Bull. Int. Assoc. Eng. Geol.*, **19**, 351÷358.

Raju, K.C.C. (1987). Geoscientific maps in land-use planning in India. In *Mineral Resources Extraction, Environmental Protection and Land Use Planning in Developing and Industrial Countries.* (Arndt, P. and Lüttig, G., eds.), E. Schweitzerbart, Stuttgart.

Rijks Geologische Dienst and Stichtung voor Bodemkartering/Dienst Grondwaterverkenning TNO, (undated). *Subsoil Uncovered.*

Robinson, G.D. and Spieker, A.M. (eds.) (1978). *Nature to be Commanded.* USGS Prof. Pap. No. 950, US Government Printing Office, (Washington, D.C.).

Ronai, A. (1979). Fundamentals of engineering geology maps. *Bull. Int. Assoc. Eng. Geol.*, **19**, 62–68.

Tönnies, G. (1987). Spatial arrangement and regional planning in the Federal German Republic. In *Mineral Resources Extraction, Environmental Protection and Land Use Planning in Industrial and Developing Countries* (Arndt, P. and Lüttig, G., eds.), E. Schweitzerbart, Stuttgart, pp. 73–82.

Viberg, L. and Adestam, L. (1980). Geotechnical terrain classification for physical planning—a Swedish research project. *Bull. Int. Assoc. Eng. Geol.*, **21**, 174–178.

Waananen, A.O., Limerinos, J.T., Kockelman, W.J., Spangle, W.E. and Blair, M.L. (1977). *Flood-prone Areas and Land Use Planning: Selected Examples from the San Francisco Bay Region, California.* USGS Prof. Pap. No. 942.

Wille, V. (1987). Example of maps in the planning process. In *Mineral Resources Extraction, Environmental Protection and Land Use Planning in Industrial and Developing Countries.* (Arndt, P. and Lüttig, G., eds.), E. Schweitzerbart, Stuttgart, 83–97.

Williams, I. (1983). Geological Reports for DOE: *Land Use Planning: the South Humberside Project.* Department of the Environment. London. Open File Reports.

Wilson, D. and Smith, M. (1985). *Thematic Geology Maps: Bridgend Area.* British Geological Survey Open File Report.

Chapter 10

Potential and Future Prospects–Earth Science Mapping for Rational Management of Natural Resources and the Environment

G. Lüttig

SITUATION AND PROBLEMS

In order to outline the potential and future prospects of earth science mapping for the rational management of natural resources and the environment discussed in the previous chapters, we must summarize the present situation and the problems apparent in this field. At the present time, there is wide realization of both the possibilities and the limitations of civilization, and an understanding that a situation has now arisen which requires not only regional and national management, but also international management. This situation is partly the result of limitations and regional disparities in the potential of the natural environment, for example in regional lack of raw materials, crop yield potential, surface and groundwater, food, as well as mineral and non-mineral energy resources. This wider awareness might also lead the way to increased exploitation of advantageous ground conditions that can be utilized by communities, for example as waste disposal sites.

Problems have also resulted from the growing need to extract resources from the natural environment, the increasing impact of such extraction, and the effects of further industrialization on the environment—seen for example in the damage to flora and fauna and the increasing problems of waste disposal and management (despite the limited amelioration potential of new recycling techniques). The cycle of interdependence in which we find ourselves is illustrated by the triangle shown in Figure 1.

Land-use planning is a tool for resolving these issues. It is capable of producing sensible measures on the basis of scientifically valid surveys and descriptions of the geo-environment, to be implemented by the relevant decision-making bodies. Such planning should incorporate cautious, technically sound, sensible and fair regulation of the natural environment, to support the continuation of human civilizaton.

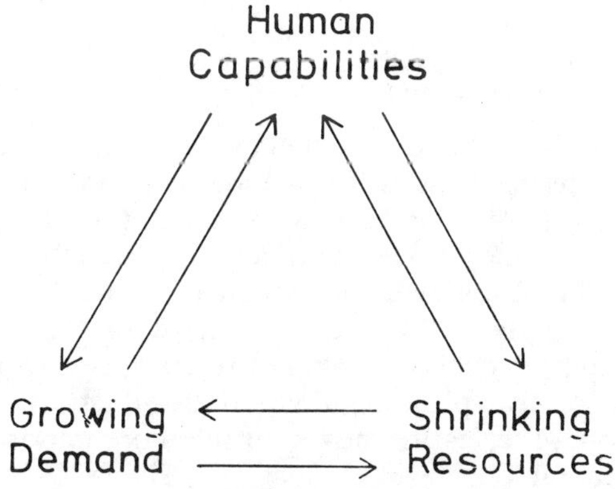

Fig. 1. The global socio-economic problem.

Since the conditions and processes of the natural environment (the 'geosphere') are subject to the laws of structure and dynamics of the 'lithosphere', land use planning must pay close attention to *geology*, and this requires constant collaboration between professional planners and geologists (in the widest sense of the term geology: i.e. including economic geology, mineralogy, geophysics, geochemistry, mining and engineering geology, soil science, hydrogeology, geomorphology, ocean-

Earth Science Mapping for Planning, Development and Conservation

ography and geography). Thus planning must be 'geo-based'. This newly-coined term denotes action on the basis of knowledge of geology *sensu lato*, and especially on the basis of earth science maps. Of course, similar involvement of other branches of natural science is equally essential to the land-use planning process,—for example, economics, medicine and pharmacology, agriculture and forestry, traffic communications, demography and social science. Despite this, on account of the close dependency of economic possibilities on raw materials resources, energy carriers, water supply, ground conditions and other geological parameters, the interaction between land-use planning and geology must, without doubt, be recognized as the essential primary step in evolving planning regulation now: and with even more urgency in the future as land becomes more scarce, populations increase and resources become more difficult to secure.

EXPECTED REQUIREMENTS OF GEOSPHERE MANAGEMENT

Land-use planning involving management of the environment takes place on several administrative levels and falls within several different categories. The regulating processes at each level are constrained by the limitations of technical procedure at that particular level. In most countries there is a hierarchy of planning levels encompassing increasingly large administrative areas. These include:

- community planning;
- district planning;
- regional planning;
- provincial planning;
- national planning.

Not all these categories exist in every country, and there may be varying degrees of legal force at the various levels.

The spatial detail and precision of definition of planning policies and allocations is normally related to the surface area of the planning division, being most precise for the smallest divisions. The scale of maps that should be produced for decision-making therefore relates to the extent of the planning division listed above. Some appropriate scales are shown for the more important of these in Table I.

Planning invariably involves effects on individual and communal rights. Classic examples of this are those mining laws which define possession of resources (landowners' rights, state rights or state ownership of underground resources), as in case of Tuscon, Arizona cited in Chapter 5. Another common case involves the compulsory purchase of private property for transportation routes, for example, the channel tunnel rail proposal in southeast England, where many householders could be affected adversely. The laws on which land-use planning is based are designed to resolve such conflicts.

In the future, especially in densely populated industrial regions, planners will be confronted by issues arising from the changing habits of civilization, and will have to approach their task in a decisive manner, solving very difficult problems. Some examples of this are:

- Increased traffic, especially motor vehicles, but also other means of transportation, changes in traffic patterns caused by changes in work and leisure activities.
- Increased leisure time and recreational needs, demands for parks, sports and recreational facilities near residential areas.
- Changes in the educational system, with greater need for higher educational facilities, and corresponding greater need for space.
- Expansion of residential and industrial areas, and road use, thus sealing off the ground surface and diminishing groundwater regeneration.
- Increased ground subsidence caused by groundwater abstraction in some areas.
- Considerable increase in *per capita* amount of solid and liquid wastes of changing composition; the need to dispose of radioactive wastes.
- Considerable increase of solid, liquid and gaseous emissions, so increasing health hazards and altering health patterns.
- Increased environmental awareness, including an increase in the need to protect nature and landscape conservation areas, sites of scientific, historic or archaeological interest, in competition with other claims for land-use (mining, agriculture, groundwater, surface housing or industrial development). Awareness of the fragility of the global biosphere and its *delicate* interdependencies.
- Particularly acute problems in the developing countries with the explosive growth of big cities. These require a major contribution to the land-use planning process by earth scientists, and this contribution is likely to increase in the future.
- The postulated rise in sea level related to the present global climatic heating up is likely to present major problems, particularly as many major centres of population are situated at or near sea level.

One could expand this list of examples at will.

Unless technology finds new answers to the means of energy production, the 21st Century will see a scarcity of fossil fuels and a shortage of energy sources in some parts

Table I Main hierarchical levels of planning and appropriate scales of mapping for land-use planning purposes

Planning Level	*Description*	*Scale*
Community planning	area utilisation plan building plan	1:1 000 1:2 500 1:5 000
Regional planning	regional land-use plan	1:25 000 1:50 000 1:100 000
Provincial planning	province land-use programme	1:200 000 1:500 000

of the globe. This must effect a radical change in the patterns of civilized activity and so of land-use planning, and may even override many of the pressures which are now regarded as key issues.

The critical point is that land-use planning has and will be forced to react decisively to those socio-economic changes. This will require adapting earth science research and development of new systems of earth science mapping related to conservation, planning and development to meet changing needs.

THE POTENTIAL OF THE EARTH SCIENCES INFORMATION

Numerical recording of facts, combined with the advantages of computerization, has sometimes produced an obsession with 'statistics' as a basis for the solution of problems. However, there is a tendency for this attitude to obscure the vision of the land-use planner and may lead to overlooking of the importance of the *map* (the author once summed this up by the remark that 'planning does not come from plannification!'—i.e. from maps).

Disregarding the conditions of the natural environment may lead to unresolvable conflicts and wrong decisions. The environment is ordered by Nature, after all, and the materials and processes, in particular, are bound to particular locations, because of the genetic processes of geology. We must consider the geological constraints as fixed obstacles which cannot be shifted but must be circumvented where necessary. For example, the dairy-owner who wishes to drill a water well in impermeable granite is ignoring the immovable geological constraints, just as a highway engineer who wishes to build a main road over a system of dolines (an area of dangerous ground subject to collapse) because he only has a limited budget. A third example relates to a case of a satellite town, where a land use planner selected a swamp site, without considering the possibility of subsidence, breaks in pipelines and cables, and risks of harmful effects on the microclimate, but was influenced solely by the cheapness of the land.

Equally, *earth resources* are fixed in the ground and are therefore vulnerable to the adverse effects of many other types of surface development. It might be folly to shut off the potential supplies for future generations.

In the case of a satellite town, the subsoil consisted of a 2-m layer of loess, with a high quality (Chernozem) soil and high reliable crop yield. This was underlain by river gravel suitable for use as construction aggregate in concrete production, and also suitable for foundations because of its load-bearing capacity. The gravel was also almost fully charged with chemically and biologically pure groundwater, and rested above several hundred metres of shale forming an aquiclude.

This locality thus provided excellent potential for several land-use interests:

- municipal engineers and architects (because of the good bearing capacity);
- gravel producers (because of the gravel quality and proximity to a large market area);
- farmers and market gardeners (because of high soil fertility of the overlying stratum and nearness to markets);
- hydrogeologists and water undertaking managers (because of the amount and quality of the groundwater resource present in the gravel aquifer).

This case provides an excellent illustration of the potential for the use of high quality earth science information, in particular maps, to assist in resolving conflicting land-use claims. The planning system would have to reject all but one claim for use of the land. It is the classic dilemma of overlapping claims for land utilization. In practice, the usual decision is that building has priority: other interested parties should look elsewhere for their resource. But is this decision justified? The future inhabitants of the satellite town might later on find themselves in need of soil, of gravel, of water. And this brings us back to the point that the geopotential is in a fixed location and cannot be moved around. So the question 'Where can the other resources be found' should be investigated *before the decision is made*, based on the facts supplied by the earth scientist and other relevant experts. Substitute resources may be a long distance away or not available at all.

THE TECHNICAL STEPS NECESSARY

As seen in preceding chapters, several earth science disciplines and subdisciplines contribute to the information useful for planning, and maps, in particular, are an important way of communicating their results.

However, a major problem is the fact that classical geological reports and maps are difficult for experts in other disciplines and the lay person to understand, especially if they have not used such material previously. It is worth remembering that geological maps are primarily prepared for geologists themselves. If others wish to obtain practical guidance from such maps on specific questions, then the geologist must be asked for a specific explanation. The same is true of soil science maps. The planner or the lay person is generally not in a position to see the relevance to planning and development of the information set before him. It is also worth bearing in mind that even a geologist may experience similar difficulty when faced by geological maps of an unfamiliar region. The geologist may be unable to give practical or technical advice on the basis of such maps—a certain rock unit may be indicated in the legend by the index fossil, the lithology and the approximate petrography, but information on soil mechanics and technical properties, of critical interest to the enquirer, are likely to be lacking.

To overcome this problem, the following steps should be taken. There should first be an improvement of the technical information content of conventional geological maps, such as increased and more complete detail on the drill holes upon which a map is based and improvements in the method of representation. Such basic improvement in information (map level 1 of Becker-Platen, Lüttig and Meine, 1979) should be accompanied by a second step, the production of interpreted and evaluated maps which show only those elements relevant to land-use planning (map level 2 of the same authors).

Mineral Resources

This was one of the first branches of geology to contribute to the development of the methods of applied geology mapping for land-use planning. While mineral resource maps existed in principle for many years, they did not constitute an adequate basis for either land-use planning or industry, although being mainly aimed at the latter. Since mineral resources at present raise great problems with regard to other land utilization claims on resource areas, it is important to communicate results adequately for planning.

- Not only geological, but also technical/economic parameters may be included in the definition of a mineral resource. A limestone bed may be shown as if it is an economic deposit throughout its outcrop length, but this may be not strictly true, for such a bed is not usually of mineable quality throughout its outcrop.
- The technical and, more specifically, the economic parameters that control potential for mining, change with time. Thus the definition of a deposit in terms of viability, or resources and reserves is subject to constant variations from year to year. Mineral resource maps require constant revision and updating.
- The value of a deposit map may depend very much on the scale used. If it is too small, the map is too imprecise for industry. If it is too big, there is a danger of incorporating unavoidable inaccuracies, because the observation points are often too far apart to determine geological boundaries accurately, and the resulting errors in interpretation may be economically misleading.

Therefore, another approach must be followed for land-use planning purposes. Since the detailed assessment of the deposit itself is not the aim of land-use planning (although it is the primary concern for industry), it is sufficient to have a relatively rough cartographical identification of the same subject on a relatively small scale (e.g. 1:50 000–1:200 000). Detailed differentiation is not necessary on the corresponding maps. The resource preservation concept can be further developed on larger scale maps by making three sub-divisions within the areas of economic resource potential:

- the main areas provided with reserves that have been well investigated and can be mined in the very near future;
- the important, well-defined resource areas which may be available for mining in the medium term;
- the priority areas to be reserved for the more distant future.

Such maps are of much more use to land-use planners than the older type of economic geology maps, but need to be re-examined and updated every 10–20 years.

Water

Numerous attempts to produce maps relevant to land-use planning are known in hydrogeology, but some of them lack unification, and are designed for different purposes. The ideal map to support land-use planning should allow the groundwater balance to be determined, and should go beyond simply describing groundwater landscapes: for example, the *International Hydrogeological Map of Europe*, 1:1.5 million (Karrenberg, 1972). The groundwater atlas of the EEC meets this requirement more closely, but unfortunately its representation in a print-out form limits interpretation of the information.

The land-use planner may wish to know the locations of:

- existing wells;
- groundwater extraction sites;
- protection areas;
- zones of natural or man-made contamination;
- regions in which reserves of groundwater are still present and can be used in the future;
- regions in which groundwater has been already depleted.

Because of the need to identify conflict areas, information is also needed on:

- areas where there is a lack of strata capable of filtering out contaminants overlying aquifers; and
- locations of past, existing and proposed waste disposal areas.

Other more far-reaching earth science factors such as fracturing patterns providing aquifers, the chemistry of the water in different aquifers, or the type of aquifer may be of only secondary interest to the land-use planner.

Contamination

Information on natural or man-made contamination is also useful, for example geological horizons or soils with high natural heavy metal contents (above ore veins, old mining heaps, contaminated surface waters, sediments and soils). Other examples are radon and other radioactive concentrations from natural rocks and methane emissions from old waste dumps. There is a whole range of contaminated industrial sites which must be avoided. Geochemistry is thus of particular value in defining areas to be avoided in land-use planning because of risk to population.

Soils

Soil science pioneered in the preparation of applied large-scale maps, but there is a tendency to use soil type terminology which is very academic and of little practical use to the user. Farmers, foresters, gardeners, vintners and fruit growers require information from maps about depth of soil, possible calcareous layers in the subsoil, 'poisonous' horizons in the soil (e.g. the 'maibolt', a clay soil rich in sulphur). They prefer maps that provide technical instructions, rather than classificatory maps. Maps on the need for soil improvement, water balance in the unsaturated soil zone, accessibility during the spring season, erosion danger, and agricultural yield potential, for example, have been widely accepted by land-use planners. The concept of geo-based land-use planning has received considerable promotion through the activities of soil scientists.

Geomorphology

Relatively simple geomorphological maps are very relevant to land-use planning, for example maps of relief expressed by the slope angles (which, in fact, set limits on agricultural cultivatability of the soil). Such maps can be of great help in deciding which type of land-use can be carried out on an area, for example:

- agriculture on slopes of less than 15% inclination;
- orchards on slopes with 15–30% inclination;
- vineyards on slopes with 15–30% inclination; and
- forestry on slopes with more than 30% inclination.

Slope angles also set limitations on, and increase costs of, construction and building.

Hazards

Maps with information on natural hazards are very important to the land-use planner, because of the risks to people, livestock and property. Important components of such maps should be active or latent tectonic disturbance zones, landslides and rock-falls, avalanches, hillside creep, and regions with a very strong valley thrust. Flood-prone areas, as well as horizons susceptible to compression and natural (subsurface erosional) or man-made (mining-related) subsidence may be usefully identified. Earthquake regions and regions endangered by volcanic hazards, tsunamis, and in coastal areas by high flood-water and storm tides, can also be identified. Distribution of permafrost (Chapter 7) is another very important problem in the sub-Arctic regions of the world.

Engineering Geology

Misgivings have sometimes been expressed about engineering geology maps which provide information concerning foundation parameters (for example, the load-bearing capacity of certain rock units). Although it is true that the ultimate decision on what construction can take place on favourable building ground and the most appropriate design will rest with the engineer, on the basis of a very detailed and exact site investigation, such thematic maps do serve a useful purpose in providing the land-use planner with a generalized picture of the distribution of good and poor foundation conditions and their quality, as well as the likely constraints on building. They also indicate the distribution of suitable rock for aggregate use and land-fill purposes in engineering. As noted in Chapter 9, all thematic maps relating to engineering geology, should clearly state that they are in no way a substitute for site investigations.

Nature Conservation

Nature conservation maps are available in an improved form today. Biotope maps and forest function maps prepared by forestry administrations in some countries, as well as maps of objects worth protecting for their natural scientific value or foreland improvement regions, assist decisions on land-uses. Some conservation maps can take earth science information into account, for example the extent and nature of wetland habitats, and some are concerned directly with the preservation of geological sites.

In every one of the subjects covered above prime consideration should be given to make the map understandable, introducing explanations and simplifying the representation. However, the user should always be able to refer to the scientific basis for the interpretation: this should not be sacrificed for the sake of simplicity. The map designer will have succeeded if the user is stimulated to study the information on which the interpretation is based.

The earth scientist or the planner can develop maps directly related to land use, among which may be listed:

- mineral or energy resource protection maps;
- maps identifying favourable areas for waste disposal sites;
- agricultural land-use maps;
- forest use maps;
- groundwater capture maps;
- maps which, in contrast, warn about certain problems (for example, those which show ground with problems for building).

The maps themselves cannot justify interference with the jurisdiction of the land-use planner. Their purpose is to allow the maximum contribution of the earth scientist towards land-use planning. The earth scientist is, in effect, saying 'If I were the land-use planner for this area, I would make this decision and no other, on the basis of my special knowledge and experience'. The earth scientist should be positive and come out in favour of certain decisions, and certainly should not sit in the 'ivory tower' of pure science and produce beautiful geological maps, leaving the planners to interpret them and decide what is the best decision, without geotechnical advice. The earth scientist must accept, however, that the planner has to consider many aspects other than those related to earth science, and may come to a different decision for very good reasons. The overall decision-making process for him is far more complex than that seen from the viewpoint of the earth science information alone.

The question of what the planner can do with these maps, whether they are understood, or whether they are even wanted, remains. There are many subjective and objective considerations involved in the whole question of acceptance. In some cases vanity and professional elitism may come into this: the belief that the planner knows all the answers and has no need for the advice of the earth scientist. Conversely, there is the point of view that the earth scientist's viewpoint is sounder than that based on other criteria—thus, the fault does not lie only with the planner. Similarly, the geologist may be reluctant to accept the need to modify and interpret sets of maps which follow a long tradition of excellent presentation and incorporate use of difficult technical terminology (jargon)—geological maps were probably the earliest maps to use colour representation (Butcher, 1983)—and from them to produce simpler plans, probably produced in black and white, and possibly produced by computerized cartography. This change requires a radical rethinking on the part of the geologist, which may not always be accepted. The production of easily used and readily understood applied geology maps can be greatly helped by the contribution of the skilled cartographer, but is possibly most effective if the users of such maps are involved in their design.

In some cases, even if planners feel the need for such maps, there may be reluctance to use them for legal and financial reasons. A mineral resource map may, for example, lead to increased land values in some areas, whereas a 'hazard' map may cause reduced property prices. The design of maps must be carried out in such a manner as to avoid such unwarranted reactions, and maps on contentious issues should only be released with discretion.

CONTRIBUTIONS OF GEOSCIENCE TO SOLVING THE PROBLEMS OF REGIONAL AND ENVIRONMENTAL PLANNING AND MANAGEMENT

As already discussed, there are many possible contributions of applied earth science to regional and environmental planning and management. We will restrict ourselves to a few broad examples, but, at the same time, will keep in mind that these represent a challenge to further activities rather than ideal solutions. In the examples, we shall proceed from sea to land.

• *Ocean Mining*: Mining is at present taking its first steps from the continents to the oceans on the basis of results obtained from prospecting for hydrocarbons, ore sludges, manganese nodules and marine gravel. Geological conditions determine the abundance here, as everywhere else. Whereas the users of minerals on the continents have for ages encountered historical boundaries, explorers in the oceans now face the new problem of the legally vague definition of deposit ownership, and thus of the disputed right to exploit deposits. In the epicontinental sea, mining and hydrocarbon exploration has led to changed territorial definitions with regard to the three (or more) mile zones of neighbouring states and the subdivision of the continental shelf. The corresponding designation of boundaries in the North Sea is shown in Fig. 2. Conflicts about the definition of boundaries are influenced by the abundance of mineral deposits in the ocean regions, as the example of the Aegean Sea clearly shows. Global territorial planning has thus acquired a clear relationship to geology; it has become *geobased*.

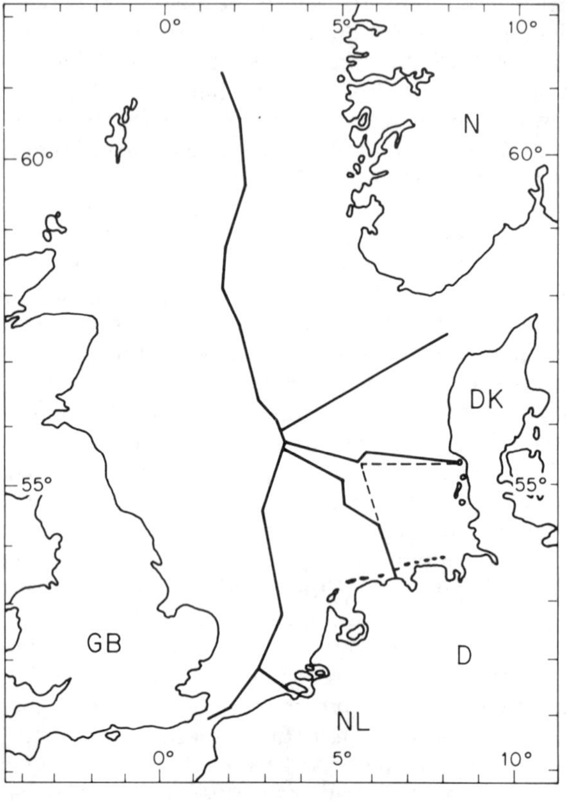

Fig. 2. North Sea—international boundaries for the continental shelf.

• *Ocean Pollution*: It was the oil industry that first exploited natural resources from the ocean, even though the oil production at Baku during the last centrury cannot quite be called 'ocean mining'. Since then, numerous stationary or permanent oil platforms have been sited at sea, producing large volumes of oil and gas: at present about 40% of the global oil and 20% of the global natural gas resources are estimated to be offshore. This ocean extraction has not been without consequences for the environment, as numerous leakages and spillages, explosions and fires have demonstrated. Regional planning has been active not only in setting national territorial boundaries and exploitation limits in the waters, but also in setting exploration concession boundaries (e.g. Fig. 3, which shows the patterns of blocks and quadrangles in the German Offshore Area of the North Sea); this is a case of 'sea-use' rather than 'land-use' planning, and planning has been and is being supported by new surveying methods.

At the same time, new laws and regulations governing pollution control have had to be formulated for environmental management of these oceanic oil-field regions. These take into account the fact that the greater part of the oceanic oil- and gas-related pollution comes not from oil production, but rather from transportation (e.g. spillage from tankers and vessels powered by oil). Lüttig (1987) estimated that in 1979 transportation accounted for 2.133 million tonnes of spillage per annum compared to an oil production spillage of 0.08 million tonnes per annum, out of a total of 4.913 million tonnes per annum, the remainder coming from rivers. However, drilling for oil also causes direct pollution in other ways, not the least of these is the reported damage to the Norwegian fishing industry due to the effect of large-scale accretion of drilling muds.

The situation in land-locked seas such as the North Sea, Mediterranean, Black Sea and Caspian Sea, with their industrial and population concentrations around the fringe and the complication of hydrocarbon exploration is very critical, and their pollution remains one of the urgent major environmental challenges of the late twentieth century.

• *The Production of Marine Sands and Gravel*: By 1970, for example, this had reached more than 12% of the total British national producton (Joliffe, 1971; Archer, 1973). Production is regulated by the use of licence boundary areas, which in most cases are in the form of blocks like those used for marine hydrocarbon exploration. At the same time, environmental management is taken into account by means of regulations to protect marine organisms, shipping and submarine cables.

• *Coastal Management*: The sensitivity of the coast to disturbance by human measures of the erosion/sedimentation equilibrium is well-documented (for example, the case cited below, p. 246). Such measures include dock construction, technical measures to protect the coasts (e.g. groyning, changing the direction and rate of currents), resource extraction (e.g. marine beach placers such as 'black sands') and tourism. As numerous seminars organized by the UN have shown, it is now well understood internationally that appropriately qualified Earth Scientists are indispensable (Lüttig, 1977; Baecker, 1975; and others) if consequent problems are to be avoided.

• *Natural Hazards and Disasters*: There have been many recent contributions on the part of geoscientists to research, cartographic documentation and technical prevention, and even predicton of natural hazards (e.g. the work of UNDRO in Geneva, see p. 245 below). Volcanology and earthquake research are very significant concerns here. Even insurance companies have made use of the advice of earth scientists.

A world map with the classes of hazards has been published by the insurance company Münchener Rückversicherung (1983), and UNESCO, UNEP, and the EEC (CORINE programme) have been active in the field of hazard management. A group of Spanish scientists has produced the first hazard map at a scale of 1:5000 (Cendrero *et al.*, 1987) and numerous publications are available for areas endangered by rock-falls and landslides, including Soviet work in the Crimea, American work in California (Brabb, 1984), and a comprehensive study of landsliding in Great Britain (Geomorphological

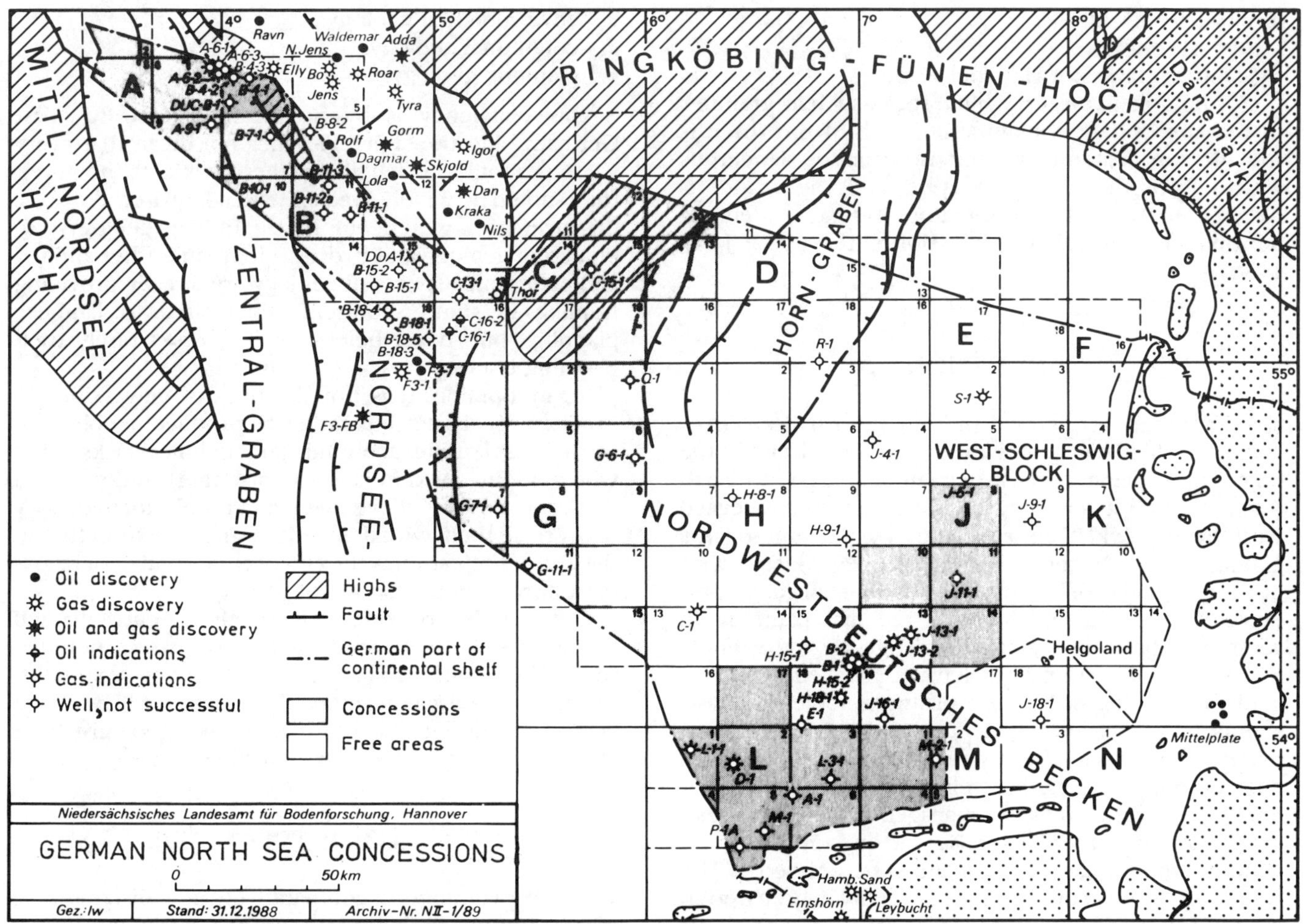

Fig. 3. West Germany offshore area—North Sea.

Services Ltd., 1986). The importance of this aspect of land-use management is evident in the light of its economic significance to the USA and Spain—in 1979 overall expenditures on hazard management in the USA were estimated to be in the range of US$179–283 billion (Kates, 1987), and in 1983 the flood catastrophe in the Basque Provinces cost US$8 billion.

- *Desert Management*: There is still much scope for the earth scientist in this field, not only in cases such as the anthropogenically accelerated desertification of the Sahel and in East Africa (in part caused by overpumping of groundwater), but also, for example, in the stabilization of migrating dunes by protection from wind-erosion; or by exploring possibilities for artificially replenishing groundwater in areas where 'fossil' groundwater is pumped. Here groundwater management is, not surprisingly, the most important branch of land-use planning and environmental management.
- *Soil Protection*: Due to the increased number of cases of soil and groundwater contamination by fertilizers, insecticides, herbicides, chemical solvents (such as chlorinated hydrocarbons), and the supposed acidification of the soil from emissions, public administration and the public at large have turned to protecting the soil in recent years. The soil possesses an important filter function against possible entry of pollutants into both vegetation and groundwater. In addition to the problems of chemical contamination, the problems of soil compaction, erosion and naturally/artificially produced salinization deserve attention. Salinization of wheat-belt land due directly to agricultural practices is a major problem, for example, in Western Australia. There is also a major problem in the continued depletion by various causes of the total global budget of agricultural land (that is, land with agricultural soils).

What is urgently needed is expert advice to planners and decision-makers from expert pedologists and hydrogeologists. Because the soil processes are difficult to understand without soil chemical, soil physical and hydrogeological expertise, it is essential that soil scientists and hydrogeologists advise planners. Again, the production of geoscientific maps is prerequisite here.

- *Soil Erosion on Slopes and in High Mountains*: In recent years there have been numerous cases of erosion-related disasters caused by heavy rainfall and characterized by flood and soil transport incidents (e.g. in the Alps—Veltlin, Savoy), and environmentalists are reviewing the role of human stimulation in high mountains soil erosion. The local impact of increased tourism and skiing and sledding areas, on slope stability, is extensive. Even 'normal' agriculture, vineyards and orchards, on slopes highly susceptible to erosion can also act as a triggering influence. Ignoring scientific advice on ways of preventing this sort of damage would be unwise, and such regions demand rigorous land-use planning.
- *Nuclear Waste Management*: In modern geotechnology geoscientists have been especially innovative in the field of identifying geological structures and formations suitable for storing medium and high-level toxic wastes, in particular radioactive wastes. Three concepts in

disposal are favoured:

(1) salt domes;

(2) thick layers of clay rocks; and

(3) homogeneous crystalline rocks such as granites. In the appropriate configuration, each of the three is suitable for guaranteeing the isolation of dangerous wastes from the biological cycle, and the reader is referred to the relevant literature for more detailed information (e.g. Faber, 1987: Hofrichter, 1982; Jahn, 1986).

EDUCATION

Some planners are trained in both the earth sciences and other allied disciplines but this is more often not the case. These planners must become familiar with earth science concepts and likewise, the geologists concerned with the new field of environmental and planning-related geology must become sufficiently acquainted with economics and econometry, statistics, human geography, ecology, land-use, and local or regional planning in addition to basic geological knowledge. Here the whole question of educational curricula is involved.

In the training of land-use planners, some emphasis should be placed on a general knowledge of geology, mineral resources, the energy industry, technology, engineering sciences, as well as on specialized but important topics like waste management and land restoration for cultivation. This raises the question, "How, in general, should land-use planners be trained"? The subject does not even exist as a university degree option in many parts of the world. In many countries, those involved in this field are not experts with adequate training. The ideal land-use planner might be a geographer with knowledge of both physical and human geography, with secondary training in socio-economics and natural sciences. In practice, land-use planners may be drawn from several branch disciplines: landscape designers, biologists of various sorts, economists, statisticians, mathematicians, industrial management experts, lawyers, sociologists and political scientists. Because of this, it is not surprising that very mixed results and much disagreement arises.

The geologist, however, should not complain about the curricula (or lack of them) of planning colleagues, but should concentrate on the role of, and the quality of training in, the earth sciences. As I have previously stated (Lüttig, 1985; 1987) this is not particularly satisfactory. In West Germany, for example, much that is required by the planner is not included in the curriculum, and much of what is included in the curriculum is not needed in practice. Necessary aspects and branches of university studies which do not receive their fair share of attention include:

- geology of unconsolidated deposits;
- Quaternary geology;
- soil science;
- irrigation and soil improvement requirements;
- water management;
- economic geology (especially applications to developing countries);
- geology of energy carriers (especially the renewable ones);
- waste management and recycling;
- geotechnology (geoscientifically based technology);
- materials science;
- environmental geology;
- communal geology;
- planning geology.

This deficiency is by no means limited to West Germany, and the above list does not mean that every student should be taught all these subjects, however, they should be widely available and adequately covered.

Universities which emphasize mining geology, engineering geology and hydrogeology, and other applied and economic aspects of geology are already providing training in some subdisciplines related to land-use planning, but, nevertheless, such courses rarely treat planning issues.

In my opinion, it is not the duty of universities to produce specialists, but rather to produce people who can solve scientific, economic and technical tasks (all of which are involved in land-use planning), and who can think and act. To achieve this, it is certainly not necessary to overload them with demands to such an extent that, at the end of their studies they incorporate and understand the entire scientific scope of geology.

After one has provided adequately for curricula and training for both planners and geologists as outlined above, then fruitful collaboration will be a more straightforward matter. Furthermore, the situation will be all the better, the earlier that the planner and geologist have met each other in the course of their studies.

THE GLOBAL PERSPECTIVE

Land-use planning is a comparatively new discipline, so planners must build on a long history of unplanned land-use. It is an idle dream to wish that the mistakes of the past had been avoided (e.g. establishing cities in the middle of spring catchment areas, wood cutting on slopes endangered by erosion, planting orchards and vineyards on volcanic slopes). The limitations imposed by its late introduction, mean that what land-use planning can achieve may commonly be limited to very modest changes in direction or the prevention of something worse.

We must not allow this limitation to cause us to regard global, regional and local environmental planning as a totally hopeless matter. The general philosophy of such planning is internationally widespread, and it is worth considering here the global perspective. It has long been realized that the narrow, interventionist philosophy of the English Reformers at the end of the nineteenth century (McLoughlin, 1973) needs modification, as the original concept does not lead to the social welfare intended by the planning philosophies and politico/economic systems. Positive or developmental actions in combination with negative or regulatory actions need constant balancing within the general framework of a long-range view (Lindblom, 1965). Otherwise, a system of short-range or incremental decisions arises, leading to incompatible results. It is in this context that the role of the international organizations, be they governmental or non-governmental, scientific or technical, is important.

While reservations have been expressed from time to time concerning the usefulness of international bodies, many have performed exceptional work throughout the world. Some examples are given below:

- The *United Nations Development Programme* (*UNDP*) is a source of numerous monographs on mineral resources, groundwater supplies, and sources of energy, referring to every continent and almost every developing country.

- The *World Energy Conference* has produced numerous reports on supply and demand of non-renewable (e.g. coal and hydrocarbons) and renewable (e.g. hydroelectric) energy.
- The *United Nations Food and Agriculture Organisation* (*FAO*) has produced numerous reports on agricultural and related land uses, including statistics on production and consumption, as well as the geological basis for food production.
- *World Health Organisation* (*WHO*) has produced fundamental national and subject-based reports on health situations, control of epidemics, nutritional state and so on, with a strong linkage, for example, to surface and groundwater quality, and problems of pollution and waste disposal.
- *United Nations Educational, Scientific and Cultural Organisation* (*UNESCO*) has produced numerous monographs and other publications on earth science topics (e.g. International Geological Map, International Hydrogeological Map) in conjunction with the International Union of Geological Scientists (IUGS) and others. It has also covered the hydrosphere (in conjunction with the IOC in the International Hydrological Decade) and the biosphere (in the Man and the Biosphere Programme). It has further engaged in social and environmental science.
- *United Nations Environmental Programme* (*UNEP*) has produced a series of monographs and country-related reports on problems of environmental research, protection and management.
- *United Nations Disaster Research Organisation* (*UNDRO*) has produced numerous expert reports, programmes and maps on natural hazards and human risk.

The publications of these and other such organisations provide an excellent foundation, and data base for rational environmental management, and much of it concerns the earth sciences. Despite this excellent foundation, there still remain the problems of its proper evaluation and conversion into practical use.

STRATEGIES AND PROBLEMS

In the future, emphasis should be given to the development of maps which are even more sophisticated than those currently used to illustrate important planning data. Furthermore, close co-operation with planners and managers should be encouraged. Such collaboration might help to resolve certain land-use conflicts. For example, a region designated for water resources might also support the development of forestry under certain conditions. Agriculture might be feasible provided certain constraints are placed on the use of fertilizers and insecticides, and, at a certain distance, temporary extraction of near-surface mineral resources might be permissable, once studies of the water-table and groundwater flow-rate have been carried out.

In contrast, there will be areas in which only one of the possible claims for land utilization can prevail. This decision should result from discussions between the planner and appropriate experts such as geologists, economists and ecologists, and the final decision will be that of the planner or of elected representatives, depending on the type of planning system used. Experience has shown, however, that incompatibility is not as common as might be supposed. Some possibilities are summarized in Fig. 4.

The final planning decision can be shown on the map of the priority use(s), which will result from weighing all

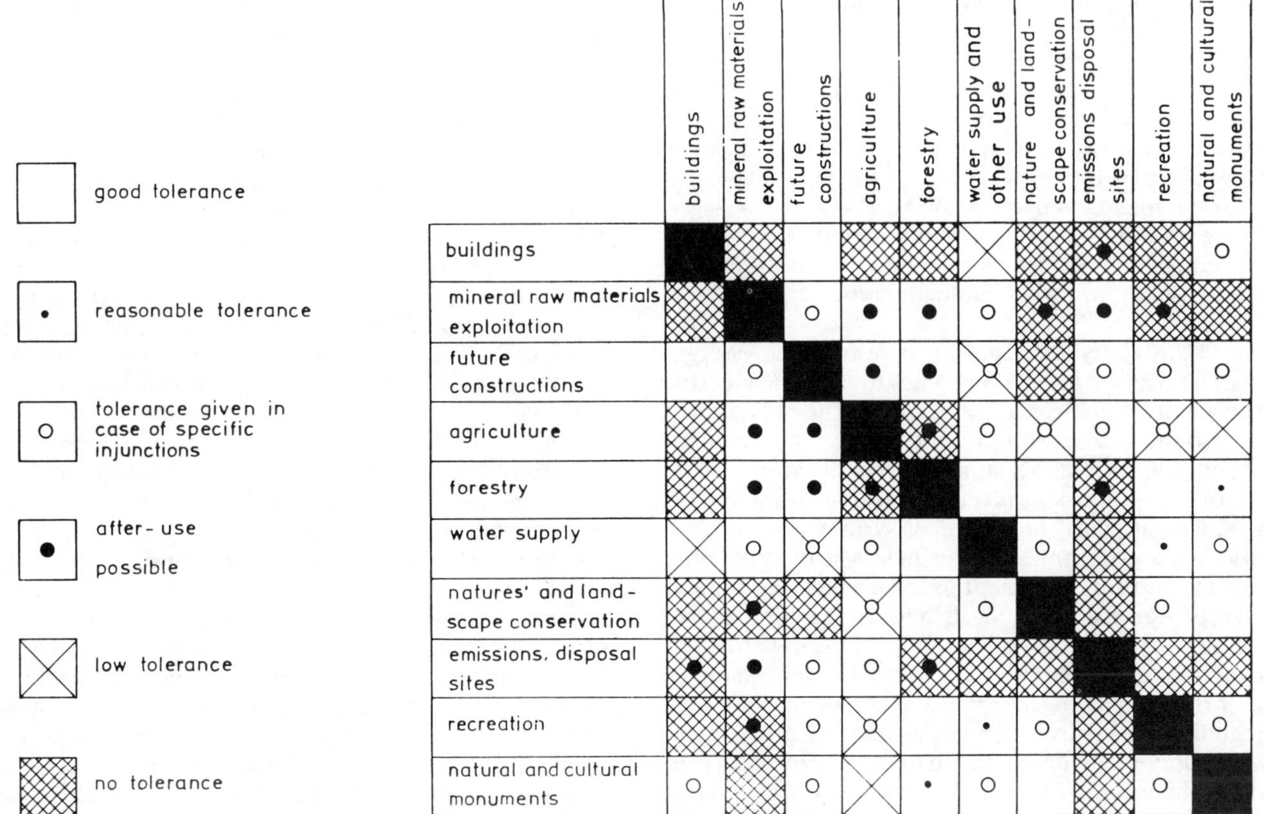

Fig. 4. Tolerance diagram for different land utilization (example FRG, from Lüttig 1988).

the different points of view; this is the desired goal of geobased land-use planning. The earth scientist can provide a priority use map which may form part of the information needed to produce the final synthesis (Becker-Platen, 1983; Becker-Platen *et al.*, 1977; Becker-Platen and Lüttig, 1980; Becker-Platen, Lüttig and Meine, 1977; Becker-Platen, Hofmeister, Klemz and Stein, 1986). The map can then be further developed by the land-use planner. Such an approach has been used, for example, in the geoscientific map of the natural environmental potential covering the whole area of Lower Saxony and Bremen at a scale of 1:200 000. At the time, it was the first complete set of maps published on the basis of Lower Saxony's land-use development plan (Becker-Platen, Dorn and Look, 1987). An accompanying resource protection map is available for this part of Germany as a sub-map on a larger scale (Stein and Hofmeister, 1977).

The methodological approach may well differ elsewhere. The approach must be adjusted to the conditions and demands of the country concerned. The amount of detail will vary considerably between maps for developing countries and industrialized nations, and will reflect quite different political, legal and planning systems. Nevertheless, in principle, there are no differences in the required methodology and possibilities.

Environmental phenomena of the earth, such as gaseous emissions, oceanic contamination, river pollution and many natural and man-made hazards, are not restricted by political boundaries. These problems cannot be solved on a national or provincial-scale, let alone on a regional- or community-scale. When considering such large-scale issues, the maxim which might be applied is that the smaller the planning area, the greater the impotence of the planner and the obstacles facing him. By reaching a consensus about the importance of general scientific thinking in planning, planning will no longer be restricted to tasks constrained by the conventional boundaries of the area, but will allow the significance of factors such as mineral resources, groundwater occurrences and soil properties, to be evaluated when appropriate from an international viewpoint. This may be an exaggeration, but co-ordinated planning over larger areas, extending across national boundaries and even on a global scale, can only be beneficial, and such coordinated effort is, perhaps, inevitable and essential if the quality of life is to be maintained and improved.

Problems that may require international and even global consideration are not limited to questions of emission, pollution or contamination; for example:

• The worldwide shortage of conventional energy sources has caused utility companies to produce electricity by means of nuclear power stations. The radioactive waste produced from this process requires waste management. The problem is supranational because of the international network involved, incorporating mining of radioactive minerals, operating electricity and waste-management companies. Nevertheless, the problems posed by further treatment, reprocessing and final storage of waste are best dealt with by locating these processes within a single community. It is not surprising that the inhabitants of such a community (for example Mol, Gorleben or Schwandorf in West Germany) are not delighted by this prospect, and this may bring the local planning bodies into conflict with the higher levels in the planning hierarchy. Much public discussion and hard political conflict may result, but the 'St. Florian principle' (in England known as the 'not in my backyard principle'), whereby a contentious development is opposed everywhere, provides no solution to a problem *which must be solved*; somewhere, a planning board (in this case a political decision-making authority, if necessary a court of law or a parliament) must decide the priority use of the potential of the natural environment (for example, a site where geology provides a favourable salt dome).

• The construction industry needs huge amounts of aggregates such as sand and gravel. Public authorities consume the largest part of this for building and civil engineering, especially for the construction of roads. The resource price is primarily determined by the transport costs of these bulk minerals and deposits as a rule extend beyond the boundaries of the planning jurisdiction areas. A gravel deposit can thus lie partly within area A and partly within area B. The extraction of the resource is, however, regulated by the administration of the individual area. In the past, the case has frequently arisen where area A behaves quite differently with regard to allocations and regulations for companies to area B. This can occur on a national scale, for example the situation which affects the border between France and the Federal German Republic, where regional disparities due to the existence of frontiers distort the economic situation by preventing sensible utilization of the geological resources. In area A, extraction may be permitted, and this region must develop and implement reasonable policies for environmental protection and waste-treatment (reclamation, minimizing groundwater impact and noise and dust pollution, provision of transportation and infrastructure). In contrast, area B may have to import the sand and gravel, almost certainly at a higher price than in region A. The elimination of the environmental consequences of extraction in area B is only a short-term apparent advantage; economically region B is at a disadvantage.

• A third example is coastal erosion. At one location on a continental coast the sediment balance (that equilibrium between erosion and accumulation which keeps the coastline stable) is disturbed, either by natural changes (e.g. if the coastal strip is rising or sinking) or by human interference (e.g. construction of sea defences, harbour dredging). Such changes may result in long stretches of damaged coasts through neighbouring countries with disastrous consequences (damage to tourist beaches, undesirable alluvial sedimentation). Even small measures are often liable to cause problems (for example construction of a mole, a dam on a river, bringing down the sand to the coast, or land reclamation extending a coastal city seawards). Since changes on the coast of one country can affect the shores of another, coastal management on a supranational scale may be necessary.

Apart from dealing with cases of planning issues that extend beyond national frontiers, an international approach also has the benefit of exchange of experience and technological know-how. In particular, interaction which involves sharing expertise between industrial and developing countries is normaly quite beneficial. This may relate to specific projects or to the general development of scientific and technical concepts at an international governmental level or between non-governmental organizations. UNESCO and UNEP have exceptional potential in this respect, and as well as from the point of view of training experts. Decision-makers would also benefit from the wider use of modern techniques and procedures, and here again international co-operation has an important role to play.

Taking for an example, coastal processes, the GEOSAT programme and other remote sensing methods have enabled us to monitor the geosphere beyond the simple evaluation provided by aerial photography. They can show changes like a moving picture, and we can record changes in the coastal processes just as we can record atmospheric changes by means of meteorological maps. Such studies should include sand balance measurements (a strenuous task and prohibitively expensive using the purely sedimentological basis of coastal maps with different dates of publication and stratigraphical data), as well as evidence of more short-term events, thus providing a better basis for predicting long-term trends.

The task of translating the enormous amount of data relevant to land-use planning into cartographic form is aided by computers and digital information systems which make the data clear and understandable. In general, the methods of information science are followed. The statistical information (e.g. the location of settlements, transportation systems, social facilities, existing land-use types, meteorological and hydrological observations, relief, soil types, rock types and vegetation) are converted from the various separate sources on to digital mesh maps. The smallest topographical units are meshes surrounded by grid lines, polygons composed of dots, lines and planes, and single dots. Thus, for example, within the Digital National Land Information in Japan (Matsuda, 1980), mesh systems of various orders for each scale of the necessary maps include all the elements necessary for planning.

A similar method has been utilized for ecological mapping carried out by the EEC (Ammer, Bechet and Klein, 1977), where the grid used 4 × 4 km polygons. At the more local level compact Geographical Information Systems are becoming commonplace in planning departments. However, many of these lack earth science elements, and until these have been added such systems are seriously inadequate.

As with coastal management problems, remote sensing has developed methods which allow yield potential and the agricultural suitability of soils to be demonstrated using false colour images on satellite photos. Soils with differences in parent rock, grain size and degree of consolidation, and depth of groundwater can be illustrated by varying grey shades in satellite photographs. These grey shades can be reinforced by converting them into brighter false colours, thus making the individual details more distinct. Although there have been claims that it is possible to make an economic geology assessment of certain bodies of rocks in the same way, without ground checks, claims spread by the media that mineral deposits have been discovered in this way are exaggerated. Detailed evaluation on the ground is still essential.

A final problem must be mentioned. The land-use planner finds the natural environment in a state which has developed historically. Efforts to regulate the use of the geosphere to suit human beings often means running against the *status quo* and existing processes and structures are hard to displace. The ability to meet the demands of both the natural environment and of human beings will depend on the administrative weight that can be wielded and the support given by laws and the decision-making bodies. In this regard there are wide variations from geographic region to geographic region, and from political system to political system. The need for planning, despite the difficulty of disturbing the *status quo* is well summed up by McLoughlin (1973):

> Spatial distributions of human activities are significant for human welfare ... human welfare can be increased by planning and controlling those spatial distributions.

The message is that in the present state of the world and increasingly in the future, it would be better to have planned land-use and some disturbance of the *status quo* than the 'law of the jungle'.

CONCLUSION

If human civilizations, despite differences in standards and systems, attempt to plan and manage the natural environment in a wise manner, the following factors must be borne in mind:

(a) Economic development, the decisive prerequisite for a more-or-less high standard of living, is best achieved if resources are available and accessible, and can be utilized close to the places of consumption. Among such natural resources are exploitable minerals and rocks, sources of energy, water above and below ground, soil and subsoil. The preservation and release of this geopotential (= the part of the natural environment's potential determined by geological conditions) plays an important part in the management of natural resources.

(b) Human exploitation of the geosphere (= lithosphere, atmosphere, hydrosphere, and to a certain extent, biosphere) affects interactive processes and causes an impact on the environment. The consumption of the potential of the natural environment should nowhere exceed the level which can be sustained within the natural laws. Excessive human influence must be brought under control, so that reason is placed above exploitation, and intelligence before lack of restraint. Compromises between resource exploitation and environmental protection are possible.

(c) Sensible, well thought-out and wisely directed planning can regulate properly, but policies and controls must take account of and be supported by sound scientific findings. The laws of the events and processes of nature should be taken into consideration. Short-term policies based on quick reactions to public opinion, particularly when that arises from hysterical or sensational accounts, should be avoided, instead, premature reactions should be replaced by considered strategies.

(d) Appropriate steps will increasingly require joint international efforts. The problems of the economic development of human civilization cannot be solved by individuals, groups, individual nations or political blocs, because the systems of the earth interweave and interact throughout the entire globe.

(e) Three major professional groups must be concerned, earth scientists, environmental scientists and land-use planners, and the following should be included in their training:

- for the earth scientists, more understanding of the national economy and the requirements of planning;
- for the environmental scientist, knowledge of all earth science factors that are relevant to the environment, to supplement their essentially biological training;
- for land-use planners, training that emphasises the essential fact that planning and development control *must* take appropriate account of earth science factors.

(f) Finally, 'Mankind's future will depend on its capacity to manage the potential of the natural environment, which should guarantee the free development of nations as well as the individual. Planning and management are impossible without *good thematic maps* and, therefore, *geoscientific mapping occupies a key position in future oriented environmental research and design.* One of the tasks of geocartography is to demonstrate this position and to use it' (Lüttig, 1987).

REFERENCES

Ammer, Bechet and Klein. (1977). *Ökologische Kartierung der EG. Leitfaden zur Erstellung der Fallstudien.* Mskrpt., 38 pp. München 1977.

Archer, A. (1973). Economics of Off-shore Exploration and Production of Solid Minerals on the Continental Shelf. *Ocean Management,* **1**, 5–40, Amsterdam.

Arndt, P. and Lüttig, G. (1987). *Mineral resources extraction, environmental protection and land-use planning in the industrial and developing countries.* Schweizerbart, Stuttgart.

Bäcker, H. (1975). Exploration of the Red Sea and Gulf of Aden during the *M.S. Valdivia* cruises "Erzschlämme A" and "Erzschlämme B". *Geol. Jb.(D)* **13**, 3–78, Hannover.

Becker-Platen, J.-D. (1983). Geowissenschaftliche Karten des Naturraumpotentials. *Forsch. dt. Landeskde., 220,* 119–164, Trier.

Becker-Platen, J.-D. *et al.* (1977). Karten des Naturraumpotentials von Niedersachsen und Bremen. *Geol. Rdsch.,* **66**, 3. 14–30, Stuttgart.

Becker-Platen, J.-D., Lüttig, G. and Meine, K.-H. (1979). Geoscientific Maps for Planning. *Natural Resources Forum* **3**, 167–177, New York.

Becker-Platen, J.-D. and Lüttig, G. (1980). Naturraumpotentialkarten als Unterlagen für Raumordnung und Landesplanung. *Arbeitsmat. Akad. Raumforsch. Landesplang.* **27**, 1–60, Hannover.

Becker-Platen, J.-D., Hofmeister, E., Klemz, B. and Stein, V. (1986). Landnutzungskarten—Ein Versuch zur Darstellung der Flächenbeanspruchung. *Raumforsch. & Raumordn,* **44**, 6, 271–234, Bonn and Hannover.

Becker-Platen, J.-D, Dorn, M. and Look, E.-R. (1987). An introduction to the legend of the Geoscientific Map of the Natural Environment's Potential (GMNEP) of Lower Saxony and Bremen. In. *Mineral resources extraction . . .* (Arndt, P. and Lüttig, G.) 119–126, Schweizerbart, Stuttgart.

Brabb, E.E. (1984). Innovative Approaches to Landslide Hazard and Risk Mapping. *Proc. 4 intern. Sympos. Landslides Toronto* 1984. **1**, 307–324, Toronto.

Butcher, N.E. (1983). The advent of colour-printed geological maps in Britain. *Proc. Roy. Inst. Great Brit.* **55**, 149–161.

Cendrero, A. *et al.* (1987). Detailed geological hazard mapping for urban and rural planning in Viscaya, Northern Spain. In *Geology for environmental planning.* Norges Geol. Unders., Spec. Publ. **2**, 25–41, Trondheim.

Faber, P. (1987). Kernenergie und nukleare Entsorgung. *Mittbl. Bund dt. Geol.* **18**, 13–18, Bonn.

Hofrichter, E. (1982). Geologische Aspekte der Endlagerung radioaktiver Abfälle. *Erzmetall* **35**, 11, 570–575, Weinheim.

Hoppe, P. (1976). Die Verfolgung küstendynamischer Prozesse an der deutschen Nordseeküste mit Hilfe von Satellitenbildern. *Geol. Jb. (A)* **33**, 49–87, 22 Abb., Hannover.

Jahn, P.G. (1986). Entsorgung radioaktiver Abfälle. *Vorlagen,* **42/43**, 1–72, Hannover, Lutherhaus.

Jolliffe, I.B. (1971). Gravel wealth from the seabed. *Geogr. Mag.* **43**, 6, 403–406, London.

Karrenberg, H. (1972). Die Internationale Hydrogeologische Karte von Europa. *Mitt. deutsch. Forschgem.,* **72/4**, 56–60, Bonn-Bad Godesberg.

Kates, R.W. (1987). Hazard assessment and management. In *Resources and World Development.* (McLaren D.J. and Skinner, B.J.). 741–752, Wiley, Chichester.

Lindblom, Ch. E. (1965). *The intelligence of democracy: decision-making through mutual adjustment.* Free Press, New York.

Lüttig, G. (1975). Geoscience and the potential of the natural environment. In *Geoscientific Studies and the Potential of the Natural Environment.* 29–40, Deutsche UNESCO-Komm., Köln.

Lüttig, G. (1977). Raw materials potential and the geodynamics of the coastal zones. In *Development and Management of Resources of Coastal Areas. Proc.* 339–344, Berlin.

Lüttig, G. (1985). Über den "Dreck, der die Geologie verhüllt" (STILLE). *Nachr. dt. geol. Ges.,* **32**, 90–105, Hannover.

Lüttig, G. (1987). Large-scale maps for detailed environmental planning. In *Geology for environmental planning.* Norges Geol. Unders., Spec. Publ. **2**, 71–76, Trondheim.

Lüttig, G. (1988). The influence of mining activities on the environment. *27th intern. geol. Congr. gen. Proc.* 167–176, Moscow. (1984).

Lüttig, G. *et al.* (1989). Geology and land-use planning. In (Kozlovsky, E.A.): Geology and environment, **3**, UNEP/UNESCO, Prague.

Matsuda, H. (1980). Digital National Land Information. *GeoJournal* **4**, 4, 313–318, Wiesbaden.

M'Bow, A.-M. (1982). World problems and the lines of emphasis of the Medium-Term Plan (1984–1989). UNESCO Gen. Conf., 4th extraord. Sess. 4 × C/4, Draft Medium-Term Plan (1984–1989). **1**, 1–38, Paris.

Mc'Loughlin, J.B. (1973). *Control and urban planning.* Faber & Faber, London.

Mc'Loughlin, J.B. (1973). Theories and techniques of spatial and environmental planning. Inform. Orts-, Regional- u. Landesplang., Dokum. *Inform. St. Planungsfr.,* **31**, 25–33, Zürich.

Münchener Rückversicherungs-Gesellschaft (1983). *World Map of Natural Hazards.* München.

Stein, V. and Hofmeister, E. 1977. Die Darstellung oberflächennaher Rohstoffvorkommen in Rohstoffsicherungskarten. *Geol. Jb. (D),* **27**, 121–132, Hannover.

Epilogue

The relevance of Earth Science information to land use planning has been demonstrated in many ways in this book. Whereas the Earth Sciences cannot resolve issues of competing requirements for land, they can materially assist decision makers in identifying rational choices. However, much work is needed to ensure that results are designed to fit the needs of users and to be readily fitted into the planning processes. This will vary from place to place and from system to system. Many of the problems which must be faced require multidisciplinary responses and these are most likely to be aimed at the right target if the planners are involved in designing their presentation and, particularly, in working on databases and maps, which are the common currency of the relevant professions.

The problems facing the land use planner in the future are likely, as Professor Lüttig has shown, to grow and increase in complexity. However, the overall problem remains:

How much more growth of population can the fragile natural environment of this planet sustain?

The land use planner is in the position of someone endeavouring to accommodate an unending line of passengers boarding a ship. At some stage the carrying limit of the ship will be reached. Though the problems can be slightly reduced by improved practices and technology, it may well be that the ability of this planet to carry more passengers, that is to provide resources for ever increasing civilised populations, without destroying the essential natural balance is fast approaching. The critical fact is that many major Earth resources are non-renewable, yet the demand for them is always increasing with the rise of population and the spread of civilised living. In the meantime, environmental planning is essential in gaining the time to find novel solutions to future difficulties, and such planning must involve the Earth Sciences as a fundamental first step in its procedures.

Glossary

Acid rock An igneous rock rich in silica (granite, etc.).
Active zone A term used in hydrogeology to denote a zone within a region subject to annual freezing and thawing, which causes disturbance to soils.
Adit An almost horizontal passage from the surface by which a mine is entered.
Aeolian An adjective applied to rocks formed of wind-blown grains.
Aggregate The mineral material, such as sand and gravel or crushed rock, which is mixed with binders to form constructional materials (e.g., with cement to make concrete or mortars, or with asphalt for road surfacing).
Airborne Multispectral Scanner A satellite-borne scanning device which operates several spectral channels in the visible and invisible (e.g. infra-red) spectrum.
Algorithm A term used in computerization to denote a sequence of instructions which give a structured solution to a particular problem.
Alluvium A general term for all granular deposits (e.g. sands, gravels, silts or clays) resulting from the operations of modern rivers, thus including sediments laid down in river beds, flood plains, lakes, fans at the foot of mountain slopes, and estuaries.
Analog (Analogue) A term used in computerization to denote data in the form of continuously variable physical quantities (contrasts with 'digital').
Andesite A very common volcanic rock (lava) type containing the minerals pyroxene, feldspar and commonly biotite and hornblende, and of 'intermediate' (moderately silicic) composition.
Anhydrite Calcium sulphate in a form which, unlike gypsum, lacks water within the crystal structure.
Aquiclude A saturated rock unit, group of rock units or part of a rock unit that yields inappreciable quantities of water to wells and springs.
Aquifer A rock unit, group of rock units or part of a rock unit that contains sufficient saturated permeable material to yield significant quantities of water to wells and springs.
Arkose A feldspar-rich sandstone, generally derived from decayed granite.
Atomic absorption spectroscopy Electrons within atoms may occupy a series of discrete energy levels which are characteristic of the type of atom concerned. These can absorb electromagnetic radiation of frequencies corresponding to these energy levels. Observation of the absorption frequencies can be used to identify and quantify the elements in the sample, which has to be tested in solution. The method is specially applicable to metals and alkalis.
Atterberg limits Three parameters widely used in engineering geology. The 'Liquid Limit' is the minimum moisture content at which a soil will flow under its own weight: it defines the change from plastic to liquid state. The 'Plastic Limit' is the minimum moisture content at which a soil can be rolled into a thread 3 mm in diameter without breaking; therefore, it defines the threshold between a soil behaving as

a plastic solid and behaving as a brittle semi-solid. The 'Shrinkage Limit' corresponds to the semi-solid to solid state. These values are determined by laboratory testing.

Auger A boring mechanism which utilizes the screw-shaped bit principle. In simple soil auguring, samples are brought to the surface adhering to the flanges of the screw-bit, picked up as it rotates and drives downwards. 'Shell and auger' drilling is a method in which a shell encases the screw-bit, retaining the sample.

Avalanche (1) Masses of snow detached from great heights in mountains which acquire fresh accumulations as they descend, acquiring great bulk, often causing great destruction in the valleys below. (2) Falling masses of rock and earth are often called rock avalanches. They are a form of landslide.

Backfilling A term used in engineering and surface mineral working for filling of a cutting with material excavated from nearby. In deep mining, it is used to denote filling up of a stope or other excavated cavity with waste rock from the mining process. This may be required to provide structural support after mining; in other cases 'open stopes' are satisfactory.

Bank storage A term used in hydrogeology to denote the volume of water temporarily stored in an aquifer adjacent to a river, filled at a time of high flow and providing return flow to a river later when its level is low.

Basalt A volcanic igneous rock (lava) composed essentially of the minerals pyroxene and feldspar. One of the most common lava types, it is of 'basic' (low silicic) composition.

Base flow The sustained flow of a stream, provided from stored sources. The flow is unrelated to a specific precipitation (rain or snowfall) event.

Base flow index A measure of the proportion of total stream flow derived from stored sources.

Base metals Any metal such as iron or lead which is altered on exposure to air (contrasts with the noble or precious metals such as gold, silver and platinum).

Basic rock Igneous rock relatively poor in silica.

Bedrock The solid rock underlying unconsolidated superficial rocks such as sands, clays and soft weathered rock.

Biochemical oxygen demand (BOD) The amount of oxygen dissolved in water used in aerobic bacterial decomposition by a sample stored for five days at 20°C.

Bit The smallest unit of information that can be held on a computer.

Bituminous coal Part of the coal series (q.v.).

'Blighting' Planning blight is a term used where land is identified in such a way that it reduces the market value of the land to the owner.

Breccia A rock composed of fragments of angular shape more than 32 mm average diameter; applied to rocks of sedimentary or volcanic origin and those produced by faulting and other tectonic processes.

Byte The smallest group of 'bits' that can be held individually by a computer. A byte generally consists of eight bits and equals one character of data (a single letter, character or symbol). It is the most common unit of computer storage capacity. Kilobyte = 2 to the power of 10(1024), but loosely used for 1000 bytes.

'CAD' Computer aided design—term used for graphics software developed to assist design, a means of producing direct computerized maps and plans.

Calcrete A crust cemented by calcite, formed by crystallization within a soil profile.

Caliche A calcitic deposit formed in the soils in certain semi-arid regions. Capilliary action draws the lime-bearing waters to the surface where, by evaporation, the lime-bearing caliche is formed. Old caliches are likely to be indurated, and therefore very hard.

Carbonate rocks Those rocks composed dominantly of carbonate minerals such as calcite ($CaCO_3$), and dolomite, the double carbonate of calcium and magnesium $(Ca, Mg)(CO_3)_2$.

Carbonatite A rare type of igneous rock, mostly intrusive and composed dominantly of carbonate minerals and not silicates.

Catena Patterns of soils related to land forms so that erosion and water movement in the soil give rise to sequences of soils developed on one type of parent material at comparable positions on slopes over a wide area of uniform climate.

Clay An earthy deposit of extremely fine texture, usually plastic when wet. Contains a proportion of colloids. Composed dominantly of hydrous alumino-silicate minerals with grain size less than 0.002 mm.

Coal Measures Those strata (of the Carboniferous System) which contain coal.

Coal series Deposits of partly decayed vegetable material may become progressively altered by increased temperature and pressure following burial. The loss of water, nitrogen and other materials gives a progressive increase in carbon content and loss of vegetable structure. The various products of the various stages in this process are known as peat, lignite (brown coal), bituminous coal and anthracite.

Colluvium Heterogeneous aggregates of rock detritus such as talus and avalanche material resulting from the transporting action of gravity downslope. A general term including several more specific types.

Compatible In computing, two devices are compatible if they can work together without special hardware or software having to be used to make this possible.

Computer graphics The processing and generation of visual information in the form of pictures by means of a computer linked to a keyboard and cathode ray tube.

Concretion A body, commonly spheroidal or discoidal, formed by the segregation and precipitation of soluble minerals (e.g., quartz, calcite) around a nucleus, which may sometimes be a fossil.

Confined aquifer An aquifer overlain by less permeable strata in which groundwater is under pressure.

Corrasion Mechanical erosion caused by loose solid material during transport, e.g. the wearing away of fragments from the beds or banks of a river by solid material carried by the river.

Corundum A mineral composed of aluminium oxide (Al_2O_3), of value as an abrasive in its common form but also forming gemstones (sapphire, ruby, emerald).

'CRT' Cathode ray tube—a device whereby electrons are sprayed onto a viewing screen under the direction of magnetic fields to form patterns (used as a synonym for Visual Display Unit).

Data Factual information.

Data base A collection of data held in files, usually for more than one purpose.

Data item The smallest unit of named data, which may consist of any number of bits or bytes. Also known as fields or data elements.

Deep mining A term used for mining at depth by means of adits and shafts (in contrast to opencast mining).

Diabase A common intrusive igneous rock occurring in dykes and intrusive sheets (sills), compositionally equivalent to basalt, and thus basic (of low silica content).

Diatomite An earthy silicic deposit formed by the accumulation of millions of internal structures of diatoms (microscopic organisms) in lakes and swamps. It is of considerable commercial value as an industrial mineral.

Digital Data in discrete quantities (contrasts with Analogue). Pertaining to data in the form of digits.

Digital terrain model A three-dimensional representation of a surface, usually the ground surface, stored as point heights in a computer, and easily reproduced by computer mapping.

Digitization To convert an analogue measurement of a physical variable into a number expressed in digital form.

Dip The angle between the bedding plane and the horizontal of a geological stratum, measured at right angles to the strike (the trace of the dipping plane on the horizontal surface).

Disk A computer memory device which looks something like an audio record ('floppy disk', 'hard disk', 'magnetic disk', etc.).

Dolomitization The process by which a calcium carbonate rock is converted, either wholly or in part, to a calcium–magnesium carbonate rock composed of dolomite $(Ca, Mg)(CO_3)_2$.

Drift Strictly denotes all superficial (rock debris) deposits deposited directly by ice or meltwater. A 'drift-map' shows the distribution of all geologically recent river, ice, gravity, coastal or wind-laid deposits.

Dyke Wall-like body of rock emplaced within the Earth's crust; usually formed by injection and subsequent cooling of molten material.

Effective precipitation The amount of precipitation remaining after the evapotranspiration has been subtracted.

Eluvial Formed by the rotting (weathering) of rock in place to varying depths below the surface.

E.M. A geophysical technique based on measuring electromagnetism of rocks and minerals in order to identify near-surface minerals, materials, strata and voids.

Epicentre That part of the Earth's surface directly above the origin (focus) of an earthquake.

Erosion The wearing away of the land surface by various natural agencies, eg. water, ice, wind, river action.

Eutrophication The process whereby a body of water received anomalously large quantities of nutrients for plants and animals. In the short term the organisms thrive, but the death and decay of these then leads to depletion of oxygen and the consequent death of most of the organisms within the lake.

Evapotranspiration The combined amount of water returned to the atmosphere by evaporation and transpiration.

Extractive industry A general term for the industry concerned with extraction of minerals.

Extrusive Igneous rocks formed from material ejected during volcanic activity.

Fault A fracture or discontinuity where the two sides have undergone differential movement parallel to the plane of the fracture or discontinuity.

Fluvial, fluviatile An adjective designating a river as forming a feature; a fluvial process is one that involves river action.

Geochemistry A subdiscipline of the earth sciences involving the application of chemical methods of analysis to rocks, minerals, water, etc.

Geomorphology A subdiscipline of the earth sciences which deals with the configuration of the Earth's surface; investigation of geological changes through interpretation of topographic forms.

Geophysics A subdiscipline of the earth sciences which is concerned with the physical properties of the Earth's interior, rocks and minerals (including economic minerals such as ore minerals and hydrocarbons). A valuable indirect method of exploring the subsurface.

Geosphere This term is used by Lüttig in Chapter 10 as a synonym for the 'Natural Environment' of the terrestrial globe, including the biosphere and atmosphere.

Geotechnical The methods of study and processes associated with quantitative engineering geology and soil mechanics.

Geothermal Pertaining to the heat of the Earth's interior.

Glacial Pertaining to periods of global glaciation, to glaciers and ice sheets.

Glaciofluvial Pertaining to melt water, streams and rivers associated with glaciers and ice sheets.

Glaciolacustrine Pertaining to glacial lakes.

Grade In economic geology denotes the quantitative index of the richness of a mineral ore.

Granite A very common intrusive igneous rock composed of quartz, alkali feldspars, and other minerals, of high silica content.

Great Oolite A stratigraphic subdivision of the Jurassic System in Britain—includes a limestone widely used for building stone.

Greenhouse effect The atmosphere receives much of its energy from short-wave solar radiation which passes through it but is absorbed and re-emitted as long-wave radiation which cannot easily escape. This leads to a nett heating. Certain gases such as carbon dioxide increase the rate of heating. The term has been applied recently to increases in global temperature due to the increased atmospheric content of CO_2 and other gases as a result of burning of fossil fuels (coal, oil, natural gas).

Greywacke A sandstone composed of 33% or more easily destroyed minerals and rock fragments together with more resistant grains such as quartz.

Ground radar More correctly 'ground impulse radar', a radar technique operated from the land surface and capable of detecting natural and artificial cavities etc. beneath the surface.

Groundwater Water occurring in the saturated zone beneath the surface.

Gypcrete A superficial crust formed by natural cementation of soils by gypsum.

Gypsum A common hydrous calcium sulphate mineral ($CaSO_4$, $2H_2O$) of value in cement and plasterboard manufacture and other industrial processes.

Hardpan A cemented gravel layer just below the topsoil, difficult to excavate.

Hard rock A loose geological term commonly used for igneous and metamorphic rocks, in distinction to sedimentary rocks. The term is also used to denote solid rock resources for aggregate use as opposed to sand and gravel.

Hardware Consists of all the physical elements of computers such as circuits, wires and keyboard.

Head An earthy deposit carrying loose stones, the product of rock decay *in situ*.

Headers Used in computerization for the primary page of data items, which summarizes most of the essentials and may lead to secondary, tertiary, etc. screen displays.

Humus The organic portion of soil, composed essentially of the products of decay of vegetable matter.

Hydraulic gradient The expression of the change in total head of groundwater, with distance, in a given direction, where the 'head' is a representation of energy per unit weight at observation points.

Hydrocarbons Compounds of hydrogen and carbon in solid, liquid or gaseous form (e.g. natural gas and petroleum).

'Hydrofracing' Creation or enlargement of fractures by injection of water at high pressures.

Hydrograph Time record of potentiometric levels in a well or the stream discharge at a given point.

Hydrology A term generally restricted to the study of surface waters, their measurement and their chemistry.

Hydrogeology The practise of geology with special reference to water. Hydrogeology embraces both surface water and underground water.

Hydrothermal A term used for mineral deposits produced by natural circulating heated water and constituents dissolved in it, in its passage through openings in the rock (characteristically forming vein deposits).

Ice wedges Wedge-shaped fissures in the ground formed by expansion of ice from freezing water, often preserved by later infilling with sediments as the ice melts.

Igneous rocks Rocks formed by solidification from a molten state; includes intrusive and extrusive rocks (the latter include lavas and pyroclasts).

Infiltration Process whereby water soaks into, or is absorbed by, and passes through the soil via pores and small openings.

Interception Precipitation intercepted by buildings or roads, and vegetation that never reaches the soil due to evaporation or surface runoff.

Interflow The lateral movement of water in the unsaturated zone beneath the ground surface during or immediately after precipitation and which discharges into a stream or lake directly, without entering the groundwater bodies.

Intrusion A body of igneous rock formed by cooling of a molten material below the surface of the Earth.

I.P. A geophysical method used mainly in the search for mineral ores, and dependent upon the injection of energy in the form of a direct current into the ground. This leads to electrical charging of some metal ores. When the current is switched off, a transient discharge lasting a few seconds can be detected. The method is useful in discovering sulphide ore bodies and gold mineralization.

Isopachyte A line drawn on a map joining places in which geological formations or strata have equal thickness.

Joint A fracture or discontinuity where there has been no differential movement parallel to the fracture surface. Joints are essentially tensional and tend to open up normal to the fracture surface. They may be of tectonic origin or be due to shrinkage on cooling of a lava or intrusive magma.

Karst topography A type of limestone plateau topography marked by sink holes interspersed with abrupt ridges and irregular protuberant rocks, caverns and underground streams. It reflects the peculiar subsurface solution and redeposition processes of groundwaters in limestone.

Keywords Selected words used in indexing which allow the reader to retrieve a data item. Keywords are used in manual data bases and also in computerized data bases, where they provide the main method of data retrieval.

Lahar A mudflow related to volcanic processes. Lahars may be hot or cold, and include rainfall lahars and lahars produced by the disruption of crater lakes or the melting of ice and snow.

Laminar flow Flow in which particles move along definite and observable paths or streamlines, generally parallel to the channel walls.

Landfill Any material used to fill up cavities and restore the original land surface after mining, quarrying or temporary excavation.

Landslide A portion of a hillside or sloping mass that becomes loosened or detached, and slips or flows downslope under the influence of gravity.

LANDSAT Imagery used in mapping and terrain interpretation (remote sensing), made by a series of satellites placed in 900 km orbits around the Earth.

Laterization Formation of residual deposits due to surface weathering *in situ* under tropical climates, composed largely of aluminium and ferric hydroxides, with variable amounts of free silica.

Lava Molten (igneous) rock poured out at the Earth's surface from a volcanic mountain or fissure.

Leachate Water which contains a large amount of dissolved solids. The term is generally used to denote the liquids seeping from a waste deposit used in landfilling or from mine wastes. Leaching occurs naturally in the zone of weathering, dissolved substances being carried downwards leaving residual crusts (e.g. laterite).

Lignite Part of the coal series (q.v.).

Limestone A general term for bedded sedimentary rocks consisting largely of calcium carbonate. A magnesian limestone contains appreciable dolomite (double carbonate of lime and magnesia).

Lineament An essentially linear geological feature, commonly resulting from a fault but not invariably so. A term often used in photo-interpretation and satellite imagery (e.g. LANDSAT) interpretation, where linear structures may be identified but their cause is not known without later examination on the ground.

Lithology The general character of rock, especially the physical properties (e.g., grain size, mineral content, colour, hardness, specific gravity, magnetism).

Lithosphere The Earth's third, outer shell or 'crust' is separated from the 'mantle' (which in turn overlies the 'core'). The lithosphere incorporates the crust and uppermost part of the Mantle and lithospheric plates move as more or less rigid entities over the rest of the Mantle beneath by the agency of Mantle convection.

Loam A detrital deposit consisting of almost equal proportions of sand, silt and clay.

Loess A fine-grained earthy deposit of silt or marl; it is buff to brownish when oxidized, porous but coherent. The finely comminuted constituents are related to the grinding action of glaciers during the Ice Age; the distribution was by wind, and the accumulation of thick deposits owes something to binding by vegetation. Loess is of restricted distribution globally, but is widespread in areas adjacent to recent or former ice sheets.

Made ground Any artificially made surface material (e.g., landfill, causeways, embankments), which overlies the natural rocks or soils.

Magma Underground liquid molten rock.

Magnetometry The measurement of the magnetic properties of rocks and minerals.

Mainframe computer A large computer with a capacity much greater than a microcomputer (or minicomputer).

Manual An adjective applied to non-computerized procedures of data handling (e.g., card indexing).

Marl A calcareous clay or loam.

Menu A list of options displayed on a monitor screen during a computer program. A choice may lead to a further menu of options.

Mercalli scale This is a scale devised by Mercalli to measure the intensity of an earthquake. The modified scale has 12 instead of 10 intensities (e.g., detected only on instruments, feeble, destructive disastrous, catastrophic). This scale is a measure of the effect experienced at a given place. The intensity decreases outwards from the epicentre, the source of the disturbance. It should not be confused with the logarithmic Richter scale of magnitude, which is a measure of the total amount of elastic energy released.

Metamorphic rock A rock derived from a pre-existing rock by mineralogical, chemical or structural change, the process being sufficiently complete to form a well-defined new rock type. Nowadays, the term is restricted to processes involving quite severe temperature or pressure effects, and 'diagenesis' is used for the less extreme changes after burial of sedimentary rocks, due to burial pressure and mineral deposition, removal and replacement by mobile solutions in the rock, processes which in general lead to consolidation of originally soft sediments. The boundary between deep diagenesis and metamorphism is not always well defined, but slates represent the incipient stage of metamorphism of sediments.

Methane 'Marsh Gas', a gaseous hydrocarbon CH_4, dangerous as 'firedamp' in coal mines and commonly produced by decay of buried landfill waste with significant organic content. Methane from such sources is a hazard in terms of fire and explosion but may be harnessed for energy purposes by the use of small plants.

Microcomputer A small computer whose processing unit is based on a microprocessor chip. It is convenient for use in table top situations in office, home, etc.

Modem A device which alters data in digital form to wave form for transmission over telephone lines (it carries the reverse process for receiving phoned data).

Moraine An accumulation of earth, stones, etc. carried by and finally deposited by a glacier or ice sheet.

Nepheline syenite An igneous rock containing the minerals orthoclase feldspar and nepheline, of alkaline chemistry, which has commercial value as an industrial mineral.

Oil shale A fine black or brown shale rich in organic material from which petroleum can be obtained by distillation.

Opencast mining Any system of mining ore minerals by means of an open pit or excavation (in contrast to deep mining). The distinction from 'quarrying' is not clear cut, but quarrying is generally applied to constructional materials and common industrial minerals such as limestone. Opencasting is more general, and includes the working of immense pits for copper, lead-zinc, and even gold and diamonds.

Overburden The soil and waste rock which overlies the good stone in a quarry, or the worthless surface material that covers a body of useful mineral, to be opencasted (for example, in the Morocco phosphate deposits, immense industrial mineral workings 30 m or more of useless soil and rock had to be removed to reach the three or more 'pay' strata).

Palaeontology The study of fossil life forms in rocks and their classification, and use for estimating the relative ages of rocks.

Parallel interface An input/output port for a computer whereby data are transmitted, several instructions at a time, through a number of channels.

Peat Dark coloured fibrous soil formed in bogs and swamps by the partial decomposition of plant material. It can be ignited for heating purposes and used as a source of energy in peat-fired power stations. It is part of the coal series (q.v.).

Pedology The study of soil from the standpoint, and by the methods, of pure science. Pedology refers to the soil in the strict sense, the regolith or superficial blanket of rock decay products, generally not more than 1–2 m deep, and relates to the agricultural soils (not the much more broadly defined 'engineering soils').

Pegmatite Very coarse acid igneous rock, product of late stage crystallization from a magma rich in dissolved gases. Individual crystals may be centimetres or even metres in dimensions.

Penetrometer An instrument used in the field by engineering geologists to measure the resistance to penetration downwards of soil material.

Percolation Downwards seepage of water under hydrostatic pressure through the zone of aeration by means of pores, cracks and fissures in the soil and underlying rocks.

Periglacial A term applied to areas close to ice caps or ice sheets, showing environmental, depositional and biological patterns strongly influenced by low temperatures and the very short summer season.

Permafrost Ground that is permanently frozen. In practice, any ground liable to remaining frozen for two or more consecutive years without seasonal melting. It is the general temperature rather than the presence or absence of ice which defines permafrost.

Permeability The ability of a material to allow fluid to pass through it, under the pressure of a hydraulic gradient.

Petrology A general term for the study of the natural history and origins of rocks. Comprises the descriptive component 'Petrography' and the interpretative component 'Petrogenesis'.

pH A measure of the acidity of a solution expressed in terms of hydrogen ion concentration. Pure water has a pH of 7, values less or greater than 7 indicate acidity or alkalinity respectively.

Photogrammetry Map production based on careful measurements made on air-photograph coverage.

Photo-interpretation The term is used for detecting and interpreting features from air-photos, and either preparing maps from them or enhancing the detail of ground surveyed maps from air-photos. In geology it may be especially useful in defining geological boundaries and structures.

Piezometric surface The level to which water will rise in a well drilled into a confined aquifer.

Pingo A relatively large mound or hillock formed by the updoming of the soil above an accreting mass of ice in a periglacial setting. It may persist for more than a single season.

Pixel An element of resolution in a satellite image or computer screen picture. A 'picture element' is the smallest portion (square) of a screen that the computer can address as an individual unit. In the satellite imagery context, if an image is progressively enlarged, it will finally emerge as a set of squares within which there is no more detail. For a particular image, the pixels will have a numerical value defined by their area.

Placer deposit A mass of gravel, sand, etc. deposited by a river and containing particles of such minerals as gold, platinum, tin, and other valuable minerals.

Plate Geology has undergone a revolution in thinking with the general acceptance that the outer layers of the Earth move laterally in a system of converging and diverging 'plates' related to ascending convection at mid-ocean ridges and descending convection at 'subduction' zones. Continental plates may collide to form mountain ranges like the Alps or Himalayas, and the interaction of oceanic and continental plates may give rise to marginal mountain chains such as the Andes.

Porosity The ratio of the volume of voids in a rock to the total volume of the rock.

Portal The surface opening to a tunnel, adit or decline.

Potentiometric surface The level at which water stands in a tightly cased well.

Precipitation Term used in meteorology for the falling of condensed moisture from the atmosphere (rain, snowfall, mist). In chemistry, the term denotes the formation of crystals from a solution with a high concentration of components of the crystalline material.

Protalus rampart A barrier-like accumulation of talus produced by alternating freezing and thawing of ground from the headwall of a hollow and emplaced by sliding to form a ridge roughly parallel with the headwall.

Pumice Extremely spongy or porous lava resembling froth. It may also occur fragmented in a pyroclastic deposit.

Pyroclastic rock A fragmental volcanic (extrusive igneous) rock formed of debris such as a volcanic ash, tuff, breccia or agglomerate (fine to coarse varieties) produced during explosive eruptions.

Quartz The most common rock mineral, composed of silica (SiO_2).

Raster A term used in computer graphics to denote a display image consisting of a matrix of pixels arranged in rows and columns.

Recharge Intake, the process by which water is absorbed and added to the zone of saturation, either directly into a formation (aquifer) or indirectly by way of another formation. Recharge potential is a complex entity, depending on rainfall, infiltration, and transmissivity. Pumping etc. from an aquifer needs to be compensated by adequate recharge, which is usually natural, but can be effected by artificial means.

Reclamation The process of reclaiming land areas from the sea, lake shores or river banks.

Regression surface A hypothetical surface estimated from the difference between observed values at various points and theoretical values calculated from mathematical formulae; thus a measure of the natural divergences from an idealized surface.

Relational data set management system Relational in computerization means that data may be retrieved by their association with a number of different data sets (e.g., keywords, map references, dates, authors names).

Remote sensing A term applied to the many developing methods of studying and analysing terrain for mineral exploration and many other purposes by means of ship, air and satellite-borne instrumental systems.

Residual deposits Those deposits accumulated *in situ* as the result of weathering, from which the more or less soluble constituents have been leached out.

Resistivity A measure of the ability of a material to retard the flow of electricity. Measured values allow deductions to be made about the nature of materials lying close to the Earth's surface or to the walls of a borehole. The geophysical method is very useful in locating mineral deposits and groundwater aquifers, also strata boundaries below the surface.

Restoration The process of treating land after mining, other than excavation, or industrial use so that it can be returned to beneficial use.

Rubric Explanatory instruction.

Salt diapirs Salt domes and related structures which have their origin in stratiform salt beds, which after deep burial suffer detachment and rise through the overlying strata, piercing them and losing their stratiform character, while adopting complex new deformation structures.

Sand Any deposit consisting of sedimentary particles averaging 0.06 mm–2 mm in diameter.

Scree A heap of loose rock debris which forms at the foot of a cliff or steep slope. Synonymous with talus which, however, is properly used to denote a type of deposit, whereas scree strictly denotes a physiographic form.

Scintillometry An instrumental technique for determining the radioactivity of a rock sample or of the ground at a test point by measuring the discharges of energy accompanying the radioactive decay of particles.

Sedimentology The study of sedimentary rocks; these consist of rock and mineral fragments laid down by the action of water, wind and glaciers.

Seismic surveying A term widely used for the geophysical method of subsurface exploration involving the recording on seismograms of artificially initiated shocks. Mainly used in hydrocarbon exploration, but also used in engineering geology and other mineral exploration.

Seismology The study of earthquakes.

Shaft A vertical or steeply inclined means of access to a deep mine.

Silcrete A silica-cemented surface crust.

Sill A sheet-like body of igneous rock emplaced within the Earth's crust by injection and subsequent cooling of molten material. A sill is conformable with (parallel to) the enclosing strata.

Silt Sediment consisting of grains mostly between 0.002 and 0.06 mm in grain size.

Sink hole A void formed by the collapse of ground into underground cavities when rocks below, soluble in fresh water (e.g., limestone and chalk), are dissolved.

Slurry A mixture of water and mud capable of flowing down slopes.

Software A term denoting all the programs, computer languages and operations used to make a computer perform a useful function. Contrasts with 'hardware'.

Soil The term is used in the agricultural sense for weathered surface materials, often with a significant organic content from decaying plants. However, in civil engineering the term is applied to all poorly consolidated sediments, even to considerable thicknesses, in distinction to the hard 'bedrock'.

Soil mechanics The study of the engineering properties and behaviour of unconsolidated earth materials.

Solifluction The slow flow from higher to lower ground of clay etc. and contained rock blocks in a water-charged state. Glacial boulder clays are commonly subject to this process.

Solution mining A modern method of mining soluble deposits such as potash by dissolving them in water at depth and piping to the surface. It can be effected from boreholes without costly shaft sinking and mine development.

Specific yield The ratio of the volume of water that will drain by gravity from saturated rock or soil to the volume of rock or soil.

Spoil dump A dump for the waste rock and tailings produced from a mining operation.

Steatite Soapstone, a massive variety of talc, a hydrous magnesian silicate. It is of considerable commercial value.

Sterilization The destruction or prevention of access to an earth resource (minerals, building land, water, good soils) by an alternative land use.

Stope A void produced during deep mining, usually in connection with metal mines, by removal of blocks from the roof and accumulating them on the floor (the ore to be hauled to the surface, the waste likewise or to be used later for backfilling). If rock left in the sides, floor and roof is sufficiently strong, the stope can be left open without backfilling. Pillars of rock may also be left unmined to prevent collapse, and finally removed allowing eventual collapse.

Stratification The original structure which divides sediments into beds, layers or strata.

Stratum A layer of rock more or less similar throughout.

Subsoil The layer of more or less decomposed and loose fragments of rock which lies between the soil and the bedrock.

Talus The sloping heap of loose rock fragments lying at the foot of a cliff or steep slope. Synonymous with 'scree' as a physiographic form, but talus is correctly applied to the type of material.

Tar Sands Sands which contain petroleum which can be extracted by distillation.

Temporary hardness of water Caused by calcium and magnesium bicarbonates which precipitate out when water is boiled.

Tectonic Pertaining to the structure of the Earth's crust and particularly to the nature of structures such as faults and folds, the large-scale Earth movements that produced them, and the patterns in which they have come to be disposed.

Thixotropy Property exhibited by some gels of becoming fluid when shaken. The change is reversible (they solidify on standing). Quicksands and quickclays are thixotropic sediments.

Till That part of glacial drift deposited from ice during melting (mainly underneath the ice), but with little transport by water.

Topsoil The uppermost layer of the soil in the pedological sense as distinct from the subsoil and weathered rock zones which are considered as soils by engineering geologists.

Transmissivity The rate at which water is transmitted through a unit width of aquifer under a unit hydraulic gradient.

Transpiration The process by which plants, having taken in moisture through their roots, return it to the atmosphere through their leaves.

Turbulent flow Flow in which the motion at any point varies rapidly in direction and magnitude, and follows no observable pattern.

Ultrabasic A term applied to igneous rocks with very low silica content, composed of minerals such as olivine, pyroxene, and serpentine, and with very minor feldspar (if any).

Ultramafic Rocks essentially composed of ferromagnesian minerals with little or no feldspar or quartz.

Unconformity An erosional break in the continuity of sedimentation. It may be 'angular' or a disconformity (without angular contrast above and below).

Vadose water Water occurring in the unsaturated zone.

Vector representation A vector is a quantity that can be expressed in terms of both quantity and direction.

Vein A tabular body of rock formed by injection of solutions rich in water or other volatile substances. Veins commonly carry quartz and calcite together with valuable ore minerals of metals such as gold, copper, lead and zinc.

Viscosity The property of a fluid describing and quantifying its resistance to flowing.

Waste tips Generally large and extensive mounds formed by tipping waste rock from mining activities or materials from industrial processes.

Work-station A computerized station including peripheral devices such as a printer, digitizer, etc.

Additional Bibliography

Anon. (1965). Geology and urban development. *Assoc. Eng. Geol. (USA), Special Publication.*

Anon. (1970). *Electric power and the environment— a report sponsored by the Energy Policy Staff.* Office of Science and Technology, US Government Printing Office, 71 pp.

Anon. (1976). Physical characteristics of the region. Service d'urbanisme, Ville de Montreal. *Bull. Technique No. 4.*

Anon. (1982). Land surface evaluation for engineering practise. *Quart. J. Geol. Soc. London,* **15**, 265–316.

Anon. (1985). Planning control over oil and gas. Department of the Environment, London, *Circular No. 2/85.*

Abad, J., Ayala, F. and Pernia Llera, J.M. (1980). Large scale 1:1000–1:5000 geotechnical mapping by the Instituto Geologico y Minero de Espana (IGME: Spanish Geological and Mining Inst. for Industrial and Urban Planning). *Bull. Int. Assoc. Eng. Geol.,* **21**, 83–91.

Abad, J., del Moral, J. and Pena Pinto, J.L. (1979). Spanish experience of geotechnical cartography in an urban area. *Bull. Int. Assoc. Eng. Geol.,* **19**, 79–84.

Abad, F.J. and Lopez-Prado Texeira, J. (1980). Geotechnical map of Spain on a scale of 1 : 1 000 000. Methodology and elaboration criteria. *Bull. Int. Assoc. Eng. Geol.,* **22**, 157–160.

Abad, F.J. and Pena, J.L. (1975). Metodologia de la Cartografia Geotecnica a E 1/200 000 ne Espana. *Mapa Geotecnico General, Techniterrae,* **3**, 44–54.

Alfors, J.T., Burnett, J.L. and Gay, T.E. (1973). Phase III Urban geology Master Plan for California. *Cal. Div. Mines and Geol. Bull. No. 198,* 98 pp.

Anderson, D.N. (1974). Problems of geothermal development in the Western United States. In *Geology, Seismicity and Environmental Impact* (Moran, D.E. *et al.*, eds.), Association of Engineering Geologists Special Publication, University Publishers, Los Angeles, 401–404.

Antoine, P. (1977). Réflexions sur la cartographie ZERMOS et bilan des expériences en cours. *Bull. BRGM 2ème Levie, Sect. III, no. 1/2,* 9–20.

Arnould, M., Deveughele, M., Monot, G. and Sanejouand, R. (1978). Projet de constitution d'un fichier données géotechniques de la Ville de Paris. *CR du 3ème Congres International de l'Asoc. Int. de Géol. de l'Ing., Session Spécial No. 4,* 108–115 (Madrid).

Artim, E.R. (1973). Geology in land use planning: some guidelines for Puget lowland. *Washington Div. Mines and Geol. Inf. Circ. No. 47.*

Baker, V.R., Craig Kochel R. and Patton, P.C. (eds.) (1988). *Flood Geomorphology,* Wiley, New York, 503 pp.

Bates, R.L. (1969). *The Geology of Industrial Rocks and Minerals.* Dover Publications, New York, 459 pp.

Barry, D.L. (1981). Prospective survey of contaminated sites. *Research Project for the Department of the Environment London, Directorate of Environmental Pollution. Atkins Research and Development, Epsom Surrey, Report 2035501/DLB/JB.*

Bell, F.G., Culshaw, M.G., and Cripps, J.C. (1987). Engineering geology of underground movements. *Preprints of papers for the 23rd Annual Conference of the Engineering Group of the Geological Society of London.*

Bini, C. and Del Sette, M. (1980). The contribution of soil cartography to land use planning. *Bull. Int. Assoc. Eng. Geol.,* **22**, 151–156.

Breuck, W. and van Burm, P. (1980). The use of historic maps for the preparation of engineering geology maps of Flanders (Belgium). *Bull. Int. Assoc. Eng. Geol.,* **21**, 99–103.

Brooks, J.W. *et al.* (1971). Environmental influences of oil and gas development in the Arctic Slope and Beaufort Sea. *US Dept. of the Interior Resource Publication No. 96,* 24 pp.

Broquet, J.F. (1978). Contribution à la cartographie géotechnique de la ville de Paris—le 13ème arrondissement. *Thèse de Doctorate d'Ingénieur,* Univ. de Paris, VI, 197 (1976).

Bromehead, E.N. (1986). *The Stability of Slopes,* Chapman and Hall, 374pp.

Brown, J. and Grave, N.A. (1979). Physical and thermal disturbance and protection of permafrost. *Cold Regions Research Engineering Laboratory, Special Report 79–5,* 42 pp.

Brown, R.D. and Kockelman, W.J. (1983). Geological principles for prudent land use: a decision-maker's guide to the San Francisco Bay region. *USGS Prof. Paper. No. 946*, 97 pp.

Buisson, J.L., Gros, G., Sanejouand, R. and Voiment, R. (1979). Computer aided updating of the engineering geology map of Rouen. *Bull. Int. Assoc. Eng. Geol.*, **19**, 303–311.

Burchell, R.W. and Listokin, D. (1975). *The Environmental Impact Handbook*. Rutgers Univ. New Jersey, Centre for Urban Policy Research.

Burgess, I.C. (1983). Thematic geological mapping, Morley. *British Geol. Surv. Open File Report and Maps, Sheet SE22NE, Morley.*

Butler, E.W. and Pick, J.B. (1982). *Geothermal Energy Development: Problems and Prospects for the Imperial Valley, California*. Plenum Press, New York, 361 pp.

Calembert, L., Fagnoul, A., Monjoie, A., Polo-Chiapolini, C., Schroeder, C., Cajot, O. and Darimont, A. (1975). *Carte Géotechnique 42.2.7 Liége*. State Geotechnical Institute.

Calembert, L., Monjoie, A. and Polo-Chiapolini, C. (1975). Carte géotechnique prototype due St. Tilman. *Annales Travaux Publics Belgique No. 5.*

Calembert, L., Monjoie, A., Polo-Chiapolini, C., Cajot, O., Fagnoul, A., Langohr, R. and Deprez, G. (1980). Utilisation de l'ordinateur pour l'élaboration des cartes géotechniques dans la region Liégeoise. *Bull. Int. Assoc. Eng. Geol.*, **21**, 159–164.

Carrara, A., Catalano, E., Somiso Valvo, M., Reali, C., Merenda, L. and Rizzo, V. (1979). Landslide and typology in two zones, Calabria, Italy. *Bull. Int. Assoc. Eng. Geol.*, **16**, 8–13.

Carrara, A., Catalano, E., Somiso Valvo, M., Reali, C. and Ossi, F. (1978). Digital terrain analysis for land evaluation. *Geologia Applicata e Idrogeologia*, **13**, 69–127.

Carrara, A. and Merenda, L. (1974). Metodologia pur un censimento degli eventi franosi in Calabria (Methodology for an inventory of slope instability events in Calabria, Southern Italy). *Geologia Applicata e Idrogeologia*, **9**, 237–255.

Carrara, A., Pugliesi-Carratelli, E. and Merenda, L. (1977). Computer based databank and statistical analysis of slope instability phenomena. *Z. Geomorph.*, **21** (2), 187–222.

Cendrero, A. (1982a). Bases fisicas para la planification provincial: el modelo de Cantabria. *XVI curso de Geologia practica, Teruel*, 325–360.

Cendrero, A. (1982b). Estudio integrado del medio fisico para la planification ambiental del Canton Guayaquil. *Revista de la Universidad de Guayaquil, No. 3*, 45–73.

Champetier de Ribes, G. and Humbert, M. (1976). Travaux préliminairies a la mise en place en France d'un plan d'étude et de cartographie des risques liés aux mouvements de terrain. *CR 2ème Congres Int. Assoc. Eng. Geol.*, Sao Paulo, 1.

Chazan, W. (1973). Le plan Ziermos: identification des zones exposées aux mouvements du sol et du sous-sol, préalable a leur prévision at a la prévention de leur effets. *Symp. Nat. S. S.S.C. t.l.*, Cannes, 93–99.

Chieregati, L.A., Casadei, S.D., Pereira Fontes, U. and Mendes Valverde, F. (1980). Mining and quarry planning in the Sao Paulo metropolitan area. *Bull. Int. Assoc. Eng. Geol*, **22**, 17–22.

Cloud, P. (1973a). Population Control—For and Against. Hart Publishing Company Inc., New York City.

Cloud, P. (1973b). Materials—their nature and nurture. *Earth Science Newsletter,* Division of Earth Sciences, National Research Council, USA.

Coates, D.R. (ed.) (1973). Environmental geomorphology and landscape conservation (3 vols). *Benchmark Papers in Geology*, Dowden, Hutchinson and Ross, Stroudsburg, Pennsylvania.

Coates, D.R. (ed.) (1976). *Glacial Foreground Geomorphology and Engineering.* Dowden, Hutchinson and Ross, Stroudsburg, Pennsylvania.

Coates, D.R. (1981). *Environmental Geology.* John Wiley, New York, 701 pp.

Collis, L. and Fox, R.A. (1985). Aggregates: sand, gravel and crushed rock aggregates for construction purposes. *Geol. Soc. London, Eng. Geol. Spec. Publ. No. 1*, 220 pp.

Cotecchia, V. (1976). Systematic reconnaissance mapping and registration of slope movements. *Bull. Int. Assoc. Eng. Geol.*, **17**, 5–37.

Cratchley, C.R., Conway, B.W., Northmore, K.J. and Denness, B. (1979). Regional geological and geotechnical survey of south Essex. *Bull. Int. Assoc. Eng. Geol.*, **19**, 30–40.

Cratchley, B.W., Northmore, K.J. and Denness, B. (1979). South Essex geological and geotechnical survey. *Proc. Int. Symposium of Eng. Geol. Mapping,* Int. Assoc. Eng. Geol., Newcastle-upon-Tyne, 30–40.

Crawford, C.B. and Legget, R.F. (1957). Ground temperature investigations in Canada. *Eng. J.*, **40**, 263–269.

Crimes, T.P. *et al.* (1988a). Assessment of sand and gravel resources in the Eastern Lleyn Peninsula, N. Gwynedd, N. Wales. *Open File Report by Liverpool University (Dept. of Geol. Sci.) for the Welsh Office.*

Crimes, T.P. et al. (1988b). Assessment of sand and gravel resources to the south of Warwick. *Open file report by Liverpool University (Dept. of Geol. Sci.), contract No. PECD/258 for the Department of the Environment, London.*

Culshaw, M.G. (1986). Planning and engineering geology. *Preprints of papers, 22nd Annual Conference, Engineering Geology Group of the Geological Society of London,* Plymouth Polytechnic, 8–12 September, 1986, 769 pp.

Culshaw, M.G., Jackson, P.D. and McCann, D.M. (1987). Geophysical mapping techniques in environmental planning. In *Planning and Environmental Geology* (Culshaw, M.G. *et al.*, eds.), Geol. Soc. London, Eng. Geol. Spec. Publ. No. 4, 171–177.

Dam, J.P., Nuyens, J., Parent, J. and Thonnard, R. (1976). *Carte Geotechnique 31.3.7 Bruxelles.* State Geotechnical Institute.

Dearman, W.R. (1987). Land evaluation and site assessment: mapping for planning purposes. In *Planning and Engineering Geology* (Culshaw, M.G. *et al.*, eds.), Geol. Soc. London, Eng. Geol. Spec. Publ. No. 4, 195–201.

Dearman, W.R., Baynes, F.J. and Pearson, R. (1977). Geophysical detection of disused mineshafts in the Newcastle-upon-Tyne area, N.E. England. *Quart. J. Eng. Geol.*, **10**, 257–269.

Dearman, W.R. and Fookes, P.G. (1974). Engineering geological mapping for civil engineering practice in the United Kingdom. *Quart. J. Eng. Geol.*, **7**, 223–256.

Dearman, W.R., Money, M.S., Coffey, J.R., Scott, P. and Wheeler, M. (1977). Engineering geological mapping of the Tyne and Wear conurbation, Northeast England. *Quart. J. Eng. Geol.*, **10**, 145–168.

Dearman, W.R., Money, M.S., Strachan, A.D., Coffey, J.R. and Marsden, A. (1979). A regional engineering geology map of the Tyne and Wear County, N.E. England. *Bull. Int. Assoc. Eng. Geol.*, **19**, 5–17.

Dearman, W.R. and Strachan, A.D. (1983). Engineering geological plans of Tyne and Wear County, N.E. England. *Bull. Int. Assoc. Eng. Geol.*, **28**, 31–41.

De Beer, E., De Breuck, W., De Moor, G., Tevernier, R., Maertens, J. and Van Burm, P. (1976). *Groundmechanische kaart 22.1.6 Gent St Pieters.* State Geotechnical Institute.

De Beer, E., De Breuck, W., Maertens, J. and Van Burm, P. (1978). *Grondmechanische kaart 15.3.8 Hoboken (Antwerpen).* State Geotechnical Institute.

Delft Geotectonics and Geological Survey of the Netherlands (1984). *Ingenieurs Geologische Kaarten van Nederland,* RGD, Haarlem

Deveughele, M. and Usseglio Polatera, J.M. (1976). Cartographie expressées aux risques liés aux dissolution de la fraction gyspsense du Marno calcaire de Saint Omen dans les 19 et 20ème arrondissements de la ville de Paris. *Symp. de L'AIGI sur la cartographie géotechnique,* **2**, Newcastle-upon-Tyne.

Dole, H. (1974). Development of oil shale in the Green River Formation. Report of the Committee on Environment and Public Planning. *The Geologist, supplement to vol. 9, no. 4,* 8 pp.

Draggan, S., Cohrssen, J.J. and Morrison, R.E. (1987). *Geochemical and Hydrological Processes and Their Protection: The Agenda for Long Term Research and Development.* Greenwood Press, 248 pp.

Echevarria, C.M.R. and Pena, J.L. (1978). La cartografie geotecnica en Espana. *Techniterrae*, **25**, 19–27.

Edwards, R. and Atkinson, K. (1986). *Ore Deposit Geology and its Influence on Mineral Exploration.* Chapman and Hall, London and New York, 466 pp.

Engineering Geology Ltd. (1989). Assessment of potentially workable sand and gravel resources in the County of Durham. *Department of the Environment Research Contract PECD 7/1/240, 5 volumes, (open file report).*

Finlayson, A. (1984). Land surface evaluation for engineering practice: applications of the Australian PUCE system for terrain analysis. *Quart. J. Eng. Geol.*, **17**, 149–158.

Forster, A., Hobbs, P.R.N., Wyatt, R.J. and Entwistle, D. (1986). Environmental geology maps of Bath and the surrounding area for engineers and planners. In *Planning and Engineering Geology* (Culshaw, M.G., ed.), *Preprint Volume, 22nd Annual Conference of the Engineering Group of the Geological Society of London*, Plymouth, 8–12 September 1986, 237–260.

Forsyth, I.H. (1983). Environmental geology mapping, Glasgow. *British Geological Survey Open File Report and Maps.*

Freshney, E.C. (1984). Geology of the Hurn-Christchurch Area. *British Geological Survey Open File Report.*

Gandemer, P.H. and Le Campian, M. (1979). Cartographie des risques ZIERMOS appliquées à des plans d'occupation des sol en Normandie *Bull. Liaison Laboratoire des Pontes et Chaussées*, **99**, ref 2277, 43–44.

Gartner, J.F. (1980). Recent Canadian experience in communicating terrain evaluation information. *Bull. Int. Assoc. Eng. Geol.*, **21**, 164–174.

Geological Society of London (1972). The preparation of maps and plans in terms of engineering geology. *Quart. J. Eng. Geol.*, **5** (4), 294–382.

Ghiste, S. (1970). Carte d'interpretation géotechnique de la région de Mons. *1re Congress IAEG, Paris, T2*, 904–915.

Ghiste, S. (1978). La cartographie géotechnique de la Province de Hainant. *3ème Congress IAEG*, Madrid.

Ghiste, S. (1980). Facteurs anthropologiques et cartographie géotechnique. *Bull. Int. Assoc. Eng. Geol.*, **22**, 125–128.

Ghiste, S. *et al.* (1969). Carte géotechnique de la région de Mons. Quatre cartes et livret explicatif de 42 pp. *ISICH*, Mons.

Ghiste, S. and Sobbe, A. (1972). Carte géotechnique de la ville de Saint Ghislain. Quatre cartes. Publication internel, *ISICH*, Mons.

Godwin, L.H., Haigler, L.B., White, D.E., Muffler, L.J.P. and Wayland, R.G. (1971). Classification of public lands valuable for geothermal steam and associated geothermal resources. *USGS Circular No. 647*, 18 pp.

Gounon, A. (1979). Intègration des risques naturels dans les documents d'urbanisme: application à la ville de Nice. *Bull. Int. Assoc. Eng. Geol.*, **19**, 126–128.

Gounon, A. (1980). La planification urbaine basée sur la géologie, l'hydrologie, la géotechnique: un example, la ville de Nice. *Bull. Int. Assoc. Eng. Geol*, **22**, 161–166.

Hake, S.S. (1987). A review of engineering, geological and geotechnical aspects of town and country planning with particular reference to minerals and the extractive process. In *Planning and Engineering Geology* (Culshaw, M.G. *et al.*, eds.), Geol. Soc. London, Eng. Geol. Spec. Publ. No. 4, 69–74.

Hasan, S.E. and West, T.R. (1982). Development of an environment geology database for land use planning. *Bull. Int. Assoc. Eng. Geol.*, **19** (2), 117–132.

Hinojosa, J.A. and Leon, C. (1978). Unstable soil mapping in Spain. *Int. Assoc. Eng. Geol., 3rd Congress, Section 1*–2, 217–227.

Holt, D.N. and Marker, B.R. (1987). Benefits of engineering geology for land use planning in areas of past metalliferous mining. In *Planning and Engineering Geology* (Culshaw, M.G. *et al.*, eds.,) Geol. Soc. London, Eng. Spec. Publ. No. 4, 75–80.

Humbert, M. (1975a). Establissement des cartes de location probable des zones exposées a des mouvements du sol/cartographie. *Zermos 75. SGN 127, AME.* BRGM, Orléans.

Humbert, M. (1975b). La cartographie des zones exposées aux risques natural liés aux mouvements du sol et du sous sol. *Bull. BRGM 2ème Série, Sect III, 1/2*, 5.

Institute of Geological Sciences, London (now BGS) (1975–1979). South Essex geological and geotechnical survey. *Open File report prepared for the Department of the Environment, London* (in twelve parts).

Kalterherberg, J. (1974). An engineering geology map of a township on the Lower Rhine. *2nd Int. Congress IAEG*, Sao Paulo.

Katz, M.B. and Langevad, E.J. (1985). Geosciences for development—the Australian role. *AGID Report Series No. 11*, 374 pp.

Kleb, B. (1979). Engineering geology mapping of a city undercut by cellular networks. *Bull. Int. Assoc. Eng. Geol.*, **19**, 128–134.

Knill, D.C. (1978). Aggregates, sand, gravel and constructional stone. In *Industrial Geology* (Knill, J.L., ed.), Oxford University Press, Oxford, 166–195.

Knill, J.L. (1978). *Industrial Geology*, Oxford University Press, Oxford.

Kockelman, W.J. (1975). Use of USGS earth science products by city planning agencies in the San Francisco Bay Region, California. *USGS Open File report 75–276*, 110 pp.

Kockelman, W.J. (1976). Use of USGS earth science products by county planning agencies in the San Francisco Bay Region, California. *USGS Open File report 76–547*, 186 pp.

Kockelman, W.J. (1979). Use of USGS earth science products by selected regional agencies in the San Francisco Bay Region, California. *USGS Open File report 79–221*, 173 pp.

Kockelman, W.J., Conomos, T.J. and Leviton, A.E. (1982). San Francisco Bay, use and protection. *Am. Assoc. for the Advancement of Sci.*, Pacific Division, 310 pp.

Kralova, Z. (1973). Engineering geology maps of the city of Prague: 1 : 5000. *Projektovy ustav dopranvnick a Inzenyrkych staveb*, Prague, CSSR.

Laird, R.T. *et al.* (1976). *Land Capability Analysis—A Method of Applying Earth Science Information to Planning and Decision Making.* Association of Bay Area Governments, San Francisco, 225 pp.

Largillier, J.F. (1971). Apport de la géomorphologie à l'étude d'un trace d'autoroute. Congres de géologie de l'ingénieur, Paris (1970), *Bull. Liaison Laboratoire des Ponts et Chaussées*, **51**, 137–146.

Larsen, J.I. (1973). Geology for planning in Lake County, Illinois. *Illinois Geol. Surv. Circular No. 481*, 43 pp.

Lee, N. and Wood, C. (1978). Environmental impact assessment of projects in EEC countries. *J. Environmental Management*, **6**, 57–71.

Legget, R.F. (1972). Duisberg harbour lowered by controlled coal mining. *Can. Geotech. J.*, **9**, 374–383.

Leggo, P.J. (1985). Ground impulse radar approach to shallow subsurface investigation. In *Prospecting in Areas of Desert Terrain*, Institution of Mining and Metallurgy, London, 95–99.

Leopold, L.B. *et al.* (1971). A procedure for evaluating environmental impact. *USGS Circular. No. 645.*

Letko, V. and Matula, M. (1977). Engineering geological conditions for the development of the Bratislava metropolitan region. *Acta Polytechnica*, **1** (2), 63–72 (in Slovak).

Little, A.D. (1975). An evaluation of the San Francisco Bay Region Evironment, Resources and Planning Study. *Report to US Department of Housing and Urban Development, Office of Policy Development and Research*, 93 pp.

Lutzen, E.E. and Rockaway, J.D. (1971). Engineering geology of St. Louis County, Missouri, USA. *Eng. Geol. Ser. No. 4, Missouri Geol. Surv. and Water Resources.*

McMillan, A.A. and Browne, M.A.E. 1987. The use and abuse of thematic mining information maps. In *Planning and Engineering Geology* (Culshaw, M.G. *et al.*, eds.), Geol. Soc. London, Eng. Geol. Spec. Publ. No. 4, 237–245.

Mahr, T. and Malgot, J. (1978). Zoning maps for regional and urban development based on slope stability. *Int. Assoc. Eng. Geol., 3rd Int. Congress, Vol 1*, 124–137.

Martin, R.P. (1977). The application of geomorphology in landslide studies. Ph.D. Thesis (unpublished), London University.

Masson, M. (1971). Cartographie géotechnique de l'agglomeration Rouenaise. *Bull. Liaison Laboratoire des Ponts et Chaussées,* **52**, 109–117.

Masson, M. (1972). Utilisation de la géomorphologie dans les études géotechniques de sites urbaniser. *Bull. Liaison Laboratoire des Ponts et Chaussées, Ref. 1242*, 59–69.

Matula, M. (1979). Regional engineering geological evaluation for planning purposes. *Bull. Int. Assoc. Eng. Geol.*, **19**, 18–24.

Meneroud, J.P. (1978). Cartographie des risques dans les Alpes Maritimes, France. *Int. Assoc. Eng. Geol., 3rd Int, Congress, 1–2*, 98–107.

Molotch, H. (1970). Santa Barbara: oil in the velvet playground In *Eco-catastrophe*, Harper and Row, New York, 85–109.

Molotch, H. (1976). Santa Barbara: oil in the velvet playground. In *Focus on Environmental Geology* (Tank, R., ed.), Oxford University Press, New York, 364–377.

Moye, D.G. (1955). Engineering geology of the Snowy Mountains Scheme. *J. Inst. Engineers of Australia*, **30**, 287–298.

Moye, D.G. (1960). Engineering geology of the Snowy Mountains Scheme. *J. Inst. Engineers of Australia*, **32**, 85–87.

National Academy of Science (USA) (1975). *Mineral Resources and the Environment.* Council on Mineral Resources and the Environment, Washington D.C., 348 pp.

National Academy Press (1988). *Estimating Probabilities of Extreme Floods: Methods and Recommended Research.* Washington D.C., 141 pp.

Nicholls, D.R. and Campbell, C.C. (eds.) (1971). Environmental Planning and Geology. US Departments of Housing and Urban Development and the Interior, *DCPD-32,* 204 pp.

Nickless, E. (1983). Environmental geology mapping. *Mineral Planning, No. 17*, September 1983, 25–29.

Philbrick, S. (1976). Kinzua Dam. In *Glacial Foreground Geomorphology and Engineering* (Coates, D.R., ed.), Dowden, Hutchinson and Ross, Stroudsburg, Pennsylvania, 195–197.

Pitts, J. (1984). A review of geology and engineering geology in Singapore. *Quart J. Eng. Geol.*, **17**, 93–101.

Polo-Chiapolini, C., Schroeder, C. and Monjoie, A. (1974). Cartographie géotechnique automatique du centre de Liége et du St. Tilman. *2ème Int. Congress IAEG*, Sao Paulo.

Price, D. *et al.* (1984–1985). Thematic geological mapping Aldridge-Brownhills. *British Geological Survey, Open File Report and Maps.*

Pryde, P.R. (1983). The decade of the environment in the USSR. *Science*, **220**, 274–279.

Rao, S.A. (1972). New York State land use and natural resources inventory. *Publication of the Water Resources and Marine Science Center*, Cornell, Ithaca, New York.

Reger, J.P. (1979). Discriminant analysis as a possible tool in landslide investigations. *Earth Science Processes*, 4, 267–273.

Rib, H.T. and Liang, H.T. (1978). Recognition and identification. In *Landslide Analysis and Control* (Schuster, R.L. and Krisek, R.J., eds.), National Academy of Science (USA), Washington D.C. 34–80.

Richards, C.A. (1974). Engineering geology aspects of petroleum in the urban environment. In *Geology, Seismicity and Environmental Impact* (Moran, D.E. *et al.*, eds.), Association of Engineering Geologists Special Publication, University Publishers, Los Angeles, 391–400.

Rivoire, M. (1979). Cartographie géotechnique pour l'amènagment. *Bull. Liaison Laboratoire des Ponts et Chausśees, 99 ref 2278*, 55–56.

Rockaway, J.D. (1974). Application of engineering geology to land use planning in the United States. *2nd Int. Congress Int. Assoc. Eng. Geol.*, Sao Paulo.

Rockaway, J.D. (1976). The influence of map scale on engineering geology mapping. *Bull. Int. Assoc. Eng. Geol.*, **14**, 119–122.

Rockaway, J.D. and Lutzen, E.E. (1971). Engineering geology of the Creve Coeur Quadrangle, St. Louis County, Missouri, USA. *Engineering Geology Series No. 2*, Missouri Geological Survey and Water Resources.

Rodrigues Ortiz, J.M., Hinojoasa, J.A. and Prieto (1978). Regional studies of mass movements in Spain. *3rd Int. Congress Int. Assoc. Eng. Geol., Sect 1*, 1, 267–277.

Roxburgh, I.S. (1987). *Geology of High-Level Nuclear Waste Disposal: An Introduction.* Chapman and Hall, 229 pp.

Sorensen, J.C. (1971). *A Framework for Identification and Control of Resource Degradation in the Multiple Use of the Coastal Zone.* Berkeley, University of California Press.

Statham, I., Golightly, C. and Terharne, G. (1987). Thematic mapping of the abandoned mining hazard: a pilot study for the South Wales Coalfield. In *Planning and Engineering Geology* (Culshaw, M.G. *et al.*, eds.), Geol. Soc. London, Eng. Geol. Spec. Publ. No. 4, 225–268.

Steinhart, C.E. and J.S. (1972). *Blow-out—a Case Study of the Santa Barbara Oil Spill.* Duxbury Press, 138 pp.

Strachan, A and Dearman, W.R. (1982). The Tyne and Wear data bank. *Bull. Int. Assoc. Eng. Geol.*, **25**, 45–51.

Szekielda, K.H. (ed.) (1986). *Satellite Remote Sensing for Resources Development.* Graham and Trotman, London, 215 pp.

Tan, B.K. (1987). Some geotechnical aspects of urban development over limestone terrain. *Bull. Int. Assoc. Eng. Geol.*, **35**, 127.

Turner, A.K. and Coffman, D.M. (1973). *Geology for Planning: a Study of Environmental Geology*, Colorado School of Mines.

Vecchia, O. (1978). A simple terrain index for the stability of hillslopes and scarps. *In Large Ground Movements and Structures* (Geddes, J.D., ed.), John Wiley and Sons, New York, 449–461.

Vidal-Font, J. (1979). Essai de valorisation géotechnique de la carte géologique de Toulouse, France, et ses environs. *Bull. Int. Assoc. Eng. Geol.*, **19**, 53–57.

Watkins, R.E. (1971). The influence of governmental regulations on the design, construction and operation of pipelines in Western Europe. *Am. Petroleum Inst. 22nd Annual Pipeline Conf.*, Dallas, Texas, April 26–28, 1971.

Wilson, A.A. (1987). Geological factors in land use planning at Aldridge–Brownhills, West Midlands. In *Planning and Engineering Geology* (Culshaw, M.G. *et al.*, eds.), Geol. Soc. London, Eng. Geol. Spec. Publ. No. 4, 87–94.

Wilson, A.A., Lowe, D.J., Price, D. and Langford, R.L. (1985) Geological reports for DOE. Land use planning at Aldridge-Brownhills. *Department of the Environment, London, Open File Report.*

Wilson, D. (1981). Geological notes and local details for sheet S098 N.W. Brierley Hills environmental geology mapping. *British Geological Survey Open File Reports.*

Wiltshire, S.E., Cowley, J. and Doornkamp, J.C. (1987). Environmental impact assessment: the background and the potential for the geologist. In *Planning and Engineering Geology* (Culshaw, M.G. *et al.*, eds.,), Geol. Soc. London, Eng. Geol. Spec. Publ. No. 4, 607–610.

Index

Aberfan Coal Tip Disaster 7
accelerated erosion 164
accessibility map 71
accuracy, survey 20–2, 27
acid rain 9, 61, 149
acid waters 68, 69
aeolian erosion 164
aerial photography 22, 86
of permafrost 168
agricultural soil surveys 122
air-borne multi-spectral scanner 159
Alaska earthquake (1964) 6
Aldridge Shopping Centre 19, 21
alluvial doline 163
alluvial fans, classification of 166
Ami 5
anticlines 25, 56, 57
applied geology mapping 210
Southampton 43–5
aquicludes 56, 61
aquifers 55–6, 59, 61
aragonite 56
area data 34
data acquisition 25–6
Areal Distribution at Present (ADP) map 108
Armenian earthquake 9
Aswan High Dam 57
atmospheric contamination 4
Aubrey, John 5
Australian Earth Science Information System 30
AutoCAD system 40
automatic contouring 27
avalanches 162–3
classification 162
main causes 162
mapping 162–3
nature of movement 162
azonal soils 121

Bahrein, soils of 123
Bangkok, threats of flooding 8
Bangladesh flooding 8, 9, 60, 166
base flow index 64–6
base metals 90–1
bearing values under static loading 183, 185
Beaumont, Elie de 5
bedrock, map showing geotechnical properties (Bath) 192, 194–7
Belmoral Gold Mine (Quebec) 87
blight 9
block diagrams 26, 29, 34, 232
block models 34
blow-holes 163
Bold Colliery 109
Bootlegger Cove Clay Formation 6
borehole data 31, 205
borehole logs 27
boundary mapping 22
British Geological Survey 30, 31, 43, 44, 45, 72
British National Geoscience Data Centre 30
Bull Piece Colliery 42

calcite 56
calcretes 120
caliche 223
canyon instability (Gabon) 187
cap-rock seals 81
carbonate rocks 56
card indexing 35
Carnilya Hill nickel deposit 29
Carnmenellis granite (Cornwall) 59
catenas 26, 128
cathodic protection 147
cave distribution in Sweden 19, 20
Central African Rail Link 151
Chacewater–St Day copper mining area 39
Check Sheet System of data collection 31
chromium 8
coal mining 81, 87–8, 109
extent and general effects under Warrington New Town 112
hazard mapping of 164
restriction imposed on mining under Warrington New Town 112
coastal erosion 164, 246
coastal flooding 166, 167
coastal management 242
Colby, Colonel T.F. 5
collapse doline 163
colour in maps, use of 27, 218, 223
common depth-point seismic reflection 109
composite capability map (Franconia) 223, 230
Computer Aided Design (CAD) 39, 40
computers
and data storage 32–3
and environmental geology maps 206–7, 234
hazard zonation maps produced by 160
role of 26–7
and soil maps 128, 144
Conshohocken Borough, soil map of 143
construction 9, 175–97, 246
design of structures 183
and excavations and tunnels 183–4
ground conditions
problems and responses 175–80
laboratory tests 182–3
map indicating preparation measures in 189–91, 193
presentation of information 184
site investigations 180–3
site preparation 185
site surveying 183
constructional minerals 97–108
contamination, *see* pollution needs
continuous permafrost 168
contour data 25
copper resources, potential for 90–1, 92–3
cost/benefit analysis 197
Cotswold Water Park 89

data
collation of 33–4
collection 30–1
handling 8, 247
locating 30
nature of 16, 29–30
retrieval 34
spatial characteristics of 34
storage 31–3
systems 8, 247
costs 41
mainframe data bases 43–8
microcomputer data bases 35–42
paper 34–5
validation of 31
decomposition 119
deep well information 31
de la Beche, Henri 5
Deltaplan 186
Department of Transport 30, 33
deposition
causes of hazards from 164
mapping 164–5
nature of hazards from 164
depth 26
derived maps 90, 204
desert management 243
development constraints map 222
development potential map 219
digital mapping 206
Digital National Land Information (Japan) 247
Digital Terrain Model 27, 45
dip 25
discontinuous permafrost 168
distomat instruments 21
disused mine openings 38–40
dolines 163
dolomite 56
potential for 91, 94–6
drainage, soil 147
drainage system maps 64, 65
Drift Maps 10, 16
Dumont, A, H, 5

earth science maps
aims of 15–16

future prospects for 237–48
contribution to problems of regional and environmental planning and management 241–4
education 244
expected requirements of geosphere management 238
global perspective 244–5
potential of earth sciences information 239
situation and problems 237–8
strategies and problems 245–7
technical steps necessary 239–41
and plans, distinction between 15
types of 9–10
using 28
earthquake hazard zoning 170, 171
earthquakes 161, 241
element maps 90, 204
EM31 conductivity meter 86
energy minerals 108–13
engineering geological thematic maps 184, 189
engineering geology mapping 4, 202, 209, 221
and land-use planning 241
engineering interpretation for a soil series 143, 146
engineering triangles 142
English Board of Agriculture 5
environmental geology mapping 9
in Africa 231
in Asia 233
commissioning agencies and contractors 204–5
data and updating 205–7
definition 202
in Europe 208–19
implementation 208–9
level of detail in presentation of results 204
limitations and liabilities 207–8
in the Middle East 231–3
with optimum planning input 209
planning and development objectives 202–4
reasons for undertaking 202
of San Francisco bay 204
themes incorporated 224
in USA 219–31
in USSR 219
erosion
causes of hazards from 164
coastal 164, 246
mapping 164–5
nature of hazards from 164
soil 9, 57, 147, 159, 164, 201, 243
by water 58
in Zimbabwe 164, 165
eskers 10
European Environmental Geology Mapping Studies 211–17
exploitability map 71

FAX 47
feature mapping 22
feldspar 120
fence diagrams 29, 105
field data sheet 31, 32, 33
flood hazard maps 157, 159
Flood Studies Report 59
flooding 59, 243
causes of hazard 166
mapping 166–7
nature of hazard 165–6
floodplain 166
flow maps 64
Forestry Commission 30
fossil pingos 16
foundation depth map (The Hague) 189, 191, 192
foundations for constructions 183, 186
frost heaving 180

gas extraction 88
geochemical tests 180
geographic information systems (GIS) 63
Geological Survey of India 233
geomorphology
and land-use planning 240
geophysics 86, 113
shallow 159
tests 184
geoscience mapping 202, 209
Geoscientific Map of Natural Environmental Potential of Lower Saxony and Bremen 218, 246
GEOSTAT programme 247
geothermal energy 8, 59
geothermal gradient 59
geothermal maps 72
of Southern England 75
of the United Kingdom 72
glacial showing map 104
gleying 120
global socio-economic problems 237
Greater Amman Master Plan 232
ground classification 220
ground stability 3
groundwater 83
quality 55
resources 69–72
as a source of heat 59
groundwater maps 66–9
groundwater vulnerability maps 72, 74
Guettard, J.E. 5
gypcretes 120

habitability maps (Morocco) 231, 233
hard rock resources 106, 110–11
hazard maps 157
aids to 158–9
application of 171
cartographical approaches to 160
and classification systems 159–60
context of 158
grid square approach to 160
matrix approach to 160
and perception 160
and plotting accuracy 160
scales 158, 159
hazards 9, 157–73
definition 157
and land-use planning 240–1
natural 157, 242–3
properties of 158
risk assessment and prediction 172
wider responses to 171–2
Heath Farm onshore drilling site (Staffordshire) 117
Hemerdon Mine 91
herbicides 62
Hitchcock, Edward 5
hydrocarbons 81, 87
hydroelectricity 8, 58
hydrogeological maps 16, 66–9
of Croatia 66
of the Rhine Valley 72
hydrological cycle 53–5
hydrological maps 64–6
hydrology 147
Hydrology of Soil Types (HOST) project 147

ice avalances 162
industrial minerals 91–6
Indus Valley (Pakistan) 61–2
instability risk map (Gabon) 188
INTERGRAPH 44
intrazonal soils 120–1
irrigation suitability map 137, 138

Kalgoorlie Goldfields 61
Kariba dam 56, 58
karst landscapes 57
Kingsley, Reverend Charles 5

lahars 170
historical 17–18
land capability classifications 134–5
land capability maps 66
land instability 201
land suitability classifications 135–8
land systems 128, 129
land systems mapping 26
land utilization map 152
land utilization, tolerance diagram for 245
landfilling 59, 61
and construction 180
landscaping and recreation 142–3
landslide failure plane 25
landslide hazard maps 157
landslides 16, 26, 170, 241, 242
classification 160, 161
distribution 18–19
main causes 161
mapping 161–2
nature of movement 160–1
landslips *see* landslides
land-use development plan (Lower Saxony and Bremen) 218, 246
land-use planning 237
definition 4
and the earth sciences 5–7
laterization 120
leaching 120
legal frameworks 4
Leiden, settlement map of 187–9
liabilities and responsibilities in mapping 9
limestone resources 91, 94–6
line data 34
data acquisition 20–2
data analysis 25
data representation 22–4
lithological contact surface 25
loess 10

Maccullough, J. 5
Maclure, William 5
Magnesian limestone 96–7
current uses of 96
physical constraints on 97

planning constraints on extraction 97
summary of maps 97
Magnesian Limestone Escarpment Local Plan for Durham (1982) 96
maibolt 240
Main Meru Lahar 17
mapping
costs and benefits of 3–4
development of 4–5
printing 27
production 27
techniques 7–8
use of colour 27, 218, 223
Mexico City earthquake (1985) 170
mineral resources 8–9
classification and settings 82–3, 84–5
constraints on extraction 88–9
consumption *per capita* 79, 81
definition 79
environmental effects of extraction 87–8
former 90
land restoration after extraction 88–9
and land-use planning 239–40
location and quantity, constraints on 81–2
mappable information in 86, 87
needs for information on 83–6
requirements for 79–81
restoration of workings and mineral tips 147–9
rock tonnages for mining 82
types of extraction 86–7
types of investigation 86–7
value of mapping 90
world production 82
Minerals Local Plan for Greater Manchester 152
mining 61, 163–4
causes of hazards from 163
mapping 163–4
nature of hazards from 163
ocean 241–2
Ministry of Agriculture, Fisheries and Food 30
Modified Mercalli Scale 159
Monnet 5
moraines 58
morphological mapping symbols 23–5

National Conservancy Council 113
National Geoscience Data Banks 46–8
National Geoscience Data Base Management 44
National Geoscience Data Centre 45
National Topographic Survey of Great Britain 5
National Trust 5
natural ground subsidence 163
nature conservation
and land-use planning 241
networks 25
nitrate pollution 62
North American Environmental Mapping Studies 225–9
North Sea international boundaries 242
Nubian Sandstone (Libya) 55
nuclear waste management 243–4

ocean mining 241–2
ocean pollution 242
oil extraction 88, 109
and ground subsidence 163
oil traps 83
olivine 57
opencast workings 87
minimizing landscape impact of 88
restoration 89
ORACLE 44, 47
Ordnance Survey 5

Parkside Colliery 109
pattern analysis 25
peat resources 98–99
pedological soil surveys 122
pedological triangles 142
pedology 119
pedon 121
penetrometer transects 151
permafrost 180, 241
causes of hazard 168
distribution in the Northern hemisphere 169
mapping 168–70
nature of hazard 168
terminology of features 168
pesticides 62, 63
petroleum 8
photogrammetry 8, 159
physiographical map (Greater Madrid) 223
pingos 168
pipelines, planning 143–7
pixels 22
platinum 81
Plymouth Limestone 35
Plymouth–Plymstock environmental geology 32, 34–8
podzolization 120
point data 34
data acquisition 16
data analysis 19–20
data representation 16–19
pollution, chemical, in soils 149–51
and land-use planning 240
pollution maps 72
polypedon 121
pot-holes 163
potential maps 90, 204
precious minerals 90–1
Property Services Agency 30, 35
protalus ramparts 16
pumice 56

quarrying 61, 87, 163–4
causes of hazards from 163
mapping 163–4
nature of hazards from 163

radioactive minerals extraction 88
radioactive waste 246
disposal 61
in water 63
radon emissions 4
Rainbow Lake Field (Canada) 109, 113–16
rainfall erosivity 164
relative ease of excavation map (Utah) 192
reliability map 71
remote sensing 86, 113–14
resistivity methods 86
ribbon diagrams 26, 29
roads, planning 143–7
rockhead contours 19–20, 21
rubification 120

Sahel famine 8
saline soils
causes of hazard 167
mapping 167–8
nature of hazard 167
salinization of soils 120
sampling 26
San Francisco earthquake (1906) 170
sand and gravel resources 98–103, 105, 106–8
marine, production of 242
sand-fills 191
sandstones 56
Santa Barbara Oil Spill 7
satellite imagery 22, 24, 86, 158–9
of permafrost 168
satellite navigation systems 21
scale 15, 22, 26, 27
of environmental geology maps 204
in land-use planning 238
Schaar Channel 187
section diagrams 26
sector collapse 170
seismic maps 159
seismological hazards 170
causes of 170
mapping 170
nature of 170
sense of faulting 25
settlement map (Leiden) 187–9, 190
sewage disposal
in soils 151
on-site 142, 144
shallow geophysics 159
Shrewsbury sand and gravel study 34
Sierra Club 5
silage 62
silcretes 120
site documentation map 205, 207
slope, changes of 22, 25
Smith, William 5
snow avalanches 162
Soil Association map 130–1
Soil Conservation Service 147
soil erodibility 164
soil costs map for housing (Utrecht) 191–2, 193
soil erosion 9, 57, 147, 164, 201
maps 159
on slopes and high mountains 243
soil neutralizing capacity 153
soil profile 121
soil resource assessment 150
soil series maps 129
soil suitability 145
soil surface properties 26
Soil Survey and Land Research Centre 72
soil surveys
and description and sampling of soils 129
intensity levels 126, 127
methods 123
compilations 126
detailed surveys 128–9
exploratory surveys 128
free survey 123
grid survey 123
physiographic survey 123
reconnaissance surveys 128

semi-detailed surveys 128–9
very detailed surveys 129–32
types of 122–3
soils 9
acidification 151
classification 120–2, 124–5, 140, 142
and construction 180
definition 119
distribution, mapping of 122–32
formation 119–20
horizons 120, 121
interpretive and specialist maps of 132–53
agriculture and forestry 134–8
conservation and creation of wildlife habitats 147–51
environmental protection 147
mapping in integrated studies 151–4
planning roads and pipelines 143–7
urban and industrial development 138–47
and land-use planning 240
pollution 243
processes 119–20
protection 243
waterlogging 120
soils mapping 5, 9, 122–32, 151–4
applications to groundwater protection 152
and construction costs 154
solid block models 29
Solid Geology Maps 9–10, 16
solution doline 163
solution pipes 163
Speede, John 5
stability factor 187
Stackyard Field (Bedfordshire) 129, 133
statements of limitations 208
statements of responsibilities 208
statistical confidence surfaces 27
storm surge barrier (Eastern Scheldt) 186–7
strike 25
subsidence
and construction 180
natural ground 163
sulphates 180
summary maps 209
surface model generation 25
surveying techniques 21
swallowholes 163
synclines 25, 56, 57

temperature-density curve of water 58
terrain classification 128
testing 26
Thrislington Quarry 96
Tibesti Mountains 55
tidal power 58–9
topsoil stone contents 140
tourism 4
Trawsfynydd nuclear power station 58
trend surface analysis 27
triangulation maps 5
tsunamis 170
tuffs 56
tunnels, construction of 183–4

underground surface maps 26
United Nations Development Programme (UNDP) 244–5
United Nations Disaster Research Organization (UNDRO) 245
United Nations Educational, Scientific and Cultural Organisation (UNESCO) 245, 246
United Nations Environmental Programme (UNEP) 245, 246
United Nations Food and Agricultural Organisation (FAO) 245
United States Geological Survey 205
uranium 8, 88
urban expansion and soils 139–43
identifying constructional problems 140–2
landscaping and recreation 142–3
protecting land from development 139–40
sewage disposal, on-site 142
urban suitability map 221
USGS Earth Science Directory 30
Ussher 32

vadose zone 54
volcanic hazards 241
causes of 171
mapping 171
nature of 170
volcanic explosion pipes 81
volcanic rocks 56
volcanoes 16
vulnerability to pollution maps 72

waste disposal 60, 61, 89
in soils 150–1
waste disposal map (Maricopa County) 223, 231
water 8
and health 62–3
importance of 53
and land uses 60–2, 240
agriculture, fish farming and forestry 61–2
industrial development 60–1
land use and water resources 62
quarrying and mining 61
urban development 60
waste disposal 61
power, uses and problems of 57–60
erosion and deposition 58
flooding 59
ground stability 59–60
quality 58
as a source of energy 58–9
weathering 57–8
quality 62–3
resources, factors affecting
climate 55
geology 55–6
geomorphology 56–7
world balance of 53–4
water pollution 60, 61, 62, 63
water-related maps 63–76
water resource potential
classes of 72
map 73
water resources maps 69–72
water-table 25
waterlogging of soils 120
wave power 58–9
Wear Catchment (Durham) 64
weathering 9, 57–8
winter rain acceptance potential (WRAP) map 147, 148
Woburn Experimental Farm (Bedfordshire) 127, 133
World Energy Conference 245
World Health Organisation 245

ZERMOS 157, 161
zonal soils 120